Seed Development, Dormancy and Germination

Seed Development, Dormancy and Germination

Edited by

KENT J. BRADFORD
Department of Plant Sciences
Seed Biotechnology Center
University of California
Davis, CA 95616-8780
USA

and

HIROYUKI NONOGAKI
Department of Horticulture
Oregon State University
Corvallis, OR 97331-7304
USA

Blackwell
Publishing

Editorial Offices:
Blackwell Publishing Ltd, 9600 Garsington Road, Oxford OX4 2DQ, UK
Tel: +44 (0)1865 776868
Blackwell Publishing Professional, 2121 State Avenue, Ames, Iowa 50014-8300, USA
Tel: +1 515 292 0140
Blackwell Publishing Asia Pty Ltd, 550 Swanston Street, Carlton, Victoria 3053, Australia
Tel: +61 (0)3 8359 1011

First published 2007 by Blackwell Publishing Ltd

ISBN-10: 1-4051-3983-8
ISBN-13: 978-14051-3983-0

Library of Congress Cataloging-in-Publication Data

Seed development, dormancy, and germination / edited by Kent Bradford and Hiroyuki Nonogaki.
p. cm. – (Annual plant reviews)
Includes bibliographical references and index.
ISBN-13: 978-1-4051-3983-0 (hardback : alk. paper)
ISBN-10: 1-4051-3983-8 (hardback : alk. paper)
1. Seeds–Development. 2. Seeds–Dormancy. 3. Germination. I. Bradford, K. J. (Kent J.) II. Nonogaki, Hiroyuki.

QK661.S415 2007
581.4′67–dc22

2006026447

A catalogue record for this title is available from the British Library

Set in 10/12 pt Times
by TechBooks, New Delhi, India
Printed and bound in Singapore
by Fabulous Printers Pte Ltd

The publisher's policy is to use permanent paper from mills that operate a sustainable forestry policy, and which has been manufactured from pulp processed using acid-free and elementary chlorine-free practices. Furthermore, the publisher ensures that the text paper and cover board used have met acceptable environmental accreditation standards.

For further information on Blackwell Publishing, visit our website:
www.blackwellpublishing.com

Contents

List of Contributors

Dr Phil S. Allen Department of Plant and Animal Sciences, Brigham Young University, 275 WIDB, Provo, UT 84602-5253, USA

Dr Diego Batlla IFEVA-Cátedra de Cerealicultura, Facultad de Agronomía, Universidad de Buenos Aires/CONICET, Av. San Martín 4453, 1417 Buenos Aires, Argentina

Dr Roberto L. Benech-Arnold IFEVA-Cátedra de Cerealicultura, Facultad de Agronomía, Universidad de Buenos Aires/CONICET, Av. San Martín 4453, 1417 Buenos Aires, Argentina

Dr Leónie Bentsink Department of Molecular Plant Physiology, Utrecht University, Padualaan 8, 3584 CH Utrecht, the Netherlands

Dr Paul C. Bethke Department of Plant and Microbial Biology, University of California, Berkeley, CA 94720-3102, USA

Professor Kent J. Bradford Department of Plant Sciences, Seed Biotechnology Center, University of California, Davis, CA 95616-8780, USA

Dr Feng Chen Department of Plant Sciences, University of Tennessee, Knoxville, TN 37996-4561, USA

Dr Isabelle Debeaujon Laboratoire de Biologie des Semences, Unité Mixte de Recherche 204 Institut National de la Recherche Agronomique/Institut National Agronomique Paris-Grignon, 78026 Versailles, France

Dr Bas J.W. Dekkers Department of Molecular Plant Physiology, University of Utrecht, Padualaan 8, 3584 CH Utrecht, the Netherlands

Dr J. Allan Feurtado Department of Biological Sciences, Simon Fraser University, Burnaby, BC, Canada V5A 1S6

Professor Ian A. Graham Department of Biology, Centre for Novel Agricultural Products, University of York, PO Box 373, York YO10 5YW, UK

Professor John J. Harada Section of Plant Biology, College of Biological Sciences, University of California, Davis, CA 95616, USA

Dr Henk W.M. Hilhorst Laboratory of Plant Physiology, Wageningen University, Arboretumlaan 4, 6703 BD Wageningen, the Netherlands

Dr Yuji Kamiya Plant Science Center, RIKEN, Growth Physiology Group, Laboratory for Cellular Growth and Development, 1-7-22 Suehirocho, Tsurumi-ku, Yokohama, 230-0045 Japan

Dr Allison R. Kermode Department of Biological Sciences, Simon Fraser University, Burnaby, BC, Canada V5A 1S6

Professor Maarten Koornneef Max Planck Institute for Plant Breeding Research, Carl-von-Linné-Weg 10, 50829 Cologne, Germany; and Laboratory of Genetics, Wageningen University, Arboretumlaan 4, 6703 BD Wageningen, the Netherlands

Professor Russell L. Jones Department of Plant and Microbial Biology, University of California, Berkeley, CA 94720-3102, USA

Dr Loïc Lepiniec Laboratoire de Biologie des Semences, Unité Mixte de Recherche 204 Institut National de la Recherche Agronomique/Institut National Agronomique Paris-Grignon, 78026 Versailles, France

Dr Igor G.L. Libourel Department of Plant Biology, Michigan State University, East Lansing, MI 48824, USA

Dr Eiji Nambara Plant Science Center, RIKEN, Growth Physiology Group, Laboratory for Cellular Growth and Development, 1-7-22 Suehirocho, Tsurumi-ku, Yokohama, 230-0045 Japan

Professor Hiroyuki Nonogaki Department of Horticulture, Oregon State University, Corvallis, OR 97331, USA

Dr Masa-aki Ohto Department of Plant Sciences, University of California, Davis, CA 95616, USA

Dr Steven Penfield Department of Biology, Centre for Novel Agricultural Products, University of York, PO Box 373, York YO10 5YW, UK

Dr Helen Pinfield-Wells Department of Biology, Centre for Novel Agricultural Products, University of York, PO Box 373, York YO10 5YW, UK

Dr Lucille Pourcel Laboratoire de Biologie des Semences, Unité Mixte de Recherche 204 Institut National de la Recherche Agronomique/Institut National Agronomique Paris-Grignon, 78026 Versailles, France

Dr Jean-Marc Routaboul Laboratoire de Biologie des Semences, Unité Mixte de Recherche 204 Institut National de la Recherche Agronomique/Institut National Agronomique Paris-Grignon, 78026 Versailles, France

Professor Sjef C.M. Smeekens Department of Molecular Plant Physiology, University of Utrecht, Padualaan 8, 3584 CH Utrecht, the Netherlands

Dr Wim Soppe Max Planck Institute for Plant Breeding Research, Carl-von-Linné-Weg 10, 50829 Cologne, Germany

Dr Camille M. Steber U.S. Department of Agriculture-Agricultural Research Service and Department of Crop and Soil Science and Graduate Program in Molecular Plant Sciences, Washington State University, Pullman, WA 99164-6420, USA

Dr Sandra L. Stone Section of Plant Biology, College of Biological Sciences, University of California, One Shields Avenue, Davis, CA 95616, USA

Dr Shinjiro Yamaguchi Plant Science Center, RIKEN, Growth Physiology Group, Laboratory for Cellular Growth and Development, 1-7-22 Suehirocho, Tsurumi-ku, Yokohama, 230-0045 Japan

Preface

The formation, dispersal, and germination of seeds are crucial stages in the life cycles of gymnosperm and angiosperm plants. The unique properties of seeds, particularly their tolerance to desiccation, their mobility, and their ability to schedule their germination to coincide with times when environmental conditions are favorable to their survival as seedlings, have no doubt contributed significantly to the success of seed-bearing plants. Humans are also dependent upon seeds, which constitute the majority of the world's staple foods (e.g., cereals and legumes), and those crops are also dependent upon seeds as propagules for establishing new fields each year. Seeds are an excellent system for studying fundamental developmental processes in plant biology, as they develop from a single fertilized zygote into an embryo and endosperm in association with the surrounding maternal tissues. As genetic and molecular approaches have become increasingly powerful tools for biological research, seeds have become an attractive system in which to study a wide array of metabolic processes and regulatory systems. The rapid pace of discovery, particularly in the model system *Arabidopsis thaliana*, and the complexity of the molecular interactions being uncovered provided the rationale for a book by leading experts to update our state of knowledge concerning seed development, dormancy, and germination.

This volume focuses on specific aspects of seed biology associated with the role of seeds as propagules. Thus, important processes in seeds, such as the accumulation of storage reserves and their subsequent mobilization during germination, are not covered in depth here. Instead, the emphasis in the development section (Chapters 1 and 2) is on the processes that contribute to seed growth and to the induction of dormancy during maturation, rather than on the very early steps of embryogenesis, which are covered in a number of other books and reviews. Dormancy is a rather mysterious physiological state in which imbibed seeds are metabolically active, yet do not progress into germination and growth. As developmental arrest is a widespread phenomenon in biology, insight into seed dormancy will have broad implications. Chapter 3 discusses the types of dormancy exhibited by seeds and the current hypotheses concerning the mechanisms by which environmental signals are transduced into regulatory mechanisms controlling dormancy. This is followed in Chapter 4 by a discussion and examples of approaches to modeling seed dormancy and germination in an ecological context. Such models have practical utility for vegetation management in both agricultural and wildland contexts, and they also identify and quantify response mechanisms for physiological investigation.

While details are still sketchy, the genetic basis of seed dormancy is being elucidated in several systems, including *Arabidopsis*, rice (*Oryza sativa*), and other cereals. Chapter 5 provides an overview and update on the genetic regulation of seed dormancy. Genes and mutations affecting dormancy and germination have identified a number of regulatory pathways, particularly those involving gibberellins (GA) and abscisic acid (ABA), that appear to be crucial for the development, maintenance, and loss of dormancy. Metabolic pathways are also involved, with lipid metabolism in particular playing an important role, as described in Chapter 6. A role for metabolic and respiratory pathways in regulating germination has been known for several decades, but new insights from work on nitric oxide discussed in Chapter 7 provide an integrating hypothesis for reinterpreting those earlier insights.

While GA and ABA are central players in regulating seed dormancy and germination, other plant hormones, including ethylene, auxin, cytokinins, and brassinosteroids, play important supporting roles. The complexity of these interacting hormonal signaling networks associated with seed dormancy is discussed in Chapter 8. Feedback loops involving hormonal synthesis, catabolism, and sensitivity govern diverse aspects of seed dormancy and initiation of germination. The specific genes encoding key enzymes in these hormonal synthesis and catabolism pathways are summarized in Chapter 9. The proteins involved in the signaling pathways through which these hormones act are also being uncovered. Chapter 10 reviews the important role of protein degradation pathways in controlling the transcription of germination-related genes. Once dormancy has been released and germination has been triggered, additional genes and mechanisms are involved in the growth of the embryo and its protrusion through any enclosing tissues – processes that are reviewed in Chapter 11. A final checkpoint appears to occur shortly after germination in the transition to seedling growth. Seeds are particularly sensitive to the effects of sugars at this stage, as described in Chapter 12.

Our goal in developing the book was to give a comprehensive look at seed biology from the point of view of the developmental and regulatory processes that are involved in the transition from a developing seed through dormancy and into germination and seedling growth. We wished to illustrate the complexity of the environmental, physiological, molecular, and genetic interactions that occur through the life cycle of seeds along with the concepts and approaches used to analyze seed dormancy and germination behavior. It has been over 10 years since a book devoted specifically to this topic has been published, and the progress made in that period is remarkable. The utility of *Arabidopsis* as a model system is evident in the focus of a number of chapters on work in this species. In addition, other chapters describe the broader implications and applications in ecological contexts of insights gained from model systems. This book provides plant developmental biologists, geneticists, plant breeders, seed biologists, graduate students, and teachers a current review of the state of knowledge on seed development, dormancy, and germination and identifies the current challenges and remaining questions for future research. The book will have been a success if it contributes to stimulating a new increment of seed biology research in the next 10 years to match or exceed that of the past decade.

We thank the distinguished group of contributors who provided authoritative reviews of their areas of expertise. Their scholarship, diligence, and responsiveness to our editorial demands made it a pleasure to work with them. We also thank Graeme MacKintosh, David McDade, Amy Brown, and their colleagues at Blackwell Publishing who offered us this opportunity and kept us on task to complete it.

Kent J. Bradford
Hiroyuki Nonogaki

Annual Plant Reviews

A series for researchers and postgraduates in the plant sciences. Each volume in this series focuses on a theme of topical importance and emphasis is placed on rapid publication.

Titles in the series:

1. **Arabidopsis**
   Edited by M. Anderson and J.A. Roberts
2. **Biochemistry of Plant Secondary Metabolism**
   Edited by M. Wink
3. **Functions of Plant Secondary Metabolites and their Exploitation in Biotechnology**
   Edited by M. Wink
4. **Molecular Plant Pathology**
   Edited by M. Dickinson and J. Beynon
5. **Vacuolar Compartments**
   Edited by D.G. Robinson and J.C. Rogers
6. **Plant Reproduction**
   Edited by S.D. O'Neill and J.A. Roberts
7. **Protein–Protein Interactions in Plant Biology**
   Edited by M.T. McManus, W.A. Laing and A.C. Allan
8. **The Plant Cell Wall**
   Edited by J.K.C. Rose
9. **The Golgi Apparatus and the Plant Secretory Pathway**
   Edited by D.G. Robinson
10. **The Plant Cytoskeleton in Cell Differentiation and Development**
    Edited by P.J. Hussey
11. **Plant–Pathogen Interactions**
    Edited by N.J. Talbot
12. **Polarity in Plants**
    Edited by K. Lindsey
13. **Plastids**
    Edited by S.G. Moller
14. **Plant Pigments and their Manipulation**
    Edited by K.M. Davies
15. **Membrane Transport in Plants**
    Edited by M.R. Blatt
16. **Intercellular Communication in Plants**
    Edited by A.J. Fleming
17. **Plant Architecture and Its Manipulation**
    Edited by C. Turnbull
18. **Plasmodesmata**
    Edited by K.J. Oparka
19. **Plant Epigenetics**
    Edited by P. Meyer

20. **Flowering and Its Manipulation**
Edited by C. Ainsworth
21. **Endogenous Plant Rhythms**
Edited by A. Hall and H. McWatters
22. **Control of Primary Metabolism in Plants**
Edited by W.C. Plaxton and M.T. McManus
23. **Biology of the Plant Cuticle**
Edited by M. Riederer
24. **Plant Hormone Signaling**
Edited by P. Hedden and S.G. Thomas
25. **Plant Cell Separation & Adhesion**
Edited by J.R. Roberts and Z. Gonzalez-Carranza
26. **Senescence Processes in Plants**
Edited by S. Gan
27. **Seed Development, Dormancy and Germination**
Edited by K.J. Bradford and H. Nonogaki
28. **Plant Proteomics**
Edited by C. Finnie
29. **Regulation of Transcription in Plants**
Edited by K. Grasser
30. **Light and Plant Development**
Edited by G. Whitelam

1 Genetic control of seed development and seed mass

Masa-aki Ohto*, Sandra L. Stone* and John J. Harada*

1.1 Introduction

Seeds are complex structures that consist of three major components, each with a distinct genotype. The embryo that will become the vegetative plant is diploid, possessing one paternal and one maternal genome equivalent. The endosperm, a structure that provides nourishment for the developing embryo and/or seedling, is triploid with two maternal and one paternal genome equivalents. Surrounding the embryo and endosperm is the testa (seed coat) that is strictly of maternal origin. The diversity of genotypes suggests that distinct genetic programs underlie the development of each seed component. Given that seed growth and development must be coordinated, communication must occur between the different components.

The ability of higher plants to make seeds has provided them with significant selective advantages that, in part, account for the success of the angiosperms (Walbot, 1978; Steeves, 1983). The seed habit has facilitated fertilization in nonaqueous environments and provided protection (ovary wall and integuments/testa) and nourishment (nucellus and endosperm) for the female gametophyte and developing embryo. The seed is also an elegantly designed dispersal unit. The desiccated embryo is metabolically quiescent enabling long-term viability, the testa serves as a permeability barrier for gases and water, and storage reserves such as lipids, proteins, and carbohydrates accumulated within the seed are a nutrient source for seedling growth. Moreover, many seeds are dormant, prohibiting reactivation of the sporophyte until conditions are appropriate for germination.

This chapter focuses on genetic mechanisms controlling seed development and seed mass. Because this review is not intended to be comprehensive, readers are referred to a number of recent reviews and to other chapters in this volume for more detailed information about specific topics (Harada, 1997; Berleth and Chatfield, 2002; Berger, 2003; Hsieh *et al.*, 2003; Gehring *et al.*, 2004; Laux *et al.*, 2004; Olsen, 2004; Vicente-Carbajosa and Carbonero, 2005).

1.2 Overview of seed development in angiosperms

Seed development is initiated with the double fertilization event in angiosperms. The haploid egg cell and the diploid central cell of the female gametophyte within

* These authors contributed equally to this chapter.

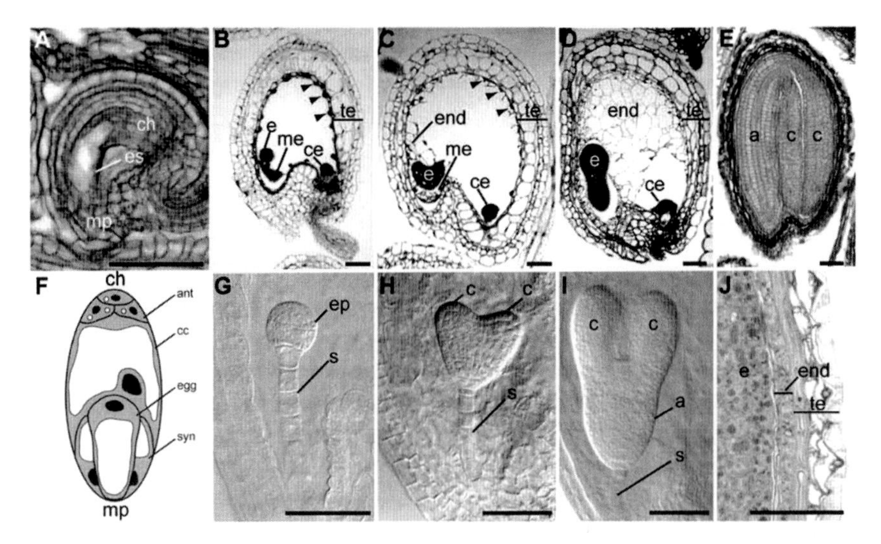

Figure 1.1 Seed development in *Arabidopsis*. (A) Section through a mature ovule. (B–E) Sections through seeds containing globular stage (B), heart stage (C), linear cotyledon stage (D), and mature green stage (E) embryos. (F) Diagram of a mature embryo sac modified from Drews *et al.* (1998). Black, gray, and white areas indicate the nucleus, cytoplasm, and vacuole, respectively. (G–I) Images of whole-mount embryos at the globular (G), heart (H), and linear cotyledon (I) stages. (J) Part of the seed containing a mature green-stage embryo with the embryo, the single layer of endosperm, and the testa. (A–E, J) Images from paraffin (A, E) and plastic (B–D, J) embedded sections through a mature ovule and developing seeds stained with (A, E) periodic acid-Schiff's, (B–D) toluidine blue O, and (J) periodic acid-Schiff's counterstained with aniline blue black. (G–I) Nomarski images of cleared embryos in seeds. Arrowhead, free endosperm nucleus; a, embryonic axis; ant, antipodal cell; c, cotyledon; cc, central cell; ce, chalazal endosperm; ch, chalazal pole; e, embryo; end, peripheral endosperm; ep, embryo proper; es, embryo sac; me, micropylar endosperm; mp, micropylar pole; s, suspensor; syn, synergid cell; te, testa. Scale bars = 50 μm.

the ovule (Figures 1.1A and 1.1F) each fuse with one sperm cell from the pollen tube to form the zygote and the endosperm cell, respectively. The zygote undergoes a series of differentiation events, resulting in the formation of the mature embryo and the suspensor, an ephemeral structure that supports the embryo physically and physiologically during early embryogenesis (Figures 1.1E, 1.1G–1.1I). The embryo consists of two major embryonic organ systems: the axis from which the body of the vegetative plant is derived and the cotyledon(s) that often functions as a storage organ for macromolecular reserves in dicotyledonous species (Figure 1.1E; reviewed by West and Harada, 1993; Goldberg *et al.*, 1994). The fertilized central cell undergoes a series of nuclear divisions without cytokinesis, resulting in the formation of a syncytium (Figures 1.1B and 1.1C) that later cellularizes (Figure 1.1D; Olsen, 2004). The endosperm is either a transient or persistent structure within the seed. In nonendospermic seeds, such as soybean (*Glycine max*) and peanuts (*Arachis hypogaea*), the endosperm is absorbed completely by the developing embryo. The endosperm is retained to varying degrees in endospermic seeds. For example, mature *Arabidopsis thaliana* seeds have a single layer of endosperm cells, whereas the endosperm makes up most of the mass of cereal seeds (Figure 1.1J). The other major component of the seed is the testa that is derived from the integuments of the ovule and, therefore, is of maternal origin (Figure 1.1J; Boesewinkel and Bouman, 1984; Bewley and Black, 1995). Many seeds also have a transient perisperm derived from the nucellar cells of the ovule. In some plants, the perisperm persists and serves as a storage organ (Bewley and Black, 1995).

1.3 Genetic control of embryo development

Embryogenesis in higher plants can be divided conceptually into two distinct phases. During the early morphogenesis phase, the basic body plan of the plant is established. Polarity is expressed with establishment of the shoot–root axis, apical–basal and radial domains from which morphological structures are derived and the embryonic organ and tissue systems are formed (Jurgens, 2001; Berleth and Chatfield, 2002; Laux *et al.*, 2004). During the maturation phase late in embryogenesis, the embryo accumulates storage macromolecules including proteins, lipids, and starch, acquires the ability to withstand the stresses of desiccation, and enters a state of developmental and metabolic quiescence as it desiccates (Harada, 1997).

Large-scale genetic screens with T-DNA or chemical mutagens have been used to identify genes that play critical roles in embryo development. Such screens have identified many mutants displaying defects in embryogenesis in *Arabidopsis*, maize (*Zea mays*), and rice (*Oyrza sativa*) (Clark and Sheridan, 1991; Meinke, 1991; Hong *et al.*, 1995). For example, screens for *Arabidopsis* mutants initially identified more than 300 *embryo-defective* (*emb*) mutations that affected embryo development (Meinke, 1991). It has been estimated that these screens have not reached saturation and that there are between 500 and 1000 *EMB* genes in *Arabidopsis* (Franzmann *et al.*, 1995; McElver *et al.*, 2001). This estimate indicates that embryo development is a complex process with hundreds, if not thousands, of required gene products. A recent report describing the nucleotide sequences of genes corresponding to 250 *emb* mutations showed that they were enriched for proteins predicted to play roles in basic cellular functions (Tzafrir *et al.*, 2004; also see www.seedgenes.org). For example, the *Arabidopsis* embryo-lethal *schleperless* (*slp*) mutant that has reduced cotyledons has a defective plastidic chaperonin 60α gene (Apuya *et al.*, 2001), the *bio* mutant that arrests at the heart or cotyledon stage is defective in the biotin synthase gene (Patton *et al.*, 1998), and the *twin2* (*twn2*) mutant that produces secondary embryos is defective in a valyl-tRNA synthetase gene (Zhang and Somerville, 1997). Although the *emb* mutations affect embryo development, many of these *EMB* gene products may be required for gametogenesis. The mutant alleles may be too weak to be lethal in gametophytes or mutant gametophytes may compensate for the lack of gene products because they are surrounded by maternal/paternal tissues heterozygous for the mutations (Springer *et al.*, 2000).

1.3.1 Central regulators of embryogenesis

As will be discussed, most regulators of processes that occur during embryogenesis function during either the morphogenesis phase or the maturation phase. However, the *LEAFY COTYLEDON* (*LEC*) genes define a small class of regulators that function during both phases (reviewed by Harada, 2001). *lec1* and *lec2* mutants were identified in screens for *emb* mutations, whereas the other *lec* mutant, *fusca3* (*fus3*), was found in screens for seeds with purple coloration. Analyses of these mutants showed that the *LEC* genes function early in embryogenesis to maintain suspensor cell fate and specify cotyledon identity. Late in embryogenesis, the *LEC* genes are required for the initiation and/or maintenance of maturation and the repression of precocious germination (Meinke, 1992; Bäumlein *et al.*, 1994; Keith *et al.*, 1994; Meinke *et al.*, 1994; West *et al.*, 1994; Lotan *et al.*, 1998; Stone *et al.*, 2001). Given their roles both early and late in embryogenesis, it has been speculated that the *LEC* genes serve to coordinate the morphogenesis and maturation phases (Harada, 2001).

Additional insight into *LEC* gene function came from ectopic expression studies. Misexpression of *LEC1*, *LEC2*, or *FUS3* confers embryonic characteristics to seedlings in that they resemble embryos morphologically and express genes encoding seed proteins such as 12S and 2S storage proteins and oleosin (Lotan *et al.*, 1998; Stone *et al.*, 2001; Gazzarrini *et al.*, 2004). Moreover, ectopic *LEC1* or *LEC2* activity is sufficient to induce somatic embryo formation from vegetative cells, suggesting that these genes enhance cellular competence to undergo embryogenesis (Lotan *et al.*, 1998; Stone *et al.*, 2001). We note that ectopic expression of *WUSCHEL* (*WUS*), a gene with a key role in establishing the shoot apical meristem in the embryo (Laux *et al.*, 1996), and of *BABYBOOM* (*BBM*), an AP2 (APETALA 2) domain protein, can also induce somatic embryogenesis in vegetative tissues (Boutilier *et al.*, 2002; Zuo *et al.*, 2002). It is not known whether *LEC*, *WUS*, and *BBM* operate in common or distinct pathways.

The *LEC* genes all encode regulatory proteins. LEC1 is a homolog of the HAP3 subunit of the CCAAT-binding transcription factor (Lotan *et al.*, 1998). LEC2 and FUS3 both possess B3 domains, a DNA-binding domain most similar to that found in the transcription factors VIVIPAROUS 1 (VP1) from maize and its apparent ortholog, ABSCISIC ACID (ABA) INSENSITIVE 3 (ABI3) from *Arabidopsis* (McCarty *et al.*, 1991; Giraudat *et al.*, 1992; Luerßen *et al.*, 1998; Stone *et al.*, 2001). Consistent with the finding that defects caused by the *lec* mutations are primarily limited to embryogenesis, all three *LEC* genes are expressed primarily during seed development (Lotan *et al.*, 1998; Luerßen *et al.*, 1998; Stone *et al.*, 2001).

1.3.2 Genes involved in the morphogenesis phase of embryo development

Genes specifically involved in establishing the embryo body plan have been identified through screens for *emb* and *seedling-defective* mutations. The rationale for the latter strategy is that mutations that cause defects in morphological development of the embryo may not cause lethality, but the defects are likely to be readily detectable during seedling development (Mayer *et al.*, 1991).

The apical–basal axis of the embryo comprises several pattern elements including the cotyledons, shoot apical meristem, hypocotyl, root, and root apical meristem (reviewed by Berleth and Chatfield, 2002). Genetic screens identified genes involved in establishing these elements. For example, *SHOOTMERISTEMLESS* (*STM*) and *WUS*, which encode different types of homeodomain proteins, play critical roles in the formation of the shoot apical meristem (Barton and Poethig, 1993; Laux *et al.*, 1996; Long *et al.*, 1996; Mayer *et al.*, 1998). *stm* mutant seeds germinate, but their seedlings do not have functional shoot apical meristems, whereas *wus* mutant seedlings have defective shoot apical meristems that can initiate only a few vegetative leaves. Recently, activities of both the *PINOID* (*PID*) and *ENHANCER OF PINOID* (*ENP*) genes have been shown to be required for the initiation of cotyledon development (Treml *et al.*, 2005). Mutations affecting patterning of the root apical meristem have also been identified. MONOPTEROS, an ARF (AUXIN RESPONSE FACTOR) transcription factor (Berleth and Jurgens, 1993; Hardtke and Berleth, 1998), and the Aux-IAA protein BODENLOS (BDL) (Hamann *et al.*, 1999, 2002) are required for the formation of the hypophysis cell of the embryo that is a precursor for the root apical meristem. Genes involved in specification of the radial axis that defines the tissue systems of the embryo have been identified. For example, *SHORT-ROOT* (*SHR*) and *SCARECROW* (*SCR*), which both encode putative GRAS family transcription factors (Benfey *et al.*, 1993; Scheres *et al.*, 1995), specify patterning of the ground tissue in the embryonic and vegetative root. Readers are referred to several excellent reviews that discuss genes involved in morphogenesis for further information (Jurgens, 2001; Berleth and Chatfield, 2002; Laux *et al.*, 2004).

1.3.3 Regulators of the maturation phase of embryo development

Many genes encoding transcription factors are required for the maturation phase of embryogenesis. The hormone ABA plays critical roles in controlling and coordinating processes that occur during maturation. Thus, it is not surprising that defects in ABA signaling, specifically mutations that confer ABA insensitivity, affect maturation processes. Such regulators include VP1 and ABI3, which are B3 domain transcription factors (McCarty *et al.*, 1991; Giraudat *et al.*, 1992), and ABI4 and ABI5, which are an AP2 domain transcription factor and a bZIP transcription factor, respectively (Finkelstein *et al.*, 1998; Finkelstein and Lynch, 2000). Although each of these genes is expressed in some vegetative tissues, all are expressed primarily during the maturation phase of embryogenesis and are required for completion of maturation (McCarty *et al.*, 1991; Giraudat *et al.*, 1992; Finkelstein *et al.*, 1998; Rohde *et al.*, 1999; Finkelstein and Lynch, 2000).

Of the *Arabidopsis* ABI transcription factors, ABI3 affects the widest range of processes during the maturation phase. Loss-of-function mutations in *ABI3*, *ABI4*, and *ABI5* cause ABA insensitivity and reduction in the levels of some seed-specific *LEA* (*LATE EMBRYOGENESIS ABUNDANT*) and *Em* (*EARLY METHIONINE-LABELED*) RNAs. However, mutations in *ABI3* have the most severe effects on seed development in that mutant embryos neither accumulate significant amounts of storage reserves nor acquire desiccation tolerance (Nambara *et al.*, 1992, 1995; Finkelstein, 1994). Additionally, *abi3* mutant seeds exhibit vivipary under humid conditions and mature seeds remain green after desiccation, suggesting that the maturation program is defective and that germinative and postgerminative development occurs prematurely. These results suggest that ABI3 is a critical positive regulator of maturation processes.

The ABI transcription factors interact with each other and with the LEC transcription factors during the maturation phase. ABI3 interacts physically with ABI5 (Nakamura *et al.*, 2001), although no other interactions among the ABI transcription factors or among the ABI and LEC transcription factors have been reported. *ABI3*, *ABI4*, and *ABI5* each interacts genetically with *LEC1* and *FUS3* to control maturation processes (Parcy *et al.*, 1997; Brocard-Gifford *et al.*, 2003). However, the precise mechanistic relationship between the ABI and LEC transcription factors remains elusive. For example, LEC1 and FUS3 have been proposed to regulate ABI3 levels (Parcy *et al.*, 1997), whereas LEC1 has been proposed to act through FUS3 and ABI3 to control processes during maturation (Kagaya *et al.*, 2005). It has also been proposed that LEC2 regulates FUS3 accumulation (Kroj *et al.*, 2003). Although the regulatory circuits controlling maturation processes remain to be defined, the ABI and LEC transcription factors clearly play key roles.

A loss-of-function mutation in the *Arabidopsis Enhanced Em Level* (*EEL*) gene that encodes a bZIP protein of the same class as ABI5 results in increased levels of *AtEm1* RNA, but has no other visible mutant seed phenotype (Bensmihen *et al.*, 2002). By contrast, *abi5* mutant seeds have decreased levels of *AtEm1* (Finkelstein and Lynch, 2000). Bensmihen *et al.* (2002) showed that EEL and ABI5 compete for the same binding sites in the *AtEm1* promoter, leading to the balanced regulation of *AtEm1* in wild-type seeds. Reduction of the expression of *EEL* alone or of *EEL* in combination with *AtbZIP67* and *AREB3* (ABA-Responsive Element Binding protein 3) that encode closely related bZIP proteins did not generate noticeable defects in seed development (Bensmihen *et al.*, 2005).

In addition to conferring insensitivity to ABA, the *abi* mutations also cause defects in sugar signaling (Arenas-Huertero *et al.*, 2000; Huijser *et al.*, 2000; Laby *et al.*, 2000; Rook *et al.*, 2001). *abi4* and, to a lesser extent, *abi3* and *abi5* mutant seeds germinate more readily than wild-type seeds in the presence of high concentrations of sugars. Consistent with this observation, ectopic expression of *ABI3*, *ABI4*, or *ABI5* causes seedling growth to be hypersensitive to sugars (Finkelstein, 2000; Brocard *et al.*, 2002; Chapter 12). Although the link between sugar and ABA signal transduction pathways remains unclear, these findings indicate the importance of sugar signaling in maturation processes, as will be discussed in greater detail subsequently.

1.4 Genetic control of endosperm development

The endosperm, formed by fertilization of the central cell of the female gametophyte with a sperm cell, initially undergoes free nuclear divisions to produce a syncytium of endosperm nuclei (Figures 1.1B and 1.1C; reviewed by Berger, 2003; Olsen, 2004). In both maize and *Arabidopsis*, as representatives of cereals and dicotyledonous oilseeds, respectively, patterning of the endosperm coenocyte (multinucleated cell) occurs with the establishment of three domains along the anterior–posterior axis: the micropylar (*Arabidopsis*) or embryo-surrounding (maize) region, the peripheral region, and the chalazal (*Arabidopsis*) or basal endosperm transfer layer (maize) region (Figures 1.1B–1.1D). Furthermore, maize possesses two additional regions: the aleurone and subaleurone layers. In maize, the basal endosperm transfer layer is involved in transferring nutrients from maternal tissues into the endosperm. The role of the micropylar or embryo-surrounding region is not known, although it has been speculated to be involved in nourishing the embryo. The peripheral region becomes the starchy endosperm in maize in which starch and proteins are stored to provide nutrients for the growing seedling. The *Arabidopsis* endosperm also appears to play a nutritional role, although it serves primarily to nourish the developing embryo rather than the seedling. Cellularization of the endosperm occurs in a wave-like manner from the micropylar endosperm through the peripheral endosperm to the chalazal endosperm (Figures 1.1C and 1.1D). The endosperm is absorbed by the growing embryo during seed development in nonendospermic seeds such as pea (*Pisum sativum*) and endospermic seeds with little persistent endosperm such as *Arabidopsis* (Figure 1.1E). By contrast, the endosperm constitutes most of the mature seed in cereals.

1.4.1 Genes required for cereal endosperm development

Genetic screens for *defective kernel* (*dek*) mutations of maize have identified genes required for seed development, and many have primary effects on the endosperm (Neuffer and Sheridan, 1980; Sheridan and Neuffer, 1980; Scanlon *et al.*, 1994). The *dek* mutants exhibit reductions in seed weight, mitotic activity, and, in most cases, endoreduplication (DNA replication without cell division) patterns (Kowles *et al.*, 1992). Some of the genes have roles in specifying cell fate. For example, aleurone cell specification requires *CRINKLY4*, which encodes a protein similar to tumor necrosis factor receptor-like receptor kinase (Becraft *et al.*, 1996). Aleurone cell fate is restricted to the outer epidermis of the endosperm by the activity of *Dek1*, which encodes a membrane protein with homology to animal calpains (Lid *et al.*, 2002). Other maize endosperm mutations have been identified that affect different parts of the endosperm. For example, the *reduced grain filling 1* (*rgf1*) mutation that affects pedicel development and the expression of transfer layer-specific markers causes final grain weight to be 30% of wild type, although the embryo is not affected by the mutation (Maitz *et al.*, 2000).

1.4.2 Genes that repress autonomous endosperm development

A group of *Arabidopsis* genes has been shown to be involved in controlling endosperm development based on their striking mutant phenotypes. Loss-of-function mutations in the *FERTILIZATION INDEPENDENT ENDOSPERM* (*FIE*), *FERTILIZATION INDEPENDENT SEED 2* (*FIS2*), *MEDEA* (*MEA*), *MULTICOPY SUPRESSOR OF IRA* (*MSI1*), *BORGIA* (*BGA*), and *RETINOBLASTOMA RELATED PROTEIN 1* (*RBR1*) genes cause the initiation of endosperm development in the absence of fertilization (Ohad *et al.*, 1996; Chaudhury *et al.*, 1997; Grossniklaus *et al.*, 1998; Kohler *et al.*, 2003a; Ebel *et al.*, 2004; Guitton *et al.*, 2004). These are gametophytic maternal-effect mutations, suggesting that the wild-type alleles of these genes operate in the female gametophyte, most likely the central cell, to control endosperm development. These genes encode proteins related to those present in polycomb group complexes that are known to be associated with gene silencing through histone modification. Therefore, these genes are likely to repress transcription at specific loci (Hsieh *et al.*, 2003; Orlando, 2003). The implication is that endosperm development is repressed prior to fertilization by polycomb group complexes that inhibit the expression of genes required for the initiation of endosperm development.

MEA encodes a SET-domain protein similar to the *Drosophila* ENHANCER OF ZESTE (Grossniklaus *et al.*, 1998; Kiyosue *et al.*, 1999; Luo *et al.*, 1999). Other SET-domain proteins have histone methyltransferase activity, indicating their importance in establishing a repressed chromatin state (Cao and Zhang, 2004). FIE, a protein with seven WD40 repeats that has similarity to *Drosophila* EXTRA SEX COMBS (Ohad *et al.*, 1999), interacts physically with MEA (Luo *et al.*, 2000; Spillane *et al.*, 2000; Yadegari *et al.*, 2000), as do their polycomb group counterparts in *Drosophila*. *FIS2* encodes a zinc-finger transcription factor similar to *Drosophila* SUPPRESSOR OF ZESTE 12 (Luo *et al.*, 1999) and may play a role in binding with the promoters of endosperm-specific genes. MSI1, a WD40 protein with similarity to the *Drosophila* retinoblastoma-binding protein P55, interacts physically with FIE and is likely to be part of the polycomb group complex (Kohler *et al.*, 2003a). RBR1 is similar to a mammalian retinoblastoma tumor suppressor protein that is a negative regulator of G1 to S phase transition of the cell cycle (Kong *et al.*, 2000). In *Drosophila*, polycomb group proteins interact with a retinoblastoma protein to form a complex that represses homeotic genes, thereby blocking mitosis (Dahiya *et al.*, 2001). The identity of the *BGA* gene is not yet known, but the mutant phenotype suggests that it encodes a polycomb group protein. A target gene repressed by this plant polycomb group complex, *PHERES1* (*PHE1*), has been identified (Kohler *et al.*, 2003b). *PHERES1* is expressed in the embryo and endosperm after fertilization and is thought to play a role in seed development.

Fertilization of ovules containing mutant alleles of the *FIE*, *FIS*, *MEA*, or *MSI1* genes causes seed development to be initiated (Ohad *et al.*, 1996; Chaudhury *et al.*, 1997; Grossniklaus *et al.*, 1998; Guitton *et al.*, 2004). However, embryos within the mutant seeds abort at about the torpedo stage and endosperm overproliferation occurs (Kiyosue *et al.*, 1999). Thus, genes encoding these plant polycomb group proteins have an additional role after fertilization to restrict the extent of endosperm growth. The effects of these mutations on seed mass will be discussed subsequently.

1.5 Genetic aspects of testa development

The testa encloses the embryo and endosperm. The testa serves a protective function against physical and ultraviolet light damage and, in some seeds, aids in seed dispersal and in the control of germination (Bewley and Black, 1995). In *Arabidopsis*, the mature ovule has inner and outer integuments, comprising three and two cell layers, respectively. After fertilization, integuments undergo anticlinal cell division and elongation that is positively regulated, in part, by *HAIKU1* (*IKU1*) and *IKU2* (Garcia *et al.*, 2003). As the testa develops, flavonoids and mucilage accumulate and specific cell layers senesce. In *Arabidopsis*, programmed cell death of the second and third layers of the innermost integument requires the activity of δVPE, a vacuolar-processing enzyme (Nakaune *et al.*, 2005).

1.5.1 Genetic regulation of flavonoid biosynthesis and accumulation

Of all aspects of testa development, most is known about flavonoid biosynthesis and accumulation. In this chapter, we will touch only briefly on selected aspects of anthocyanin and condensed tannin or proanthocyanidin biosynthesis and refer readers to recent reviews for additional details (Marles *et al.*, 2003; Broun, 2005; Chapter 2). Anthocyanins are pigments that are important, in part, for providing color to petals and UV protection in aerial organs and seeds (Chapple *et al.*, 1994). Proanthocyanidins (or condensed tannins) are soluble and colorless, accumulate in flowers, leaves, and testa of many plant species, and are agriculturally and environmentally important for reducing bloating in ruminants (Lees, 1992).

In *Arabidopsis*, anthocyanins and proanthocyanidins accumulate specifically in the endothelium, or innermost inner integument layer of the testa (Beeckman *et al.*, 2000). The *BANYULS* (*BAN*) gene controls the branch point between anthocyanins and proanthocyanidins (see Figure 2.4. for the pathway). *BAN* encodes an anthocyanidin reductase that converts anthocyanidin, a common precursor of both anthocyanins and proanthocyanidins, selectively to the latter (Xie *et al.*, 2003). RNA localization experiments showed that *BAN* is transiently expressed in seeds specifically in the endothelium after fertilization until the preglobular stage (Devic *et al.*, 1999). In loss-of-function *ban* mutants, anthocyanin accumulation in the testa is no longer restricted to the later part of seed development but also occurs in the endothelium during the early morphogenesis phase (Albert *et al.*, 1997). As a result, proanthocyanidins are reduced in *ban* seeds (Devic *et al.*, 1999). Members of the *TRANSPARENT TESTA* (*TT*) and *TRANSPARENT TESTA GLABRA* (*TTG*) mutant gene classes have been identified as regulators of proanthocyanidin and anthocyanin biosynthesis (Marles *et al.*, 2003; Broun, 2005).

1.5.2 Regulators of mucilage biosynthesis and accumulation

Mucilage is present in the testa of members of the Brassicaceae, Solanaceae, Linaceae, and Plantaginaceae families and plays an important role in seed hydration under natural conditions (Grubert, 1981). In *Arabidopsis*, mucilage accumulation occurs only in the outermost layer of the outer integument (Windsor *et al.*, 2000). The mucilage is primarily composed of pectin rhamnogalacturonan I (Western *et al.*, 2004). Starting in seeds at the torpedo stage and continuing during maturation, large amounts of mucilage accumulate in the outermost cells of the outer integument and are secreted in a ring between the plasma membrane and the cell wall (Beeckman *et al.*, 2000; Windsor *et al.*, 2000). As this occurs, the remaining cytoplasm is forced into a columnar shape (columella). After the mucilage is deposited and the columella is formed, a secondary cell wall is deposited external to the columella. During late maturation, the testa, including the mucilage, desiccates. When the mature seed imbibes, the mucilage is extruded from the testa to form a halo. Lack of mucilage extrusion is associated with reduced germination and seedling establishment under dry conditions (Penfield *et al.*, 2001), suggesting that mucilage is an important factor for seedling establishment under adverse conditions.

It has been proposed that there are at least three pathways controlling mucilage biosynthesis in the *Arabidopsis* testa (Western *et al.*, 2004). Two of these pathways involve a complex composed of AP2 and TTG1 and a bHLH (basic helix-loop-helix) protein, possibly ENHANCER OF GLABRA 3 (EGL3) and/or TT8, and a tissue-specific MYB transcription factor. In one pathway, this complex activates TTG2, which acts through a yet unknown factor to enhance mucilage biosynthesis and accumulation. In the second pathway, the complex activates GL2, which then activates *MUCILAGE MODIFIED 4* (*MUM4*), a putative pectin biosynthetic gene. In a third pathway, MYB61 is required for mucilage biosynthesis and accumulation but is not regulated by AP2 and TTG1.

1.6 Control of seed mass

The mass of a seed is determined by the number and size of cells within tissues that constitute its three components: embryo, endosperm, and testa. As will be discussed, interactions among the different components ultimately determine the size of the seed. In this section, we focus on genetic and physiological factors that influence seed mass.

1.6.1 Genetic factors affecting seed mass

Different accessions or ecotypes of *Arabidopsis* exhibit variations in seed mass. For example, seeds of the Cape Verde Island (Cvi) ecotype are almost double the weight of those of L*er* (Landsberg *erecta*) (Alonso-Blanco *et al.*, 1999). This natural variation in seed mass results from differences in both the number and size of cells within the embryo and testa. Reciprocal crosses between the two ecotypes showed that the variation in cell number is controlled primarily by maternal factors, whereas cell size is affected by nonmaternal factors. A number of quantitative trait loci (QTL) affecting seed mass co-localize with QTL affecting ovule and/or seed number, fruit size, and total leaf number. These traits emphasize the importance of maternal factors in controlling seed mass. In many species, there exists a negative correlation between seed mass and the number of seeds produced, which is attributed to limitations in resources from the mother plant (Harper *et al.*, 1970). Consistent with this finding, plants in which fruit number is controlled either by selected pollination of male sterile mutants or by manual removal of flowers produce seeds that are heavier than those of control plants (Jofuku *et al.*, 2005; Ohto *et al.*, 2005). Late-flowering mutants of *Arabidopsis* produce more leaves than wild type, and they set more seeds (Alonso-Blanco *et al.*, 1999). Moreover, seeds from late-flowering mutants are slightly larger than those from wild-type plants. This finding suggests that the increase in resource availability due to higher leaf number may overcome the negative effects of seed number on seed mass (Alonso-Blanco *et al.*, 1999). These correlations indicate that genetic factors affecting seed mass operate primarily through maternal tissues.

1.6.2 Testa development and seed mass

The testa is a maternal component of the seed that affects its mass. Differentiation of ovule integuments into the testa occurs initially through cell divisions, which in *Arabidopsis* end by about 4 days after pollination (DAP). This is followed by a period of cell elongation that continues until the seed achieves its maximal length at about 6–7 DAP (Alonso-Blanco *et al.*, 1999; Western *et al.*, 2000; Garcia *et al.*, 2005). Although the seed has reached maximal size by 6–7 DAP, the embryo is only at the bent-cotyledon stage and has not yet filled the seed, suggesting that testa growth may be a determinant of seed mass. The *ttg2* mutation that affects pigment production in the testa is a maternal mutation that affects seed mass (Garcia *et al.*, 2005). This mutation causes reduced growth of integument cells, suggesting that *TTG2* may regulate integument cell elongation. Similarly, seed mass is also affected by many *tt* mutations that cause defects in flavonoid synthesis in the seed coat (Debeaujon *et al.*, 2000; Chapter 2). It is possible that some intermediates or by-products of the proanthocyanidin biosynthesis pathway accumulate in integument cell walls and change the cells' competence to elongate.

1.6.3 Endosperm development and seed mass

The endosperm has a major influence on seed mass (Scott *et al.*, 1998; Garcia *et al.*, 2003). Because the endosperm makes up most of the cereal seed, genes that affect endosperm growth have direct effects on seed size. The endosperm also affects the mass of nonendospermic seeds and of endospermic seeds with small amounts of residual endosperm. In such seeds, the endosperm proliferates extensively and serves as a sink for nutrients supplied by the maternal plant during early seed development to provide nutrients to sustain embryo growth later in embryogenesis. In addition, endosperm growth in coordination with testa growth early during seed development may have a primary role in determining the size of the postfertilization embryo sac, which, in turn, appears to influence embryo and seed size.

Endosperm growth is influenced by parent-of-origin effects. For example, interploidy crosses between diploid females and tetraploid males produce offspring with an excess of paternal genomes (Figure 1.2). This imbalance between parental genomes results in overproliferation of endosperm nuclei, a delay in the onset of endosperm cellularization, and overgrowth of the chalazal endosperm involved in transferring nutrient resources from the mother plant (Scott *et al.*, 1998). Because of these changes in endosperm development, seeds with an excess of paternal genomes are larger than progeny from diploid parents. Reciprocal crosses, resulting in an excess of maternal genomes, have the opposite effect on endosperm development and give rise to smaller seeds (Figure 1.2; Scott *et al.*, 1998). The parental conflict theory

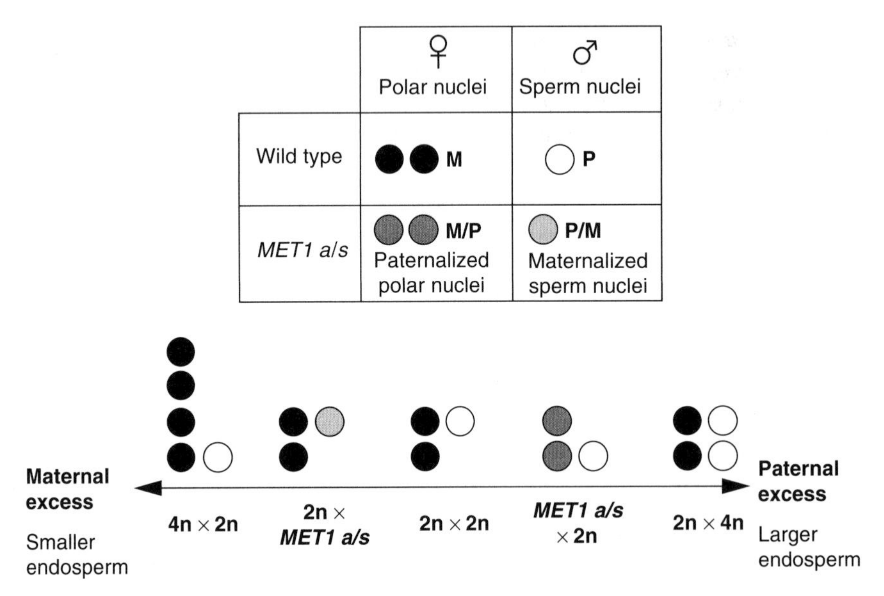

Figure 1.2 Parent-of-origin effects on endosperm size. Endosperm nuclei resulting from self-pollination of a diploid contain two maternal genomes, contributed by polar nuclei from the female gametophyte, and one paternal genome, contributed by the sperm. Interploidy crosses produce seeds with an excess of either maternal or paternal genomes relative to progeny from self-fertilized diploid plants. Maternal or paternal excesses result in small seeds with small endosperms or large seeds with large endosperms, respectively. The *MET1 antisense* (*a*/*s*) transgene induces DNA hypomethylation. Reciprocal crosses between diploid *MET1 a/s* plants with nontransgenic diploids reproduce the parent-of-origin effects observed in interploidy crosses; *MET1 a*/*s* in maternal or paternal genomes results in large seeds with large endosperm or small seeds with small endosperm, respectively. These observations suggest that *MET1 a/s* transgene activates imprinted loci that are normally silenced by DNA methylation. (Modified from Adams *et al.*, 2000.)

has been proposed to explain these parent-of-origin effects on seed size (Haig and Westoby, 1989, 1991). This theory proposes that the paternal plant will attempt to channel maternal resources primarily to its progeny, whereas the maternal plant will try to allocate resources equally among all progeny.

The parent-of-origin effects on seed size observed in interploidy crosses can be phenocopied by crossing plants with defects in DNA methylation. METHYLTRANSFERASE 1 (MET1) is a maintenance methylase that catalyzes the covalent attachment of methyl groups to cytosine residues in DNA (Finnegan and Dennis, 1993). Plants expressing a *MET1* antisense gene (*MET1 a/s*) have reduced levels of DNA methylation, resulting in the ectopic activation of genes that are normally silenced (Finnegan *et al.*, 1996). Crosses between nontransgenic females and hypomethylated males containing *MET1 a/s* produce seeds that exhibit underproliferation of endosperm nuclei, precocious endosperm cellularization, reduction in chalazal endosperm growth, and smaller seed size, similar to progeny from interploidy crosses that have an excess of maternal genomes (Adams *et al.*, 2000). Crosses between hypomethylated females and normal males produce progeny with the opposite endosperm phenotype and produce larger seeds. Thus, hypomethylation of female or male gametophyte genomes causes paternalization or maternalization of the endosperm, respectively. These results have been interpreted to indicate that imprinting underlies parent-of-origin effects (Berger, 2003; Gehring *et al.*, 2004; Autran *et al.*, 2005). That is, DNA methylation is thought to silence parent-specific loci that would otherwise promote or inhibit endosperm proliferation and seed growth.

As described above, some endosperm mutants are also defective in embryo development. Seed abortion occurs when ovules containing *fie*, *fis2*, and *mea* mutant alleles are self-fertilized or pollinated with wild-type pollen (Ohad *et al.*, 1996; Chaudhury *et al.*, 1997; Grossniklaus *et al.*, 1998; Guitton et al., 2004). However, mutant ovules can produce viable seeds if they are fertilized with pollen with hypomethylated paternal genomes (Vielle-Calzada *et al.*, 1999; Luo *et al.*, 2000; Vinkenoog *et al.*, 2000). Thus, activation of paternal alleles thought to be silenced by DNA methylation rescues embryos from abortion caused by mutations in *FIE*, *FIS2*, and *MEA* genes. With the *fie* mutant, a wild-type *FIE* allele in the hypomethylated paternal genome is required for seed rescue (Vinkenoog *et al.*, 2000). By contrast, it is not clear whether it is the wild-type *FIS2* and *MEA* genes or other silenced loci that must be activated in pollen to produce viable seeds in *fis2* and *mea* mutants (Berger, 2003). In addition to rescuing seed viability, pollen with hypomethylated paternal genomes also enhanced endosperm growth in seeds with a *fie* mutant maternal allele. Rescued seeds with maternal *fie* mutant alleles were 1.5-fold larger than wild-type self-fertilized seeds (Vinkenoog *et al.*, 2000). Developing seeds displayed characteristics similar to those caused by a paternal genome excess in interploidy crosses (Vinkenoog *et al.*, 2000). Together, these findings suggest that *FIE*, *FIS2*, and *MEA* genes are essential for endosperm and embryo development but negatively influence endosperm growth.

Arabidopsis HAIKU1 (*IKU1*) and *IKU2* genes influence seed mass (Garcia *et al.*, 2003). Endosperm size is decreased in *iku* mutants as early as the globular stage, and cellularization occurs prematurely as compared to wild-type seeds. Thus, *iku* seeds have reduced mass. Although the *iku* mutations affect endosperm development as do the gametophytic maternal effects of *fie*, *fis2*, and *mea* mutations, *IKU* genes act sporophytically to control endosperm development. The mechanistic relationship between the *IKU* and *FIE/FIS2/MEA* genes is not understood. Effects of the *iku* mutations on seed mass result from a coordinated reduction in endosperm size, embryo growth, and the elongation of integument cells. These defects occur even when the homozygous *iku* endosperm and embryo are present within a testa heterozygous for the recessive *iku* mutation (Garcia *et al.*, 2003). This finding suggests that the endosperm produces signals that regulate integument cell elongation to coordinate overall seed growth (Garcia *et al.*, 2003, 2005).

1.6.4 Sugar transport and metabolism during seed development

Sugar metabolism has been implicated to be an important determinant in the control of seed mass. Studies of sugar transport, metabolism, and sensing in faba bean (*Vicia faba*) have uncovered potential mechanisms by which changes in sugar composition

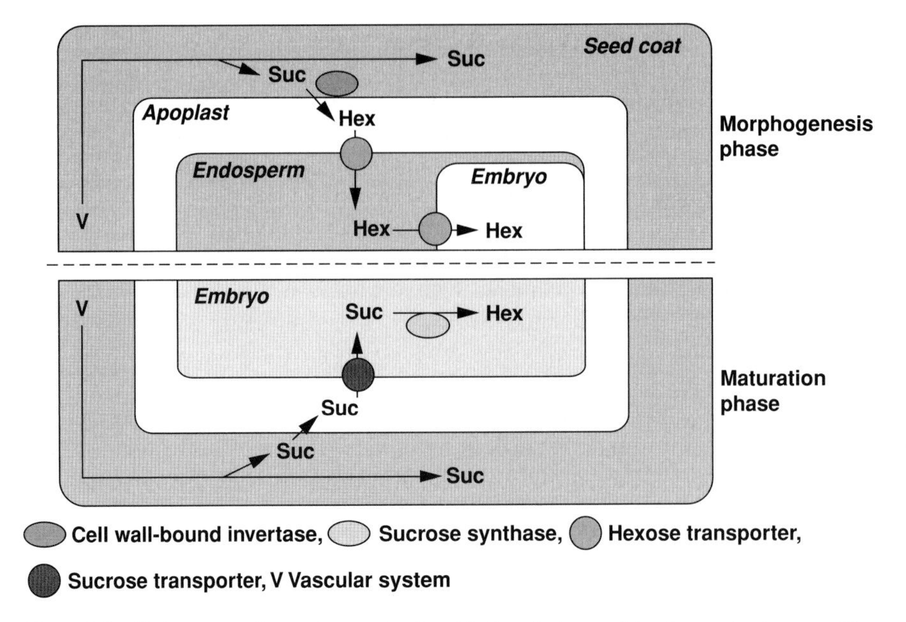

Figure 1.3 Major routes of sugar transport and metabolism during seed development in legumes. The upper half of the diagram represents the major pathway of sucrose import and metabolism during the morphogenesis phase of seed development. Sucrose imported from the mother plant through the vascular system is distributed in testa tissue via plasmodesmata. Testa-associated cell wall-bound invertase activity is high during this period of seed development, causing sucrose to be converted into hexoses. The hexoses move across the apoplastic space and are transported into the embryo via hexose transporters. The lower half of the diagram represents the major pathway of sucrose import and metabolism during the maturation phase. Because invertase activity is low during this period, sucrose is taken up directly by the embryo. Sucrose synthase catalyzes the breakdown of sucrose into hexoses.

and concentration may control endosperm and embryo growth (reviewed by Weber *et al.*, 1997b; Wobus and Weber, 1999; Borisjuk *et al.*, 2003, 2004). Before discussing these studies, we first briefly review sugar transport and metabolism in seeds.

Sucrose is the primary form of sugar transported to developing seeds. Sucrose is transported to testa parenchyma tissue by symplastic phloem unloading, and it is metabolized in one of two ways, depending upon the developmental status of the seed (Figure 1.3; Weber *et al.*, 1997b). During the morphogenesis phase of faba bean seed development, most of the sucrose is hydrolyzed to the hexoses, glucose and fructose, by cell wall-bound acid invertases (cwbINV) localized at the innermost tissue of the testa, the thin-walled parenchyma (Weber *et al.*, 1995). The hexoses produced move across the apoplastic space into the endosperm and embryo via transporters located in epidermal cells (Weber *et al.*, 1997a) and move symplastically to the inner tissues. Sucrose can also be transported directly into the embryo and endosperm through sucrose transporters. Some of this transported sucrose is hydrolyzed into hexoses by vacuolar soluble invertases in the endosperm or by sucrose synthase (SuSy) in the embryo (Weber *et al.*, 1995, 1996b, 1997a). During early seed development, hexoses are the primary sugars that accumulate in the embryo.

During the maturation phase, apoplastic conversion of sucrose to hexoses declines because cwbINV activity decreases as the thin-walled parenchyma of the testa is crushed by the enlarging embryo (Weber *et al.*, 1995). The decrease in cwbINV activity is the primary reason why the hexoses to sucrose ratio is low in the maturation phase. Instead, sucrose is transported predominantly directly into the embryo from the testa via a sucrose transporter at the epidermis of the embryo (Weber *et al.*, 1997a) and is metabolized by SuSy (Weber *et al.*, 1996b). By this stage of seed development, the endosperm in faba bean has largely been absorbed by the developing embryo. Thus, there is a developmentally regulated shift in the ratio of hexoses to sucrose; this ratio is high in the morphogenesis phase and low in the maturation phase. Similar shifts in sugar composition have also been observed in *Arabidopsis*, pea and canola (*Brassica napus*) (Focks and Benning, 1998; Borisjuk *et al.*, 2002; Hill *et al.*, 2003), suggesting that basic mechanisms of sugar transport and metabolism may be conserved in these species. In pea, sucrose transporters in developing embryos are localized in both epidermal cells and inner storage parenchyma cells (Tegeder *et al.*, 1999). In canola, soluble acid invertases in the endosperm may make a greater contribution to sucrose metabolism than seed coat-associated acid invertases during the morphogenesis phase, whereas alkaline invertases may contribute to sucrose metabolism during the maturation phase (Hill *et al.*, 2003).

1.6.5 Metabolic control of seed development and size

Variations in seed mass in faba bean, pea, and cereals are strictly correlated with differences in cell number and not cell size (Wobus and Weber, 1999). Therefore, mechanisms that dictate seed size must control the extent of cell proliferation in the embryo, endosperm, and/or testa. Changes in sugar composition that occur during seed development are thought to control the shift from cell division to seed filling. A high hexoses to sucrose ratio is correlated positively with cell divisions (Wobus and Weber, 1999). Histographical mapping of glucose concentrations in faba bean showed that cotyledon regions with high mitotic activities are characterized by high hexose levels (Borisjuk *et al.*, 1998). Furthermore, hexose transporter genes are expressed in epidermal cells over mitotically active regions of the cotyledons (Weber *et al.*, 1997a). By contrast, a low hexoses to sucrose ratio has a different effect on embryo differentiation. High sucrose levels are correlated with seed filling regardless of whether the embryo stores starch or oils (Weber *et al.*, 1995; Focks and Benning, 1998; Hill *et al.*, 2003). In faba bean embryos, sucrose transporter genes are uniformly expressed in epidermal cell layers during the morphogenesis phase but are more strongly expressed in regions of the epidermis overlying storage parenchyma cells with high storage activity during the maturation phase (Weber *et al.*, 1997a). Histographical imaging revealed that sucrose was concentrated in regions of the cotyledon with cells that are elongating and accumulating starch (Borisjuk *et al.*, 2002).

In vitro culture of faba bean cotyledons provided evidence to support a role for sugar composition in controlling seed development (Weber *et al.*, 1996a). Cotyledons from mitotically active embryos were cultured in medium containing either a high or a low ratio of hexoses to sucrose. Cotyledons cultured in high hexoses medium maintained mitotic activity, whereas cotyledons cultured with high sucrose medium showed a significant decrease in mitotic activity. Cotyledons in the high sucrose medium displayed an increase in cell size and nuclear diameter, and they started to accumulate starch and storage proteins. Conversely, cotyledons from maturation-phase embryos exhibited a decrease in storage protein gene expression when cultured with high hexoses medium (Weber *et al.*, 1996a).

Cell wall-bound invertases have been implicated to control seed mass through their role in modulating changes in sugar composition during seed development. For example, two faba bean genotypes that produce seeds of different sizes were compared (Weber *et al.*, 1996a). Seed coats from the large-seeded genotype have a much larger region of thin-walled parenchyma that is the site at which sucrose is hydrolyzed by cwbINV. This seed coat tissue remained intact in the large-seeded genotype for a longer period of seed development than it did in small-seeded genotypes. Thus, extension of the period of cwbINV activity correlates with a prolonged period of cell division and a larger seed. A similar mechanism to control cell proliferation in seed tissues through cwbINV also appears to operate in cereals. Maize *miniature1* (*mn1*) mutant seeds are smaller than wild type; endosperm weight is approximately 20% of normal (Vilhar *et al.*, 2002). The *Mn1* gene encodes an endosperm-specific cwbINV that normally accumulates in the basal endosperm transfer layer (Cheng *et al.*, 1996). Thus, *mn1* mutant endosperm has reduced cwbINV activity that may cause reduced mitotic activity.

In contrast to the role of hexoses produced by cwbINV in cell proliferation, sucrose is related to seed filling. The role of sucrose in the control of seed size was tested in *Vicia narbonesis*, a species closely related to faba bean. A secreted form of yeast invertase expressed during the maturation phase under the control of the *legumin B4* promoter acts in the apoplast of cotyledons (Weber *et al.*, 1998; Wobus and Weber, 1999). The accumulation of starch and some storage proteins was reduced in transgenic cotyledon cells. These transgenic cells also had large and persistent vacuoles instead of protein bodies, and hexoses were present at high levels in the vacuoles. As a result, mature dried seeds were smaller than wild-type seeds. The *shrunken1* (*sh1*) mutant of maize also emphasizes the importance of sucrose in seed filling. *sh1* seeds are defective in the accumulation of storage starch in endosperm and are smaller than wild-type seeds (Chourey and Nelson, 1976). Because *sh1* encodes SuSy, mutant seeds are unable to utilize sucrose efficiently for further metabolism (Echt and Chourey, 1985).

Seeds from the *Arabidopsis* floral homeotic mutant *ap2* are larger than wild type and they accumulate more storage proteins and lipids (Jofuku *et al.*, 2005; Ohto *et al.*, 2005). Although the *ap2* mutation causes reduced fertility due to defects in flower structure, the mutation appears to have direct effects on seed mass that go beyond the positive effects that result from a reduction in seed number. Similar to the effects of natural variation observed among *Arabidopsis* ecotypes, the *ap2* mutation affects seed mass by increasing cell number and size (Ohto *et al.*, 2005). Sugar composition differed between developing *ap2* mutant seeds and wild-type seeds. The ratio of hexoses to sucrose remains high in *ap2* mutants for an extended period of time as compared to wild type (Ohto *et al.*, 2005). This defect in sugar metabolism may explain the increase in cell number in *ap2* mutant seeds; the prolonged period in which the hexoses to sucrose ratio remains high may allow cell division to continue for a longer period of time in *ap2* mutants. Maternal tissues appear to be responsible for these changes in sugar composition. The recessive *ap2* mutation affects seed mass through the maternal sporophyte and endosperm genomes (Jofuku *et al.*, 2005). In addition, the *ap2* mutant testa has irregularly shaped epidermal cells that lack an epidermal plateau and produces reduced amounts of mucilage (Jofuku *et al.*, 1994). Thus, it is possible that the *ap2* mutation may induce defects, perhaps involving testa-associated cwbINV, that extend the period of seed development during which the ratio of hexoses to sucrose remains high.

1.7 Perspective

In the past 15 years, significant progress has been made in defining regulatory genes that control aspects of seed development. A major challenge is to define the gene networks that operate during seed development. Identifying all of the genes that are expressed in the different compartments of the seed and defining their functions will allow for dissection of the regulatory circuitries controlling seed development and the definition of gene products that underlie the processes that occur during seed development. Initial progress toward this goal has been made with descriptions of RNA populations in developing seeds and/or embryos (Girke *et al.*, 2000; Hennig *et al.*, 2004; Casson *et al.*, 2005; Grimanelli *et al.*, 2005; Ma *et al.*, 2005). Although much work remains to be done, the tools are in place to accomplish these goals. Ultimately, this knowledge, integrated with our understanding of the physiological, biochemical, and morphological processes in seeds, will provide a comprehensive understanding of seed development.

Acknowledgements

We thank Sandra Floyd and Samantha Duong for help in preparing some of the micrographs used in Figure 1.1 and Siobhan Braybrook for her comments about the manuscript. Research cited in this review from our laboratory was supported by grants from the National Science Foundation and the United States Department of Energy.

References

S. Adams, R. Vinkenoog, M. Spielman, H. Dickinson and R. Scott (2000) Parent-of-origin effects on seed development in *Arabidopsis thaliana* require DNA methylation. *Development* **127**, 2493–2502.

S. Albert, M. Delseny and M. Devic (1997) *BANYULS*, a novel negative regulator of flavonoid biosynthesis in the *Arabidopsis* seed coat. *The Plant Journal* **11**, 289–299.

C. Alonso-Blanco, H. Blankestijn-de Vries, C.J. Hanhart and M. Koornneef (1999) Natural allelic variation at seed size loci in relation to other life history traits of *Arabidopsis thaliana*. *Proceedings of the National Academy of Sciences of the United States of America* **96**, 4710–4717.

N.R. Apuya, R. Yadegari, R.L. Fischer, J.J. Harada, J.L. Zimmerman and R.B. Goldberg (2001) The *Arabidopsis* embryo mutant *schlepperless* has a defect in the *chaperonin-60α* gene. *Plant Physiology* **126**, 717–730.

F. Arenas-Huertero, A. Arroyo, L. Zhou, J. Sheen and P. Leon (2000) Analysis of *Arabidopsis* glucose-insensitive mutants, *gin5* and *gin6*, reveals a central role of the plant hormone ABA in the regulation of plant vegetative development by sugar. *Genes and Development* **14**, 2085–2096.

D. Autran, W. Huanca-Mamani and J.-P. Vielle-Calzada (2005) Genomic imprinting in plants: the epigenetic version of an Oedipus complex. *Current Opinion in Plant Biology* **8**, 19–25.

M.K. Barton and R.S. Poethig (1993) Formation of the shoot apical meristem in *Arabidopsis thaliana*: an analysis of development in the wild type and in the *shoot meristemless* mutant. *Development* **119**, 823–831.

H. Bäumlein, S. Miséra, H. Luerßen, *et al.* (1994) The *FUS3* gene of *Arabidopsis thaliana* is a regulator of gene expression during late embryogenesis. *The Plant Journal* **6**, 379–387.

P.W. Becraft, P.S. Stinard and D.R. McCarty (1996) CRINKLY4: A TNFR-like receptor kinase involved in maize epidermal differentiation. *Science* **273**, 1406–1409.

T. Beeckman, R. De Rycke, R. Viane and D. Inze (2000) Histological study of seed coat development in *Arabidopsis thaliana*. *Journal of Plant Research* **113**, 139–148.

P.N. Benfey, P.J. Linstead, K. Roberts, J.W. Schiefelbein, M.-T. Hauser and R.A. Aeschbacher (1993) Root development in *Arabidopsis*: four mutants with dramatically altered root morphogenesis. *Development* **119**, 57–70.

S. Bensmihen, J. Giraudat and F. Parcy (2005) Characterization of three homologous basic leucine zipper transcription factors (bZIP) of the ABI5 family during *Arabidopsis thaliana* embryo maturation. *Journal of Experimental Botany* **56**, 597–603.

S. Bensmihen, S. Rippa, G. Lambert, *et al.* (2002) The homologous ABI5 and EEL transcription factors function antagonistically to fine-tune gene expression during late embryogenesis. *The Plant Cell* **14**, 1391–1403.

F. Berger (2003) Endosperm: the crossroad of seed development. *Current Opinion in Plant Biology* **6**, 42–50.

T. Berleth and S. Chatfield (2002) Embryogenesis: pattern formation from a single cell. In: *The Arabidopsis Book* (eds C.R. Somerville and E.M. Meyerowitz). American Society of Plant Biologists, Rockville, MD. http://www.aspb.org/publications/arabidopsis/.

T. Berleth and G. Jurgens (1993) The role of the monopteros gene in organising the basal body region of the *Arabidopsis* embryo. *Development* **118**, 575–587.

J.D. Bewley and M. Black (1995) *Seeds: Physiology of Development and Germination*. Plenum Press, New York.

F.D. Boesewinkel and F. Bouman (1984) The seed: structure. In: *Embryology of Angiosperms* (ed. B.M. Johri), pp. 567–610. Springer-Verlag, Berlin.

L. Borisjuk, H. Rolletschek, R. Radchuk, W. Weschke, U. Wobus and H. Weber (2004) Seed development and differentiation: a role for metabolic regulation. *Plant Biology* **6**, 375–386.

L. Borisjuk, H. Rolletschek, U. Wobus and H. Weber (2003) Differentiation of legume cotyledons as related to metabolic gradients and assimilate transport into seeds. *Journal of Experimental Botany* **54**, 503–512.

L. Borisjuk, S. Walenta, H. Rolletschek, W. Mueller-Klieser, U. Wobus and H. Weber (2002) Spatial analysis of plant metabolism: sucrose imaging within *Vicia faba* cotyledons reveals specific developmental patterns. *The Plant Journal* **29**, 521–530.

L. Borisjuk, S. Walenta, H. Weber, W. Mueller-Klieser and U. Wobus (1998) High-resolution histographical mapping of glucose concentrations in developing cotyledons of *Vicia faba* in relation to mitotic activity and storage processes: glucose as a possible developmental trigger. *The Plant Journal* **15**, 583–591.

L. Borisjuk, T.L. Wang, H. Rolletschek, U. Wobus and H. Weber (2002) A pea seed mutant affected in the differentiation of the embryonic epidermis is impaired in embryo growth and seed maturation. *Development* **129**, 1595–1607.
K. Boutilier, R. Offringa, V.K. Sharma, *et al.* (2002) Ectopic expression of *BABY BOOM* triggers a conversion from vegetative to embryonic growth. *The Plant Cell* **14**, 1737–1749.
I.M. Brocard, T.J. Lynch and R.R. Finkelstein (2002) Regulation and role of the *Arabidopsis ABSCISIC ACID-INSENSITIVE 5* gene in abscisic acid, sugar, and stress response. *Plant Physiology* **129**, 1533–1543.
I.M. Brocard-Gifford, T.J. Lynch and R.R. Finkelstein (2003) Regulatory networks in seeds integrating developmental, abscisic acid, sugar, and light signaling. *Plant Physiology* **131**, 78–92.
P. Broun (2005) Transcriptional control of flavonoid biosynthesis: a complex network of conserved regulators involved in multiple aspects of differentiation in *Arabidopsis*. *Current Opinion in Plant Biology* **8**, 272–279.
R. Cao and Y. Zhang (2004) The functions of E(Z)/EZH2-mediated methylation of lysine 27 in histone H3. *Current Opinion in Genetics and Development* **14**, 155–164.
S. Casson, M. Spencer, K. Walker and K. Lindsey (2005) Laser capture microdissection for the analysis of gene expression during embryogenesis of *Arabidopsis*. *The Plant Journal* **42**, 111–123.
C.C.S. Chapple, B.W. Shirley, M. Zook, R. Hammerschmidt and S.C. Sommerville (1994) Secondary metabolism in *Arabidopsis*. In: *Arabidopsis* (eds M.E. Meyerowitz and C.R. Sommerville), pp. 989–1030. Cold Spring Harbor Laboratory Press, Cold Spring Harbor, NY.
A.M. Chaudhury, L. Ming, C. Miller, S. Craig, E.S. Dennis and W.J. Peacock (1997) Fertilization-independent seed development in *Arabidopsis thaliana*. *Proceedings of the National Academy of Sciences of the United States of America* **94**, 4223–4228.
W.H. Cheng, E.W. Taliercio and P.S. Chourey (1996) The MINIATURE1 seed locus of maize encodes a cell wall invertase required for normal development of endosperm and maternal cells in the pedicel. *The Plant Cell* **8**, 971–983.
P.S. Chourey and O.E. Nelson (1976) The enzymatic deficiency conditioned by the *shrunken 1* mutations in maize. *Biochemical Genetics* **14**, 1041–1055.
J.K. Clark and W.F. Sheridan (1991) Isolation and characterization of 51 embryo-specific mutations of maize. *The Plant Cell* **3**, 935–951.
A. Dahiya, S. Wong, S. Gonzalo, M. Gavin and D.C. Dean (2001) Linking the Rb and polycomb pathways. *Molecular Cell* **8**, 557–568.
I. Debeaujon, K.M. Leon-Kloosterziel and M. Koornneef (2000) Influence of the testa on seed dormancy, germination, and longevity in *Arabidopsis*. *Plant Physiology* **122**, 403–414.
M. Devic, J. Guilleminot, I. Debeaujon, *et al.* (1999) The *BANYULS* gene encodes a DFR-like protein and is a marker of early seed coat development. *The Plant Journal* **19**, 387–398.
G.N. Drews, D. Lee and C.A. Christensen (1998) Genetic analysis of female gametophyte development and function. *The Plant Cell* **10**, 5–18.
C. Ebel, L. Mariconti and W. Gruissem (2004) Plant retinoblastoma homologues control nuclear proliferation in the female gametophyte. *Nature* **429**, 776–780.
C.S. Echt and P.S. Chourey (1985) A comparison of two sucrose synthetase isozymes from normal and *shrunken-1* maize. *Plant Physiology* **79**, 530–536.
R.R. Finkelstein (1994) Mutations at two new *Arabidopsis* ABA response loci are similar to the *abi3* mutations. *The Plant Journal* **5**, 765–771.
R.R. Finkelstein and T.J. Lynch (2000) The *Arabidopsis* abscisic acid response gene *ABI5* encodes a basic leucine zipper transcription factor. *The Plant Cell* **12**, 599–610.
R.R. Finkelstein, M.L. Wang, T.J. Lynch, S. Rao and H.M. Goodman (1998) The *Arabidopsis* abscisic acid response locus *ABI4* encodes an APETALA 2 domain protein. *The Plant Cell* **10**, 1043–1054.
E.J. Finnegan and E.S. Dennis (1993) Isolation and identification by sequence homology of a putative cytosine methyltransferase from *Arabidopsis thaliana*. *Nucleic Acids Research* **21**, 2383–2388.
E.J. Finnegan, W.J. Peacock and E.S. Dennis (1996) Reduced DNA methylation in *Arabidopsis thaliana* results in abnormal plant development. *Proceedings of the National Academy of Sciences of the United States of America* **93**, 8449–8454.

N. Focks and C. Benning (1998) *wrinkled1*: A novel, low-seed-oil mutant of *Arabidopsis* with a deficiency in the seed-specific regulation of carbohydrate metabolism. *Plant Physiology* **118**, 91–101.

L.H. Franzmann, E.S. Yoon and D.W. Meinke (1995) Saturating the genetic map of *Arabidopsis thaliana* with embryonic mutations. *The Plant Journal* **7**, 341–350.

D. Garcia, J.N. Fitz Gerald and F. Berger (2005) Maternal control of integument cell elongation and zygotic control of endosperm growth are coordinated to determine seed size in *Arabidopsis*. *The Plant Cell* **17**, 52–60.

D. Garcia, V. Saingery, P. Chambrier, U. Mayer, G. Jurgens and F. Berger (2003) *Arabidopsis haiku* mutants reveal new controls of seed size by endosperm. *Plant Physiology* **131**, 1661–1670.

S. Gazzarrini, Y. Tsuchiya, S. Lumba, M. Okamoto and P. McCourt (2004) The transcription factor *FUSCA3* controls developmental timing in *Arabidopsis* through the hormones gibberellin and abscisic acid. *Developmental Cell* **7**, 373–385.

M. Gehring, Y. Choi and R.L. Fischer (2004) Imprinting and seed development. *The Plant Cell* **16**, S203–S213.

J. Giraudat, B.M. Hauge, C. Valon, J. Smalle, F. Parcy and H.M. Goodman (1992) Isolation of the Arabidopsis *ABI3* gene by positional cloning. *The Plant Cell* **4**, 1251–1261.

T. Girke, J. Todd, S. Ruuska, J. White, C. Benning and J. Ohlrogge (2000) Microarray analysis of developing *Arabidopsis* seeds. *Plant Physiology* **124**, 1570–1581.

R.B. Goldberg, G. de Paiva and R. Yadegari (1994) Plant embryogenesis: zygote to seed. *Science* **266**, 605–614.

D. Grimanelli, E. Perotti, J. Ramirez and O. Leblanc (2005) Timing of the maternal-to-zygotic transition during early seed development in maize. *The Plant Cell* **17**, 1061–1072.

U. Grossniklaus, J.-P. Vielle-Calzada, M.A. Hoeppner and W.B. Gagliano (1998) Maternal control of embryogenesis by *MEDEA*, a *polycomb* group gene in *Arabidopsis*. *Science* **280**, 446–450.

M. Grubert (1981) *Mucilage or Gum in Seeds and Fruits of Angiosperms: A Review*. Minerva-Publikation, Munich.

A.-E. Guitton, D.R. Page, P. Chambrier, *et al.* (2004) Identification of new members of FERTILIZATION INDEPENDENT SEED polycomb group pathway involved in the control of seed development in *Arabidopsis thaliana*. *Development* **131**, 2971–2981.

D. Haig and M. Westoby (1989) Parent-specific gene expression and the triploid endosperm. *American Naturalist* **134**, 147–155.

D. Haig and M. Westoby (1991) Genomic imprinting in endosperm: its effect on seed development in crosses between species, and between different ploidies of the same species, and its implications for the evolution of apomixis. *Philosophical Transactions of the Royal Society of London. Series B: Biological Sciences* **333**, 1–13.

T. Hamann, E. Benkova, I. Baurle, M. Kientz and G. Jurgens (2002) The *Arabidopsis BODENLOS* gene encodes an auxin response protein inhibiting MONOPTEROS-mediated embryo patterning. *Genes and Development* **16**, 1610–1615.

T. Hamann, U. Mayer and G. Jurgens (1999) The auxin-insensitive *bodenlos* mutation affects primary root formation and apical–basal patterning in the *Arabidopsis* embryo. *Development* **126**, 1387–1395.

J.J. Harada (1997) Seed maturation and control of germination. In: *Advances in Cellular and Molecular Biology of Plants, Vol. 4: Cellular and Molecular Biology of Seed Development* (eds B.A. Larkins and I.K.Vasil), pp. 545–592. Kluwer Academic Publishers, Dordrecht.

J.J. Harada (2001) Role of *Arabidopsis LEAFY COTYLEDON* genes in seed development. *Journal of Plant Physiology* **158**, 405–409.

C.S. Hardtke and T. Berleth (1998) The *Arabidopsis* gene *MONOPTEROS* encodes a transcription factor mediating embryo axis formation and vascular development. *The EMBO Journal* **17**, 1405–1411.

J.L. Harper, P.H. Lovell and K.G. Moore (1970) The shapes and sizes of seeds. *Annual Review of Ecology and Systematics* **1**, 327–356.

L. Hennig, W. Gruissem, U. Grossniklaus and C. Kohler (2004) Transcriptional programs of early reproductive stages in *Arabidopsis*. *Plant Physiology* **135**, 1765–1775.
L.M. Hill, E.R. Morley-Smith and S. Rawsthorne (2003) Metabolism of sugars in the endosperm of developing seeds of oilseed rape. *Plant Physiology* **131**, 228–236.
S.K. Hong, T. Aoki, H. Kitano, H. Satoh and Y. Nagato (1995) Phenotypic diversity of 188 rice embryo mutants. *Developmental Genetics* **16**, 298–310.
T.F. Hsieh, O. Hakim, N. Ohad and R.L. Fischer (2003) From flour to flower: how polycomb group proteins influence multiple aspects of plant development. *Trends in Plant Science* **8**, 439–445.
C. Huijser, A. Kortstee, J. Pego, P. Weisbeek, E. Wisman and S. Smeekens (2000) The *Arabidopsis SUCROSE UNCOUPLED-6* gene is identical to *ABSCISIC ACID INSENSITIVE-4*: involvement of abscisic acid in sugar responses. *The Plant Journal* **23**, 577–585.
K.D. Jofuku, B. Boer, M.V. Montagu and J.K. Okamuro (1994) Control of *Arabidopsis* flower and seed development by the homeotic gene *APETALA2*. *The Plant Cell* **6**, 1211–1225.
K.D. Jofuku, P.K. Omidyar, Z. Gee and J.K. Okamuro (2005) Control of seed mass and seed yield by the floral homeotic gene *APETALA2*. *Proceedings of the National Academy of Sciences of the United States of America* **102**, 3117–3122.
G. Jurgens (2001) Apical–basal pattern formation in *Arabidopsis* embryogenesis. *The EMBO Journal* **20**, 3609–3616.
Y. Kagaya, R. Toyoshima, R. Okuda, H. Usui, A. Yamamoto and T. Hattori (2005) LEAFY COTYLEDON1 controls seed storage protein genes through its regulation of *FUSCA3* and *ABSCISIC ACID INSENSITIVE3*. *Plant and Cell Physiology* **46**, 399–406.
K. Keith, M. Kraml, N.G. Dengler and P. McCourt (1994) *fusca3*: a heterochronic mutation affecting late embryo development in Arabidopsis. *The Plant Cell* **6**, 589–600.
T. Kiyosue, N. Ohad, R. Yadegari, *et al.* (1999) Control of fertilization-independent endosperm development by the *MEDEA* polycomb gene in *Arabidopsis*. *Proceedings of the National Academy of Sciences of the United States of America* **96**, 4186–4191.
C. Kohler, L. Hennig, R. Bouveret, J. Gheyselinck, U. Grossniklaus and W. Gruissem (2003a) *Arabidopsis* MSI1 is a component of the MEA/FIE *polycomb* group complex and required for seed development. *The EMBO Journal* **22**, 4804–4814.
C. Kohler, L. Hennig, C. Spillane, S. Pien, W. Gruissem and U. Grossniklaus (2003b) The *polycomb*-group protein MEDEA regulates seed development by controlling expression of the MADS-box gene *PHERES1*. *Genes and Development* **17**, 1540–1553.
L.-J. Kong, B.M. Orozco, J.L. Roe, *et al.* (2000) A geminivirus replication protein interacts with the retinoblastoma protein through a novel domain to determine symptoms and tissue specificity of infection in plants. *The EMBO Journal* **19**, 3485–3495.
R.V. Knowles, M.D. McMullen, G. Yerk, R.L. Phillips, S. Kraemer and F. Srienc (1992) Endosperm mitotic activity and endoreduplication in maize affected by defective kernel mutations. *Genome* **35**, 68–77.
T. Kroj, G. Savino, C. Valon, J. Giraudat and F. Parcy (2003) Regulation of storage protein gene expression in *Arabidopsis*. *Development* **130**, 6065–6073.
R.J. Laby, M.S. Kincaid, D. Kim and S.I. Gibson (2000) The *Arabidopsis* sugar-insensitive mutants *sis4* and *sis5* are defective in abscisic acid synthesis and response. *The Plant Journal* **23**, 587–596.
T. Laux, K.F.X. Mayer, J. Berger and G. Juergen (1996) The *WUSCHEL* gene is required for shoot and floral meristem integrity in *Arabidopsis*. *Development* **122**, 87–96.
T. Laux, T. Wurschum and H. Breuninger (2004) Genetic regulation of embryonic pattern formation. *The Plant Cell* **16**, S190–S202.
G.L. Lees (1992) Condensed tannins in some forage legumes: their role in the prevention of ruminant pasture bloat. In: *Plant Polyphenols, Basic Life Sciences*, Vol. 59 (eds R.W. Hemingway and P.E. Laks), pp. 915–934. Plenum Press, New York.
S.E. Lid, D. Gruis, R. Jung, *et al.* (2002) The *defective kernel 1* (*dek1*) gene required for aleurone cell development in the endosperm of maize grains encodes a membrane protein of the calpain gene superfamily. *Proceedings of the National Academy of Sciences of the United States of America* **99**, 5460–5465.

J.A. Long, E.I. Moan, J.I. Medford and M.K. Barton (1996) A member of the KNOTTED class of homeodomain proteins encoded by the *STM* gene of *Arabidopsis*. *Nature* **379**, 66–69.

T. Lotan, M.-A. Ohto, K.M. Yee, *et al.* (1998) *Arabidopsis* LEAFY COTYLEDON1 is sufficient to induce embryo development in vegetative cells. *Cell* **93**, 1195–1205.

H. Luerßen, V. Kirik, P. Herrmann and S. Miséra (1998) *FUSCA3* encodes a protein with a conserved VP1/ABI3-like B3 domain which is of functional importance for the regulation of seed maturation in *Arabidopsis thaliana*. *The Plant Journal* **15**, 755–764.

M. Luo, P. Bilodeau, E.S. Dennis, W.J. Peacock and A. Chaudhury (2000) Expression and parent-of-origin effects for *FIS2*, *MEA*, and *FIE* in the endosperm and embryo of developing *Arabidopsis* seeds. *Proceedings of the National Academy of Sciences of the United States of America* **97**, 10637–10642.

M. Luo, P. Bilodeau, A. Koltunow, E.S. Dennis, W.J. Peacock and A.M. Chaudhury (1999) Genes controlling fertilization-independent seed development in *Arabidopsis thaliana*. *Proceedings of the National Academy of Sciences of the United States of America* **96**, 296–301.

L. Ma, N. Sun, X. Liu, Y. Jiao, H. Zhao and X.W. Deng (2005) Organ-specific expression of *Arabidopsis* genome during development. *Plant Physiology* **138**, 80–91.

M. Maitz, G. Santandrea, Z. Zhang, *et al.* (2000) *rgf1*, a mutation reducing grain filling in maize through effects on basal endosperm and pedicel development. *The Plant Journal* **23**, 29–42.

M.A.S. Marles, H. Ray and M.Y. Gruber (2003) New perspectives on proanthocyanidin biochemistry and molecular regulation. *Phytochemistry* **64**, 367–383.

K.F.X. Mayer, H. Schoof, A. Haecker, M. Lenhard, G. Jurgens and T. Laux (1998) Role of *WUSCHEL* in regulating stem cell fate in the *Arabidopsis* shoot meristem. *Cell* **95**, 805–815.

U. Mayer, R.A. Torres Ruiz, T. Berleth, S. Misera and G. Jurgens (1991) Mutations affecting body organization in the *Arabidopsis* embryo. *Nature* **353**, 402–407.

D.R. McCarty, T. Hattori, C.B. Carson, V. Vasil, M. Lazar and I.K. Vasil (1991) The *Viviparous-1* developmental gene of maize encodes a novel transcriptional activator. *Cell* **66,** 895–906.

J. McElver, I. Tzafrir, G. Aux, *et al.* (2001) Insertional mutagenesis of genes required for seed development in *Arabidopsis thaliana*. *Genetics* **159**, 1751–1763.

D.W. Meinke (1991) Embryonic mutants of *Arabidopsis thaliana*. *Developmental Genetics* **12**, 382–392.

D.W. Meinke (1992) A homoeotic mutant of *Arabidopsis thaliana* with leafy cotyledons. *Science* **258**, 1647–1650.

D.W. Meinke, L.H. Franzmann, T.C. Nickle and E.C. Yeung (1994) Leafy cotyledon mutants of *Arabidopsis*. *The Plant Cell* **6**, 1049–1064.

S. Nakamura, T.J. Lynch and R.R. Finkelstein (2001) Physical interactions between ABA response loci of *Arabidopsis*. *The Plant Journal* **26**, 627–635.

S. Nakaune, K. Yamada, M. Kondo, *et al.* (2005) A vacuolar processing enzyme, δVPE, is involved in seed coat formation at the early stage of seed development. *The Plant Cell* **17**, 876–887.

E. Nambara, K. Keith, P. McCourt and S. Naito (1995) A regulatory role for the *ABI3* gene in the establishment of embryo maturation in *Arabidopsis thaliana*. *Development* **121**, 629–636.

E. Nambara, S. Naito and P. McCourt (1992) A mutant of *Arabidopsis* which is defective in seed development and storage protein accumulation is a new *abi3* allele. *The Plant Journal* **2**, 435–441.

M.G. Neuffer and W.F. Sheridan (1980) Defective kernel mutants of maize. I: Genetic and lethality studies. *Genetics* **95**, 929–944.

N. Ohad, L. Margossian, Y.-C. Hsu, C. Williams, P. Repetti and R.L. Fischer (1996) Mutation that allows endosperm development without fertilization. *Proceedings of the National Academy of Sciences of the United States of America* **93**, 5319–5324.

N. Ohad, R. Yadegari, L. Margossian, *et al.* (1999) Mutations in *FIE*, a WD polycomb group gene, allow endosperm development without fertilization. *The Plant Cell* **11**, 407–416.

M.-A. Ohto, R.L. Fischer, R.B. Goldberg, K. Nakamura and J.J. Harada (2005) Control of seed mass by *APETALA2*. *Proceedings of the National Academy of Sciences of the United States of America* **102**, 3123–3128.

O.-A. Olsen (2004) Nuclear endosperm development in cereals and *Arabidopsis thaliana. The Plant Cell* **16**, S214–S227.

V. Orlando (2003) Polycomb, epigenomes, and control of cell identity. *Cell* **112**, 599–606.

F. Parcy, C. Valon, A. Kohara, S. Misera and J. Giraudat (1997) The *ABSCISIC ACID-INSENSITIVE3, FUSCA3*, and *LEAFY COTYLEDON1* loci act in concert to control multiple aspects of *Arabidopsis* seed development. *The Plant Cell* **9**, 1265–1277.

D.A. Patton, A.L. Schetter, L.H. Franzmann, K. Nelson, E.R. Ward and D.W. Meinke (1998) An embryo-defective mutant of *Arabidopsis* disrupted in the final step of biotin synthesis. *Plant Physiology* **116**, 935–946.

S. Penfield, R.C. Meissner, D.A. Shoue, N.C. Carpita and M.W. Bevan (2001) *MYB61* is required for mucilage deposition and extrusion in the *Arabidopsis* seed coat. *The Plant Cell* **13**, 2777–2791.

A. Rohde, M. Van Montagu and W. Boerjan (1999) The *ABSCISIC ACID-INSENSITIVE 3* (*ABI3*) gene is expressed during vegetative quiescence processes in *Arabidopsis*. *Plant, Cell and Environment* **22**, 261–270.

F. Rook, F. Corke, R. Card, G. Munz, C. Smith and M.W. Bevan (2001) Impaired sucrose-induction mutants reveal the modulation of sugar-induced starch biosynthetic gene expression by abscisic acid signaling. *The Plant Journal* **26**, 421–433.

M.J. Scanlon, P.S. Stinard, M.G. James, A.M. Myers and D.S. Robertson (1994) Genetic analysis of 63 mutations affecting maize kernel development isolated from Mutator stocks. *Genetics* **136**, 281–294.

B. Scheres, L. Di Laurenzio, V. Willemsen, *et al.* (1995) Mutations affecting the radial organisation of the *Arabidopsis* root display specific defects throughout the embryonic axis. *Development* **121**, 53–62.

R. Scott, M. Spielman, J. Bailey and H. Dickinson (1998) Parent-of-origin effects on seed development in *Arabidopsis thaliana*. *Development* **125**, 3329–3341.

W.F. Sheridan and M.G. Neuffer (1980) Defective kernel mutants of maize. II: Morphological and embryo culture studies. *Genetics* **95**, 945–960.

C. Spillane, C. MacDougall, C. Stock, *et al.* (2000) Interaction of the *Arabidopsis* polycomb group proteins FIE and MEA mediates their common phenotypes. *Current Biology* **10**, 1535–1538.

P. Springer, D. Holding, A. Groover, C. Yordan and R. Martienssen (2000) Parent-of-origin effects on seed development in *Arabidopsis thaliana*. *Development* **127**, 1815–1822.

T.A. Steeves (1983) The evolution and biological significance of seeds. *Canadian Journal of Botany* **61**, 3550–3560.

S.L. Stone, L.W. Kwong, K.M. Yee, *et al.* (2001) *LEAFY COTYLEDON2* encodes a B3 domain transcription factor that induces embryo development. *Proceedings of the National Academy of Sciences of the United States of America* **98**, 11806–11811.

M. Tegeder, X.-D. Wang, W.B. Frommer, C.E. Offler and J.W. Patrick (1999) Sucrose transport into developing seeds of *Pisum sativum* L. *The Plant Journal* **18**, 151–161.

B.S. Treml, S. Winderl, R. Radykewicz, *et al.* (2005) The gene *ENHANCER OF PINOID* controls cotyledon development in the *Arabidopsis* embryo. *Development* **132**, 4063–4074.

I. Tzafrir, R. Pena-Muralla, A. Dickerman, *et al.* (2004) Identification of genes required for embryo development in *Arabidopsis*. *Plant Physiology* **135**, 1206–1220.

J. Vicente-Carbajosa and P. Carbonero (2005) Seed maturation: developing an intrusive phase to accomplish a quiescent state. *International Journal of Developmental Biology* **49**, 645–651.

J.-P. Vielle-Calzada, J. Thomas, C. Spillane, A. Coluccio, M.A. Hoeppner and U. Grossniklaus (1999) Maintenance of genomic imprinting at the *Arabidopsis medea* locus requires zygotic *DDM1* activity. *Genes and Development* **13**, 2971–2982.

B. Vilhar, A. Kladnik, A. Blejec, P.S. Chourey and M. Dermastia (2002) Cytometrical evidence that the loss of seed weight in the *miniature1* seed mutant of maize is associated with reduced mitotic activity in the developing endosperm. *Plant Physiology* **129**, 23–30.

R. Vinkenoog, M. Spielman, S. Adams, R.L. Fischer, H.G. Dickinson and R.J. Scott (2000) Hypomethylation promotes autonomous endosperm development and rescues postfertilization lethality in *fie* mutants. *The Plant Cell* **12**, 2271–2282.

V. Walbot (1978) Control mechanisms for plant embryogeny. In: *Dormancy and Developmental Arrest* (ed. M.E. Clutter), pp. 113–166. Academic Press, New York.

H. Weber, L. Borisjuk, U. Heim, P. Buchner and U. Wobus (1995) Seed coat-associated invertases of fava bean control both unloading and storage functions: cloning of cDNAs and cell type-specific expression. *The Plant Cell* **7**, 1835–1846.

H. Weber, L. Borisjuk, U. Heim, N. Sauer and U. Wobus (1997a) A role for sugar transporters during seed development: molecular characterization of a hexose and a sucrose carrier in fava bean seeds. *The Plant Cell* **9**, 895–908.

H. Weber, L. Borisjuk and U. Wobus (1996a) Controlling seed development and seed size in *Vicia faba*: a role for seed coat-associated invertases and carbohydrate state. *The Plant Journal* **10**, 823–834.

H. Weber, L. Borisjuk and U. Wobus (1997b) Sugar import and metabolism during seed development. *Trends in Plant Science* **2**, 169–174.

H. Weber, P. Buchner, L. Borisjuk and U. Wobus (1996b) Sucrose metabolism during cotyledon development of *Vicia faba* L. is controlled by the concerted action of both sucrose-phosphate synthase and sucrose synthase: expression patterns, metabolic regulation and implications for seed development. *The Plant Journal* **9**, 841–850.

H. Weber, U. Heim, S. Golombek, L. Borisjuk, R. Manteuffel and U. Wobus (1998) Expression of a yeast-derived invertase in developing cotyledons of *Vicia narbonensis* alters the carbohydrate state and affects storage functions. *The Plant Journal* **16**, 163–172.

M.A. West and J.J. Harada (1993) Embryogenesis in higher plants: an overview. *The Plant Cell* **5**, 1361–1369.

M.A.L. West, K.L. Matsudaira Yee, J. Danao, *et al.* (1994) LEAFY COTYLEDON1 is an essential regulator of late embryogenesis and cotyledon identity in *Arabidopsis*. *The Plant Cell* **6**, 1731–1745.

T.L. Western, D.J. Skinner and G.W. Haughn (2000) Differentiation of mucilage secretory cells of the *Arabidopsis* seed coat. *Plant Physiology*, **122**, 345–356.

T.L. Western, D.S. Young, G.H. Dean, W.L. Tan, A.L. Samuels and G.W. Haughn (2004) *MUCILAGE-MODIFIED4* encodes a putative pectin biosynthetic enzyme developmentally regulated by *APETALA2*, *TRANSPARENT TESTA GLABRA1*, and *GLABRA2* in the *Arabidopsis* seed coat. *Plant Physiology* **134**, 296–306.

J.B. Windsor, V.V. Symonds, J. Mendenhall and A.M. Lloyd (2000) *Arabidopsis* seed coat development: morphological differentiation of the outer integument. *The Plant Journal* **22**, 483–493.

U. Wobus and H. Weber (1999) Sugars as signal molecules in plant seed development. *Biological Chemistry* **380**, 937–944.

D.-Y. Xie, S.B. Sharma, N.L. Paiva, D. Ferreira and R.A. Dixon (2003) Role of anthocyanidin reductase, encoded by *BANYULS*, in plant flavonoid biosynthesis. *Science* **299**, 396–399.

R. Yadegari, T. Kinoshita, O. Lotan, *et al.* (2000) Mutations in the *FIE* and *MEA* genes that encode interacting polycomb proteins cause parent-of-origin effects on seed development by distinct mechanisms. *The Plant Cell* **12**, 2367–2382.

J.Z. Zhang and C.R. Somerville (1997) Suspensor-derived polyembryony caused by altered expression of valyl-tRNA synthetase in the *twn2* mutant of *Arabidopsis*. *Proceedings of the National Academy of Sciences of the United States of America* **94**, 7349–7355.

J. Zuo, Q.-W. Niu, G. Frugis and N.-H. Chua (2002) The *WUSCHEL* gene promotes vegetative-to-embryonic transition in *Arabidopsis*. *The Plant Journal* **30**, 349–359.

2 Seed coat development and dormancy

Isabelle Debeaujon, Loïc Lepiniec, Lucille Pourcel
and Jean-Marc Routaboul

2.1 Introduction

Two major types of dormancy mechanisms exist: embryo dormancy where the agents inhibiting germination are inherent to the embryo and coat-imposed dormancy where inhibition is conferred by the seed envelopes (Bewley, 1997). Generally, complex interactions between the embryo and covering structures determine whether a seed will germinate. As a consequence, many intermediate situations are encountered due to varying contributions by the embryo and envelopes to dormancy. Seed dormancy is a typical quantitative genetic character involving many genes and being substantially influenced by environmental effects (Koornneef *et al.*, 2002; Alonso-Blanco *et al.*, 2003). It is an adaptative trait allowing germination to occur during the most suitable period for seedling establishment and life cycle completion.

Embryo growth potential and characteristics of the seed envelopes that determine the intrinsic capacity of a seed to germinate are established during development. The purpose of this review is to analyze the role of the seed envelopes, particularly the testa (seed coat), in dormancy and germination. The developmental events leading to the formation of the testa in *Arabidopsis* are presented. Special attention is paid to the roles played by flavonoids, particularly proanthocyanidins (PAs), in determining the physicochemical characteristics of the testa that influence seed dormancy, germination, and longevity in various species. In particular, the recent progress made in this field using the model plant *Arabidopsis*, which also illustrates the power of molecular genetics combined with physiology, is emphasized to dissect the mechanisms of seed coat-imposed dormancy.

2.2 Development and anatomy of the seed coat

2.2.1 *The seed envelopes*

In seed plants, the ovule consists of the embryo sac surrounded by the nucellus and integument(s) (Schneitz *et al.*, 1995). After fertilization, the ovule develops into a seed, which contains an embryo embedded in nutritive tissues such as the endosperm (angiosperms) or the megagametophyte (gymnosperms). The embryo and the nutritive tissues are surrounded by the testa. In the family Gramineae to which the cereals such as wheat (*Triticum aestivum*), rice (*Oryza sativa*), and maize (*Zea mays*) belong, the 'seed' is a caryopsis (called a grain), i.e., a dry and indehiscent

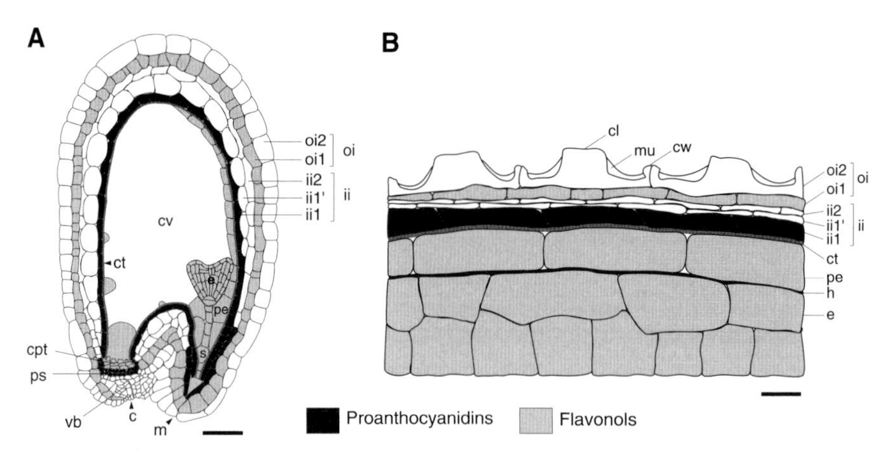

Figure 2.1 Testa structure and flavonoid localization in *Arabidopsis* seed. (A) Anatomy of a developing seed at the heart stage of embryo development (longitudinal section). Cells accumulating either proanthocyanidins or flavonols are highlighted in black or gray, respectively. The integument layers are labeled according to Beeckman *et al.* (2000); the endothelium corresponds to the ii1 layer (adapted from Pourcel *et al.*, 2005). (B) Cross section of the mature testa. Abbreviations: c, chalaza; cl, columella; cpt, chalazal-proliferating tissue (nucellus); ct, cuticle; cv, central vacuole; cw, cell wall; e, embryo; h, hyaline layer; ii, inner integument; m, micropyle; mu, mucilage; oi, outer integument; pe, peripheral endosperm (aleurone layer); ps, pigment strand; s, suspensor; vb, vascular bundle. Bar = 40 μm in (A) and 7 μm in (B).

fruit containing one seed in which the pericarp is fused with the thin testa. In angiosperm species, the nutritive tissue of mature seeds can be either the endosperm or the perisperm. Alternatively, reserves can be stored in the cotyledons (embryonic leaves), thus distinguishing among endospermic, perispermic, and nonendospermic seed types (Werker, 1997). The testa and perisperm are maternal tissues derived from the differentiation of the integument(s) and nucellus, respectively. The endosperm and embryo are of both maternal and paternal origin because they originate from double fertilization. In mature *Arabidopsis thaliana* and rapeseed (*Brassica napus*) seeds, the endosperm is reduced to one cell layer, which is tightly associated with the testa (Iwanowska *et al.*, 1994; Debeaujon and Koornneef, 2000) (Figure 2.1B). It corresponds to the peripheral endosperm and is sometimes called the aleurone layer, by structural analogy to the aleurone layer of cereal seeds (Olsen, 2004). The cells of the aleurone layer remain alive at seed maturity. In mature *Arabidopsis* seeds, a thin hyaline layer without pectin also surrounds the embryo (Debeaujon *et al.*, 2000).

The testa includes the integument(s) and the chalazal tissues (Figure 2.1A). The testa consists of several layers of specialized cell types originating from the differentiation of ovular integuments that is triggered by fertilization. They develop from the epidermis of the ovule primordium that derives ontogenetically from the meristematic L1 cell layer (Schneitz *et al.*, 1995). The number of ovule integuments varies depending on plant species. Most monocotyledons (e.g., wheat) and dicotyledons (e.g., *Arabidopsis*, bean [*Phaseolus vulgaris*]) have two integuments (bitegmic ovules). A single integument (unitegmic ovules) is mainly found in the Rosidae, Ericales, Asteridae, and Solanaceae that includes tomato (*Lycopersicon esculentum*), petunia (*Petunia* sp.), and tobacco (*Nicotiana* sp.) (Boesewinkel and Bouman, 1995; Angenent and Colombo, 1996). The micropyle is a pore formed by the integument(s) as an entrance for the pollen tube. The chalazal region is also an important part of the testa where the connection of the vascular tissues of the maternal funiculus to the seed ends. The scar where the funiculus was connected that remains after seed detachment is called the hilum. The differentiation of testa tissues involves important cellular changes generally ending with programmed cell death (PCD).

2.2.2 *The* Arabidopsis *testa*

The *Arabidopsis* testa involves two integuments, the inner integument (ii) having three cell layers and the outer integument (oi) being two-layered (Figure 2.1; Schneitz *et al.*, 1995; Beeckman *et al.*, 2000). In the first few days after fertilization (daf), integument growth proceeds through both cell division and expansion (Haughn and Chaudhury, 2005). At the heart stage of embryo development, the cellular organization of testa tissues becomes evident and distinguishable (Figure 2.1A), although further modifications are necessary to lead to the formation of the mature testa structure (Figure 2.1B).

During seed development, the individual integumentary cell layers follow different fates. The innermost cell layer (ii1), also called the endothelium, specializes in PA biosynthesis (Devic *et al.*, 1999; Debeaujon *et al.*, 2003). PAs are flavonoid compounds also known as condensed tannins (Marles *et al.*, 2003; Dixon *et al.*, 2005). They accumulate in vacuoles of the endothelium cells as colorless compounds during the early stages of seed development (Figure 2.2). PA biosynthesis starts very early (around 1–2 daf) in the micropylar region and the deposition progresses toward the chalaza until around 5–6 daf. Their oxidation during the course of seed desiccation leads to the formation of brown pigments that confer the color of mature seeds (Figure 2.2; Stafford, 1988; Debeaujon *et al.*, 2003; Marles *et al.*, 2003; Pourcel *et al.*, 2005). With the exception of a few ii2 cells at the micropyle that can accumulate PAs (Figure 2.1A), the cells of the two other ii layers do not differentiate further from their original parenchymatic stage and are crushed at seed maturity (Figure 2.1B; Debeaujon *et al.*, 2003).

Cells of oi1 and oi2 layers both first produce and degrade starch granules and afterward fulfill different developmental fates. The oi2 layer differentiates into the surface cells known as columellae containing mucilage, thickened radial cell walls, and central elevations (Western *et al.*, 2000; Windsor *et al.*, 2000; Haughn and Chaudhury, 2005). The mucilage accumulates in the apoplastic space of oi2 cells around the columellae. Its major component is pectin, a highly hydrophilic polysaccharide that has a gel-like consistency when hydrated (Goto, 1985; Western *et al.*, 2000). Gibberellins (GAs) present in oi2 cells induce α-amylase for starch degradation preceding mucilage formation (Kim *et al.*, 2005). The subepidermal oi1 layer produces a thickened wall on the inner tangential side of the cell, forming

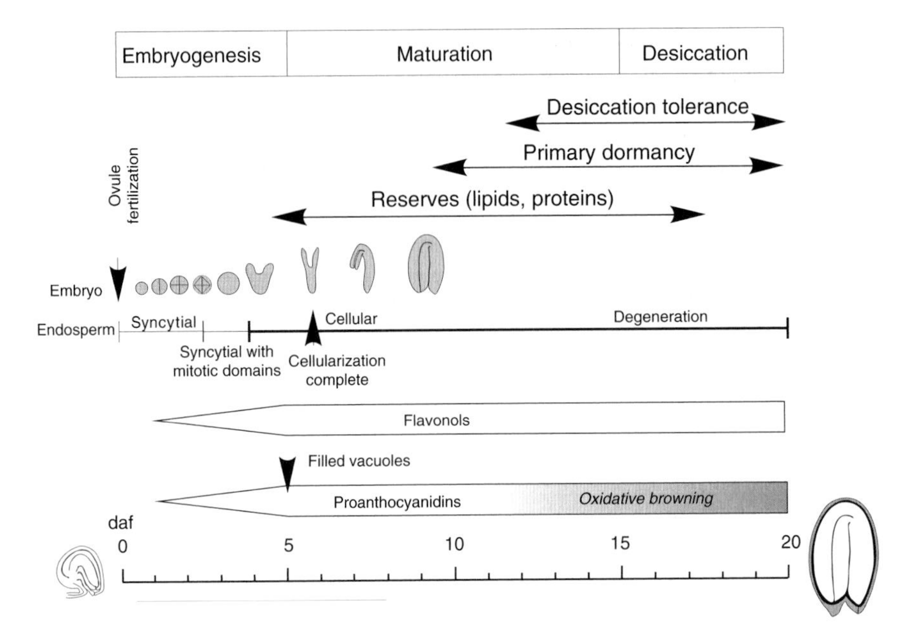

Figure 2.2 Overview of flavonoid biosynthesis and dormancy induction during seed development in *Arabidopsis*. Double fertilization results in the formation of a diploid embryo and a triploid endosperm. Embryo organization is completed during the embryogenesis phase. Reserve accumulation and primary dormancy establishment take place mainly during the maturation phase, which is followed by desiccation. Flavonoids (flavonols, proanthocyanidins) are synthesized early during development. During the desiccation phase, proanthocyanidins are oxidized to give brown derivatives that confer mature seed color (adapted from Baud *et al.*, 2002; Bentsink and Koornneef, 2002; Debeaujon *et al.*, 2003; Lepiniec *et al.*, 2005; Pourcel *et al.*, 2005; Routaboul *et al.*, 2006). Abbreviation: daf, days after fertilization.

the palisade layer (Goto, 1985; Western *et al.*, 2000). It also accumulates flavonols, which are colorless to pale yellow flavonoids (Figure 2.1; Pourcel *et al.*, 2005).

The cell walls of endothelial cells (ii1) facing the aleurone layer are bordered by an electron-dense layer that reacts positively to osmium tetroxide, suggesting its lipidic nature (Figure 2.1; Beeckman et al., 2000). The same layer also appears strongly refringent in seed confocal sections (Garcia *et al.*, 2003). This layer corresponds to the cuticle (ct) originally present in the ii during development (Beeckman *et al.*, 2000). Such a cuticle was found only at the surface of cells facing the endosperm and not on the other integument layers (Figure 2.1). Some cells in the chalazal region (pigment strand) of the *Arabidopsis* testa are also able to accumulate PAs, which creates a continuum of tannin-producing cells with a bulb-like shape at the chalazal pole, just above the end of the vascular bundle (vb) from the funiculus (Figure 2.1A), corresponding to the pigment strand in cereals (Zee and O'brien, 1970; Debeaujon *et al.*, 2003). There is also a very small amount of remaining nucellus (chalazal-proliferating tissue) at the chalazal pole of the seed (Figure 2.1A; Beeckman *et al.*, 2000).

Testa growth and differentiation proceed coordinately with the development of the endosperm and embryo. However, as the integuments are not directly involved in fertilization, some signal(s) produced during or following fertilization must co-ordinate the differentiation of the testa concomittantly with embryo and endosperm development (Haughn and Chaudhury, 2005). Garcia *et al.* (2003) presented genetic evidence that the *Arabidopsis* HAIKU protein is an endosperm-derived signal stimulating elongation (but not division) of integument cells, together with endosperm and embryo proliferation and growth. On the other hand, prevention of cell elongation in the integuments by mutations in the *TRANSPARENT TESTA GLABRA2* (*TTG2*) gene restricts endosperm and seed growth (Garcia *et al.*, 2005). The authors suggest that regulatory cross talk between the integuments and the endosperm is the primary regulator of the coordinated control of seed size in *Arabidopsis* (see Chapter 1).

The *Arabidopsis* testa cells undergo PCD during seed development and maturation (Haughn and Chaudhury, 2005). The first cell layers that initiate this process are the parenchymatic ii2 and ii1' layers from the ii (Figure 2.1). Nakaune *et al.* (2005) have found a correlation between the presence of VPE (VACUOLAR PROCESSING ENZYME), a cysteine proteinase with caspase-like activity, in these cell layers and their subsequent PCD. It is not known whether similar mechanisms promote PCD in the other cell layers. PCD involving a cysteine proteinase (BnCysP1) was also reported to occur in the ii of rapeseed testa, which suggests that this process is conserved at least among members of the Brassicaceae family (Wan *et al.*, 2002).

Testa structure is used in determining taxonomic relationships among members of the Brassicaceae family (Vaughan and Whitehouse, 1971; Bouman, 1975). In contrast to *Arabidopsis*, the mature testa of rapeseed lacks mucilage in the oi and, for most varieties, exhibits a brown to black color mainly due to the presence of PAs (Marles and Gruber, 2004). It also has a strong palisade layer composed of cells with thickened radial walls (stellar cells) that are impregnated with phenolics at maturity. Seeds at the heart stage of embryo development exhibit a four-layered oi and a five- to seven-layered ii. PAs are detected in the innermost layer of the ii, which therefore may be functionally homologous to the *Arabidopsis* endothelium (Iwanowska *et al.*, 1994; Naczk *et al.*, 1998).

2.3 Role of the seed coat in seed dormancy and germination

2.3.1 Constraints imposed by the seed coat

Germination begins with water uptake by the quiescent dry seed and is completed by radicle protrusion through the tissues surrounding the embryo. It occurs when the growth potential of the embryo can overcome the constraints imposed by the covering structures (Bewley and Black, 1994). An intact viable seed is considered to be dormant when it is unable to germinate under environmental conditions that are

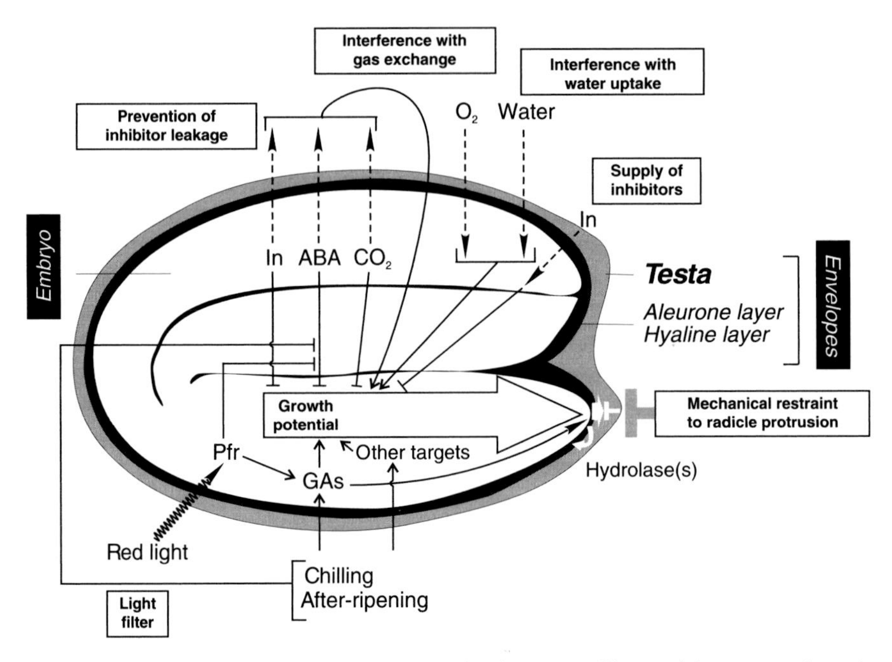

Figure 2.3 Interactions between the envelopes and embryo controlling seed dormancy and germination. Radicle protrusion occurs when embryo growth potential overcomes the constraints imposed by the envelopes. The main mechanisms through which the testa can influence embryo growth potential are mentioned in boxes. Hydrolase(s) secreted by the endosperm may contribute to the rupture of micropylar endosperm and testa. Full lines represent an action, and dashed lines indicate diffusion or leakage. Sharp and blunt arrows stand for promotive and inhibitory actions, respectively. Abbreviations: ABA, abscisic acid; GAs, gibberellins; In, inhibitor; Pfr, far-red light photoreceptor phytochrome. (Adapted from Bewley and Black, 1994; Debeaujon and Koornneef, 2000; Bentsink and Koornneef, 2002; Leubner-Metzger, 2002.)

appropriate for germination (Bewley, 1997). To understand dormancy mechanisms, it is necessary to know what constraints the envelopes impose and why the embryo cannot overcome them (Bewley and Black, 1994). The main effects exerted by the tissues surrounding the embryo are (1) interference with water uptake; (2) mechanical restraint to radicle protrusion; (3) interference with gas exchange, particularly oxygen and carbon dioxide; (4) prevention of inhibitor leakage from the embryo; (5) supply of inhibitors to the embryo; and (6) light filtration (Figure 2.3; Kelly *et al.*, 1992; Bewley and Black, 1994; Werker, 1997, 1980/81).

Many studies have demonstrated that phenolic compounds, particularly flavonoids, contribute to the germination-inhibiting effects mentioned above (Figure 2.4). Several other physicochemical characteristics of the envelopes other than phenolics, such as specific structural elements and the presence of mucilage, cutin or callose, also influence coat-imposed dormancy (Kelly *et al.*, 1992; Werker, 1997, 1980/81). Here, we will illustrate examples of the contribution of phenolic compounds to seed dormancy in several plant families.

2.3.2 Flavonoids in Arabidopsis *seeds*

2.3.2.1 Main flavonoid end-products present in seeds

Arabidopsis seeds accumulate only flavonols and PAs (Figure 2.4; Routaboul *et al.*, 2006). Flavonols are present mainly as glycoside derivatives and are found in the testa (essentially the oi1 layer), endosperm, and embryo (Figure 2.1; Pourcel

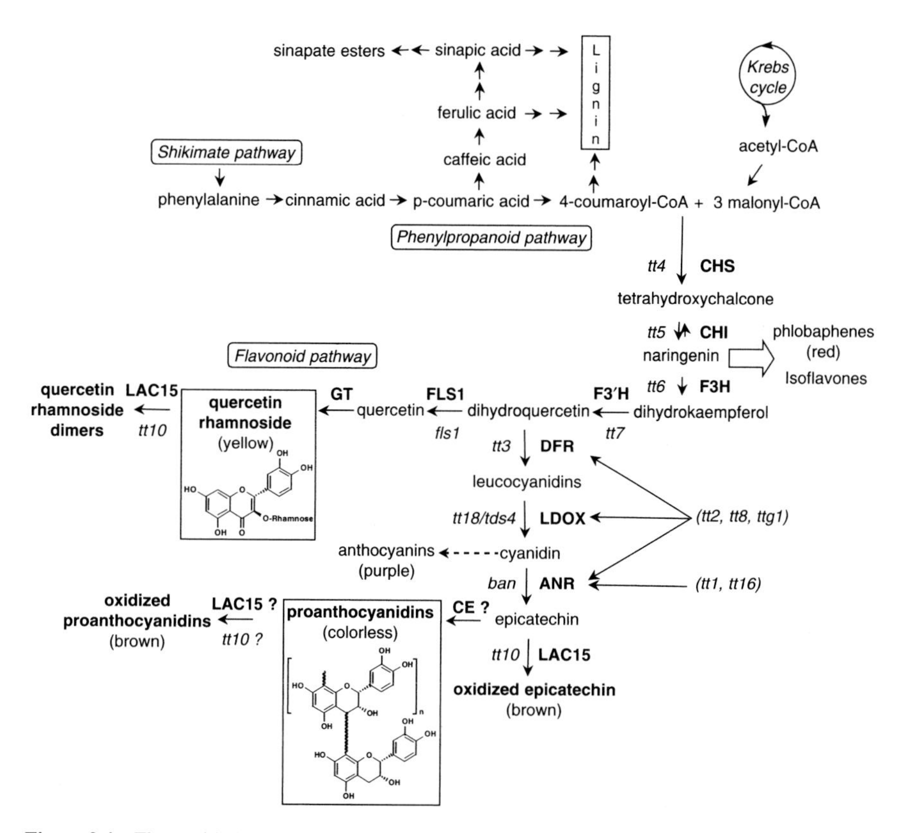

Figure 2.4 Flavonoid biosynthetic pathway and the upstream enzymatic steps. Flavonoids are plant-specific secondary metabolites derived from 4-coumaroyl-CoA and acetyl CoA, formed through the phenylpropanoid pathway and the Krebs cycle, respectively. In *Arabidopsis* seeds, the flavonoid pathway leads to the formation of two major end-products: PAs that become brown after oxidation, and flavonol glycosides (yellow). Anthocyanins (purple) are found only in the *banyuls* (*ban*) mutant testa (see Table 2.1), where they replace PAs. Red phlobaphenes are found in the seed-covering tissues in species such as wheat and rice, and isoflavones are present essentially in *Leguminosae* (adapted from Winkel-Shirley, 1998, 2002b; Pourcel *et al.*, 2005; Routaboul *et al.*, 2006; Lepiniec *et al.*, 2006). The dashed arrow indicates that this step takes place in vegetative tissues. Enzymes are represented in uppercase and bold letters, the corresponding mutants in lowercase and italics, and the regulatory mutants in brackets. Abbreviations: ANR, anthocyanidin reductase; CE, condensing enzyme; CHI, chalcone isomerase; CHS, chalcone synthase; DFR, dihydroflavonol reductase; F3H, flavanone 3-hydroxylase; F3′H, flavanone 3′-hydroxylase; FLS, flavonol synthase; GT, glycosyltransferase; LAC, laccase; LDOX, leucoanthocyanidin dioxygenase; *tannin-deficient seed* (*tds*); *ttg*, *transparent testa glabra*.

et al., 2005; Routaboul *et al.*, 2006). In mature seeds, the major flavonol is quercetin-3-*O*-rhamnoside (Q-3-O-R). Recently, a novel group of biflavonols (quercetin-rhamnoside dimers) has been detected. Both Q-3-O-R and biflavonols are found mainly in the testa (Routaboul *et al.*, 2006).

PAs are present specifically in the testa (Figure 2.1). In most plant species, PAs are generally polymers of the two flavan-3-ol stereoisomers epicatechin (EC, 2-3-*cis*) and catechin (C, 2-3-*trans*) (Dixon *et al.*, 2005). However, *Arabidopsis* seeds accumulate only EC polymers. The mean degree of polymerization varies between 5 and 8, depending on the accession (Abrahams *et al.*, 2003; Routaboul *et al.*, 2006). Interestingly, natural variation in the quantity of PAs also occurs among accessions (Lepiniec *et al.*, 2006). A detailed characterization of flavonoid metabolome in seeds of *Arabidopsis* mutants and natural accessions using liquid chromatography–tandem mass spectrometry (LC-MS) and nuclear magnetic resonance (NMR) methods has been undertaken with the aims of drawing correlations between germination/dormancy behaviors and specific flavonoid products and isolating novel flavonoid mutants that may not have clear color phenotypes (Routaboul *et al.*, 2006).

Purple anthocyanins are also present in *Arabidopsis*, but only in vegetative parts (Shirley *et al.*, 1995). They are conspicuous in 4-day-old seedlings and in aging plants exhibiting chlorophyll degradation.

2.3.2.2 Molecular genetics of flavonoid metabolism

The mutants. Most *Arabidopsis* mutants impaired in flavonoid biosynthesis have been identified through visual screenings of various collections for altered seed coat pigmentation (Koornneef, 1990; Shirley *et al.*, 1995; Lepiniec *et al.*, 2006). As such, they are all affected in PA metabolism. Mutant seed colors range from pale yellow to pale brown or gray, with the chalaza and micropyle remaining normally pigmented in some of them. The *fls1* (*flavonol synthase 1*) mutant, obtained through reverse genetics (Wisman *et al.*, 1998), lacks only flavonols because of a mutation in the *FLS1* gene (Figure 2.4) and produces brown seeds like the wild type (WT) due to the presence of oxidized PAs (Routaboul *et al.*, 2006). No mutant seeds whose testa color is significantly darker than WT have been reported in *Arabidopsis*.

Twenty-three genetic complementation groups of testa mutants have been identified. Many mutants have been collectively called *transparent testa* (*tt*) mutants (*tt1* to *tt19*), or *tt glabra* (*ttg1* and *ttg2*) (Koornneef, 1990; Shirley *et al.*, 1995; Lepiniec *et al.*, 2006). The *banyuls* (*ban*) mutant is unique in the fact that it accumulates anthocyanins in place of PAs (Albert *et al.*, 1997). The *tt10* mutant is also exceptional because it does not affect the biosynthesis of PAs but their subsequent oxidative browning (Pourcel *et al.*, 2005). Six independent mutations affecting PA biosynthesis (*tannin-deficient seed*; *tds1* to *tds6*) have been reported (Abrahams *et al.*, 2002). It is not known whether these represent additional loci except for *tds4*, which is allelic to *tt18* (Abrahams *et al.*, 2003). In addition, the *tt11* and *tt14* mutants appear to be allelic to *tt18* and *tt19*, respectively (I. Debeaujon and M. Koornneef, unpublished results). Seed pigmentation mutants may be deprived of anthocyanins in vegetative parts if the mutation affects the general core flavonoid pathway (Figure 2.4). A

Table 2.1 Classification of testa mutants based on their flavonoid composition

Mutant (or WT)	Seed			Plant	
	PAs	Flavonols	Anthocyanins	Flavonols	Anthocyanins
WT, *tt7*[a], *tt10*[b]	+	+	−	+	+
tt1, tt2, tt9, tt12, tt13, tt15, tt16, ttg2, aha10, tds1, tds3, tds5, tds6	−	+	−	+	+
tt3, tt8, tt17, tt18[c], *tt19*[bd], *ttg1, tds2*	−	+	−	+	−
tt4, tt5, tt6	−	−	−	−	−
ban	−	+	+	+	+
fls1	+	−	−	−	+

Note: PAs and flavonols present in seeds were assessed either by histochemistry (vanillin or dimethylamino-cinnamaldehyde staining for PAs, and diphenylboric acid-2-aminoethyl ester for flavonols) or by liquid chromatography-tandem mass spectrometry (LC-MS). Presence or absence/reduction are denoted by + and −, respectively. Anthocyanins were detected by visually examining for purple color in seeds or plants. Data from various published and unpublished works are summarized (Shirley *et al.*, 1995; Albert *et al.*, 1997; Wisman *et al.*, 1998; Focks *et al.*, 1999; Debeaujon *et al.*, 2001; Abrahams *et al.*, 2002; Johnson *et al.*, 2002; Nesi *et al.*, 2002; Bharti and Khurana, 2003; Shikazono *et al.*, 2003; Baxter *et al.*, 2005; Pourcel *et al.*, 2005; Routaboul *et al.*, 2006; L. Pourcel, unpublished results).
[a]Flavonols and anthocyanins in *tt7* are kaempferol derivatives in place of quercetin derivatives.
[b]Seeds brownish with storage time.
[c]Allelic to *tt11* and *tds4*.
[d]Allelic to *tt14*.

classification of the 23 mutants on the basis of their flavonoid composition in seed and vegetative tissues is presented in Table 2.1.

The proteins. Twenty genes affecting flavonoid metabolism have already been characterized at the molecular level (see Lepiniec *et al.*, 2006, for a detailed review). Nine genes encode biosynthetic enzymes (chalcone synthase [CHS], chalcone isomerase [CHI], flavanone 3-hydroxylase [F3H] , flavanone 3′-hydroxylase [F3′H], dihydroflavonol reductase [DFR], leucoanthocyanidin dioxygenase [LDOX], glycosyltransferase [GT], flavonol synthase 1 [FLS1], and anthocyanidin reductase [ANR]) and one gene encodes a modification enzyme (laccase 15 [LAC15]) (Figure 2.4). In addition to FLS1, five other genes (FLS2 to FLS6) exhibit sequence homology with flavonol synthases (Alerding *et al.*, 2005), although it is not yet known whether they are also involved in flavonol biosynthesis. The *TT10* gene has been cloned and shown to encode a polyphenol oxidase of the laccase type that belongs to a multigene family containing 17 members (Pourcel *et al.*, 2005). TT10 (LAC15) is involved in the formation of epicatechin quinones that spontaneously polymerize into brown derivatives (Figure 2.4). It may catalyze the oxidative browning of colorless PAs, which is consistent with the fact that *tt10* has normal PA levels but yellow seeds at harvest. TT10/LAC15 also catalyzes the formation of biflavonols from quercetin rhamnoside, probably in the oil layer (see below the description of *TT10/LAC15* promoter activity). Interestingly, *TT10/LAC15* expression is lower in the Landsberg *erecta* (L*er*) accession than in Wassilewskija (Ws-2) and Columbia (Col), consistent with a reduction in enzyme activity because L*er* exhibits twice the amount of soluble PAs (corresponding to less oxidized PAs) compared to Ws-2 and Col (Pourcel *et al.*, 2005; Routaboul *et al.*, 2006). The biological role of TT10/LAC15 during seed development may be to strengthen the testa to protect the embryo and endosperm from biotic and abiotic stresses.

The genes encoding *Desmodium uncinatum* leucoanthocyanidin reductase (LAR) and *Arabidopsis* and *Medicago truncatula* ANR are involved specifically in PA biosynthesis (Tanner *et al.*, 2003; Xie *et al.*, 2003; Dixon *et al.*, 2005). The corresponding recombinant proteins catalyze the formation of *trans*-flavan-3-ols (e.g., catechin) and *cis*-flavan-3-ols (e.g., epicatechin), respectively (Tanner *et al.*, 2003; Xie *et al.*, 2004). In *Arabidopsis*, ANR is encoded by the *BANYULS* (*BAN*) gene (Devic *et al.*, 1999). However, no sequence with significant homology to LAR enzymes has been found in the *Arabidopsis* genome. This is consistent with the fact that only epicatechin is synthesized in this species (Abrahams *et al.*, 2002; Routaboul *et al.*, 2006).

Six loci in *Arabidopsis* (*TT1*, *TT2*, *TT8*, *TT16*, *TTG1*, and *TTG2*) have a regulatory function in PA biosynthesis. *TT2*, *TT8*, and *TTG1* encode a R2R3-MYB protein (Nesi *et al.*, 2001), a basic helix-loop-helix (bHLH) protein (Nesi *et al.*, 2000), and a WD40 protein (Walker *et al.*, 1999), respectively. The corresponding mutants produce seeds with no PAs. The *ttg1* and *tt8* (but not *tt2*) mutants are also affected in anthocyanin biosynthesis in vegetative tissues (Koornneef, 1981; Shirley *et al.*, 1995). This is consistent with the specific expression of *TT2* in PA-producing cells (Nesi *et al.*, 2001; Debeaujon *et al.*, 2003), whereas *TT8* and *TTG1* are expressed also in vegetative parts (Walker *et al.*, 1999; Nesi *et al.*, 2001; Baudry *et al.*, 2004). *TT1 and TT16/ABS* (*CARABIDOPSIS BSISTER*) code a zinc-finger protein and a 'B-sister' group MADS box protein, respectively (Becker *et al.*, 2002; Nesi *et al.*, 2002; Sagasser *et al.*, 2002). Both are necessary for the pigmentation in the endothelium, but not in the micropyle and chalazal areas. *TTG2* encodes a WRKY transcription factor and mutation of this gene leads to the formation of completely yellow seeds (Johnson *et al.*, 2002). Epistatic relationships suggest that TTG2 acts downstream of TTG1 in regulating PA accumulation.

Three proteins are probably involved in flavonoid compartmentation (epicatechin transport into vacuoles): a MATE secondary transporter, a H^+-ATPase, and a glutathione-S-transferase encoded by *TT12*, *AHA10* (*AUTO-INHIBITED* H^+-*ATPase*) and *TT19*, respectively (Debeaujon *et al.*, 2001; Kitamura *et al.*, 2004; Baxter *et al.*, 2005). Expression of the *TT12* (MATE secondary transporter), *BAN* (ANR), *TT3* (DFR), and *TT18* (LDOX) genes was absent in mutants defective in *TT2* (MYB), *TT8* (bHLH), and *TTG1* (WD40), suggesting regulatory roles for these transcription factors in the induction of PA biosynthesis and transport enzymes (Nesi *et al.*, 2000, 2001; Debeaujon *et al.*, 2003).

Regulation of BANYULS and TT10 gene expression. Promoter activity of both *BAN* and *TT10/LAC15* were analyzed using a promoter:reporter approach to obtain more insights into the tissue specificity of their transcriptional regulation. *BAN* promoter activity was detected mainly in cells accumulating PAs (Debeaujon *et al.*, 2003; Sharma and Dixon, 2005). The *TT10/LAC15* promoter exhibited a more complex pattern of regulation than that of the *BAN* promoter, appearing first in the endothelium and then in the oi1 cell layer. Expression was correlated with PA- and flavonol-producing cells (Figure 2.1), which is consistent with the role for TT10/LAC15 in flavonoid metabolism in the testa. Characterization of *BAN*- and *TT10/LAC15*-promoter:reporter constructs in the the *tt2*, *tt8*, and *ttg1* regulatory mutants demonstrated that *TT2*, *TT8*, and *TTG1* are necessary for expression of *BAN* but not *TT10/LAC15* (Debeaujon *et al.*, 2003; Pourcel *et al.*, 2005). These promoter:reporter constructs provide valuable markers for the PA- and flavonol-accumulating cells in the *Arabidopsis* testa.

2.3.2.3 Effects of flavonoids on seed dormancy and germination

Flavonoids, particularly PAs, have been shown to reinforce coat-imposed dormancy by increasing testa thickness and mechanical strength. Indeed, cell layers containing pigments generally do not crush as dramatically as the nonpigmented ones (Debeaujon *et al.*, 2000). Moreover, during oxidation PAs have a tendency to cross-link with proteins and carbohydrates in cell walls, thus reinforcing testa structure and also modifying its permeability properties (Marles *et al.*, 2003; Marles and Gruber, 2004). This was demonstrated using *Arabidopsis* mutants affected in flavonoid metabolism in the testa (Léon-Kloosterziel *et al.*, 1994; Focks *et al.*, 1999; Debeaujon and Koornneef, 2000; Debeaujon *et al.*, 2000, 2001). Freshly harvested mutant seeds lacking PAs germinate faster than the corresponding WT brown seeds (Figure 2.5A). Moreover, their mature testa is thinner and more permeable to tetrazolium salts (Debeaujon *et al.*, 2000). Permeability to water, an important parameter for germination, could not be examined with *tt* mutants by the traditional seed water content analysis that depends on weighing seeds during imbibition because of the presence of a hydrophilic mucilage excreted by the oi. NMR imaging may represent an alternative method to analyze water distribution inside seeds, as demonstrated in tobacco (*Nicotiana tabacum*) and white pine (*Pinus monticola*) (Manz *et al.*, 2005; Terskikh *et al.*, 2005).

Reduced seed dormancy observed in *tt* mutant seeds is controlled maternally based on germination of F1 seeds resulting from reciprocal crosses between *tt* mutants and WT (Figure 2.5B). This is consistent with the fact that the testa derives from the integuments, which are maternal tissues. Little difference in germination between *tt* mutant and WT seeds is observed when they are after-ripened, suggesting that the 'seed coat' effect is particularly conspicuous under physiological conditions unfavorable for germination such as embryo dormancy (Debeaujon *et al.*, 2000). Seeds of the *Arabidopsis ga1* mutant are deficient in GAs and thus unable to germinate in the absence of additional GAs. The *tt4* mutant (CHS deficient; Figure 2.4) is deprived of flavonoids and exhibits reduced testa inhibition of germination. Genetically reducing the testa inhibition in *ga1* seeds (in a *ga1 tt4* double mutant) enabled germination of nondormant seeds without GA requirement (Figure 2.5C). This experiment also showed that both cold and light may in part be able to stimulate germination independently from GAs (Debeaujon and Koornneef, 2000).

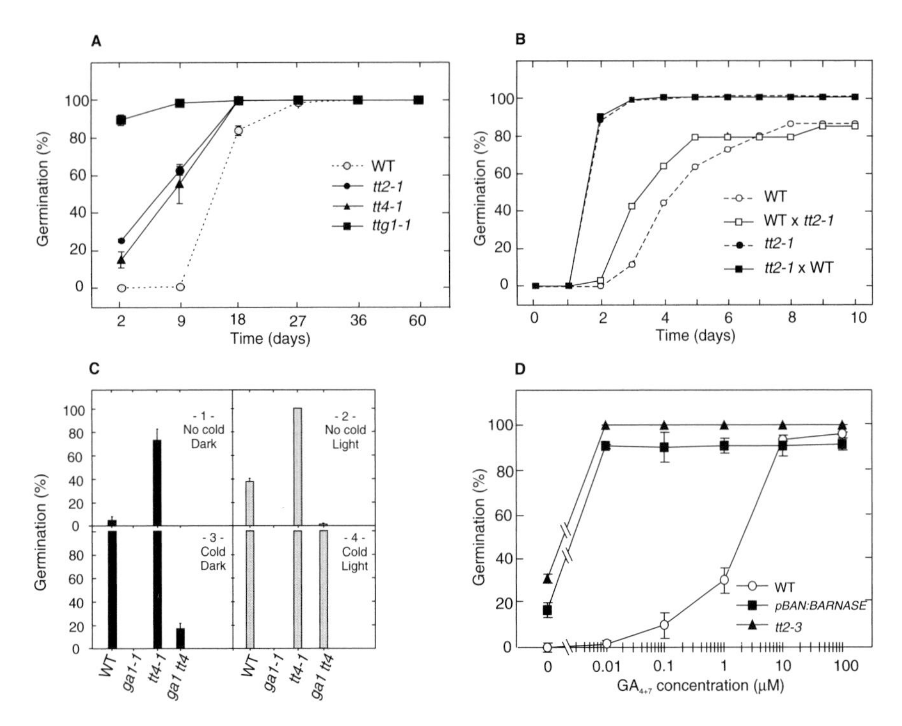

Figure 2.5 Effect of *transparent testa* mutations on dormancy and germination of *Arabidopsis* seeds. (A) Effect of dry storage (after-ripening) on dormancy release (adapted from Debeaujon *et al.*, 2000). (B) Maternal control of seed dormancy in the *tt2-1* mutant. The germination behaviors of F1 seed progenies from reciprocal crosses between *tt2-1* and WT are shown and compared to WT and the mutant parent. The parent line indicated first in the cross (e.g., 'WT' in 'WT × *tt2-1*') was used as a female. The time course of germination after 16 days of storage is presented (adapted from Debeaujon *et al.*, 2000). (C) Influence of testa mutation on germination behavior of the gibberellin-deficient mutant *ga1-1*. The *ga1-1* mutation was introduced into a *tt4-1* background. The effect of light and cold stratification (prechilling) on dormancy breakage and germination of WT, single and double mutants are compared (adapted from Debeaujon and Koornneef, 2000). (D) Permeability of the testa to GA and GA biosynthesis inhibitor. Germination of WT seeds was examined in the presence of 100 μM paclobutrazol (GA biosynthesis inhibitor) and various concentrations of GA, and compared to the germination of typical *transparent testa* mutant seeds (*tt2-3*) and seeds of a transgenic line deprived of proanthocyanidin-producing cells (*pBAN:BARNASE*) through genetic ablation (see text for details). Nondormant seeds were used (adapted from Debeaujon *et al.*, 2003).

The permeability of the *Arabidopsis* testa to exogenous GAs and GA biosynthesis inhibitors (tetcyclacis [TET] and paclobutrazol [PAC]) was also increased in *tt* mutants (Figure 2.5D; Debeaujon and Koornneef, 2000). Application of 100 μM PAC reduces germination of both *tt2-3* and WT seeds, which is gradually rescued by increasing concentrations of exogenous GAs. Relatively lower concentrations of GAs (0.01 μM and above) enables *tt2-3* seeds to germinate in the presence of PAC, while much higher concentrations of GAs (10 μM and above) are required to recover PAC-inhibited WT seed germination (Figure 2.5D). The greater response (permeability) of seeds to GAs compared to WT was also observed in pigmentless seeds produced by the *pBAN::BARNASE* line (Debeaujon *et al.*, 2003). In this line, the cytotoxic gene *BARNASE* was expressed using the *BAN* gene promoter which drove specific expression in PA-producing cells, leading to their genetic ablation and as a consequence to the formation of pigment-less seeds. These seeds were capable of germinating in relatively lower concentrations of GAs in the presence of PAC, in the same way as typical *tt* seeds (Figure 2.5D). Therefore, flavonoids and no other physicochemical factor present in PA-producing cells are significantly responsible for the *tt* germination phenotype.

Abscisic acid (ABA) is synthesized *de novo* in primary dormant *Arabidopsis* seeds during imbibition (Debeaujon and Koornneef, 2000; Ali-Rachedi *et al.*, 2004). Freshly harvested *ga1 tt4* seeds were able to germinate to 50% in the presence of carotenoid synthesis inhibitor norflurazon, which inhibits ABA biosynthesis, while *ga1* seeds exhibited only 5% germination (Debeaujon and Koornneef, 2000).

Together, these data suggest that GAs are necessary during *Arabidopsis* germination essentially to overcome the restraints to radicle protrusion imposed by the testa and *de novo* synthesized ABA. Removal of seed envelopes (testa and endosperm) allows germination of both *ga1* seeds and freshly harvested Cape Verde Islands (Cvi) seeds, which have a strong primary dormancy (Debeaujon and Koornneef, 2000; Alonso-Blanco *et al.*, 2003). It is possible that coat removal enables ABA leakage from dormant seeds, thus releasing dormancy, but this still remains to be investigated. Analysis of natural allelic variation between the Cvi and L*er* accessions revealed a quantitative trait locus (QTL) with a maternal effect (*Delay of Germination2* [*DOG2*]). The authors hypothesized that *DOG2* may affect seed coat-imposed dormancy through the genetic structure of the testa, or a factor imported from the mother plant (Alonso-Blanco *et al.*, 2003; see Chapter 5). The *tt* mutants exhibit normal ABA sensitivity, although some of them showed a slight decrease in sensitivity (I. Debeaujon and M. Koornneef, unpublished results). This may indicate that flavonoids do not affect permeability of the testa to ABA as much as they affect GA permeability. PAs have been shown to act as GA antagonists in pea (*Pisum sativum*) (Corcoran *et al.*, 1972; Green and Corcoran, 1975). Similarly, (+)-catechin extracted from bean seeds was shown to act as an inhibitor of GA biosynthesis, blocking the conversion of GA_{12}-aldehyde to GA_{12}; the catechin was localized mainly in the testa and not in the embryo during seed maturation (Kwak *et al.*, 1988). Moreover, Buta and Lusby (1986) observed that catechin and epicatechin inhibited *Lespedeza* seed germination and seedling growth. These data suggest that PAs and their precursors may inhibit germination not only by influencing the structural properties of the testa at maturity, but also by acting as biochemical inhibitors of GA metabolism and action when released from the testa during seed imbibition. Whether this biological effect is specific to GA metabolism or common to other biosynthetic pathways remains to be investigated more thoroughly.

Testa permeability is an important parameter to consider when undertaking germination experiments to test the effect of any exogenous substances on embryo behavior. For example, Rajjou *et al.* (2004) used *tt2* mutant seeds totally lacking PAs in the testa to test the effect of the transcription inhibitor α-amanitin on seed germination. In preliminary experiments the authors had established that *tt2* seeds were far more permeable to the molecule than WT seeds. Therefore, they were able to conclude that the inability of α-amanitin to block radicle protrusion was not likely due to inability of this substance to cross the testa and reach the embryo.

The *tt12* mutant came out of a reduced dormancy screen and was isolated on the basis of a faster seed germination of freshly harvested seeds on water (Debeaujon *et al.*, 2001). Recently, several *tt* mutants were isolated from a screen for fast seed germination at 10°C (Salaita *et al.*, 2005). These additional examples confirm that testa flavonoid defects enable the seed to germinate quicker in unfavorable conditions.

Desiccation plays an important role in switching seeds from a developmental mode to germinative mode. Premature drying can redirect metabolism from a developmental to a germination program (Kermode, 1995). In *tt* mutant seeds that have more permeable testae, the embryo might dehydrate quicker than WT embryos and therefore shift earlier from the developmental to germinative mode. The earlier loss of dormancy and shift to the germinative mode could cause precocious germination, as was observed in drying siliques of the *tt16* mutant (Nesi *et al.*, 2002).

2.3.3 Flavonoids in seed dormancy and germination of various species

2.3.3.1 Solanaceae

Mutants affected in flavonoid metabolism were also isolated in tomato. The mutants *anthocyaninless of Hoffmann* (*ah*), *anthocyanin without* (*aw*), and *baby lea syndrome* (*bls*) contain drastically reduced or no PAs in the endothelium. Absence of PAs in the testa probably enhances its permeability to water, since the mutant seeds exhibit increased seed weight (or water content) during imbibition (Atanassova *et al.*, 2004). Moreover, seeds of the three mutants germinate faster than the corresponding WT lines not only in optimal but also in stress conditions such as high and low temperatures (33 and 13°C, respectively), high salinity (120 mM NaCl), and high osmoticum (15% polyethyleneglycol). Mutations *ah* and *bls* both resulted in coordinated reduction in CHS, F3H, and DFR activities (Figure 2.4), while *aw* completely lacked DFR activity (Atanassova *et al.*, 1997a,b). On the other hand, *brownseed* (*bs1* to *bs4*) and *blackseed* (*bks1*) mutants are tomato mutants with darker testae than WT, which display poor germination rates and percentages (Downie *et al.*, 2004). They accumulate an additonal dark pigment in the cell layers surrounding the endothelium, which itself contains normal levels of PAs. In *bks1* seeds, the black pigment is a melanic substance that also enhances the mechanical strength of the testa. The phenotypes of the *bs* and *bks* mutations are determined by the embryo and the endosperm, in contrast to the *ah*, *aw*, and *bls* mutants that exhibit the typical maternal control. These results suggest that the endosperm and embryo may secrete a factor that influences testa characteristics (Downie *et al.*, 2003). Seeds of the *bs1* and *bs4* mutants have increased catalase activity, and *bs4* seeds also exhibit increased peroxidase activity (Downie *et al.*, 2004). Likewise, transgenic maize grains that overexpress a gene encoding a fungal laccase-type polyphenol oxidase become brown and exhibit poor germination (Hood *et al.*, 2003).

2.3.3.2 Water permeability of testae in Leguminosae and other species

The fact that flavonoids increase coat-imposed dormancy by restricting the permeability of the testa to water is best exemplified by so-called hard seed, which have been abundantly described in the Leguminosae family (Wyatt, 1977; Legesse and Powell, 1992, 1996; Serrato-Valenti *et al.*, 1994; Kantar *et al.*, 1996). In the genus *Pisum*, seeds with impermeable testae, such as the wild pea (*P. elatius*), have high phenolic contents and catechol oxidase activity, while species having seeds with permeable testae, such as the cultivated pea (*P. sativum*), do not. It is hypothesized that during seed desiccation, oxidation of phenolic compounds into the corresponding quinones by catechol oxidase in the presence of molecular oxygen may trigger testa impermeability. Permeability is inversely proportional to the phenolics content and their degree of oxidation, probably through some tanning reaction (Marbach and Mayer, 1974, 1975; Stafford, 1974; Werker *et al.*, 1979). Gillikin and Graham (1991) proposed that an anionic peroxidase may play a role in the hardening of soybean (*Glycine max*) seed coats. In this species, the majority of peroxidase activity detected in seeds is localized in the testa. In cotton (*Gossypium hirsutum*) seed also, oxidation of seed coat tannins during ripening causes seed coloration and reduction of testa permeability to water (Halloin, 1982).

2.3.3.3 Flavonoids and other phenolics as direct and indirect germination inhibitors

Flavonoids (PAs, catechin, epicatechin), phenolic acids (caffeic, *p*-coumaric, ferulic, sinapic, vanillic acids), and lignans have been considered as possible germination inhibitors. Seed germination is reduced in the presence of exogenous phenolics in a dose-dependent manner (Buta and Lusby, 1986; Reigosa *et al.*, 1999; Gatford *et al.*, 2002; Basile *et al.*, 2003; Cutillo *et al.*, 2003). The presence of phenolic compounds in developing grains is correlated with the prevention of preharvest sprouting (PHS; i.e., germination on the ear in high relative humidity) in cereals (Weidner *et al.*, 2002). Wrobel *et al.* (2005) have shown that the amount of tannins and phenolic acids present in mature dry seeds of riverbank grape (*Vitis riparia*) was reduced during cold stratification (imbibition in cold conditions), possibly by leaching of these compounds into the surrounding medium. Zobel *et al.* (1989) followed the changes in phenolic localization in rapeseed seeds during imbibition and found that after 3 h of imbibition, part of the phenolic compounds originally found in the testa leached onto the embryo surface, where they could potentially exert an inhibitory effect on radicle protrusion.

Tissues surrounding the embryo might interfere with seed germination by impeding oxygen entry or escape of carbon dioxide, which could inhibit respiration (Bewley and Black, 1994). Flavonoids are efficient antioxidants (Rice-Evans *et al.*, 1997). When present in seed coats, they may fix molecular oxygen through reactions catalyzed by polyphenol oxidases and peroxidases and also through nonenzymatic oxidation, therefore limiting oxygen availability for the embryo. Similarly, the dormancy-imposing glumellae of barley (*Hordeum vulgare*) grains also consume oxygen (Lenoir *et al.*, 1986). Restricted oxygen diffusion through the seed coat and an increased sensitivity of the embryo to hypoxia cause a coat-imposed dormancy of muskmelon (*Cucumis melo*) seeds at low temperature (Edelstein *et al.*, 1995). Porter and Wareing (1974) suggested that the presence of germination inhibitors in *Xanthium pennsylvanicum* seeds results in a high oxygen requirement because the removal of inhibitors occurs by oxidation. This is supported by experiments using beechnut (*Fagus sylvatica*) seeds, where covering structures prevent germination by interfering not only with water uptake but also with oxygen availability. In this case, oxygen was demonstrated to be involved in oxidative degradation of ABA using (+)-[^{3}H]ABA (Barthe *et al.*, 2000). The testa may impose an indirect restraint to radicle protrusion by impeding ABA leakage from the embryo in yellow cedar (*Chamaecyparis nootkatensis*) seeds (Ren and Kermode, 1999).

Flavonoid pigments in the testa are likely to act as filters modifying the spectrum of light received by the embryo. Flavonols present in the testa efficiently absorb ultraviolet (UV) light, protecting the embryo from radiation damage (Winkel-Shirley, 2002a; Griffen *et al.*, 2004). However, no data are available for absorption of other light wavelengths, particularly red light known to induce GA biosynthesis through phytochrome action (Yamaguchi *et al.*, 1998).

2.3.3.4 Pre-harvest sprouting (PHS) in cereals

Important crop species such as wheat, barley, rice, and sorghum (*Sorghum bicolor*) exhibit low dormancy during grain development, leading to a susceptibility to PHS. Wheat grain dormancy is controlled both by maternally expressed *R* (*Red grain color*) genes conferring red pericarp pigmentation and by other genes such as *Phs*, which has a major effect in the embryo. Therefore, wheat PHS is regulated both by coat-imposed and embryonic pathways controlled by separate genetic systems (Flintham, 2000; Himi *et al.*, 2002; Mares *et al.*, 2002). Dominant alleles of *R* promote the biosynthesis of red phlobaphenes (Figure 2.4). Recently, the wheat *R* gene was cloned and shown to encode a MYB-type transcription factor (Himi and Noda, 2005). This protein is involved in activation of the early biosynthetic genes CHS, CHI, F3H, and DFR (see Figure 2.4 for the pathway), which is consistent with its role in phlobaphene biosynthesis (Himi *et al.*, 2005). Genetic resolution from the QTL approach is insufficient to determine whether the *R* gene increases dormancy by itself or is linked to another dormancy-promoting locus. Analyzing dormancy of white-grained wheats overexpressing *R* in the pericarp could help answer this question. In weedy rice, which is far more dormant than cultivated rice, red or black pericarp and red or black hull (palea and lemma) were correlated with deep seed dormancy. QTL have been found for these characters (Gu *et al.*, 2005). Synteny among cereal species enables comparison of wheat PHS loci to maize and rice seed dormancy loci, allowing application of the available rice genomic DNA sequence to PHS research in other cereals (Gale *et al.*, 2002).

2.3.3.5 Heteromorphism and physiological heterogeneity among seeds

Heteromorphism (heteroblasty) is caused by maternal factors within an individual plant, such as the position of the seed in the fruit or in the inflorescence, that influence the color, shape, or size and therefore germination capacity of seeds. The resulting physiological heterogeneity provides a very important ecological advantage, especially under extreme climates (Gutterman, 2000; Matilla *et al.*, 2005). Individual siliques of *Brassica rapa* contain seeds that differ in seed color: black, dark brown, and light brown. Light-brown seeds are more water-permeable than black seeds, which is correlated to a faster germination. They are also more responsive to exogenous ethylene than black or dark-brown seeds (Puga-Hermida *et al.*, 2003).

Heterogeneity in seed dormancy can also be generated by the light environment. Dormancy of mature *Chenopodium album* seeds is influenced by the photoperiod that the mother plant experiences during seed development. Seeds obtained from plants grown under long days (LD, 18 h light per day) produce small dormant seeds with thick and black testae, whereas plants grown under short days (SD, 8 h light per day) produce large nondormant seeds with thinner brownish testae. It is generally known that photoperiodic plants monitor the length of the dark period. Consequently, a short (1 h) exposure of red light to interrupt the long night under short-day conditions (termed SDR by the author) mimicked long-day conditions and made the maternal plants produce dormant seeds (Karssen, 2002). However, these seeds did not have black testae. Dormancy of SDR seeds was released by 3-months of after-ripening, while the authentic black testa-dormant seeds produced under LD were still dormant after 3 months. Together, these data suggested that two types of dormancy are present in this species – embryo dormancy and seed coat-imposed dormancy, and that total light energy received by the plant is important for determining the second type (Karssen, 2002).

2.3.3.6 Interactions with endosperm

The impact of the endosperm on *Arabidopsis* seed germination is still a matter of debate. Interactions between the testa and the endosperm may influence germination behavior. It is possible that enzymes secreted by endosperm cells may hydrolyze some components of testa cell walls and thus reduce their resistance (Dubreucq *et al.*, 2000; Leubner-Metzger, 2002). In turn, hydrolase inhibitors may leach from the testa during imbibition. Another hypothesis would be that hydrolytic enzymes weaken the endosperm itself on the model of the tomato endosperm, which was proposed to be a more important obstacle to radicle protrusion than the testa (Groot *et al.*, 1988; Chen *et al.*, 2002). No mutant affected specifically in the structure of the peripheral endosperm has been recovered until now that could shed some light on its contribution to *Arabidopsis* seed germination.

2.4 Link between seed coat-imposed dormancy and longevity

When seeds deteriorate, they lose vigor, become more sensitive to stresses during germination, and finally become unable to germinate. The rate of aging is strongly influenced by storage temperature, seed moisture content, and seed quality (Walters, 1998). The seed coat performs important functions to protect the embryo and seed reserves (Mohamed-Yasseen *et al.*, 1994; Boesewinkel and Bouman, 1995), and as such, seed coat-imposed dormancy and longevity are directly related. Indeed, the physicochemical characteristics of the seed coat that determine the level of coat-imposed dormancy are also instrumental in protecting the seeds from stressful environmental conditions during storage and upon germination. For instance, lignin content in the soybean testa correlates with seed resistance to mechanical damage (Capeleti *et al.*, 2005). Flavonols present in the testa of *B. rapa* protect the embryo from UV-B radiation (Griffen *et al.*, 2004). PAs were shown to deter, poison, or starve bruchid larvae feeding on cowpea (*Vigna unguiculata*) seeds (Lattanzio *et al.*, 2005). Defense-related proteins such as chitinases, polyphenol oxidases, and peroxidases are prevalent in testae of *Arabidopsis* and soybean (Gillikin and Graham, 1991; Gijzen *et al.*, 2001; Pourcel *et al.*, 2005). Flavonoids present in the testa, particulary PAs, provide a chemical barrier against infections by fungi due to their antimicrobial properties (Scalbert, 1991; Skadhauge *et al.*, 1997; Islam *et al.*, 2003; Aveling and Powell, 2005). They also limit imbibitional damage due to solute leakage by decreasing testa permeability, thus controlling the rate of water uptake (Kantar *et al.*, 1996). Oxidative degradation of proteins was shown to occur during development, germination, and aging of *Arabidopsis* seeds (Job *et al.*, 2005). It is possible that flavonoids have a beneficial effect on seed longevity by scavenging free radicals. Yellow-seeded flax (*Linum usitatissimum*) showed a higher tendency of germination loss compared to dark seeds after accelerated aging (Diederichsen and Jones-Flory, 2005). Germination of *Arabidopsis* mutant seeds exhibiting testa defects, such as *tts* and *aberrant testa shape* (*ats*), was reduced more compared to WT after both long-term ambient storage and controlled deterioration, confirming the importance of seed coat integrity for seed longevity (Debeaujon *et al.*, 2000; Clerkx *et al.*, 2004).

2.5 Concluding remarks

The contribution of seed envelopes, particularly the testa, to the level of seed dormancy and germination is important and needs to be appreciated to have a complete and integrative understanding of seed dormancy. This requires anatomical, histochemical, and chemical analysis of the developing testa until maturation to identify the factors playing roles in dormancy. Moreover, the physiological response of the seed to the environmental conditions prevailing at imbibition must be dissected. A better understanding of the genetic and molecular events during testa development and differentiation not only improves our fundamental knowledge on the important contribution of this multifunctional organ in seed biology, but also may open the way toward: (1) the discovery of molecular markers linked to precise testa quality parameters, which can be used in plant breeding, and (2) the genetic engineering of these testa characters to fulfill requirements for seed quality, which includes characters influencing not only seed dormancy and germination but also longevity. Fundamental knowledge obtained on flavonoid metabolism in *Arabidopsis* testa will speed up the improvement of seed quality in crop plants, such as rapeseed.

References

S. Abrahams, E. Lee, A.R. Walker, G.J. Tanner, P.J. Larkin and A.R. Ashton (2003) The *Arabidopsis TDS4* gene encodes leucoanthocyanidin dioxygenase (LDOX) and is essential for proanthocyanidin synthesis and vacuole development. *The Plant Journal* **35**, 624–636.

S. Abrahams, G.J. Tanner, P.J. Larkin and A.R. Ashton (2002) Identification and biochemical characterization of mutants in the proanthocyanidin pathway in *Arabidopsis*. *Plant Physiology* **130**, 561–576.

S. Albert, M. Delseny and M. Devic (1997) *BANYULS*, a novel negative regulator of flavonoid biosynthesis in the *Arabidopsis* seed coat. *The Plant Journal* **11**, 289–299.

A.B. Alerding, D.K. Owens, J.H. Westwood and B.S.J. Winkel (2005) Flavonol synthases in *Arabidopsis*: gene-specific responses to developmental and biotic signals. In: *Poster Abstract No 583: Proceedings of the 16th International Conference on Arabidopsis Research*. University of Wisconsin, Madison, USA.

S. Ali-Rachedi, D. Bouinot, M.H. Wagner, *et al.* (2004) Changes in endogenous abscisic acid levels during dormancy release and maintenance of mature seeds: studies with the Cape Verde Islands ecotype, the dormant model of *Arabidopsis thaliana*. *Planta* **219**, 479–488.

C. Alonso-Blanco, L. Bentsink, C.J. Hanhart, H. Blankestijn-de Vries and M. Koornneef (2003) Analysis of natural allelic variation at seed dormancy loci of *Arabidopsis thaliana*. *Genetics* **164**, 711–729.

G.C. Angenent and L. Colombo (1996) Molecular control of ovule development. *Trends in Plant Science* **1**, 228–232.

B. Atanassova, L. Shtereva, Y. Georgieva and E. Balatcheva (2004) Study on seed coat morphology and histochemistry in three anthocyaninless mutants in tomato (*Lycopersicon esculentum* Mill.) in relation to their enhanced germination. *Seed Science and Technology* **32**, 79–90.

B. Atanassova, L. Shtereva and E. Molle (1997a) Effect of three anthocyaninless genes on germination in tomato (*Lycopersicon esculentum* Mill). 1: Seed germination under optimal conditions. *Euphytica* **95**, 89–98.

B. Atanassova, L. Shtereva and E. Molle (1997b) Effect of three anthocyaninless genes on germination in tomato (*Lycopersicon esculentum* Mill.). 2: Seed germination under stress conditions. *Euphytica* **97**, 31–38.

T.A.S. Aveling and A.A. Powell (2005) Effect of seed storage and seed coat pigmentation on susceptibility of cowpeas to pre-emergence damping-off. *Seed Science and Technology* **33**, 461–470.

P. Barthe, G. Garello, J. Bianco-Trinchant and M.T. le Page-Degivry (2000) Oxygen availability and ABA metabolism in *Fagus sylvatica* seeds. *Plant Growth Regulation* **30**, 185–191.

A. Basile, S. Sorbo, J.A. Lopez-Saez and R.C. Cobianchi (2003) Effects of seven pure flavonoids from mosses on germination and growth of *Tortula muralis* HEDW. (*Bryophyta*) and *Raphanus sativus* L. (*Magnoliophyta*). *Phytochemistry* **62**, 1145–1151.

S. Baud, J.P. Boutin, M. Miquel, L. Lepiniec and C. Rochat (2002) An integrated overview of seed development in *Arabidopsis thaliana* ecotype WS. *Plant Physiology and Biochemistry* **40**, 151–160.

A. Baudry, M.A. Heim, B. Dubreucq, M. Caboche, B. Weisshaar and L. Lepiniec (2004) TT2, TT8, and TTG1 synergistically specify the expression of *BANYULS* and proanthocyanidin biosynthesis in *Arabidopsis thaliana*. *The Plant Journal* **39**, 366–380.

I.R. Baxter, J.C. Young, G. Armstrong, *et al.* (2005) A plasma membrane H^+-ATPase is required for the formation of proanthocyanidins in the seed coat endothelium of *Arabidopsis thaliana*. *Proceedings of the National Academy of Sciences of the United States of America* **102**, 2649–2654.

A. Becker, K. Kaufmann, A. Freialdenhoven, *et al.* (2002) A novel MADS-box gene subfamily with a sister-group relationship to class B floral homeotic genes. *Molecular Genetics and Genomics* **266**, 942–950.

T. Beeckman, R. De Rycke, R. Viane and D. Inzé (2000) Histological study of seed coat development in *Arabidopsis thaliana*. *Journal of Plant Research* **113**, 139–148.

L. Bentsink and M. Koornneef (2002) Seed dormancy and germination. In: *The Arabidopsis Book* (eds C.R. Somerville and E.M. Meyerowitz). American Society of Plant Biologists, Rockville, MD. http://www.aspb.org/publications/arabidopsis/.

J.D. Bewley (1997) Seed germination and dormancy. *The Plant Cell* **9**, 1055–1066.

J.D. Bewley and M. Black (1994) *Seeds: Physiology of Development and Germination*. Plenum Press, New York.

A.K. Bharti and J.P. Khurana (2003) Molecular characterization of *transparent testa* (*tt*) mutants of *Arabidopsis thaliana* (ecotype Estland) impaired in flavonoid biosynthetic pathway. *Plant Science* **165**, 1321–1332.

F.D. Boesewinkel and F. Bouman (1995) The seed: structure and function. In: *Seed Development and Germination* (eds J. Kigel and G. Galili), pp. 1–24. Marcel Dekker, New York.

F. Bouman (1975) Integument initiation and testa development in some *Cruciferae*. *Botanical Journal of the Linnean Society* **70**, 213–229.

J.G. Buta and W.R. Lusby (1986) Catechins as germination and growth inhibitors in *Lespedeza* seeds. *Phytochemistry* **25**, 93–95.

I. Capeleti, M.L.L. Ferrarese, F.C. Krzyzanowski and O. Ferrarese (2005) A new procedure for quantification of lignin in soybean (*Glycine max* (L.) Merrill) seed coat and their relationship with the resistance to mechanical damage. *Seed Science and Technology* **33**, 511–515.

F. Chen, H. Nonogaki and K.J. Bradford (2002) A gibberellin-regulated xyloglucan endotransglycosylase gene is expressed in the endosperm cap during tomato seed germination. *Journal of Experimental Botany* **53**, 215–223.

E.J.M. Clerkx, H. Blankestijn-De Vries, G.J. Ruys, S.P.C. Groot and M. Koornneef (2004) Genetic differences in seed longevity of various *Arabidopsis* mutants. *Physiologia Plantarum* **121**, 448–461.

M.R. Corcoran, T.A. Geissman and B.O. Phinney (1972) Tannins as gibberellin antagonists. *Plant Physiology* **49**, 323–330.

F. Cutillo, B. D'Abrosca, M. DellaGreca, A. Fiorentino and A. Zarrelli (2003) Lignans and neolignans from *Brassica fruticulosa*: effects on seed germination and plant growth. *Journal of Agricultural and Food Chemistry* **51**, 6165–6172.

I. Debeaujon and M. Koornneef (2000) Gibberellin requirement for *Arabidopsis* seed germination is determined both by testa characteristics and embryonic abscisic acid. *Plant Physiology* **122**, 415–424.

I. Debeaujon, K.M. Léon-Kloosterziel and M. Koornneef (2000) Influence of the testa on seed dormancy, germination, and longevity in *Arabidopsis*. *Plant Physiology* **122**, 403–413.

I. Debeaujon, N. Nesi, P. Perez, *et al.* (2003) Proanthocyanidin-accumulating cells in *Arabidopsis* testa: regulation of differentiation and role in seed development. *The Plant Cell* **15**, 2514–2531.

I. Debeaujon, A.J. Peeters, K.M. Léon-Kloosterziel and M. Koornneef (2001) The *TRANSPARENT TESTA12* gene of *Arabidopsis* encodes a multidrug secondary transporter-like protein required for flavonoid sequestration in vacuoles of the seed coat endothelium. *The Plant Cell* **13**, 853–871.

M. Devic, J. Guilleminot, I. Debeaujon, *et al.* (1999) The *BANYULS* gene encodes a DFR-like protein and is a marker of early seed coat development. *The Plant Journal* **19**, 387–398.

A. Diederichsen and L.L. Jones-Flory (2005) Accelerated aging tests with seeds of 11 flax (*Linum usitatissimum*) cultivars. *Seed Science and Technology* **33**, 419–429.

R.A. Dixon, D.Y. Xie and S.B. Sharma (2005) Proanthocyanidins – a final frontier in flavonoid research? *New Phytologist* **165**, 9–28.

A.B. Downie, L.M.A. Dirk, Q.L. Xu, *et al.* (2004) A physical, enzymatic, and genetic characterization of perturbations in the seeds of the *brownseed* tomato mutants. *Journal of Experimental Botany* **55**, 961–973.

A.B. Downie, D.Q. Zhang, L.M.A. Dirk, *et al.* (2003) Communication between the maternal testa and the embryo and/or endosperm affect testa attributes in tomato. *Plant Physiology* **133**, 145–160.

B. Dubreucq, N. Berger, E. Vincent, *et al.* (2000) The *Arabidopsis AtEPR1* extensin-like gene is specifically expressed in endosperm during seed germination. *The Plant Journal* **23**, 643–652.

M. Edelstein, F. Corbineau, J. Kigel and H. Nerson (1995) Seed coat structure and oxygen availability control low-temperature germination of melon (*Cucumis melo*) seeds. *Physiologia Plantarum* **93**, 451–456.

J.E. Flintham (2000) Different genetic components control coat-imposed and embryo-imposed dormancy in wheat. *Seed Science Research* **10**, 43–50.

N. Focks, M. Sagasser, B. Weisshaar and C. Benning (1999) Characterization of *tt15*, a novel transparent testa mutant of *Arabidopsis thaliana* (L.) Heynh. *Planta* **208**, 352–357.

M.D. Gale, J.E. Flintham and K.M. Devos (2002) Cereal comparative genetics and preharvest sprouting. *Euphytica* **126**, 21–25.

D. Garcia, J.N. Fitz Gerald and F. Berger (2005) Maternal control of integument cell elongation and zygotic control of endosperm growth are coordinated to determine seed size in *Arabidopsis*. *The Plant Cell* **17**, 52–60.

D. Garcia, V. Saingery, P. Chambrier, U. Mayer, G. Jürgens and F. Berger (2003) *Arabidopsis haiku* mutants reveal new controls of seed size by endosperm. *Plant Physiology* **131**, 1661–1670.

K.T. Gatford, R.F. Eastwood and G.M. Halloran (2002) Germination inhibitors in bracts surrounding the grain of *Triticum tauschii*. *Functional Plant Biology* **29**, 881–890.

M. Gijzen, K. Kuflu, D. Qutob and J.T. Chernys (2001) A class I chitinase from soybean seed coat. *Journal of Experimental Botany* **52**, 2283–2289.

J.W. Gillikin and J.S. Graham (1991) Purification and developmental analysis of the major anionic peroxidase from the seed coat of *Glycine max*. *Plant Physiology* **96**, 214–220.

N. Goto (1985) A mucilage polysaccharide secreted from testa of *Arabidopsis thaliana*. *Arabidopsis Information Service* **22**, 143–145.

F.B. Green and M.R. Corcoran (1975) Inhibitory action of five tannins on growth induced by several gibberellins. *Plant Physiology* **56**, 801–806.

L.R. Griffen, A.M. Wilczek and F.A. Bazzaz (2004) UV-B affects within-seed biomass allocation and chemical provisioning. *New Phytologist* **162**, 167–171.

S.P.C. Groot, B. Kieliszewska-Rokicka, E. Vermeer and C.M. Karssen (1988) Gibberellin-induced hydrolysis of endosperm cell walls in gibberellin-deficient tomato seeds prior to radicle protrusion. *Planta* **174**, 500–504.

X.Y. Gu, S.F. Kianian, G.A. Hareland, B.L. Hoffer and M.E. Foley (2005) Genetic analysis of adaptive syndromes interrelated with seed dormancy in weedy rice (*Oryza sativa*). *Theoretical and Applied Genetics* **110**, 1108–1118.

Y. Gutterman (2000) Maternal effects on seeds during development. In: *Seeds: The Ecology of Regeneration in Plant Communities*, 2nd edn (ed. M. Fenner), pp. 59–84. CAB International, Wallingford, UK.

J.M. Halloin (1982) Localization and changes in catechin and tannins during development and ripening of cottonseed. *New Phytologist* **90**, 641–657.

G. Haughn and A. Chaudhury (2005) Genetic analysis of seed coat development in *Arabidopsis*. *Trends in Plant Science* **10**, 472–477.

E. Himi, D.J. Mares, A. Yanagisawa and K. Noda (2002) Effect of grain colour gene (R) on grain dormancy and sensitivity of the embryo to abscisic acid (ABA) in wheat. *Journal of Experimental Botany* **53**, 1569–1574.

E. Himi, A. Nisar and K. Noda (2005) Colour genes (R and Rc) for grain and coleoptile upregulate flavonoid biosynthesis genes in wheat. *Genome* **48**, 747–754.

E. Himi and K. Noda (2005) Red grain colour gene (R) of wheat is a Myb-type transcription factor. *Euphytica* **143**, 239–242.

E.E. Hood, M.R. Bailey, K. Beifuss, *et al.* (2003) Criteria for high-level expression of a fungal laccase gene in transgenic maize. *Plant Biotechnology Journal* **1**, 129–140.

F.M.A. Islam, J. Rengifo, R.J. Redden, K.E. Basford and S.E. Beebe (2003) Association between seed coat polyphenolics (tannins) and disease resistance in common bean. *Plant Foods for Human Nutrition* **58**, 285–297.

A. Iwanowska, T. Tykarska, M. Kuras and A.M. Zobel (1994) Localization of phenolic compounds in the covering tissues of the embryo of *Brassica napus* (L.) during different stages of embryogenesis and seed maturation. *Annals of Botany* **74**, 313–320.

C. Job, L. Rajjou, Y. Lovigny, M. Belghazi and D. Job (2005) Patterns of protein oxidation in *Arabidopsis* seeds and during germination. *Plant Physiology* **138**, 790–802.

C.S. Johnson, B. Kolevski and D.R. Smyth (2002) *TRANSPARENT TESTA GLABRA2*, a trichome and seed coat development gene of *Arabidopsis*, encodes a WRKY transcription factor. *The Plant Cell* **14**, 1359–1375.

F. Kantar, C.J. Pilbeam and P.D. Hebblethwaite (1996) Effect of tannin content of faba bean (*Vicia faba*) seed on seed vigour, germination and field emergence. *Annals of Applied Biology* **128**, 85–93.

C.M. Karssen (2002) Germination, dormancy and red tape. *Seed Science Research* **12**, 203–216.

K.M. Kelly, J. van Staden and W.E. Bell (1992) Seed coat structure and dormancy. *Plant Growth Regulation* **11**, 213–222.

A.R. Kermode (1995) Regulatory mechanisms in the transition from seed development to germination: interactions between the embryo and the seed environment. In: *Seed Development and Germination* (eds J. Kigel and G. Galili), pp. 273–332. Marcel Dekker, New York.

Y.C. Kim, M. Nakajima, A. Nakayama and I. Yamaguchi (2005) Contribution of gibberellins to the formation of *Arabidopsis* seed coat through starch degradation. *Plant and Cell Physiology* **46**, 1317–1325.

S. Kitamura, N. Shikazono and A. Tanaka (2004) TRANSPARENT TESTA 19 is involved in the accumulation of both anthocyanins and proanthocyanidins in *Arabidopsis*. *The Plant Journal* **37**, 104–114.

M. Koornneef (1981) The complex syndrome of *ttg* mutants. *Arabidopsis Information Service* **18**, 45–51.

M. Koornneef (1990) Mutations affecting the testa colour in *Arabidopsis*. *Arabidopsis Information Service* **27**, 1–4.

M. Koornneef, L. Bentsink and H. Hilhorst (2002) Seed dormancy and germination. *Current Opinion in Plant Biology* **5**, 33–36.

S.S. Kwak, Y. Kamiya, A. Sakurai and N. Takahashi (1988) Isolation of a gibberellin biosynthesis inhibitor from testas of *Phaseolus vulgaris* L. *Agricultural and Biological Chemistry* **52**, 149–152.

V. Lattanzio, R. Terzano, N. Cicco, A. Cardinali, D.D. Venere and V. Linsalata (2005) Seed coat tannins and bruchid resistance in stored cowpea seeds. *Journal of the Science of Food and Agriculture* **85**, 839–846.

N. Legesse and A.A. Powell (1992) Comparisons of water uptake and imbibition damage in 11 cowpea cultivars. *Seed Science and Technology* **20**, 173–180.

N. Legesse and A.A. Powell (1996) Relationship between the development of seed coat pigmentation, seed coat adherence to the cotyledons and the rate of imbibition during the maturation of grain legumes. *Seed Science and Technology* **24**, 23–32.

C. Lenoir, F. Corbineau and D. Côme (1986) Barley (*Hordeum vulgare*) seed dormancy as related to glumella characteristics. *Physiologia Plantarum* **68**, 301–307.

K.M. Léon-Kloosterziel, C.J. Keijzer and M. Koornneef (1994) A seed shape mutant of *Arabidopsis* that is affected in integument development. *The Plant Cell* **6**, 385–392.

L. Lepiniec, I. Debeaujon, J.-M. Routaboul, *et al.* (2006) Genetics and biochemistry of seed flavonoids. *Annual Review of Plant Biology*, **57**, 405–430.

L. Lepiniec, M. Devic and F. Berger (2005) Genetic and molecular control of seed development in *Arabidopsis*. In: *Plant Functional Genomics* (ed. D. Leister), pp. 511–564. Haworth Press, Inc., New York.

G. Leubner-Metzger (2002) Seed after-ripening and over-expression of class I beta-1,3-glucanase confer maternal effects on tobacco testa rupture and dormancy release. *Planta* **215**, 959–968.

B. Manz, K. Muller, B. Kucera, F. Volke and G. Leubner-Metzger (2005) Water uptake and distribution in germinating tobacco seeds investigated *in vivo* by nuclear magnetic resonance imaging. *Plant Physiology* **138**, 1538–1551.

I. Marbach and A.M. Mayer (1974) Permeability of seed coats to water as related to drying conditions and metabolism of phenolics. *Plant Physiology* **54**, 817–820.

I. Marbach and A.M. Mayer (1975) Changes in catechol oxidase and permeability to water in seed coats of *Pisum elatius* during seed development and maturation. *Plant Physiology* **56**, 93–96.
D. Mares, K. Mrva, M.K. Tan and P. Sharp (2002) Dormancy in white-grained wheat: progress towards identification of genes and molecular markers. *Euphytica* **126**, 47–53.
M.A.S. Marles and M.Y. Gruber (2004) Histochemical characterisation of unextractable seed coat pigments and quantification of extractable lignin in the Brassicaceae. *Journal of the Science of Food and Agriculture* **84**, 251–262.
M.A. Marles, H. Ray and M.Y. Gruber (2003) New perspectives on proanthocyanidin biochemistry and molecular regulation. *Phytochemistry* **64**, 367–383.
A. Matilla, M. Gallardo and M.I. Puga-Hermida (2005) Structural, physiological and molecular aspects of heterogeneity in seeds: a review. *Seed Science Research* **15**, 63–76.
Y. Mohamed-Yasseen, S.A. Barringer, W.E. Splittstoesser and S. Costanza (1994) The role of seed coats in seed viability. *Botanical Review* **60**, 426–439.
M. Naczk, R. Amarowicz, A. Sullivan and F. Shahidi (1998) Current research developments on polyphenolics of rapeseed/canola: a review. *Food Chemistry* **62**, 489–502.
S. Nakaune, K. Yamada, M. Kondo, *et al.* (2005) A vacuolar processing enzyme, δVPE, is involved in seed coat formation at the early stage of seed development. *The Plant Cell* **17**, 876–887.
N. Nesi, I. Debeaujon, C. Jond, *et al.* (2002) The *TRANSPARENT TESTA16* locus encodes the ARABIDOPSIS BSISTER MADS domain protein and is required for proper development and pigmentation of the seed coat. *The Plant Cell* **14**, 2463–2479.
N. Nesi, I. Debeaujon, C. Jond, G. Pelletier, M. Caboche and L. Lepiniec (2000) The *TT8* gene encodes a basic helix-loop-helix domain protein required for expression of *DFR* and *BAN* genes in *Arabidopsis* siliques. *The Plant Cell* **12**, 1863–1878.
N. Nesi, C. Jond, I. Debeaujon, M. Caboche and L. Lepiniec (2001) The *Arabidopsis TT2* gene encodes an R2R3 MYB domain protein that acts as a key determinant for proanthocyanidin accumulation in developing seed. *The Plant Cell* **13**, 2099–2114.
O.A. Olsen (2004) Nuclear endosperm development in cereals and *Arabidopsis thaliana*. *The Plant Cell* **16** (Suppl), S214–S227.
N.G. Porter and P.F. Wareing (1974) The role of the oxygen permeability of the seed coat in the dormancy of seeds of *Xanthium pennsylvanicum* Wallr. *Journal of Experimental Botany* **25**, 583–594.
L. Pourcel, J.-M. Routaboul, L. Kerhoas, M. Caboche, L. Lepiniec and I. Debeaujon (2005) *TRANSPARENT TESTA10* encodes a laccase-like enzyme involved in oxidative polymerization of flavonoids in *Arabidopsis* seed coat. *The Plant Cell* **17**, 2966–2980.
M.I. Puga-Hermida, M. Gallardo, M.D. Rodriguez-Gacio and A.J. Matilla (2003) The heterogeneity of turnip-tops (*Brassica rapa*) seeds inside the silique affects germination, the activity of the final step of the ethylene pathway, and abscisic acid and polyamine content. *Functional Plant Biology* **30**, 767–775.
L. Rajjou, K. Gallardo, I. Debeaujon, J. Vandekerckhove, C. Job and D. Job (2004) The effect of α-amanitin on the *Arabidopsis* seed proteome highlights the distinct roles of stored and neosynthesized mRNAs during germination. *Plant Physiology* **134**, 1598–1613.
M.J. Reigosa, X.C. Souto and L. Gonzalez (1999) Effect of phenolic compounds on the germination of six weeds species. *Plant Growth Regulation* **28**, 83–88.
C.W. Ren and A.R. Kermode (1999) Analyses to determine the role of the megagametophyte and other seed tissues in dormancy maintenance of yellow cedar (*Chamaecyparis nootkatensis*) seeds: morphological, cellular and physiological changes following moist chilling and during germination. *Journal of Experimental Botany* **50**, 1403–1419.
C.A. Rice-Evans, J. Miller and G. Paganga (1997) Antioxidant properties of phenolic compounds. *Trends in Plant Science* **2**, 152–159.
J.-M. Routaboul, L. Kerhoas, I. Debeaujon, L. Pourcel, M. Caboche, J. Einhorn and L. Lepiniec (2006) Flavonoid diversity and biosynthesis in seed of *Arabidopsis thaliana*. *Planta*, **224**, 96–107.
M. Sagasser, G. Lu, K. Hahlbrock and B. Weisshaar (2002) *A. thaliana* TRANSPARENT TESTA 1 is involved in seed coat development and defines the WIP subfamily of plant zinc finger proteins. *Genes and Development* **16**, 138–149.

L. Salaita, R.K. Kar, M. Majee and A.B. Downie (2005) Identification and characterization of mutants capable of rapid seed germination at 10°C from activation-tagged lines of *Arabidopsis thaliana*. *Journal of Experimental Botany* **56**, 2059–2069.

A. Scalbert (1991) Antimicrobial properties of tannins. *Phytochemistry* **30**, 3875–3883.

K. Schneitz, M. Hulskamp and R.E. Pruitt (1995) Wild-type ovule development in *Arabidopsis thaliana* – a light microscope study of cleared whole-mount tissue. *The Plant Journal* **7**, 731–749.

G. Serrato-Valenti, L. Cornara, P. Ghisellini and M. Ferrando (1994) Testa structure and histochemistry related to water uptake in *Leucaena leucocephala* Lam (Dewit). *Annals of Botany* **73**, 531–537.

S.B. Sharma and R.A. Dixon (2005) Metabolic engineering of proanthocyanidins by ectopic expression of transcription factors in *Arabidopsis thaliana*. *The Plant Journal* **44**, 62–75.

N. Shikazono, Y. Yokota, S. Kitamura, *et al.* (2003) Mutation rate and novel *tt* mutants of *Arabidopsis thaliana* induced by carbon ions. *Genetics* **163**, 1449–1455.

B.W. Shirley, W.L. Kubasek, G. Storz, *et al.* (1995) Analysis of *Arabidopsis* mutants deficient in flavonoid biosynthesis. *The Plant Journal* **8**, 659–671.

B. Skadhauge, K.K. Thomsen and D. Von Wettstein (1997) The role of the barley testa layer and its flavonoid content in resistance to *Fusarium* infections. *Hereditas* **126**, 147–160.

H.A. Stafford (1974) The metabolism of aromatic compounds. *Annual Review of Plant Physiology* **25**, 459–486.

H.A. Stafford (1988) Proanthocyanidins and the lignin connection. *Phytochemistry* **27**, 1–6.

G.J. Tanner, K.T. Francki, S. Abrahams, J.M. Watson, P.J. Larkin and A.R. Ashton (2003) Proanthocyanidin biosynthesis in plants. Purification of legume leucoanthocyanidin reductase and molecular cloning of its cDNA. *Journal of Biological Chemistry* **278**, 31647–31656.

V.V. Terskikh, J.A. Feurtado, C.W. Ren, S.R. Abrams and A.R. Kermode (2005) Water uptake and oil distribution during imbibition of seeds of western white pine (*Pinus monticola* Dougl. ex D. Don) monitored *in vivo* using magnetic resonance imaging. *Planta* **221**, 17–27.

J.G. Vaughan and J.M. Whitehouse (1971) Seed structure and the taxonomy of the Cruciferae. *Botanical Journal of the Linnean Society* **64**, 383–409.

A.R. Walker, P.A. Davison, A.C. Bolognesi-Winfield, *et al.* (1999) The *TRANSPARENT TESTA GLABRA1* locus, which regulates trichome differentiation and anthocyanin biosynthesis in *Arabidopsis*, encodes a WD40 repeat protein. *The Plant Cell* **11**, 1337–1350.

C. Walters (1998) Understanding the mechanisms and kinetics of seed aging. *Seed Science Research* **8**, 223–244.

L. Wan, Q. Xia, X. Qiu and G. Selvaraj (2002) Early stages of seed development in *Brassica napus*: a seed coat-specific cysteine proteinase associated with programmed cell death of the inner integument. *The Plant Journal* **30**, 1–10.

S. Weidner, U. Krupa, R. Amarowicz, M. Karamac and S. Abe (2002) Phenolic compounds in embryos of triticale caryopses at different stages of development and maturation in normal environment and after dehydration treatment. *Euphytica* **126**, 115–122.

E. Werker (1980/81) Seed dormancy as explained by the anatomy of embryo envelopes. *Israel Journal of Botany* **29**, 22–44.

E. Werker (1997) *Seed Anatomy – Encyclopedia of Plant Anatomy*. Gebrüder Borntraeger, Berlin.

E. Werker, I. Marbach and A.M. Mayer (1979) Relation between the anatomy of the testa, water permeability and the presence of phenolics in the genus *Pisum*. *Annals of Botany* **43**, 765–771.

T.L. Western, D.J. Skinner and G.W. Haughn (2000) Differentiation of mucilage secretory cells of the *Arabidopsis* seed coat. *Plant Physiology* **122**, 345–355.

J.B. Windsor, V.V. Symonds, J. Mendenhall and A.M. Lloyd (2000) *Arabidopsis* seed coat development: morphological differentiation of the outer integument. *The Plant Journal* **22**, 483–493.

B. Winkel-Shirley (1998) Flavonoids in seeds and grains: physiological function, agronomic importance and the genetics of biosynthesis. *Seed Science Research* **8**, 415–422.

B. Winkel-Shirley (2002a) Biosynthesis of flavonoids and effects of stress. *Current Opinion in Plant Biology* **5**, 218–223.

B. Winkel-Shirley (2002b) A mutational approach to dissection of flavonoid biosynthesis in *Arabidopsis*. In: *Phytochemistry in the Genomics and Post-genomics Eras, Recent Advances in*

Phytochemistry, Vol. 36 (eds J.T. Romeo and R.A. Dixon), pp. 95–110. Pergamon–Elsevier Science, Amsterdam.

E. Wisman, U. Hartmann, M. Sagasser, *et al.* (1998) Knock-out mutants from an En-1 mutagenized *Arabidopsis thaliana* population generate phenylpropanoid biosynthesis phenotypes. *Proceedings of the National Academy of Sciences of United States of America* **95**, 12432–12437.

M. Wrobel, M. Karamac, R. Amarowicz, E. Frqczek and S. Weidner (2005) Metabolism of phenolic compounds in *Vitis riparia* seeds during stratification and during germination under optimal and low temperature stress conditions. *Acta Physiologiae Plantarum* **27**, 313–320.

J.E. Wyatt (1977) Seed coat and water absorption properties of seed of near-isogenic snap bean lines differing in seed coat color. *Journal of the American Society for Horticultural Science* **102**, 478–480.

D.Y. Xie, S.B. Sharma and R.A. Dixon (2004) Anthocyanidin reductases from *Medicago truncatula* and *Arabidopsis thaliana*. *Archives of Biochemistry and Biophysics* **422**, 91–102.

D.Y. Xie, S.B. Sharma, N.L. Paiva, D. Ferreira and R.A. Dixon (2003) Role of anthocyanidin reductase, encoded by *BANYULS* in plant flavonoid biosynthesis. *Science* **299**, 396–399.

S. Yamaguchi, M.W. Smith, R.G.S. Brown, Y. Kamiya and T.P. Sun (1998) Phytochrome regulation and differential expression of gibberellin 3β-hydroxylase genes in germinating *Arabidopsis* seeds. *The Plant Cell* **10**, 2115–2126.

S.Y. Zee and T.P. O'Brien (1970) Studies on the ontogeny of the pigment strand in the caryopsis of wheat. *Australian Journal of Biological Sciences* **23**, 1153–1171.

A. Zobel, M. Kuras and T. Tykarska (1989) Cytoplasmic and apoplastic location of phenolic compounds in the covering tissue of the *Brassica napus* radicles between embryogenesis and germination. *Annals of Botany* **64**, 149–157.

3 Definitions and hypotheses of seed dormancy

Henk W.M. Hilhorst

3.1 Introduction

The term *dormancy* refers to a state of a whole plant or plant organ that is generally characterized by a temporary arrest in growth and development. Dormancy can be found among most forms of plant life and may occur in seeds, bulbs, tubers, buds, and whole plants. Dormancy is a trait that has likely been acquired during evolution by selection for the ability to survive in adverse environments, such as heat, cold, and drought. There is evidence that the evolutionary origin of dormancy is related to climatic changes during the earth's history and the spread of plant species around the globe. The number of plant species with dormancy tends to increase with the geographical distance from the equator, i.e., as seasonal variation in precipitation and temperature increases (Baskin and Baskin, 1998). Dormancy has significantly contributed to the development of new species and the successful dispersion of already existing species (Baskin and Baskin, 1998). Several reviews discuss the ecology and regulation mechanisms of seed dormancy (Bewley and Black, 1994; Hilhorst, 1995, 1998; Hilhorst *et al.*, 2006; Bewley, 1997; Baskin and Baskin, 1998; Koornneef *et al.*, 2002). As a part of the mechanism to survive adverse conditions, dormancy provides a strategy for seeds to spread germination in time in order to reduce the risk of premature death by catastrophe. This review discusses seed dormancy classifications and definitions and focuses primarily on recent research related to the regulation of dormancy. Results are discussed in relation to existing theories. Emphasis will be on the model plant *Arabidopsis thaliana*, as this species has yielded the most significant research progress over the last decade.

3.2 Classifications of dormancy

A number of classification schemes have been proposed to describe seed dormancy. Most classifications that are in use (Table 3.1; Baskin and Baskin, 1998, 2004) are derived from the one that was proposed by Nikolaeva (1977) on the basis of seed morphology. An additional classification is based on the timing of the occurrence of dormancy: *primary dormancy* refers to the type of dormancy that occurs prior to dispersal as part of the seed developmental program, whereas *secondary dormancy* refers to the acquisition of dormancy in a mature seed after imbibition as a result of the lack of proper conditions for germination (Amen, 1968; see Table 3.1). Dormancy may be located in the embryo or imposed by the tissues that surround the

Table 3.1 A simplified classification of dormancy

	Primary dormancy		Secondary dormancy	
	Exogenous	Endogenous	Combinatorial	Endogenous
Location of block	• Maternal tissues (testa) including perisperm or endosperm enclosing embryo	• Immature embryo (morphological dormancy) • Metabolic blocks (physiological dormancy) • Morpho-physiological dormancy (combined)	• Combination of exogenous and endogenous dormancy	• Metabolic blocks
Mechanism	• Inhibition of water uptake ('physical dormancy') • Mechanical restraint preventing embryo expansion ('mechanical dormancy') • Modification of gas exchange • Prevention of leaching of inhibitors from embryo • Supplying inhibitors to embryo and endosperm ('chemical dormancy')	• Embryo in mature seed has to complete development prior to germination • Physiological mechanisms largely unknown		• Physiological mechanisms largely unknown

Source: Adapted from Baskin and Baskin (1998).

embryo. In a number of species, dormancy is imposed by both the embryo and the tissues enclosing it. The completion of germination (radicle protrusion) is the net result of the opposing forces: the 'thrust' of the embryo and the restraints by the surrounding tissues. In the case of embryo dormancy the properties of the embryo are of principal importance. In coat-imposed dormancy, the properties of the covering tissues are the determinants, including mechanical, chemical, and permeability features, all of which may interfere with the successful completion of germination. For example, many seeds possess a testa (seed coat) that poses a mechanical restraint to embryonic growth and that may also contain chemical inhibitors, such as phenolic compounds, that prevent embryo growth (mechanical and chemical dormancy) (see Chapter 2). Endosperm tissue may restrict embryo growth until the thick endosperm cell walls are degraded by hydrolytic enzymes that can be induced by factors (e.g., plant hormones) derived from the embryo (physiological/mechanical dormancy) (see Chapter 11). Both embryo and coat-imposed dormancy are common and there does not seem to be a preference for a specific category or type of dormancy among plant families or genera (Baskin and Baskin, 1998).

The two principal classes of dormancy, physiological and morphophysiological dormancy, have been subdivided into several levels and types based on morphological characteristics and requirements for dormancy release (Baskin and Baskin, 2004). Although this subdivision is very useful for phylogenetic, biogeographic, and evolutionary studies, it is not suitable for physiological and molecular studies since this subdivision is largely arbitrary, as it does not distinguish underlying mechanisms and uses subjective criteria for dormancy intensity. Here, I will provide a short overview of the different kinds of dormancy. For an extensive listing and discussion of dormancy classifications, see Baskin and Baskin (1998, 2004).

3.2.1 Endogenous dormancy

In some plant families, such as the Orchidaceae and Orobanchaceae, seeds contain undifferentiated embryos usually consisting of only 2–100 cells (Watkinson and Welbaum, in press). Unlike other seeds with rudimentary embryos where a substantial endosperm provides nutrition, these seeds do not contain sufficient food reserves to complete embryo development and germination. They require exogenous sources of nutrition for embryo growth, such as from higher plants (parasitic) or from mycoheterotrophs (symbiotic). Developmental arrest in these seeds occurs at an early stage of embryonic histodifferentiation. Seeds with undifferentiated embryos are not dormant in a strict sense and are therefore not included in the most current dormancy classifications (Baskin and Baskin, 2004).

Other seeds with small embryos do complete their morphogenetic phase and have a fully differentiated embryo in terms of structure, but these embryos do not enter the maturation phase; i.e., they do not expand and accumulate food reserves. Seeds of this type usually contain relatively large amounts of endosperm tissue, which often entirely surrounds the small embryo. These embryos have to grow inside the dispersed seed prior to germination. This has been studied particularly well in coffee (*Coffea arabica*) and celery (*Apium graveolens*). In both species the surrounding endosperm is digested during embryo growth. Radicle protrusion does not occur until the embryo has attained a predefined length and the micropylar endosperm is sufficiently digested (Van der Toorn and Karssen, 1992; Da Silva *et al.*, 2004).

Physiological dormancy, which is exhibited by the fully developed and matured embryo, represents a reversible block to germination; i.e., embryo dormancy could be reinduced after the release of primary dormancy. In contrast, coat-imposed dormancy is irreversible: it can be released but not induced again. The reversible nature of physiological dormancy allows the occurrence of repeated dormancy cycles in seeds in the soil, following seasonal changes in temperature (Hilhorst *et al.*, 1996). Dormancy cycling is crucial to the establishment and survival of many plant communities.

3.2.2 *Exogenous dormancy*

Embryo coverings include seed and fruit tissues that may be dead or alive. Their contribution to the level of dormancy may be of a physical, chemical, or mechanical nature, or combinations of these. Many seeds contain an endosperm that is of both maternal (2/3) and paternal (1/3) genetic origin. The endosperm usually functions as a storage tissue. In some species (e.g., in cereals), it is predominantly dead tissue that is filled with storage reserves, such as starch granules and protein bodies. In other cases the endosperm is a living tissue that could contribute significantly to the dormancy and germination behavior of the seed. The relative contribution of the endosperm to seed size varies greatly, from only a few cell layers surrounding the embryo to an almost complete occupation of the seed. Very often endosperm tissue contains thickened cell walls. This provides considerable mechanical protection to the embryo and is also a source of reserve carbohydrates in the form of hemicelluloses. In a number of species, the primary seed storage tissue is the perisperm, a diploid endosperm-like tissue derived from the maternal nucellus.

The testa derived from the integuments and/or pericarp derived from fruit tissues may also delay germination. Physical dormancy is usually the result of impermeability of the testa to water or oxygen. In these cases, the seed coat requires chemical modification or mechanical scarification before water and oxygen can enter the embryo. A range of chemical compounds, including phenolics, derived from seeds or dispersal units may inhibit germination. Such chemicals have been extracted from all seed and fruit parts. The plant hormone abscisic acid (ABA) seems to be omnipresent in seed and fruit tissues. ABA present in mature seeds is generally considered to be a remnant of the ABA pools that are involved in the regulation of seed development (Hilhorst, 1995). If the ABA content in the seed is high enough to inhibit germination, it has to leach out or be catabolized to allow completion of germination. However, the specific role of these compounds in the inhibition of germination is not clear. In most cases a direct relationship between ABA content and its physiological action in seed dormancy is lacking. On the other hand, germination of seeds containing inhibitory chemicals can be accelerated by extensively rinsing the seeds with water prior to germination. These chemicals could be important not only for germination regulation but also for other survival strategies of seeds. It is possible that many of these inhibitory substances are present primarily to avoid predation or microbial infections (Baskin and Baskin, 1998). Seeds that combine elements of both endogenous and exogenous dormancy are classified in the group of combinatorial dormancy (Table 3.1).

Tissues surrounding the embryo will impose a certain mechanical restraint to expansion of the embryo. Removal of all or part of the covering tissues often leads to normal embryo growth, indicating that the block to germination is located in the covering tissues. The embryo can overcome the mechanical restraint of the enclosing tissues by generating a thrust sufficient to penetrate them. Alternatively, the restraint may be reduced by weakening of the tissues by an active process such as enzyme secretion, which in turn is controlled by the embryo, without apparent changes in the embryonic growth potential. This has been demonstrated for a number of species in which the endosperm is digested by hydrolytic enzymes, e.g., tomato (*Lycopersicon esculentum*; Groot and Karssen, 1987), *Datura ferox* (Sánchez *et al.*, 1990), coffee (Da Silva *et al.*, 2004), and tobacco (*Nicotiana tabacum*; Leubner-Metzger, 2003; see Chapter 11).

3.3 Definitions of dormancy

Seed dormancy results in the absence of germination. However, absence of germination can have several causes: the seed can be nonviable, the environment may simply be limiting germination, or some factors in the seed or dispersal unit itself may block germination. Hence, the frequently used definition of seed dormancy is "*the absence of germination of a viable seed under conditions that are favorable to germination*" (Harper, 1959). However, this definition is not perfect since it could lead to different interpretations. A first caveat relates to the determination of 'absence of germination'. Many species display very slow germination; e.g., coffee seeds may take up to 50 days before radicle protrusion occurs under the optimal growing conditions (Da Silva *et al.*, 2004). Observation before 50 days would classify the seeds as dormant whereas observation after 50 days would classify them as nondormant. In addition, the time required for germination can vary depending upon the degree or depth of dormancy. An arbitrary germination time to distinguish between dormant and non-dormant seeds has obvious practical advantages for dormancy classification but is not useful in physiological and molecular studies of seed dormancy and germination. A second caveat refers to 'conditions that are favorable to germination'. Less dormant seeds germinate in a wider range of conditions than more dormant ones (e.g., Karssen, 1982). In the case of temperature, dormant seeds germinate over a narrower range of temperatures whereas nondormant seeds are more tolerant to a wider range of temperatures (see Chapter 4). Consequently, dormancy release is associated with a widening of the germination temperature 'window' and, vice versa, when dormancy is induced, the germination temperature window becomes narrower. Obviously, fully dormant seeds will not germinate at any temperature. This makes it clear that seed dormancy is not an all-or-nothing property but a relative phenomenon the expression of which varies with the environment. This implies that when seeds are moved into a suitable temperature range, they appear as nondormant and will germinate. This does not necessarily mean that dormancy is completely absent in these seeds. There may still be differences in germination rate (speed), for example, which is a more sensitive indicator of dormancy. In other words, the expression of seed dormancy, but not dormancy as such, depends on the environment. This seems to contradict with the fact that dormancy is regarded as an adaptive seed trait and not related to the environment (Simpson, 1990). It should be noted, however, that conditions other than the germination environment influence the depth of dormancy. For example, dry after-ripening widens the germination temperature window whereas the actual germination temperature only determines whether dormancy is expressed.

The root of the problem is that, as yet, we can only measure dormancy *post facto* by testing germination, which is the result of dormancy release. As germination (in a given time for an individual seed) is an all-or-nothing property, it cannot be avoided that dormancy of an individual seed is also considered as such. Although on a population level intermediate dormancy categories can be found, it is technically not possible at present to quantify the degree of dormancy in an individual seed prior to its germination.

The second cause of the confusion is that the release of dormancy and initiation of germination are often regarded as equivalent phenomena. It is important to consider dormancy release and initiation of germination as distinct properties within a seed (Karssen, 1982; Hilhorst, 1995; Vleeshouwers *et al.*, 1995; Baskin and Baskin, 1998). One set of conditions will release dormancy whereas another set of conditions may be required for subsequent germination. For example, cold imbibition may release dormancy, but light is required to initiate germination. This has led to an alternative definition of seed dormancy: "*Dormancy is a seed characteristic, the degree of which defines what conditions should be met to make the seeds germinate*" (Vleeshouwers *et al.*, 1995). Within this context 'absence of germination' is equivalent to the term 'quiescence', which is commonly used to denote the prevention of germination because of a lack of water (Bewley and Black, 1994). The definition by Vleeshouwers *et al.* (1995) clearly states that germination occurs when internal requirements and external factors overlap (Figure 3.1).

There is some controversy among seed biologists as to which factors control the induction and breaking of dormancy (Vleeshouwers *et al.*, 1995). For example, the effects of light, nitrate, and temperature on dormancy are difficult to distinguish from their effects on the process of germination itself, as discussed in the previous section. This apparent continuum between dormancy and germination is a major constraint to studying dormancy (Cohn, 1996). This is particularly true in crop species where dormancy is usually minimal and may occur only under stress conditions. This feature can make crop species unsuitable for studies of dormancy responses as they cannot be distinguished from germination-related processes (Figure 3.2). For such studies it is essential to have a well-characterized and unambiguous experimental system. Annual soil seed-bank species are better subjects than domesticated crops and have particular features that make them suitable as model species for physiological, genetic, and molecular studies of seed dormancy (Cohn, 1996; Hilhorst, 1997). For example, germination of seeds of *Sisymbrium officinale* is

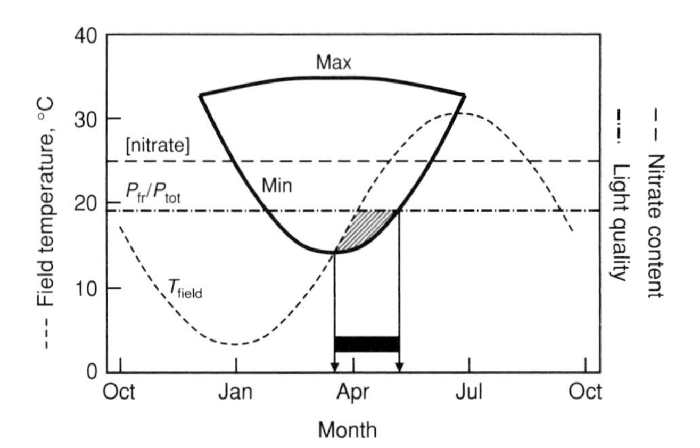

Figure 3.1 Model of dormancy cycling and potential timing of seedling emergence of summer annuals in temperate regions. Central to the model is the increase in sensitivity of the seeds to temperature, light, and nitrate, indicated by max/min threshold temperature and concentrations. The dotted lines indicate field temperature (T_{field}), light quality (P_{fr}/P_{tot}), and nitrate content of the soil. Germination is induced only when the thresholds required for germination fall below *all* the prevalent field conditions, which is indicated by the shaded area. Emergence occurs in the period indicated between the two arrows. Although light quality and nitrate content are represented by straight lines for simplicity, they are seasonally variable and highly unpredictable. Thus, the suggested sequence in fulfillment of requirements for light, nitrate, and temperature is coincidental.

absolutely dependent on stimulants such as light and nitrate, whereas the dormancy releasing conditions (e.g., chilling) increase the seeds' sensitivity to these stimulants. These characteristics provide the tools to separate the state of nondormancy from germination (reviewed by Hilhorst, 1997). Both *S. officinale* and the model species *A. thaliana* are members of the Brassicaceae family. However, most known

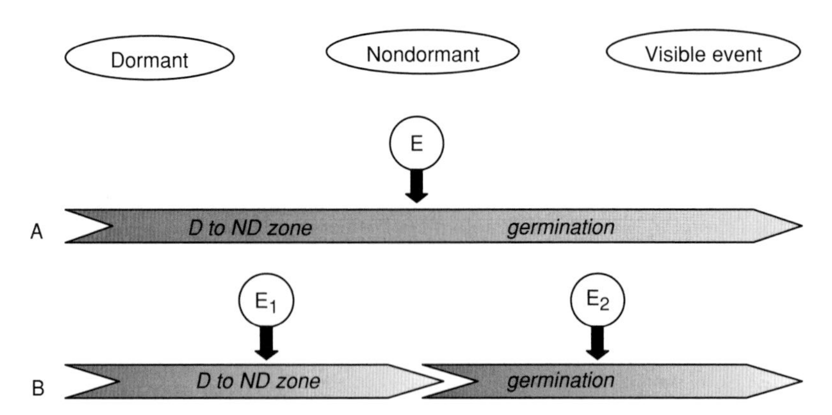

Figure 3.2 Two alternatives for the transition from dormancy to germination: (A) If only one set of environmental factors (E) is required to break dormancy and induce germination, the two events cannot be separated. (B) If the set of environmental factors that break dormancy (E_1) is different from the set that induces germination (E_2), these events can be separated and studied individually. (Based on Cohn, 1996.)

accessions of *Arabidopsis* have only a shallow dormancy that is released by relatively short periods of dry after-ripening. This makes the species less suitable for dormancy research. However, the Cvi accession, which originates from the Cape Verde Islands, is considered deeply dormant as it requires several months of after-ripening to release its dormancy. This accession is now in use in a number of genetic and molecular studies of dormancy (Alonso-Blanco *et al.*, 2003; Chapter 5).

3.4 Primary dormancy

3.4.1 Induction of primary dormancy

3.4.1.1 Role of ABA in dormancy induction

Primary dormancy is commonly associated with the transient increase in ABA content during seed development. In most of the species studied, ABA levels increase during the first half of seed development and decline during late maturation concomitantly with the decline in seed water content. ABA has been detected in all seed and fruit tissues examined and has been related to a number of developmental processes, including synthesis of storage proteins and late embryogenesis-abundant proteins, suppression of precocious germination, and induction of desiccation tolerance (for reviews, see McCarty, 1995; Holdsworth *et al.*, 1999; Finkelstein *et al.*, 2002; Chapter 1). Studies with ABA-deficient mutants of *Arabidopsis* and tomato have demonstrated that defects in ABA synthesis during seed development result in the formation of nondormant seeds (Karssen *et al.*, 1983; Koornneef *et al.*, 1984, 1985; Groot and Karssen, 1992). In tomato and *Arabidopsis*, dormancy was induced only when the embryo contained the normal *ABA* allele and was capable of synthesizing ABA. Crosses between wild-type and *aba* (*Arabidopsis*) or *sitiens* (sit^{w}, tomato) mutant plants indicated that maternal ABA (i.e., located in testa and fruit tissues) does not affect dormancy. Thus, a transient increase in embryonic ABA content during seed development is required to induce dormancy. Further support for this conclusion was provided by experiments in tobacco where overexpression of zeaxanthin oxidase, one of the key enzymes in ABA synthesis, resulted in more dormant seed phenotypes than wild type, whereas silencing of this gene yielded less dormant seed phenotypes (Frey *et al.*, 1999).

Sensitivity to ABA plays an equally important role to ABA content in the induction of dormancy. The ABA-insensitive *abi1*, *abi2*, and *abi3* mutants display variable reductions in seed dormancy (Koornneef *et al.*, 1984). Conversely, seeds from the ABA-hypersensitive *era1* (*enhanced response to ABA 1*) mutant show an enhanced seed dormancy phenotype (Cutler *et al.*, 1996). Several cultivars of wheat (*Triticum aestivum*) and maize (*Zea mays*) that exhibit vivipary, also known as preharvest sprouting, have a reduced sensitivity to ABA. Analysis of *Arabidopsis* seeds and maize kernels displaying vivipary led to the identification of the ABA-responsive genes *ABI3* and *VP1*, respectively, which are responsible for these phenotypic characteristics (see Chapters 1 and 8). These orthologous genes encode transcription factors of the B3 domain family, which activate the transcription of ABA-inducible genes that play roles in seed development (Finkelstein *et al.*, 2002). They are involved in maintaining the developmental state of seeds and suppressing the transition to the vegetative or growth stage. Thus, embryos of mutants defective in these genes typically exhibit little or no dormancy and often germinate precociously.

3.4.1.2 Developmental programs and dormancy induction

Mutant analyses have identified a number of genes that altered seed dormancy without affecting ABA contents throughout seed development, including *Arabidopsis abi3*, *leafy cotyledon1* (*lec1*), *lec2*, and *fusca3* (*fus3*) mutants. These mutants all exhibit phenotypes that are characteristic of the vegetative state: e.g., reduced desiccation tolerance, active meristems, expression of germination-related genes, and absence of dormancy (Parcy *et al.*, 1997; Chapter 1). Apparently, these genes have partially overlapping functions in the overall control of seed maturation, leading to mutants defective in many aspects of seed maturation (Parcy *et al.*, 1997). *LEC1* and *FUS3* loci probably regulate developmental arrest, as mutations in these genes cause a continuation of embryo growth. *ABI3* is also active during vegetative quiescence processes in other parts of the plant in which it suppresses meristematic activity (Rohde *et al.*, 2000). These results suggest that *ABI3*, *LEC1*, and *FUS3* are required for the induction and/or the maintenance of dormancy.

Because the *abi3*, *lec1*, *lec2*, and *fus3* mutants are defective in seed maturation, no dormancy is observed in the mutant seeds, which germinate precociously (Nambara *et al.*, 2000). Detailed analyses of the *abi3*, *lec1*, and *fus3* mutants have revealed that they differ in the time of occurrence of precocious germination (Raz *et al.*, 2001). *LEC1* and *FUS3* probably function earlier to switch off the developmental program and then *ABI3* functions at a later time point to induce dormancy (Raz *et al.*, 2001). Double mutants between a gibberellin (GA)-deficient mutant and each of the three genes also behave differently. For example, GA dependency of germination is maintained in the *fus3* mutant but not in *lec1*, suggesting that these mutants affect the germination potential of seeds in different ways. Comparison of *LEC1*, *FUS3*, and *ABI3* gene expression during seed development and dormancy release in several accessions of *Arabidopsis* varying in degree of dormancy demonstrated that there was no relationship between expression of these genes and the degree of dormancy (Baumbusch *et al.*, 2004). Thus, it is unlikely that these genes directly control the establishment of dormancy in *Arabidopsis*.

Similarly, in tomato it has been suggested that developmental arrest and dormancy are separate phenomena. After the completion of embryonic histodifferentiation both wild-type (cv. Moneymaker) and ABA-deficient *sit*^{w} mutant seeds enter a state of developmental arrest. In wild-type seeds this is followed by the induction of dormancy, consistent with a transient rise in ABA content (Hilhorst and Karssen, 1992; Hilhorst, 1995). In the *sit*^{w} mutant there is no induction of dormancy and hence developmental arrest is not maintained, resulting in viviparous germination within overripe fruits. However, at this stage ABA is also undetectable in wild-type seeds, yet germination is suppressed. The osmotic environments of the locular tissues of fruits of both genotypes are comparable (Berry and Bewley, 1992; Liu *et al.*, 1996). Apparently, the mutant seeds possess a greater 'growth potential', as is evident *in vitro* in their stronger resistance to osmotic inhibition of germination (Groot and Karssen, 1992). These results could be explained by aberrations in the testae of the mutant seeds, which contain only one cell layer while wild-type testae have four to five cell layers (Hilhorst and Downie, 1996). This is a pleitropic effect of the mutation in abscisic aldehyde oxidase (AAO3), which catalyzes the final step in ABA synthesis (Nambara and Marion-Poll, 2005; Chapter 9). Numerous mutants in *Arabidopsis* that alter testa properties are known to affect seed dormancy (Léon-Kloosterziel *et al.*, 1994; see Chapter 2).

In addition to genetic factors, the environment has a profound influence on the acquisition of dormancy during seed development. So far, no generalizations have been made as to which environmental factors result in consistent effects on dormancy over a range of species. Diverse effects of a variety of environmental factors on the development of seed dormancy have been reported, including day length, light quality, mineral nutrition, competition, temperature, physiological age of plants and position of seeds on the parent plant. A summary of these factors and their effects on dormancy can be found in Baskin and Baskin (1998).

3.4.2 Release of primary dormancy

3.4.2.1 After-ripening

Many species require variable periods of dry after-ripening or moist chilling (or cold stratification) to release their dormancy (Bewley and Black, 1994). Under natural conditions, dry after-ripening may occur in winter annuals in which dormancy is released by high summer temperatures in order to make the seeds germinable in the fall. The rate of dry after-ripening increases with temperature. Moist chilling is effective in many summer annuals to release dormancy during cold winter months (Baskin and Baskin, 1998). However, this distinction is not absolute. For example, dormancy in *Arabidopsis* Cvi, a winter annual, can also be released by a short period of cold stratification (Ali-Rachedi *et al.*, 2004).

The requirement for dry after-ripening of dormant *Arabidopsis* seeds depends on the conditions during seed development. Derkx and Karssen (1993) found that seeds harvested in a greenhouse between September and March were much more sensitive to light (less dormant) than seeds harvested between April and June. Sensitivity to GAs showed very similar patterns (higher GA sensitivity in seeds harvested between September and March). The sensitivity to GAs and light increased further during after-ripening, indicative of dormancy release. A decrease in ABA content of *Arabidopsis* seeds also occurred during dry after-ripening (Ali-Rachedi *et al.*, 2004). In seeds of *Nicotiana plumbaginifolia* that were after-ripened for 10–12 months, ABA content dropped to approximately 40% of that found in dormant seeds with a concurrent increase in germination rate (Grappin *et al.*, 2000). A similar decrease was found in tomato seeds after 1 year of dry storage (Groot and Karssen, 1992). The sensitivity to ABA also decreased in tobacco seeds during after-ripening (Grappin *et al.*, 2000; Leubner-Metzger and Meins, 2000). As ABA is a potent inhibitor of seed germination, these changes in ABA content or sensitivity are likely related to the increasing germination potential of after-ripened seeds.

Our understanding of the processes that occur in dry seeds during after-ripening is very limited. The constraint that limits study of the processes in dry systems is that any treatment with aqueous solutions may initiate a rapid activation of preformed enzymes leading to an alteration in their activities or in the amounts of their substrates and products. Also, it is not always clear how much water the seed contains or whether the water is evenly distributed among the different seed tissues. Therefore, results should be interpreted with caution. Nevertheless, there are convincing indications that processes related to dormancy release occur in dry seeds. The reduction of ABA content, as described above, is an example. In wild oat (*Avena fatua*), transcripts encoding avenin (*AV10*), puroindoline-like (*AV1*), and serine proteinase inhibitor-like (*Z1*) proteins were reported to be preferentially degraded during after-ripening relative to other mRNAs that were present (Johnson and Dyer, 2000). During after-ripening of *N. tabacum* seeds containing 8–13% overall moisture content, a transient increase in β-1,3-glucanase (βGlu1) mRNA abundance was observed, which was followed by an increase in βGlu enzyme activity in the following 10 days (Leubner-Metzger, 2005). βGlu activity is a prerequisite for loss of dormancy in tobacco seeds (Leubner-Metzger, 2003). The βGlu enzyme activity was not present in antisense transgenic seeds that did not express the β*Glu1* gene. Those seeds did not show the after-ripening effects. These results suggest that βGlu1 is transiently expressed during dry after-ripening and, hence, that transcription and translation can occur in seeds with very low moisture content. However, magnetic resonance imaging revealed that the small amount of seed moisture is unevenly distributed over the different seed tissues, resulting in areas or 'pockets' of high water content (Leubner-Metzger, 2005). Thus it cannot be precluded that the events occurred in these areas of high water content (see also Chapter 11).

3.4.2.2 Regulation of dormancy in imbibed seeds

The release of dormancy in imbibed seeds by moist chilling has been studied more extensively than the changes during after-ripening. Early studies showed that chilling increases the seed sensitivity to environmental factors, such as light and nitrate, and also to exogenously applied GAs (for reviews, see Hilhorst and Karssen, 1992; Bewley and Black, 1994; Hilhorst, 1995). More recent studies have focused on ABA and GA biosynthesis and catabolism, including expression analyses of the genes involved. Chilling of beechnut (*Fagus sylvatica*) embryos, which is required to release dormancy, does not significantly reduce ABA content during chilling treatment. However, when embryos were transferred to optimal conditions for germination after chilling, the germinable seeds displayed a rapid decline in ABA content, which was not observed in the dormant embryos kept in the same condition without chilling. The authors suggested that the cold treatment decreased the ABA-biosynthesis capacity at the optimal germination temperature, which resulted in a change of the equilibrium from biosynthesis to catabolism (Le Page-Degivry *et al.*, 1997). In accordance with this is a similar observation in *Arabidopsis* Cvi seeds in which seed ABA contents did not change during the cold treatment but decreased only after transfer to the germination conditions (Ali-Rachedi *et al.*, 2004). The decrease in ABA content is not caused by simple leakage from the seed but by active

ABA catabolism (Toyomasu *et al.*, 1994; Ali-Rachedi *et al.*, 2004). During dormancy release of yellow cedar (*Chamaecyparis nootkatensis*) seeds by a combined warm and cold treatment, ABA content of the embryo decreased by approximately 50%. However, the decrease in ABA content alone was not sufficient for dormancy release; a concurrent decrease in sensitivity to ABA was also required (Schmitz *et al.*, 2000). In beechnut seeds, expression of an ABA-inducible protein phosphatase type-2C (*PP2C*) gene increased during the first weeks of cold stratification (González-García *et al.*, 2003). Overexpression of the beechnut *PP2C* in *Arabidopsis* resulted in less dormant seeds with a markedly reduced ABA sensitivity. These results suggest that the PP2C protein is a negative regulator of ABA signaling and that phosphorylation/dephosphorylation may be involved in seed dormancy release during chilling. From these studies it can be concluded that the reduction of ABA content and/or sensitivity is of great importance in dormancy release during both after-ripening and cold stratification.

In *Arabidopsis* seeds ABA content decreases at the end of seed maturation, suggesting that after the onset of dormancy endogenous ABA is no longer required for its maintenance (Karssen *et al.*, 1983). However, the carotenoid biosynthesis inhibitors norfluorazon or fluridone (which therefore inhibit ABA biosynthesis) promote germination in dormant *Arabidopsis* seeds (Debeaujon and Koornneef, 2000). ABA is synthesized *de novo* in imbibed dormant seeds of *N. plumbaginifolia*, suggesting that ABA synthesis is required for the maintenance of dormancy (Grappin *et al.*, 2000). In barley (*Hordeum vulgare*), ABA content does not decrease upon imbibition in dormant seeds whereas it decreases rapidly in nondormant seeds (Jacobsen *et al.*, 2002).

A more detailed analysis of the kinetics of ABA synthesis and catabolism during imbibition of *Arabidopsis* Cvi seeds showed that ABA content decreased to a similar extent in both dormant and nondormant seeds (Ali-Rachedi *et al.*, 2004). However, the same study showed that after the time when the nondormant seeds started to germinate (after approximately 3 days), the ABA content of the dormant seeds almost doubled within 3 days of further incubation. These results strongly suggest that the maintenance of dormancy in imbibed seeds may be an active process involving *de novo* ABA synthesis. It has been shown in a number of species including barley (Jacobsen *et al.*, 2002) and lettuce (*Lactuca sativa*) (Toyomasu *et al.*, 1994) that ABA catabolism through oxidative degradation to 8′-hydroxy-ABA is the principal regulatory pathway in the decline of ABA content in seeds during imbibition. In lettuce, light and GAs promote the degradation of ABA (Toyomasu *et al.*, 1994), and light has the additional effect to promote GA synthesis by induction of a GA 3-oxidase gene (Toyomasu *et al.*, 1998). Recently, a genome-wide analysis of the *Arabidopsis* transcriptome concluded that a delicate interplay between GA and ABA biosynthesis and catabolism determines whether the GA-ABA balance would favor dormancy or germination (Cadman *et al.*, 2006). The key genes in this balance were suggested to be the *NCED* (9-*cis*-epoxycarotenoid dioxygenase) and *CYP707A* (ABA 8′-hydroxylase) gene families for ABA synthesis and catabolism, respectively, and *GA3ox1* (GA 3-oxidase) and *GA2ox2* (GA 2-oxidase) for GA-synthesis and degradation, respectively. These examples of interaction between the synthesis and catabolism of ABA and GA during imbibition strongly suggest that an ABA–GA balance is operational. The results from the studies described above are in sharp contrast to the revised hormone balance theory (Karssen and Laçka, 1986), which implicated ABA only in seed development and excluded its role during imbibition and dormancy maintenance, which were postulated to be sensitive only to the presence or absence of GA.

There is strong evidence that GAs are an absolute requirement for germination. GA-deficient mutants of tomato (*gib1*) and *Arabidopsis* (*ga1*) require exogenous GA to complete germination (Koornneef and Van der Veen, 1980; Groot and Karssen, 1987). As indicated above, the sensitivity to GAs may increase with the progress of dormancy release (Hilhorst *et al.*, 1986; Karssen *et al.*, 1989). An increase in the content of bioactive GAs during cold stratification in *Arabidopsis* seeds has been reported, but did not show a close correlation with the control of dormancy (Derkx *et al.*, 1994). Yamauchi *et al.* (2004) showed that an increase in bioactive GA_1 and GA_4 occurred during cold stratification of *Arabidopsis* seeds. The major increase in GA_4 content was mediated by enhanced expression of the genes encoding GA 3-oxidase (*GA3ox1* and *GA3ox2*), the enzyme that catalyzes the conversion of GA_9 to GA_4. Furthermore, red light irradiation during the cold treatment further elevated *GA3ox1* transcript accumulation (Yamauchi *et al.*, 2004). Although all the requirements for dormancy release, namely the increase in GA sensitivity and content as well as light sensitivity, may be met during the cold treatment, seeds may not (or not within the time frame of observation) germinate at this low temperature. It was suggested that cell-type specificity of *GA3ox1* mRNA accumulation depends on the environment, including temperature (Yamauchi *et al.*, 2004). It is possible that a permissive temperature for germination is essential for spatial changes in GA synthesis or of its secondary messenger across different cell layers, although another possibility is that the responses of cells to GA (e.g., cell elongation) are inhibited at cold temperature.

The results from these recent studies seem to contradict the long-held model of dormancy release separated from, and followed by, the induction of germination, as described in a previous section (Figure 3.2B). The synthesis of GAs has been commonly associated with the promotion of germination rather than the release from dormancy (Bewley and Black, 1994). Observations that GAs induce germination of dormant seeds were interpreted as indicating that GAs may bypass the requirements for dormancy release, rather than being integral directly to dormancy release. Similarly, in the previous model, light was not considered to affect dormancy but was recognized to be active only during germination, after the seeds had been sensitized to light and GAs by cold treatment (Hilhorst *et al.*, 1986; Hilhorst and Karssen, 1992). Evidence of the induction of GA biosynthetic enzymes by chilling and their sensitivity to light during chilling may require reexamination of this model.

Similarly, recent studies have shed new light on the role of nitrate in seed dormancy and germination. *S. officinale* seed germination is absolutely dependent on light and nitrate and both factors are active only in the completion of germination after dormancy-releasing cold stratification has enhanced the seed sensitivity to both exogenous and endogenous nitrate (Hilhorst *et al.*, 1986; Hilhorst, 1990). However, in the case of dormant *Arabidopsis* Cvi seeds, nitrate could substitute for the long period (7–12 months) of dry storage or several days of cold stratification required to release dormancy (Ali-Rachedi *et al.*, 2004) and therefore can be regarded as a dormancy-releasing agent. So far, very little is known about the nitrate signaling pathways that lead to germination, although there is evidence in *S. officinale* that nitrate is a signaling molecule by itself without being reduced to nitrite or NO (Hilhorst and Karssen, 1989; see Chapter 7 for NO signaling). Seeds of an *Arabidopsis* nitrate reductase [NADH] 1 and 2 double mutant *nia1 nia2*, which retains only 0.5% of the nitrate reductase activity of the wild type, are much less dormant than the wild-type seeds. The mutant seeds contained 100-fold more nitrate than the wild type, probably because the mutant accumulates more nonreduced nitrate that could be transported to and stored in the seeds (Alboresi *et al.*, 2005). It has been proposed that nitrate signaling may have cross talk with the ABA or GA pathways, but there is only circumstantial evidence for this hypothesis (Ali-Rachedi *et al.*, 2004; Alboresi *et al.*, 2005).

3.5 Secondary dormancy

The term *secondary dormancy* is used for the type of dormancy that is imposed after seeds have lost primary dormancy. Secondary dormancy may be the result of a prolonged inhibition of germination. The inhibition may be due to active factors (e.g., endogenous ABA or secondary metabolites) or passive ones (e.g., the lack of proper conditions for germination) (reviewed by Hilhorst, 1998). The occurrence of secondary dormancy is highly relevant to seed behavior in the soil seed bank as it is central to so-called dormancy cycling. The concept of dormancy cycling has been developed to explain seasonal flushes of seedling emergence of annual temperate species (Karssen, 1982; Hilhorst *et al.*, 1996; Baskin and Baskin, 1998). Dormancy cycling involves repeated induction and termination of dormancy parallel to seasonal variations in temperature (Figure 3.1). In this way seeds avoid germination during short favorable spells within the unfavorable season and germinate just prior to the favorable season for plant growth (Vleeshouwers *et al.*, 1995).

There is no conclusive evidence that ABA is involved in the acquisition of secondary dormancy, although it is very likely. ABA synthesis is required for the maintenance of primary dormancy in *Arabidopsis* Cvi and *N. plumbaginifolia* (Grappin *et al.*, 2000; Ali-Rachedi *et al.*, 2004). In the case of *Arabidopsis*, the conditions that maintain ABA synthesis and seed dormancy are also effective in inducing secondary dormancy (Cadman *et al.*, 2006). Induction of secondary dormancy in *Brassica napus* seeds by prolonged incubation in an osmoticum was associated with an increase in ABA content and maintenance of ABA sensitivity (Gulden *et al.*, 2004). Moreover, in *Arabidopsis* seeds possessing secondary dormancy, ABA biosynthesis genes show significantly higher expression than in nondormant seeds (Cadman *et al.*, 2006).

A long-standing question is whether secondary dormancy differs mechanistically or physiologically from primary dormancy. Only recently it was shown that there are significant differences in primary and secondary dormancy of *Arabidopsis* seeds (Cadman *et al.*, 2006). Dormancy was induced in Cvi seeds by applying several different temperature regimes of cold (5°C) and warm (20°C) incubation, which made it possible for seeds to go through two dormancy loss and induction cycles. Primary and secondary dormant seeds required light and nitrate to germinate, but seeds that had undergone two cycles (*tertiary dormancy*) did not respond to light and nitrate, although they were still viable. This is clearly indicative of a changed physiology in tertiary dormant seeds. In addition, there were substantial differences in the transcriptomes of primary, secondary, and tertiary dormant seeds (Cadman *et al.*, 2006). However, it remains to be shown that the different forms of secondary dormancy also occur in the field and in other species.

3.6 Signaling in dormancy

3.6.1 Stress signaling

Several studies aiming at the discovery of dormancy-associated genes have identified stress-related genes. From differential screening of dormant and nondormant *A. fatua* cDNA libraries, several genes were found to be expressed in the dormant but not in the nondormant seeds (Li and Foley, 1995, 1996). However, when nondormant seeds were stressed by exposure to elevated temperatures, expression of the 'dormancy genes' was increased. Many of these dormancy-associated genes encode the late embryogenesis abundant family proteins, which are thought to play a protective role during seed desiccation (Philips *et al.*, 2002). Other studies have also identified stress-related genes, such as the barley *Per1* and *Arabidopsis AtPer1*, which encode peroxiredoxin (Prx) (Aalen, 1999). Prx is an enzyme that plays a role in protecting seeds from damage due to exposure to reactive oxygen species. *Per1* expression is induced by osmotic stress and ABA in immature barley embryos (Aalen *et al.*, 1994). The closely related *pBS128* transcript is upregulated in both dormant and nondormant imbibed mature embryos of *Bromus secalinus* (Goldmark *et al.*, 1992). Prx was suggested to play a role in dormancy (Goldmark *et al.*, 1992; Haslekas *et al.*, 1998). However, more conclusive evidence argues against the involvement of Prx in the regulation of dormancy. Overexpression and RNAi experiments with *Arabidopsis AtPER1* found no correlation between Prx levels and the maintenance of dormancy (Haslekas *et al.*, 2003). Furthermore, *AtPER1* expression in the nondormant *Arabidopsis aba1* mutant was similar to that in wild-type seeds, which suggests that expression of the *AtPer1* transcript alone is not sufficient to maintain dormancy (Haslekas *et al.*, 1998).

More information is required to draw firm conclusions on cross talk between stress- and dormancy-related signaling. Evidently, ABA is a common denominator in these signaling networks since it is central to stress responses in plants (Finkelstein *et al.*, 2002; Xiong *et al.*, 2002). Overlapping stress and dormancy responses could be related to the ecological relevance of dormancy. Seeds maintain their dormancy during unfavorable periods for growth, during which they are exposed to cold, heat, or drought stress. Thus, seeds have to be equipped to successfully tolerate these stresses while dormant.

3.6.2 Signaling networks

As expression of a large number of genes is affected by ABA, it is virtually certain that ABA signaling consists of complex signaling networks. *Cis*-acting elements appear to play crucial roles as nodes within these networks (Nakabayashi *et al.*, 2005), and various combinations of *cis*-acting elements have been identified that are involved in ABA, stress, and other hormone responses. For example, the promoters of the most abundant transcripts present in dry *Arabidopsis* Cvi seeds were more likely to contain multiple ABREs (ABA-responsive elements), a combination of ABRE with CE (coupling element), or a combination of ABRE with RY/Sph (seed-specific motif) (Nakabayashi *et al.*, 2005). Similarly, of the more than 400 genes associated with various dormancy states in *Arabidopsis* Cvi seeds, including many related to stress, a statistically greater fraction contained one or more of these *cis*-elements as compared to their occurrence in the whole genome (Cadman *et al.*, 2006). A greater specificity of ABA action may be attained by the binding of certain transcription factors with specific combinations of *cis*-acting elements to activate a smaller number of promoters as compared to the number of genes potentially regulated by ABA, as described for osmotic and cold stress responses (Yamaguchi-Shinozaki and Shinozaki, 2005). Furthermore, posttranscriptional and posttranslational control mechanisms are likely to be involved in further fine-tuning of these responses.

3.6.3 Environmental signals

Although rapid progress is now being made in unraveling hormonal signaling in dormancy regulation, especially the mechanisms downstream of ABA and GA perception, little attention has been paid to the perception of the environmental signals that evoke the changes in seed dormancy. The perception of light through phytochrome in initiating germination has been relatively well studied (Casal and Sánchez, 1998). With the aid of single, double, and triple mutants of phytochromes A to E in *Arabidopsis*, it was shown that phytochromes B, D, and E are required for the photoreversible regulation of germination by red (R) and far-red (FR) light (Casal and Sánchez, 1998; Hennig *et al.*, 2002). Light, through phytochrome, also increases bioactive GA content, as described above.

Nitrate is another important environmental signal in seed dormancy. Since nitrate reductase is not involved in the nitrate-mediated germination of *S. officinale* and *Arabidopsis* (see above), other primary receptors must be involved in these species. Based on detailed dose–response analysis of nitrate-induced germination of *S. officinale*, Hilhorst (1990) hypothesized that a low- and high-affinity nitrate receptor mediates nitrate signaling. A biphasic germination response to nitrate indicated that a high-affinity response occurred at 1 mM nitrate whereas a low-affinity response occurred at 10 mM nitrate. The *Arabidopsis* chlorate resistant (*chl1-5*) mutant is defective in the dual high/low affinity NRT1.1 (CHL1) nitrate transporter. Mutant seeds failed to germinate at 1 mM nitrate but did germinate at 10 mM nitrate (Alboresi *et al.*, 2005). Thus, the nitrate receptor (Hilhorst, 1990) hypothesized to be involved in nitrate perception and signaling in seed germination may be the dual affinity nitrate transporter NRT1.1.

While it can be debated whether light and nitrate act on dormancy release or germination induction, there is little doubt that temperature is an environmental cue that has a profound influence on the dormancy status of seeds. The changes in seasonal temperatures drive dormancy cycling of seeds in the soil. Seeds in the soil seed bank not only perceive the changes in temperature but also 'store' this information in order to monitor the season they are in. A role for membranes in dormancy regulation has been postulated in the literature for a long time, mainly because of the effect of anesthetics on dormancy and germination (reviewed by Hilhorst, 1998). Membranes have been suggested to be the primary target for temperature perception at the cellular level (Murata and Los, 1997) and therefore might mediate temperature signaling in the regulation of dormancy. On the basis of a consilience of circumstantial evidence, Hilhorst (1998) proposed a model in which temperature alters membrane fluidity, which in turn results in altered conformation of membrane proteins (e.g., receptors) and/or membrane permeability, and ultimately results in a change in dormancy. The degree of unsaturation of the membrane phospholipids, which is sensitive to temperature, could function as 'temperature memory'. However, in this model the connection between changes in membrane fluidity and the downstream signaling events was missing.

Evidence is now becoming available that changes in membrane fluidity may indeed result in altered gene expression. For example, in cyanobacteria *Synechocystis*, rigidification of the membranes by cold induces the transient expression of genes encoding desaturases that are required to increase the degree of unsaturation in the membrane phospholipids and, hence, membrane fluidity (Mikami and Murata, 2003). *HISTIDINE KINASE 33* (*HIK33*) regulates expression of a large number of cold-inducible genes in *Synechocystis*. HIK33 proteins appear to recognize a decrease in membrane fluidity, which invokes their dimerization and the subsequent activation of the kinase (Inaba *et al.*, 2003). It is not known yet which domain(s) in the protein perceive these changes in membrane fluidity. A similar mechanism has been demonstrated in plants. In cell suspension cultures of *Medicago sativa*, activity of mitogen-activated protein kinases (MAPKs) is induced by heat and cold stress. Heat specifically activated a heat-shock activated MAPK (HAMK) and cold activated a distinct stress-activated MAPK (SAMK). The activation of these kinases was directly mediated by a decrease (SAMK) or an increase (HAMK) in membrane fluidity (Sangwan *et al.*, 2002). Although the identities of the cascades activated by SAMK and HAMK remain to be elucidated, this would be an exciting area for dormancy research since putatively similar cascades are activated by opposite changes in membrane fluidity. This could explain why low temperatures release dormancy of seeds from summer annuals but induce dormancy in winter annuals.

3.7 Challenges for the future

The now widely accessible transcriptomics technology allows rapid progress in the elucidation of signaling networks involved in the regulation of dormancy and germination. Complete genome sequences have become available for a growing number of species and gene functional analysis by reverse genetics is a realistic option in seed science. The functions of specific genes on a seed phenotype (e.g., dormancy) can be rapidly determined by characterizing the phenotype of putative gene knockout plants, which can be electronically identified in the available databases or by screening the available DNA pools from the mutagenized plant populations. Similarly, proteome analysis will also be a robust tool for seed science, which includes analysis of protein structure–function relationships, regulation by posttranslational modification, and protein–protein, protein–DNA/RNA, or protein–small ligand interactions. Dormancy is an adaptive trait and highly dependent on the environment. Forward genetics, particularly QTL analysis, in *Arabidopsis* and cereals has identified many loci that account for the natural variation in dormancy of several accessions (e.g. Alonso-Blanco *et al.*, 2003; Chapter 5). However, the subsequent fine-mapping to identify the genes involved is tedious and time-consuming. Recently, a reverse genetics method called Ecotilling has been proposed to identify polymorphisms in natural populations (Comai *et al.*, 2004). This method holds the promise to more rapidly identify genes that contribute to natural variation in dormancy. However, reverse genetics can only be powerful if efficient and reliable high-throughput phenotyping is available. In addition to the technology, this requires the input of expertise on plant and seed physiology and morphology. Functional genomics will undoubtedly boost our knowledge of signaling pathways and networks that are involved in the seed's 'decision' to germinate or remain dormant and likely change current models and theories of the regulation of dormancy. Only a combination of multiple disciplines will enable us to address the fundamental questions in seed science.

References

R.B. Aalen, H.-G. Opsahl-Ferstad, C. Linnestad and O.-A. Olsen (1994) Transcripts encoding an oleosin and a dormancy-related protein are present in both the aleurone layer and the embryo of developing barley (*Hordeum vulgare* L.) seeds. *The Plant Journal* **5**, 385–396.

R.B. Aalen (1999) Peroxiredoxin antioxidants in seed physiology. *Seed Science Research* **9**, 285–295.

A. Alboresi, C. Gestin, M.-T. Leydecker, M. Bedu, C. Meyer and H.-N. Truong (2005) Nitrate, a signal relieving seed dormancy in *Arabidopsis*. *Plant, Cell and Environment* **28**, 500–512.

S. Ali-Rachedi, D. Bouinot, M.-H. Wagner, *et al.* (2004) Changes in endogenous abscisic acid levels during dormancy release and maintenance of mature seeds: studies with the Cape Verde Islands ecotype, the dormant model of *Arabidopsis thaliana*. *Planta* **219**, 479–488.

C. Alonso-Blanco, L. Bentsink, C.J. Hanhart, H. Blankestijn–de Vries and M. Koornneef (2003) Analysis of natural allelic variation at dormancy loci of *Arabidopsis thaliana*. *Genetics* **164**, 711–729.

R.D. Amen (1968) A model of seed dormancy. *Botanical Review* **34**, 1–31.

C.C. Baskin and J.M. Baskin (1998) *Seeds: Ecology, Biogeography and Evolution of Dormancy and Germination*. Academic Press, San Diego.

C.C. Baskin and J.M. Baskin (2004) A classification system for seed dormancy. *Seed Science Research* **14**, 1–16.

L.O. Baumbusch, D.W. Hughes, G.A. Galau and K.S. Jakobsen (2004) *LEC1*, *FUS3*, *ABI3* and *Em* expression reveals no correlation with dormancy in *Arabidopsis*. *Journal of Experimental Botany* **55**, 77–87.

T. Berry and J.D. Bewley (1992) A role for the surrounding fruit tissues in preventing the germination of tomato (*Lycopersicon esculentum*) seeds. *Plant Physiology* **100**, 951–957.

J.D. Bewley (1997) Seed germination and dormancy. *The Plant Cell* **9**, 1055–1066.

J.D. Bewley and M. Black (1994) *Seeds: Physiology of Development and Germination*. Plenum Press, New York.

C.S.C. Cadman, P.E. Toorop, H.W.M. Hilhorst and W.E. Finch-Savage (2006) Gene expression profiles of *Arabidopsis* Cvi seeds during cycling through dormant and non-dormant states indicate a common underlying dormancy control mechanism. *The Plant Journal*, **46**, 805–822.

J.J. Casal and R.A. Sánchez (1998) Phytochromes and seed germination. *Seed Science Research* **8**, 317–329.

M.A. Cohn (1996) Operational and philosophical decisions in seed dormancy research. *Seed Science Research* **6**, 147–153.

L. Comai, K. Young, B.J. Till, *et al.* (2004) Efficient discovery of DNA polymorphisms in natural populations by ecotilling. *The Plant Journal* **37**, 778–786.

S. Cutler, M. Ghassemian, D. Bonetta, S. Cooney and P. McCourt (1996) A protein farnesyl transferase involved in abscisic acid signal transduction in *Arabidopsis*. *Science* **273**, 1239–1241.

E.A.A. Da Silva, P.E. Toorop, A.C. van Aelst and H.W.M. Hilhorst (2004) Abscisic acid controls embryo growth potential and endosperm cap weakening during coffee (*Coffea arabica* cv. Rubi) seed germination. *Planta* **220**, 251–261.

I. Debeaujon and M. Koornneef (2000) Gibberellin requirement for *Arabidopsis thaliana* seed germination is determined both by testa characteristics and embryonic ABA. *Plant Physiology* **122**, 415–424.

M.P.M. Derkx and C.M. Karssen (1993) Effects of light and temperature on seed dormancy and gibberellin-stimulated germination in *Arabidopsis thaliana*: studies with gibberellin-deficient and -insensitive mutants. *Physiologia Plantarum* **89**, 360–368.

M.P.M. Derkx, E. Vermeer and C.M. Karssen (1994) Gibberellins in seeds of *Arabidopsis thaliana*: biological activities, identification and effects of light and chilling on endogenous levels. *Plant Growth Regulation* **15**, 223–234.

R.R. Finkelstein, S.S.L. Campala and C.D. Rock (2002) Abscisic acid signaling in seeds and seedlings. *The Plant Cell* **14**, S15–S45.

A. Frey, C. Audran, E. Marin, B. Sotta and A. Marion-Poll (1999) Engineering seed dormancy by the modification of zeaxanthin epoxidase gene expression. *Plant Molecular Biology* **39**, 1267–1274.

P.J. Goldmark, J. Curry, C.F. Morris and M.K. Walker-Simmons (1992) Cloning and expression of an embryo-specific mRNA up-regulated in hydrated dormant seeds. *Plant Molecular Biology* **19**, 433–441.

M.P. González-García, D. Rodríguez, C. Nicolás, P.L. Rodríguez, G. Nicolás and O. Lorenzo (2003) Negative regulation of abscisic acid signalling by the *Fagus sylvatica* FsPP2C1 plays a role in seed dormancy regulation and promotion of seed germination. *Plant Physiology* **133**, 135–144.

P. Grappin, D. Bouinot, B. Sotta, E. Miginiac and M. Julien (2000) Control of seed dormancy in *Nicotiana plumbaginifolia*: post-imbibition abscisic acid synthesis imposes dormancy maintenance. *Planta* **210**, 279–285.

S.P.C. Groot and C.M. Karssen (1987) Gibberellins regulate seed germination in tomato by endosperm weakening: a study with gibberellin-deficient mutants. *Planta* **171**, 525–531.

S.P.C. Groot and C.M. Karssen (1992) Dormancy and germination of abscisic acid-deficient tomato seeds. Studies with the *sitiens* mutant. *Plant Physiology* **99**, 952–958.

R.H. Gulden, S. Chiwocha, S. Abrams, I. McGregor, A. Kermode and S. Shirtliffe (2004) Response to abscisic acid application and hormone profiles in spring *Brassica napus* seed in relation to secondary dormancy. *Canadian Journal of Botany* **82**, 1618–1624.

J.L. Harper (1959) The ecological significance of dormancy and its importance in weed control. In: *Proceedings of the IVth International Congress of Crop Protection*, Hamburg, September 1957, Vol. 4, pp. 415–420.

C. Haslekas, R.A.P. Stacy, V. Nygaard, F.A. Culianez-Macia and R.B. Aalen (1998) The expression of a peroxiredoxin antioxidant gene, *AtPer1*, in *Arabidopsis thaliana* is seed specific and related to dormancy. *Plant Molecular Biology* **36**, 833–845.

C. Haslekas, M.K. Viken, P.E. Grini, *et al.* (2003) Seed 1-cysteine peroxiredoxin antioxidants are not involved in dormancy, but contribute to inhibition of germination during stress. *Plant Physiology* **133**, 1148–1157.

L. Hennig, W.M. Stoddart, M. Dieterle, G.C. Whitelam and E. Schäfer (2002) Phytochrome E controls light-induced germination of *Arabidopsis*. *Plant Physiology* **128**, 194–200.

H.W.M. Hilhorst (1990) Dose–response analysis of factors involved in germination and secondary dormancy of seeds of *Sisymbrium offinicale*. 2: Nitrate. *Plant Physiology* **94**, 1096–1102.

H.W.M. Hilhorst (1995) A critical update on seed dormancy. I: Primary dormancy. *Seed Science Research* **5**, 61–73.

H.W.M. Hilhorst (1997) Seed dormancy. *Seed Science Research* **7**, 221–223.

H.W.M. Hilhorst (1998) The regulation of secondary dormancy. The membrane hypothesis revisited. *Seed Science Research* **8**, 77–90.

H.W.M. Hilhorst, L. Bentsink and M. Koornneef (2006) Dormancy and germination. In: *Handbook of Seed Science and Technology* (ed. A. Basra). The Haworth Press, Binghamton, New York 271–299.

H.W.M. Hilhorst, M.P.M. Derkx and C.M. Karssen (1996) An integrating model for seed dormancy cycling; characterization of reversible sensitivity. In: *Plant Dormancy: Physiology, Biochemistry and Molecular Biology* (ed. G. Lang), pp. 341–360. CAB International, Wallingford, UK.

H.W.M. Hilhorst and B. Downie (1996) Primary dormancy in tomato (*Lycopersicon esculentum* cv. Moneymaker) – studies with the *sitiens* mutant. *Journal of Experimental Botany* **47**, 89–97.

H.W.M. Hilhorst and C.M. Karssen (1989) Nitrate reductase independent stimulation of seed germination in *Sisymbrium officinale* L. (hedge mustard) by light and nitrate. *Annals of Botany* **63**, 131–138.

H.W.M. Hilhorst and C.M. Karssen (1992) Seed dormancy and germination: the role of abscisic acid and gibberellins and the importance of hormone mutants. *Plant Growth Regulation* **11**, 225–238.

H.W.M. Hilhorst, I. Smitt and C.M. Karssen (1986) Gibberellin biosynthesis and sensitivity mediated stimulation of seed germination of *Sisymbrium officinale* by red light and nitrate. *Physiologia Plantarum* **67**, 285–290.

M. Holdsworth, S. Kurup and R. McKibbin (1999) Molecular and genetic mechanisms regulating the transition from embryo development to germination. *Trends in Plant Science* **4**, 275–280.

M. Inaba, I. Suzuki, B. Szalontai, *et al.* (2003) Gene-engineered rigidification of membrane lipids enhances the cold inducibility of gene expression in *Synechocystis*. *Journal of Biological Chemistry* **278**, 12191–12198.

J.V. Jacobsen, D.W. Pearce, A.T. Poole, R.P. Pharis and L.N. Mander (2002) Abscisic acid, phaseic acid and gibberellin contents associated with dormancy and germination in barley. *Physiologia Plantarum* **115**, 428–441.

R.R. Johnson and W.E. Dyer (2000) Degradation of endosperm mRNAs during dry afterripening of cereal grains. *Seed Science Research* **10**, 233–241.

C.M. Karssen (1982) Seasonal patterns of dormancy in weed seeds. In: *The Physiology and Biochemistry of Seed Development, Dormancy and Germination* (ed. A.A. Khan), pp. 243–270. Elsevier Biomedical Press, Amsterdam.

C.M. Karssen, D.L.C. Brinkhorst-van der Swan, A.E. Breekland and M. Koornneef (1983) Induction of dormancy during seed development by endogenous abscisic acid: studies on abscisic acid deficient genotypes of *Arabidopsis thaliana* (L.) Heynh. *Planta* **157**, 158–165.

C.M. Karssen and E. Laçka (1986) A revision of the hormone balance theory of seed dormancy: studies on gibberellin- and/or abscisic acid-deficient mutants of *Arabidopsis thaliana*. In: *Plant Growth Substances 1985* (ed. M. Bopp), pp. 315–323. Springer-Verlag, Berlin.

C.M. Karssen, S. Zagórski, J. Kepczynski and S.P.C. Groot (1989) Key role for endogenous gibberellins in the control of seed germination. *Annals of Botany* **63**, 71–80.

M. Koornneef, L. Bentsink and H.W.M. Hilhorst (2002) Seed dormancy and germination. *Current Opinion in Plant Biology* **5**, 33–36.

M. Koornneef, J.W. Cone, C.M. Karssen, R.E. Kendrick, J.H. Van der Veen and J.A.D. Zeevaart (1985) Plant hormone and photoreceptor mutants in *Arabidopsis* and tomato. In: *Plant Genetics, UCLA Symposia on Molecular and Cellular Biology, New Series*, Vol. 35 (ed. M. Freeling), pp. 1–12. Alan Liss, New York.

M. Koornneef, G. Reuling and C.M. Karssen (1984) The isolation and characterization of abscisic acid-insensitive mutants of *Arabidopsis thaliana. Physiologia Plantarum* **61**, 377–383.

M. Koornneef and J.H. Van der Veen (1980) Induction and analysis of gibberellin sensitive mutants in *Arabidopsis thaliana* (L.) Heynh. *Theoretical and Applied Genetics* **58**, 257–263.

K.M. Léon-Kloosterziel, C.J. Keijzer and M. Koornneef (1994) A seed shape mutant of *Arabidopsis* that is affected in integument development. *The Plant Cell* **6**, 385–392.

M.T. Le Page-Degivry, G. Garello and P. Barthe (1997) Changes in abscisic acid biosynthesis and catabolism during dormancy breaking in *Fagus sylvatica* embryo. *Journal of Plant Growth Regulation* **16**, 57–61.

G. Leubner-Metzger (2003) Functions and regulation of β-1,3-glucanase during seed germination, dormancy release and after-ripening. *Seed Science Research* **13**, 17–34.

G. Leubner-Metzger (2005) β-1,3-Glucanase gene expression in low-hydrated seeds as a mechanism for dormancy release during tobacco after-ripening. *The Plant Journal* **41**, 133–145.

G. Leubner-Metzger and F. Meins, Jr. (2000) Sense transformation reveals a novel role for class I β-1,3-glucanase in tobacco seed germination. *The Plant Journal* **23**, 215–221.

B. Li and M. Foley (1995) Cloning and characterization of differentially expressed genes in imbibed dormant and afterripened *Avena fatua* embryos. *Plant Molecular Biology* **29**, 823–831.

B. Li and M. Foley (1996) Transcriptional and posttranscriptional regulation of dormancy-associated gene expression by afterripening in wild oat. *Plant Physiology* **110**, 1267–1273.

Y. Liu, R.J. Bino, C.M. Karssen and H.W.M. Hilhorst (1996) Water relations of GA- and ABA-deficient tomato mutants during seed and fruit development and their influence on germination. *Physiologia Plantarum* **96**, 425–432.

D.R. McCarty (1995) Genetic control and integration of maturation and germination pathways in seed development. *Annual Review of Plant Physiology and Plant Molecular Biology* **46**, 71–93.

K. Mikami and N. Murata (2003) Membrane fluidity and the perception of environmental signals in cyanobacteria and plants. *Progress in Lipid Research* **42**, 527–543.

N. Murata and D.A. Los (1997) Membrane fluidity and temperature perception. *Plant Physiology* **115**, 875–879.

K. Nakabayashi, M. Okamoto, T. Koshiba, Y. Kamiya and E. Nambara (2005) Genome-wide profiling of stored mRNA in *Arabidopsis thaliana* seed germination: epigenetic and genetic regulation of transcription in seed. *The Plant Journal* **41**, 697–709.

E. Nambara, R. Hayama, Y. Tsuchiya, *et al.* (2000) The role of *ABI3* and *FUS3* loci in *Arabidopsis thaliana* on phase transition from late embryo development to germination. *Developmental Biology* **220**, 412–423.

E. Nambara and A. Marion-Poll (2005) Abscisic acid biosynthesis and catabolism. *Annual Review of Plant Biology* **56**, 165–185.

M.G. Nikolaeva (1977) Factors controlling the seed dormancy pattern. In: *The Physiology and Biochemistry of Seed Dormancy and Germination* (ed. A.A. Khan), pp. 51–74. North-Holland, Amsterdam.

F. Parcy, C. Valon, A. Kohara, S. Miséra and J. Giraudat (1997) The *ABSCISIC ACID-INSENSITIVE3*, *FUSCA3*, and *LEAFY COTOLEDON1* loci act in concert to control multiple aspects of *Arabidopsis* seed development. *The Plant Cell* **9**, 1265–1277.

J.R. Philips, M.J. Oliver and D. Bartels (2002) Molecular genetics of desiccation tolerant systems. In: *Desiccation and Survival in Plants: Drying without Dying* (eds M. Black and H.W. Pritchard), pp. 319–341. CAB International, Wallingford, UK.

V. Raz, J.H.W. Bergervoet and M. Koornneef (2001) Sequential steps for developmental arrest in *Arabidopsis* seeds. *Development* **128**, 243–252.
A. Rohde, S. Kurup and M. Holdsworth (2000) ABI3 emerges from the seed. *Trends in Plant Science* **5**, 418–419.
R.A. Sánchez, L. Sunell, J.M. Labavitch and B.A. Bonner (1990) Changes in the endosperm cell walls of two *Datura* species before radicle protrusion. *Plant Physiology* **93**, 89–97.
V. Sangwan, B.L. Orvar, J. Beyerly, H. Hirt and R.S. Dhindsa (2002) Opposite changes in membrane fluidity mimic cold and heat stress activation of distinct plant MAP kinase pathways. *The Plant Journal* **31**, 629–638.
N. Schmitz, S.R. Abrams and A.R. Kermode (2000) Changes in abscisic acid content and embryo sensitivity to (+)-abscisic acid during the termination of dormancy of yellow cedar seeds. *Journal of Experimental Botany* **51**, 1159–1162.
G.M. Simpson (1990) *Seed Dormancy in Grasses*. Cambridge University Press, Cambridge, UK.
T. Toyomasu, H. Kawaide, W. Mitsuhashi, Y. Inoue and Y. Kamiya (1998) Phytochrome regulates gibberellin biosynthesis during germination of photoblastic lettuce seeds. *Plant Phsiology*, **118**, 1517–1523.
T. Toyomasu, H. Yamane, N. Murofushi and Y. Inoue (1994) Effects of exogenously applied gibberellin and red light on the endogenous levels of abscisic acid in photoblastic lettuce seeds. *Plant and Cell Physiology* **35**, 127–129.
P. Van der Toorn and C.M. Karssen (1992) Analysis of embryo growth in mature fruits of celery (*Apium graveolens*). *Physiologia Plantarum* **84**, 593–599.
L.M. Vleeshouwers, H.J. Bouwmeester and C.M. Karssen (1995) Redefining seed dormancy: an attempt to integrate physiology and ecology. *Journal of Ecology* **83**, 1031–1037.
J.O. Watkinson and G.E. Welbaum (in press) Seeds of the family Orchidaceae. In: *The Encyclopedia of Seeds* (eds D. Bewley, M. Black and P. Halmer). CABI International, Wallingford, UK.
L. Xiong, H. Lee, M. Ishitani and J.-K. Zhu (2002) Regulation of osmotic stress-responsive gene expression by the *LOS6/ABA1* locus in *Arabidopsis*. *Journal of Biological Chemistry* **277**, 8588–8596.
K. Yamaguchi-Shinozaki and K. Shinozaki (2005) Organization of *cis*-acting regulatory elements in osmotic- and cold-stress responsive promoters. *Trends in Plant Science* **10**, 88–94.
Y. Yamauchi, M. Ogawa, A. Kuwahara, A. Hanada, Y. Kamiya and S. Yamaguchi (2004) Activation of gibberellin biosynthesis and response pathways by low temperature during imbibition of *Arabidopsis thaliana* seeds. *The Plant Cell* **16**, 367–378.

4 Modeling of seed dormancy

Phil S. Allen, Roberto L. Benech-Arnold,
Diego Batlla and Kent J. Bradford

4.1 Introduction

Understanding seed dormancy sufficiently to predict germination and seedling emergence from natural seed banks has long been a goal of both seed ecologists and agriculturalists. These predictions require more complex models than for germination of agronomic species that lack appreciable seed dormancy, but offer the potential for significant economic and environmental benefits where such models succeed in predicting seed outcomes. For example, models to predict the timing of weed seed germination would be useful for scheduling seedling control efforts (Benech-Arnold and Sánchez, 1995; Grundy, 2003). Seed dormancy models could also aid in the design of practices for managing native or introduced plant populations. Although several empirical models have been developed that predict germination and emergence with some success, one of the most significant limitations to the advancement of such models is the existence of dormancy in seed populations (Forcella *et al.*, 2000; Vleeshouwers and Kropff, 2000).

A detailed understanding of how environmental factors regulate dormancy status is needed in order to develop an adequate theoretical framework for the construction of predictive models that address dormancy changes in seed banks (see also Chapter 3). Several environmental factors associated with dormancy in the field are identified and their relationships diagrammed in Figure 4.1. While our focus in this chapter is primarily on those stages most closely responsible for changes in dormancy status, it is important to recognize that predictions related to seed germination can span all the processes that begin with inputs related to weather and soil and end with outcomes related to seedling establishment and survival. Since field measurements relating to specific variables of interest can be difficult to obtain, we often accept less accurate or less precise measurements for practical reasons (e.g., atmospheric temperature and precipitation instead of seed zone temperature and water potential). It is also generally easier to measure seedling emergence from the soil rather than germination per se. The decision whether to accept a particular measure as a proxy for a variable that is theoretically more desirable but in practice more difficult to obtain often depends on the complexity of interactions among environmental factors and the desired level of accuracy for model predictions. In many cases a model must be constructed and tested in order to determine whether particular variables lead to sufficiently accurate predictions related to dormancy and germination. A further challenge in modeling of seed dormancy is that the level of dormancy (i.e., the range of conditions under which germination can proceed) is not the same for all individuals within a seed population, and models need to explicitly account for this variation.

In the sections below we discuss environmental factors affecting dormancy in natural seed banks according to the general framework presented in Figure 4.1, with an emphasis on classifying and understanding the effects of these factors on seed dormancy changes under field situations. The aim of this classification is to facilitate conceptualization of dormancy systems for modeling purposes. Later sections are devoted to presenting some examples of population-based models that relate seed dormancy behavior to the effects of those environmental factors.

4.2 Types and phenology of seed dormancy

According to Benech-Arnold *et al.* (2000), 'Dormancy is an internal condition of the seed that impedes its germination under otherwise adequate hydric, thermal and gaseous conditions'. This implies that once the impedance has been removed, seed germination would proceed under a wide range of environmental conditions. Karssen (1982) proposed that dormancy could be classified into primary and secondary dormancy. Primary dormancy refers to the innate dormancy possessed by seeds when they are dispersed from the mother plant. Numerous types and patterns of primary dormancy have been documented and classified by Baskin and Baskin (2004). Secondary dormancy refers to a dormant state that is induced in nondormant seeds by unfavorable conditions for germination, or reinduced in once-dormant seeds after primary dormancy has been alleviated. Thus, primary and secondary dormancy refer only to the timing of occurrence, not to the mechanisms by which dormancy is imposed or released, although knowledge of the mechanistic basis of both types of dormancy is still rudimentary (see Chapter 3). Release from primary dormancy followed by subsequent entrance into secondary dormancy may lead to dormancy cycling, indicated in Figure 4.1 by solid and dashed arrows alternating between low and high dormancy levels. Evidence for dormancy cycling has been obtained for seeds of many species (Fenner, 2000). For species that exhibit dormancy cycling under natural field conditions, the transitions into and out of dormancy may continue for several years before the seeds germinate, decay, or are otherwise lost from the soil seed bank (Karssen, 1980/81; Baskin and Baskin, 1985; Meyer *et al.*, 1997). Overlapping additions to and losses from the soil seed bank and varying levels of dormancy within the seed population complicate attempts to model germination and emergence cycles in such species.

In general, seeds are released from dormancy preceding a season with favorable conditions for seedling emergence and survival, whereas dormancy is induced prior to a period with adverse conditions for seedling survival. For example, several summer annual species exhibit a high dormancy level in autumn, undergo dormancy release during winter and, for the fraction of seeds that fail to germinate in spring, show an increase in dormancy again during the following summer. In contrast, winter annual species show an opposite seasonal dormancy pattern. Therefore, seasonal patterns of seed dormancy are of high survival value to species by constraining

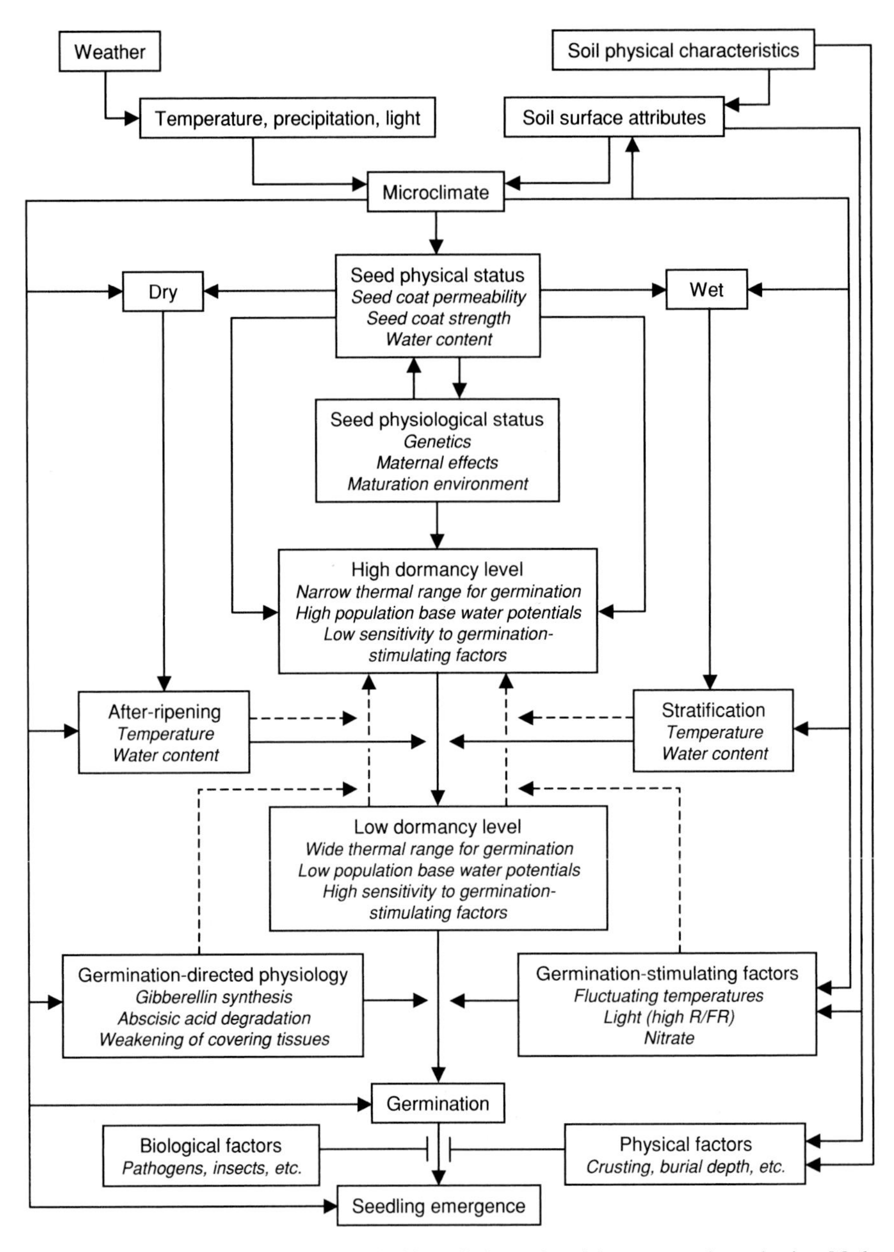

Figure 4.1 Targeted variables associated with predictions of seed dormancy and germination. Modeling efforts may focus on any of several processes. Weather, soil surface attributes (e.g., presence or absence of litter layer, topography), and soil physical characteristics determine seed zone microclimate. Microclimate controls seed dormancy status and germination-directed physiology through its effect on seed physical status, primarily moisture content and temperature. Physical attributes of the seed, such as seed coat permeability, can mitigate these microclimate effects. In addition, (*Continued over*)

germination to periods when environmental conditions favor seedling growth and survival (Karssen, 1982; Donohue *et al.*, 2005).

Dormancy is also a relative rather than an absolute condition. The concept of degrees of relative dormancy was first introduced by Vegis (1964) following observations that as dormancy is released, the temperature range permissive for germination widens until it is maximal. Conversely, as dormancy is induced the range of temperatures over which germination can proceed narrows until germination is no longer possible at any temperature. On this basis, Karssen (1982) proposed that seasonal periodicity in field emergence of annual species is the combined result of seasonal periodicity in soil temperatures and physiological changes within seeds that alter the temperature range permissive for germination. Germination in the field is therefore restricted to periods when the soil temperature and the temperature range over which germination can proceed overlap.

The concept of relative dormancy can be expanded to include seed responses to environmental factors other than temperature. For example, dormancy release or imposition can also be correlated with changes in sensitivity to water potential, light, or nitrate (Hilhorst, 1990; Bradford, 2002; Batlla and Benech-Arnold, 2005). In some species, the transition from dormancy to nondormancy is relatively continuous, and is exhibited by increases in germination percentage and rate under a given set of conditions. In other species, the conditions for releasing dormancy may be distinct from those required for germination, and the latter must be stimulated by factors such as light, nitrate, or fluctuating temperatures after dormancy per se has been alleviated (see also Chapter 3). In these cases, the loss of dormancy is associated with an increasing sensitivity or responsiveness of seeds to these germination-stimulating (or dormancy-terminating) factors (Benech-Arnold *et al.*, 2000).

Seedling emergence of a particular species in the field occurs only when all environmental conditions are within the ranges permissive for seed germination. Since the permissive ranges of conditions change with dormancy levels of seed populations, models for predicting germination or emergence need to incorporate

Figure 4.1 (*Continued*) the physiological status of the seed, including its genetic background and maternal and environmental effects during development and maturation, influences dormancy level. Dormancy-releasing processes in dry seeds (after-ripening) differ qualitatively from those that occur in imbibed seeds (stratification). Interactions between environmental factors and seed dormancy status are shown via arrows. A high dormancy level (whether primary or secondary) is associated with a narrow thermal range permissive for germination, relatively high (more positive) population base water potentials, and a low sensitivity to germination-stimulating factors. Conversely, a low dormancy level is associated with a wider temperature range permissive for germination, lower (more negative) population base water potentials, and increased sensitivity to germination stimulating factors. In some cases following dormancy release through temperature and moisture conditions, germination may need to be stimulated by factors such as light, nitrate, or fluctuating temperatures. Dashed arrows indicate that when environmental conditions are not conducive to dormancy-breakage or germination, seeds can revert to a higher dormancy level (secondary dormancy). Even after germination has occurred, additional biotic and abiotic stresses can reduce the number of seedlings that successfully emerge from the soil. (Expanded from Allen and Meyer, 1998, and Benech-Arnold *et al.*, 2000.)

both changes over time of the seeds' sensitivity to environmental conditions and their subsequent responses to those conditions as dormancy levels vary. It is therefore necessary to identify the environmental factors that modify dormancy levels of seed populations and to establish functional relationships between these factors and the rates of dormancy release or induction.

4.3 Environmental control of dormancy

4.3.1 Factors affecting dormancy levels of seed populations

4.3.1.1 Temperature

Dormancy cycles are often regulated by temperature in temperate environments where water is not seasonally restricted (Baskin and Baskin, 1977, 1984; Kruk and Benech-Arnold, 1998). For example, in some summer annual species dormancy is released when imbibed seeds experience low temperatures during winter (stratification), while the dormancy level is increased by high temperatures experienced during summer (Bouwmeester and Karssen, 1992, 1993). Winter annual species show the reverse dormancy pattern, where high temperatures during summer result in lower dormancy levels and low temperatures during winter induce secondary dormancy (Baskin and Baskin, 1976; Karssen, 1982; Probert, 1992). As noted in Figure 4.1, these changes in dormancy level are expressed through a narrowing and widening of the temperature range permissive for germination. Germination occurs in the field when soil temperature is in the permissive range characterized by two thermal parameters determined by the seed dormancy level, the low limit temperature (T_l) and the high limit temperature (T_h) (Figure 4.2). T_l and T_h vary among seeds within the population, consistent with the idea that dormancy level is different for individual seeds (Kruk and Benech-Arnold, 2000; Batlla and Benech-Arnold, 2003; Batlla *et al.*, 2004). For example, $T_l(50)$ and $T_h(50)$ represent the temperatures below and above which dormancy is expressed for 50% of the seed population. These distributions of lower and higher temperature limits shift downward and upward in relation to changes in dormancy level of the population (Figure 4.2). Changes in the dormancy level (or variations in the thermal range where germination can occur) in summer species are due to increases or decreases in T_l, while in winter species changes in dormancy level result from fluctuations in T_h (Figure 4.2; Baskin and Baskin, 1980).

Although soil temperature is generally considered the primary environmental factor regulating dormancy levels in seed bank populations, the effect of temperature on dormancy release and induction may be modulated by soil moisture (Adamoli *et al.*, 1973; Reisman-Berman *et al.*, 1991; Christensen *et al.*, 1996; Bauer *et al.*, 1998; Bair *et al.*, 2006; Batlla and Benech-Arnold, 2006). For example, in *Polygonum aviculare* seeds dormancy release occurs most rapidly when seeds are moist-chilled (stratified) at 4°C, but dry storage at 4°C also releases seed dormancy, though at a much slower rate (Kruk and Benech-Arnold, 1998). *P. aviculare* seeds buried in the field under contrasting soil water content conditions show different

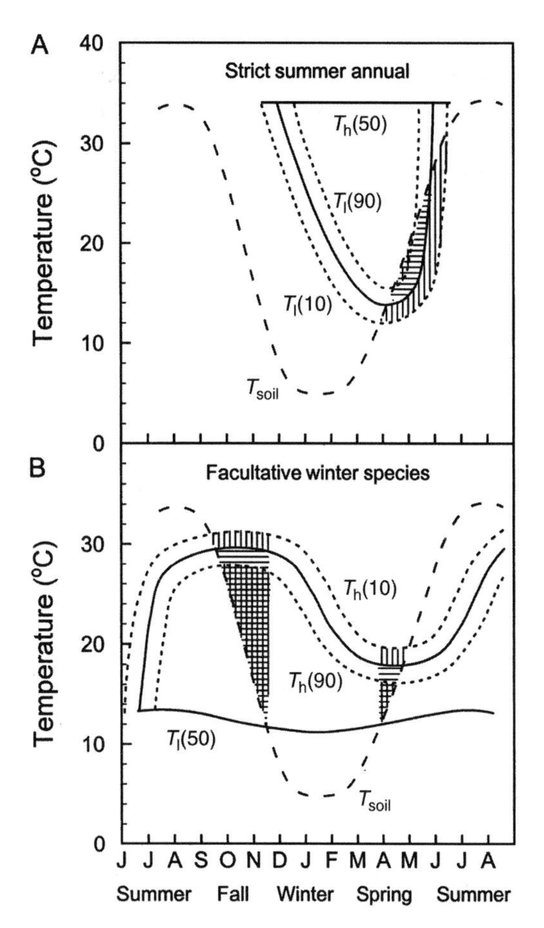

Figure 4.2 Seasonal changes in the permissive germination thermal range and its relation with soil temperature dynamics. Solid lines indicate median lower ($T_l(50)$) and higher ($T_h(50)$) limit temperatures allowing germination. Dotted lines indicate the T_l and T_h values at which 10 or 90% of the seed population could germinate due to variation in these limit temperatures within the seed population. Dashed lines show mean daily maximum soil temperatures (T_{soil}). The vertically shaded areas represent the periods when between 10 and 50% of the seed population could germinate due to the overlap of the permissive and actual temperatures. The horizontally shaded regions indicate the periods when between 50 and 90% of the seeds could germinate, and the hatched regions indicate where over 90% of the population could germinate. (A) Germination of strict summer annual species is controlled primarily by a lowering of T_l during the winter and spring. (B) Germination of facultative winter species is controlled primarily by seasonal variation in T_h except in the winter when T_l limits prevents germination. (Adapted from Benech-Arnold *et al.*, 2002.)

annual patterns of changes in sensitivities to light and to alternating temperatures and in the range of temperatures permissive for germination (Batlla *et al.*, 2003; Batlla and Benech-Arnold, 2004, 2005, 2006). In the winter annual *Bromus tectorum*, which requires dry after-ripening for dormancy release, very low soil moisture slows the rate at which dormancy is lost (Bair *et al.*, 2006).

In the field, induction of secondary dormancy can occur at temperatures that are within the range permissive for germination. Possible explanations include an inhibition of germination *per se* (e.g., inhibition of germination by low water potentials or reduced light under leaf canopies) or from a situation in which conditions that normally stimulate germination are absent (e.g., loss of sensitivity to light in light-requiring seeds held in darkness; loss of sensitivity to fluctuating temperatures in seeds held at low thermal amplitudes) (Benech-Arnold *et al.*, 2000). Induction of secondary dormancy involves a narrowing of the range of conditions suitable for germination (Karssen, 1982).

4.3.1.2 After-ripening

For species that produce seeds in spring/summer and whose seedlings emerge in autumn, a period of prolonged desiccation often leads to a loss of primary seed dormancy that is present when mature seeds are shed. Populations of recently harvested seeds may show some germination, particularly at low temperatures, but such seeds germinate slowly and nonuniformly. Although slow germination is often interpreted as an indication of low vigor, this is clearly not the case for these dormant seeds because they will eventually acquire the ability to germinate quickly and nearly completely (Figure 4.3). Termed 'dry after-ripening', this dormancy-breaking process occurs in the dry state at a rate that is highly temperature dependent. Hydration of seeds often prevents after-ripening and maintains dormancy (e.g., Allen and Meyer, 1998; Bair *et al.*, 2006). Mechanisms responsible for after-ripening have been difficult to unravel, in part because the desiccated state does not lend itself to easy study. Dry seeds lack sufficient hydration to carry out normal metabolism, i.e., 'integrated pathways of enzyme-catalyzed reactions that are regulated in rate and direction and that contribute to the maintenance of the cells in which they occur' (Lynch and Clegg, 1986), and it is unlikely that there is enough water in dry seed tissues to support enzyme-mediated activities (Vertucci and Farrant, 1995; Walters *et al.*, 2001). Although a recent report claimed that both transcription and translation could take place in air-dry seeds and were involved in after-ripening (Leubner-Metzger, 2005), it is difficult to conceive how these processes, and their associated respiratory energy requirements, could be active at the low water contents at which after-ripening can occur. A more likely explanation is that dormancy loss through after-ripening is associated with a breakdown of dormancy-imposing factors, possibly through nonenzymatic oxidative processes (Leopold *et al.*, 1988; Esashi *et al.*, 1993; Foley, 1994; Corbineau and Côme, 1995).

Many species with an after-ripening requirement are facultatively autumn-germinating. Seeds germinate in response to autumn rains, postpone germination until winter or early spring, or remain dormant across years in the soil seed bank where they undergo dormancy cycling as previously discussed. From an ecological perspective, the after-ripening requirement prevents seeds from germinating during the hottest periods of summer, when the risk of seedling death due to rainfall inadequate to permit seedling establishment is highest. The importance of germination timing in semi-arid habitats is illustrated by results from a 38-year simulation of

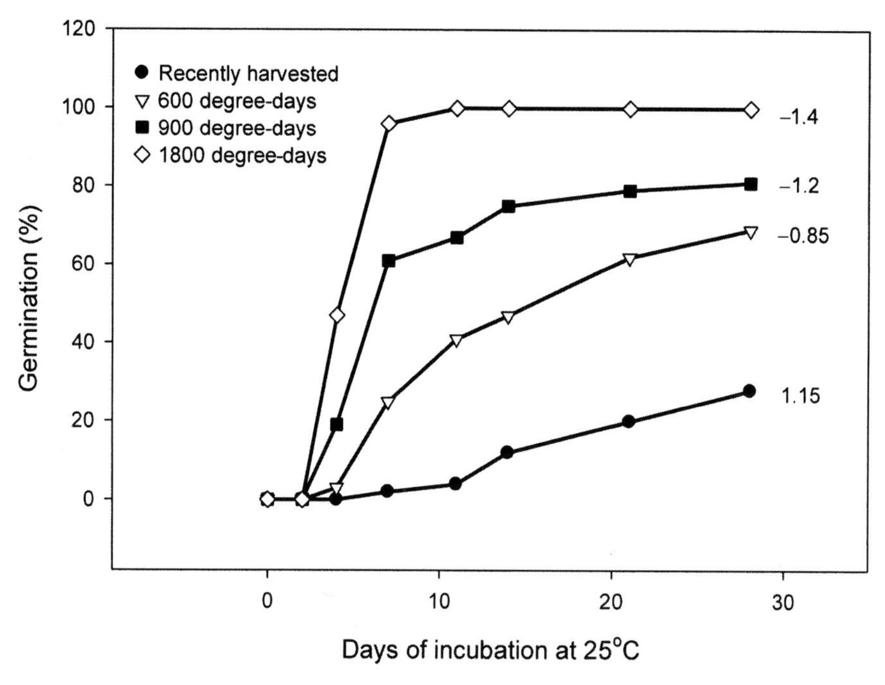

Figure 4.3 Selected cumulative time courses for germination of the perennial grass *Elymus elymoides* at different stages of dormancy loss through dry storage. As after-ripening progresses, as indicated by increasing thermal after-ripening times, seeds germinate to higher percentages and eventually reach maximum rate and uniformity. Number to the right of each line indicates the population mean base water potential ($\psi_b(50)$, MPa), which decreases (becomes more negative) during after-ripening. (Previously unpublished data of P.S. Allen and S.E. Meyer.)

seed zone temperature and moisture conditions which showed that, on average, only 30 days per year had conditions favorable for seedling establishment of wildland grasses (Hardegree *et al.*, 2003).

4.3.1.3 Stratification

In a high proportion of species from temperate regions, particularly those adapted to spring emergence, exposure to cold temperatures under moist conditions releases seed dormancy (Baskin and Baskin, 1985, 1998; Probert, 1992). This stimulatory effect of exposing imbibed seeds to low temperatures has been recognized for centuries. More than 300 years ago, J. Evelyn (1664) recommended placing seeds of forestry species in moist sand or soil and holding them outdoors during the winter prior to spring sowing in order to obtain better seed germination and plant establishment. This method is a long-standing practice used to break primary dormancy in many forestry and horticultural seeds, and is generally referred to as stratification because seeds were stored arranged in layers (i.e., 'stratified') in moist sand or vermiculite for exposure to low temperatures prior to planting. However, other terms such as moist chilling and cold after-ripening are also used to refer to dormancy release by cold temperatures in hydrated seeds. Results from many ecophysiological studies of the emergence patterns of buried seed populations indicate that stratification is a significant process controlling the dormancy status and timing of emergence of wild species in temperate regions (Probert, 1992; Meyer *et al.*, 1995). The requirement for low temperatures prevents seedling establishment and growth under conditions unfavorable for plant survival (i.e., during winter), while allowing germination and establishment under environmental conditions that favor species growth and perpetuation (i.e., during spring) (Karssen, 1982).

In contrast with after-ripening, for stratification to be effective seeds must be at least partially hydrated and have access to oxygen (Stokes, 1965; Bewley and Black, 1982; Vertucci and Leopold, 1986). The period required for stratification may range from a few days to several months, with differences known to exist among species, populations within a species, and individual seeds within a single seed population (Allen and Meyer, 1998; Baskin and Baskin, 2004; Batlla *et al.*, 2004). The optimum temperatures for dormancy release through stratification are generally in the range of 1–5°C, although seeds of some species can be stratified at temperatures as high as 16–17°C (Roberts and Smith, 1977; Batlla and Benech-Arnold, 2003). Prolonged exposure to temperatures near the upper limit for stratification can lead to secondary dormancy (e.g., Totterdell and Roberts, 1979). In some species, after-ripening must precede stratification in order for dormancy loss to occur for at least a fraction of the seed population (Baskin and Baskin, 1984; Garvin and Meyer, 2003).

4.3.2 Factors that stimulate germination

Promotion of germination following dormancy release often requires exposure to specific environmental stimuli (Figure 4.1). For example, fluctuating temperatures and light result in dormancy loss for many weed seeds (Benech-Arnold *et al.*, 1990b; Scopel *et al.*, 1991; Ghersa *et al.*, 1992), although other factors (i.e., CO_2, NO_3^-, O_2, and ethylene) can be involved in the termination of dormancy under field conditions (Corbineau and Côme, 1995; Benech-Arnold *et al.*, 2000). Not all seeds require exposure to germination-stimulating factors in order for germination to occur. For example, seeds within a population may proceed to germinate immediately after dormancy is released as long as water and temperature conditions are adequate. In these cases, there is a gradual transition from dormancy to nondormancy to germination, and it is difficult to distinguish specific stages (see Chapter 3). Whether efforts to model seed dormancy require the inclusion of additional germination-stimulating factors to predict germination outcomes depends on the specific requirements of a species.

4.3.2.1 Fluctuating temperature

In addition to signaling changes in season, fluctuating temperatures likely provide seeds with an indication of burial depth. Several characteristics of diurnal temperature cycles could be responsible for stimulating germination (Roberts and Totterdell,

1981). For example, thermal amplitude is important in species such as *Chenopodium album*, where dormancy-releasing effectiveness improved as the amplitude of temperature fluctuations increased from 2.4 to about 15°C (Murdoch *et al.*, 1988). In addition, the response to a given amplitude was greater with a higher mean temperature (i.e., average of lower and upper temperature) up to an optimum of about 25°C. In some cases, diurnal temperature cycles tend to be additive in their effect. For example, in *Sorghum halepense* seeds, 10 stimulatory cycles released twice the proportion of the population from dormancy as was released with only 5 cycles (Benech-Arnold *et al.*, 1990a).

4.3.2.2 Light

In many species, germination is stimulated when hydrated seeds are exposed to light, which is perceived through photoreceptors, particularly those from the phytochrome family. Phytochromes have two mutually photoconvertible forms: Pfr (considered the active form) with maximum absorption at 730 nm (far-red) and Pr with maximum absorption at 660 nm (red). Phytochromes are synthesized as Pr and the proportion of the pigment population (P) in the active form (Pfr/P) in a particular tissue depends on the light environment to which the seeds are exposed. Exposure of seeds to light with a high red (R) to far-red (FR) ratio (R/FR) leads to larger Pfr/P, which stimulates germination. Phytochrome-mediated responses can be classified physiologically into three 'action modes' (Kronenberg and Kendrik, 1986). Two of these action modes – the very low fluence responses (VLFR) and the low fluence responses (LFR) – are characterized by the correlation between the intensity of the effect and the level of Pfr that is established by the light environment. They differ in that extremely low levels of Pfr saturate the VLFR and the response is often not reversible by FR light, while higher Pfr levels are necessary to elicit LFR and the response to R is often reversible if followed immediately by FR light (Casal and Sánchez, 1998). High irradiance responses (HIR) are the third action mode, which show no simple relationship to Pfr levels, require extended illumination, and may involve additional components of the phytochrome system (Heim and Schäfer, 1982, 1984).

A number of species with light-mediated seed dormancy are agricultural weeds. Crop or pasture leaf canopies typically reduce the R/FR ratio perceived by seeds located at the soil surface (Smith, 1982; Pons, 1992). This change in light quality leads to a low Pfr/P, preventing seed germination. The LFR and/or HIR modes of action can mediate this type of inhibition by canopy presence (Deregibus *et al.*, 1994; Batlla *et al.*, 2000). Reductions in canopy density, for example by grazing, result in increases in R/FR and Pfr/P ratios, promoting germination of seeds present on the soil surface via the LFR response (Deregibus *et al.*, 1994; Insausti *et al.*, 1995). A VLFR response is typically observed when soil is disturbed by agricultural practices. Seeds of several species acquire a high sensitivity to light as dormancy is released through a period of after-ripening or stratification in the soil (Scopel *et al.*, 1991; Batlla and Benech-Arnold, 2005). These seeds can respond to exposures on the order of sub-milliseconds of sunlight, resulting in significant germination from the seed bank when the soil is disturbed by tillage practices.

4.3.2.3 Nitrate

The widespread stimulation of germination by nitrate is illustrated by the large number of species reported to respond to low doses of this chemical during routine seed testing (International Seed Testing Association, 2004). Although mechanisms through which nitrate stimulates germination are largely unknown, they have been hypothesized to act on cellular membranes or transporters (Karssen and Hilhorst, 1992; Vleeshouwers and Bouwmeester, 2001; see Chapter 3 for possible molecular mechanisms). As discussed previously for fluctuating temperatures and light, seasonal changes in dormancy of buried seeds may coincide with changes in seed responsiveness to nitrate (Derkx and Karssen, 1993). However, experimental evidence supporting an effect of soil nitrate on dormancy and germination of natural seed banks under field situations is inconsistent (Egley, 1986).

4.3.3 Conceptual scheme of dormancy and its relationship to modeling

The framework presented in Figure 4.1 and discussed above provides a brief overview of how environmental factors control seed dormancy. In order to develop models for describing seed dormancy processes in natural soil seed banks, it is first necessary to evaluate the independent variables that impact dormancy levels for a particular species and determine which ones to measure quantitatively. Functional relationships between these variables and the rates of processes they control then need to be identified and quantified. Finally, to advance beyond purely empirical approaches to predicting seed dormancy and germination outcomes, models need to relate various stages (arrows in Figure 4.1) to the controlling mechanisms likely to be involved. A wealth of information on the way seeds respond to various light, temperature, and water potential conditions has been obtained from laboratory experiments, and simulation modeling allows the best test of whether seed responses in the laboratory can be used to make meaningful predictions about dormancy loss in the field. For recent examples of efforts to incorporate controlling mechanisms into modeling of processes diagrammed in Figure 4.1, see Bauer *et al.* (1998) for seed zone microclimate; Benech-Arnold *et al.* (2000), Meyer *et al.* (2000), Allen and Meyer (2002), and Bradford (2002) for dormancy; Grundy *et al.* (2000), Alvarado and Bradford (2002), and Rowse and Finch-Savage (2003) for germination; and Shrestha *et al.* (1999), Vleeshouwers and Kropff (2000), and Finch-Savage (2004) for seedling establishment.

4.4 Approaches to modeling seed dormancy

Many models of dormancy and seedling emergence have been developed, and several reviews are available (Benech-Arnold and Sanchez, 1995; Allen and Meyer, 1998; Forcella *et al.*, 2000; Vleeshouwers and Kropff, 2000; Batlla *et al.*, 2004; Finch-Savage, 2004). Among these models, at least two general approaches can be distinguished. A pragmatic approach seeks to predict seasonal seedling emergence patterns primarily as a function of environmental data such as temperature, moisture, cultivation, soil type, solar radiation, etc. (e.g., Forcella, 1998; Kebreab and Murdoch, 1999a; Vleeshouwers and Bouwmeester, 2001). These models are primarily empirically based and use commonly available environmental data (soil temperature, rainfall, soil moisture, etc.) to predict the likely periods when conditions are conducive to emergence of a particular species, given its specific requirements to lose dormancy and complete germination. This 'engineering' approach prefers continuous functions across the range of inputs, may use numerous parameters or variables to fit different components of the model, and is less concerned about the biological relevance of the parameters. These approaches have value for developing broadly applicable predictive forecasts based on meteorological inputs and information about the composition of the soil seed bank in a particular location. However, as Forcella *et al.* (2000) concluded, 'the most critical need for improving models of seedling emergence in the future is mechanistic integration of microclimate and management variables with the rates of dormancy alleviation/induction, germination and seedling elongation. Equally important is the integration of these latter three components of emergence'.

An alternative (or complementary) approach is to attempt to characterize on a physiological basis how seeds respond to environmental conditions, and then use that understanding to model how the seeds will respond to the environment that they have experienced. While in some ways a more challenging approach to developing predictive models, physiologically based models can provide testable hypotheses regarding potential biological mechanisms underlying seed dormancy and germination behavior. A more complete understanding of those mechanisms should eventually allow for improved predictions of seed behavior. In particular, these models often explicitly incorporate variation in dormancy levels among individual seeds in the population, which can be linked directly to proportions of the seed bank in a germinable state in a given environment. Physiological models can also be incorporated as subcomponents of empirical models, such as in anticipating the rates of dormancy loss, which are often included as 'black boxes' in such models. Vleeshouwers and Kropff (2000) examined the component processes in a sophisticated emergence model and concluded, 'Research to improve the prediction of weed seedling emergence in the field should . . . focus on improving the simulation of seasonal changes in dormancy of buried seeds'. This may be a suitable target for physiologically based models. Two common features of these models are that they account for threshold-type behavior in response to regulatory factors, and they explicitly address the range of dormancy/germination behavior within seed populations.

4.4.1 Temperature response models and thermal time

Germination responses to temperature are generally characterized by the three 'cardinal temperatures' – the minimum, optimum, and maximum. The minimum or base temperature (T_b) is the lowest temperature at which germination will occur, regardless of how long the seeds are incubated. Similarly, the maximum or ceiling temperature (T_c) is the highest temperature at which seeds will germinate. The optimum temperature (T_o) is the temperature at which germination is most rapid; in some cases, there is an optimal range of temperatures rather than a single optimum (Labouriau and Osborn, 1984; Orozco-Segovia *et al.*, 1996). This is evident when the time to germination (t_g) for a specific fraction or percentage (g) of the seed population is plotted as a function of temperature. The slower germination at higher and lower temperatures results in broad U-shaped 'germination characteristic' curves that have a wide optimum range and sharp increases in t_g at the extremes (see, e.g., Hilhorst, 1998). However, these responses are more readily analyzed if the inverse of time to germination for a given percentage of the seed population ($1/t_g$), or the germination rate (GR_g), is plotted versus temperature. This results in an inverted V-shaped curve with two linear components between T_b and T_o and between T_o and T_c (Figure 4.4A) (Labouriau, 1970; Bierhuizen and Wagenvoort, 1974).

For the suboptimal temperature range (between T_b and T_o) this relationship can be described mathematically as:

$$\theta_T(g) = (T - T_b)t_g \tag{4.1}$$

or

$$GR_g = 1/t_g = (T - T_b)/\theta_T(g) \tag{4.2}$$

where $\theta_T(g)$ is the thermal time to germination of fraction or percentage g, T is the germination temperature, T_b is the base temperature, and GR_g is the germination rate of percentage g. Because $\theta_T(g)$ is a constant for a given seed fraction g, as the difference $T - T_b$ increases, t_g decreases proportionately. Thus, the time required for germination is the same at all suboptimal temperatures for a given seed fraction when expressed in terms of thermal time (i.e., degree-hours or degree-days). T_b sets a minimum threshold temperature below which germination will not occur, and the rate of germination increases linearly above T_b with a slope of $1/\theta_T(g)$ (Figure 4.4A). The value of T_b can be estimated graphically in this way (e.g., Steinmaus *et al.*, 2000) or by probit regression analyses that combine data from all tested temperatures (Covell *et al.*, 1986; Dahal *et al.*, 1990). The variation or spread in germination times among individual seeds is accounted for by the variation in thermal time requirements for different seed fractions ($\theta_T(g)$, or the inverses of the slopes in Figure 4.4A). The values of $\theta_T(g)$ in the seed population can be characterized by their mean and standard deviation (Figure 4.4A, inset 1). When the GR_g versus T plots for different fractions of the population have a common intercept on the T-axis (Figure 4.4A), it indicates that T_b is the same for all seeds in the population, as is often the case (Covell *et al.*, 1986; Dahal *et al.*, 1990; Kebreab and Murdoch, 1999b). In other cases, the apparent T_b varies among seeds in the population (e.g., Labouriau and Osborn, 1984; Fyfield and Gregory, 1989; Phelps and Finch-Savage, 1997; Kebreab and Murdoch, 2000), particularly when dormancy is a factor, and other models may better account for seed germination behavior in those cases (see below). In general, the thermal time approach to modeling germination rates at suboptimal temperatures has been quite useful (e.g., Garcia-Huidobro *et al.*, 1982; Covell *et al.*, 1986; Dahal *et al.*, 1990; Dahal and Bradford, 1994), and most predictive models for germination or seedling emergence use a thermal timescale

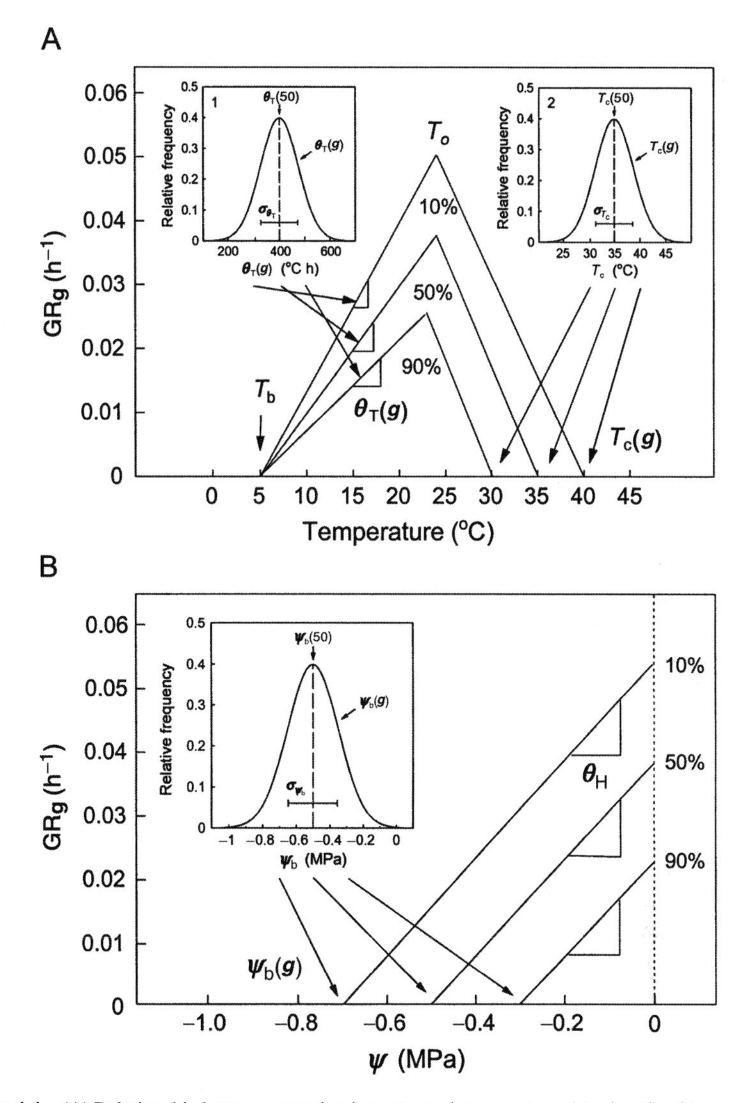

Figure 4.4 (A) Relationship between germination rates and temperature. At suboptimal temperatures, germination rates ($GR_g = 1/t_g$) for different fractions (percentages) of the seed population increase linearly with temperature above a common base temperature (T_b). The slopes of the lines are equal to the inverses of the thermal times to germination ($1/\theta_T(g)$), which vary among individual seeds in a normal distribution (inset 1; in some cases, $\theta_T(g)$ is normally distributed on a logarithmic time scale, not shown.) The maximum GR_g occurs at the optimum temperature (T_o), and above this temperature GR_g decreases linearly. The ceiling temperatures for germination ($T_c(g)$) vary among seed fractions in a normal distribution (inset 2). (B) Relationship between germination rates and water potential (ψ). As ψ decreases, germination rates for different percentages decrease linearly with a common slope of $1/\theta_H$, intercepting the x-axis at different threshold or base water potential values ($\psi_b(g)$), which are normally distributed among seeds in the population (inset). (From Bradford, 2002, with permission of the Weed Science Society of America.)

(degree-days) to normalize for temperature variation over time in the field (Forcella, 1998; Vleeshouwers and Kropff, 2000).

According to Equations 4.1 and 4.2, t_g should continue to decrease (GR_g should continue to increase) indefinitely with increasing temperature, but this is not the case. Instead, GR_g increases to a maximum at T_o, and then decreases approximately linearly at higher temperatures until it intersects the x-axis at T_c (Figure 4.4A). At supraoptimal temperatures, the slopes of these lines for different germination fractions are often parallel, resulting in different intercepts on the x-axis (different T_c values) (Figure 4.4A) (Covell *et al.*, 1986; Ellis *et al.*, 1987; Ellis and Butcher, 1988), although common T_c values for all seeds in a population have been reported as well (Garcia-Huidobro *et al.*, 1982; Orozco-Segovia *et al.*, 1996). When T_c values differ among seeds in the population, germination time courses in the supraoptimal temperature range can be analyzed by holding thermal time constant (θ_{Tc}) and allowing T_c to vary among different seed fractions (i.e., $T_c(g)$) (Ellis *et al.*, 1986, 1987; Ellis and Butcher, 1988):

$$\theta_{Tc} = (T_c(g) - T)t_g \tag{4.3}$$

or

$$GR_g = 1/t_g = (T_c(g) - T)/\theta_{Tc} \tag{4.4}$$

Thus, the thermal time required for completion of germination (θ_{Tc}) is the same for all seeds in the supraoptimal temperature range, but differences in germination rates are due to different T_c values (Figure 4.4A). That is, seeds having higher T_c values will accumulate thermal time ($T_c - T$) more rapidly and will therefore complete germination more quickly than seeds having lower T_c values. The seed population can then be characterized in terms of the distribution (i.e., the mean and standard deviation) of $T_c(g)$ values in the supraoptimal range (Figure 4.4A, inset 2).

The presence of dormancy can affect germination responses to temperature. Stratification of *Aesculus hippocastanum* seeds resulted in a progressive lowering of T_b, such that seeds could eventually germinate at temperatures that initially prevented germination (Steadman and Pritchard, 2004). Apparent ceiling temperatures for lettuce (*Lactuca sativa*) seed germination can be altered by ethylene (Dutta and Bradford, 1994; Nascimento *et al.*, 2000). An alternative approach to analyzing and modeling this situation was proposed by Benech-Arnold and coworkers (Kruk and Benech-Arnold, 1998, 2000; Batlla and Benech-Arnold, 2003) based on the early work of Washitani (1987). This approach assumes that the actual T_b and T_c for germination do not change due to dormancy, but rather that there is in addition a lower (T_l) and upper (T_h) temperature limit for germination that varies according to the depth of dormancy. When T is lower than T_l of a certain fraction of the population, dormancy is expressed in individuals belonging to that fraction but not in seeds from fractions with a T_l lower than T. As dormancy is alleviated by the effect of stratification, for example, the entire T_l distribution is displaced toward lower temperatures and, consequently, additional fractions of the population will have a T_l lower than T, and will be able to germinate. Thermal time required for germination is accumulated relative to T_b, but only once T exceeds T_l for a given seed fraction, as T_l and T_h both vary among seeds (Figure 4.2). This approach allows changes in temperature responses due to dormancy to be modeled without altering the fundamental cardinal temperatures for a species. Its application will be illustrated subsequently.

Both approaches explicitly account for variation in temperature response parameters among individual seeds, which results in the overall population behavior as the temperature sensitivity shifts during dormancy release or imposition. The consequences of both approaches are similar, in that they account for shifts in the ability of seeds to germinate at low or high temperatures due to varying degrees of dormancy by a corresponding physiological shift in the threshold or temperature limit values.

4.4.2 Water potential responses and hydrotime models

Seeds cannot germinate without water, and seed germination rates and percentages are closely attuned to water availability (Bradford, 1995). As for temperature, there is a minimum water potential (ψ) that must be exceeded in order for seeds to complete germination, and seeds in a population vary in the value of this minimum or base ψ (ψ_b) (Hegarty, 1978). In analogy with the thermal time concept, Gummerson (1986) proposed that the time to germination for a given seed fraction (g) was inversely related to the difference between the current seed ψ and the ψ_b for that fraction ($\psi_b(g)$). He defined this hydrotime model as:

$$\theta_H = (\psi - \psi_b(g))t_g \tag{4.5}$$

or

$$GR_g = 1/t_g = (\psi - \psi_b(g))/\theta_H \tag{4.6}$$

where θ_H is the hydrotime constant in MPa-hours or MPa-days. If GR_g values determined by incubating seeds at different constant ψ are plotted as a function of ψ, the resulting relationships are linear for different seed fractions (g), having a common slope and intercepting the ψ-axis at different $\psi_b(g)$ values (Figure 4.4B). The common slopes indicate that the total accumulated hydrotime (MPa-hours or MPa-days) to radicle emergence is the same for all seeds in the population, while the different intercepts show that individual seeds vary in their ψ_b thresholds at which radicle emergence is prevented. As θ_H is a constant, t_g must increase proportionately as ψ is reduced and approaches $\psi_b(g)$ (see Bradford, 1997, 2002, for illustrations). In most cases, $\psi_b(g)$ values vary among seeds in the population in a normal or Gaussian distribution (e.g., Gummerson, 1986; Bradford, 1990; Dahal and Bradford, 1990), although this is not essential to the model. If the distribution is normal, it can be defined by its mean ($\psi_b(50)$) and standard deviation ($\sigma_{\psi b}$) (Figure 4.4B, inset). $\psi_b(50)$ is more properly the median, since it refers to the 50th percentile of the population rather than to the average value, but for a normal distribution, the mean and the median coincide, and so we will refer to it here as the mean base water potential. As for temperature, both graphical and probit analysis methods for estimating the hydrotime parameters from original germination time course data at different water potentials have been developed (Gummerson, 1986; Bradford, 1990, 1995).

The hydrotime model functions well in matching both the timing and the percentage of germination of seed populations in relation to their ψ environment (Figure 4.5). In addition, the parameters of the hydrotime model can be used to

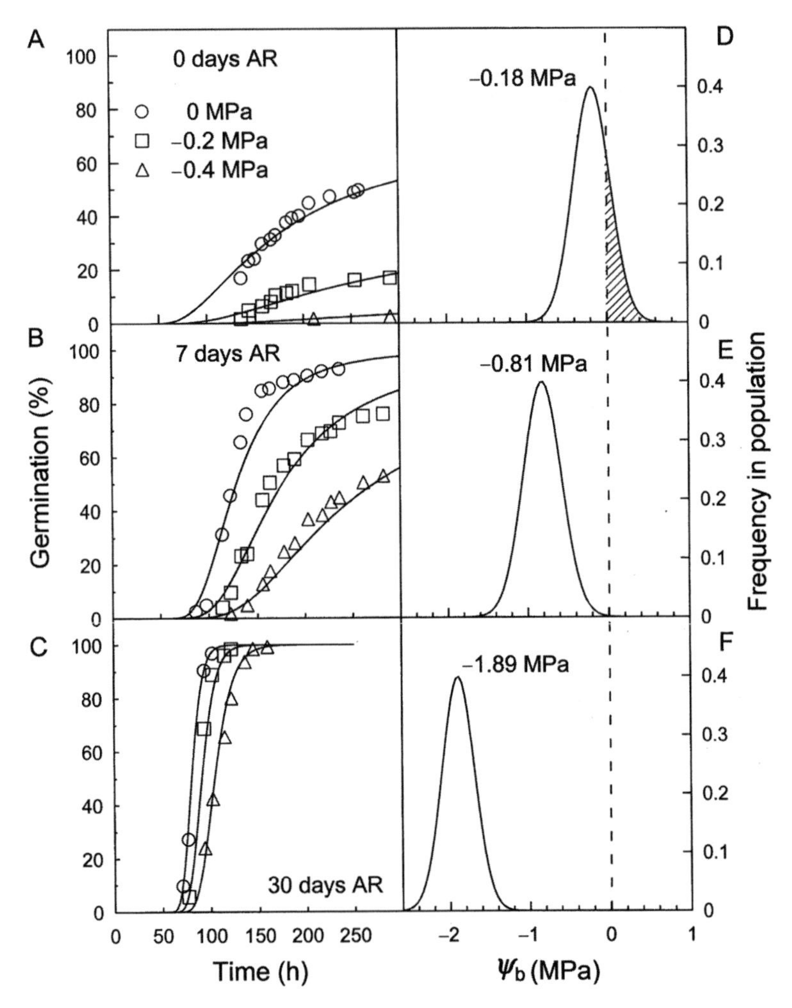

Figure 4.5 Germination of true potato seeds at 14–15°C and different ψ (0, −0.2, −0.4 MPa) as affected by after-ripening for (A) 0, (B) 7, or (C) 30 days at 4% seed moisture content and 37°C. The symbols are the experimental data, and the curves are predicted from the hydrotime model based upon the $\psi_b(g)$ distributions shown in panels D, E, and F. Initially, the ψ_b distribution was high and a fraction of the seed population was unable to germinate even in water (represented by shaded area under curve to the right of 0 MPa line in panel D). During after-ripening, the $\psi_b(g)$ distributions shifted to lower values (panels E and F), corresponding to the more rapid germination and decreased sensitivity to inhibition by low ψ (panels B and C). (Modified from Alvarado and Bradford, 2005.)

characterize the properties of seed populations. The θ_H value quantifies the inherent speed of germination, the $\psi_b(50)$ value is an indication of the average water stress tolerance of the population, and the standard deviation of $\psi_b(g)$ ($\sigma_{\psi b}$) indicates the uniformity of threshold values which corresponds to the synchrony in germination timing among seeds in the population. These three parameters are sufficient to predict complete germination time courses at any ψ at a constant T (Bradford, 1995, 2002). In addition, the hydrotime model is insensitive to whether the ψ of the environment changes or the physiological $\psi_b(g)$ thresholds are altered. That is, equivalent effects on germination timing result either from variation in environmental ψ or from physiological adjustments in the water potential thresholds of seed populations (Bradford, 2002). Changes in $\psi_b(g)$ distributions have proven to be a useful way to characterize and model increases or decreases in seed dormancy levels (Figures 4.5D–4.5F).

A limitation of the hydrotime model as presented above is that it does not account for metabolic advancement toward germination that can occur at water potentials below ψ_b. The practice of seed priming, in which seeds are incubated at $\psi < \psi_b(g)$ for extended periods of time and subsequently exhibit more rapid germination rates, indicates that seeds can accumulate hydrotime at water potentials below ψ_b. There is a limit or minimum ψ below which metabolic advancement does not occur (ψ_{min}), and advancement in germination is often linear at water potentials between ψ_{min} and ψ_b (Tarquis and Bradford, 1992; Bradford and Haigh, 1994; Finch-Savage *et al.*, 2005). Modifications to Equation 4.6 have been suggested that can empirically account for this effect on germination timing (Cheng and Bradford, 1999). An alternative approach developed by Rowse and colleagues (Rowse *et al.*, 1999; Rowse and Finch-Savage, 2003), termed the 'virtual osmotic potential' or VOP model, directly accounts for advancement toward germination at all water potentials above ψ_{min}. While sharing the concept of a distribution of $\psi_b(g)$ thresholds for germination, the VOP model assumes that progress toward germination is due to either accumulation of osmotic solutes to lower embryonic water potential or a reduction in the restraint by enclosing tissues or both. As the VOP model incorporates the effects of water potentials between ψ_{min} and ψ_b as well as between ψ_b and ψ, it may be more amenable to predicting germination times under fluctuating ψ conditions than the standard hydrotime model (Finch-Savage, 2004; Finch-Savage *et al.*, 2005).

4.4.3 *Interactions of temperature and water potential*

The combined effects of suboptimal temperatures and reduced water potentials on germination can be modeled by using thermal time instead of actual time in the following equation (Gummerson, 1986; Bradford, 1995):

$$\theta_{HT} = (\psi - \psi_b(g))(T - T_b)t_g \tag{4.7}$$

This hydrothermal model has successfully described germination time courses at constant suboptimal temperatures in the relatively high ψ range (Gummerson, 1986; Dahal and Bradford, 1994; Alvarado and Bradford, 2002, 2005). This equation assumes that $\psi_b(g)$ is constant and independent of temperature, and that T_b is independent of ψ, which is not always the case, particularly when the incubation temperatures affect seed dormancy (Bradford and Somasco, 1994; Dahal and Bradford, 1994; Kebreab and Murdoch, 1999b). In those cases, however, the effects on dormancy can be quantified in terms of their influence on the parameters of Equation 4.7. For example, seeds of lettuce and of many species that normally germinate in the fall or winter are prevented from germinating at warm temperatures (Baskin and Baskin, 1998). In lettuce seeds, this can be attributed to an increase in the $\psi_b(g)$ distribution (to less negative values) as T increases (Bradford and Somasco, 1994). Several studies have found that $\psi_b(g)$ values are at a minimum around the optimum T and become higher at supraoptimal T (Christensen *et al.*, 1996; Kebreab and Murdoch, 1999b; Meyer *et al.*, 2000; Alvarado and Bradford, 2002; Rowse and Finch-Savage, 2003). While different species may have similar germination time courses in water, dramatic differences in germination rates are frequently observed when those seeds are incubated at reduced water potentials, as will be reflected in their θ_{HT} and $\psi_b(g)$ values (Allen *et al.*, 2000). Analysis of hydrothermal time parameters has led to important ecological interpretations regarding germination for a wide variety of species (Allen *et al.*, 2000).

4.4.4 Modeling responses to other factors affecting dormancy and germination

Population-based threshold models have been applied to describe or predict seed germination behavior in response to a number of factors. Hormonal (e.g., gibberellin [GA], abscisic acid [ABA], and ethylene) effects on germination of tomato (*Lycopersicon esculentum*) and lettuce seeds were quantified using this approach (Ni and Bradford, 1992, 1993; Dutta and Bradford, 1994). Fennimore and Foley (1998) quantitatively described how GA sensitivity of wild oat (*Avena fatua*) seeds changed during loss of dormancy by after-ripening. Bradford (1996, 2005) has suggested that seed responses to light and nitrate would also be amenable to modeling by this approach.

The concept of degrees of relative dormancy described previously presupposes the existence of environmental thresholds for dormancy expression. Depending on the dormancy level of the population, there will be a temperature above which (T_b or T_l) or below which (T_c or T_h) dormancy is not expressed. Similarly, there will be a threshold water potential (ψ_b) above which germination can proceed. While the thermal time and hydrotime models described above assume a linear response to a given factor above the threshold value, this is not the only possibility. With respect to light, for example, the phytochrome LFR response has long been known to exhibit reciprocity; i.e., a low dose over a longer time is equivalent to a high dose over a shorter time. This is strictly analogous to the threshold models described above. However, whether or not a seed is responsive to light at all may be an all-or-none or quantal response, possibly based upon whether the seed expresses a specific member of the phytochrome gene family (Casal and Sánchez, 1998). Individual seeds may move from a nonresponsive fraction of the population to a responsive fraction as dormancy is lost. As sensitivity thresholds for various factors shift with changes in the dormancy level of the population, the range of conditions under which germination can proceed will widen if dormancy is released or, conversely, will narrow when dormancy is induced. Since the dormancy level can be different for each individual seed within the population, these thresholds cannot be regarded as unique for the whole population; on the contrary, their values can be assumed to be distributed within the population, with a mean and a standard deviation. Hence, not only the mean value of the threshold(s) but also the entire accompanying distribution is shifting with changes in the dormancy level of the population (e.g., Figures 4.5D–4.5F).

4.5 Examples of seed dormancy models

To illustrate the application of seed dormancy models, some cases that have been investigated in depth will be presented. As there are many unique types of dormancy, different approaches may be better suited to particular situations and species. The examples below illustrate methods that can be used to quantify and predict seed dormancy behavior in relation to the principal factors that influence the rates of dormancy release.

4.5.1 *Solanum tuberosum*

Recently harvested true (botanical) potato (*Solanum tuberosum*) seeds exhibit dormancy that can be alleviated by after-ripening. After relatively short after-ripening periods, they exhibit conditional dormancy in which they are capable of germination at low $T (<17°C)$, but are increasingly inhibited at higher T; further after-ripening expands the upper temperature limit for germination (Pallais, 1995). These changes in germination behavior in relation to after-ripening and germination temperature can be modeled using the hydrotime and hydrothermal time models (Alvarado and Bradford, 2002, 2005). The dormancy of recently harvested potato seeds is evident in their delayed initiation of germination, low final germination percentages, and high sensitivity to inhibition by reduced ψ even at relatively low temperatures (14°C) (Figure 4.5A). After only 7 days of dry storage at 37°C, germination was much improved, and after 30 days of after-ripening, germination was rapid and relatively insensitive to reduced ψ (Figures 4.5B and 4.5C). These changes in germination behavior as dormancy was relieved can be described by associated shifts in the $\psi_b(g)$ distributions of the seed population to lower values (Figures 4.5D–4.5F). The hydrotime constant (θ_H) also increased during after-ripening, and when combined with the $\psi_b(g)$ distribution, the complete germination time courses at all ψ can be predicted with high precision (Figures 4.5A–4.5C). These relationships held across the entire range of suboptimal temperatures, illustrating that a major consequence of after-ripening is a lowering of the $\psi_b(g)$ distribution (Alvarado and Bradford,

2005). Similarly, the effects of fluridone (an inhibitor of ABA synthesis), GA, and moist-chilling in releasing dormancy were accompanied by reductions in mean ψ_b values (Alvarado and Bradford, 2005). An interesting aspect of the hydrotime model is that while the variation among seeds in their ψ_b values was relatively unaffected by the loss of dormancy, variation in actual times to germination (i.e., the time from first to last germinant) was much greater in dormant than in after-ripened seeds (Figures 4.5A–4.5C). An automatic consequence of the threshold model is that as the difference between the seed ψ and $\psi_b(g)$ increases, the times to germination and the variation in germination times decrease. In addition, the model also automatically predicts the final germination percentages, as the fraction of the $\psi_b(g)$ distribution that extends above the current ψ is prevented from completing germination.

Even after dormancy is largely released, potato seeds are still sensitive to germination inhibition at high temperatures. In contrast to the decrease in mean ψ_b values during after-ripening, $\psi_b(g)$ distributions increase (become less negative) as temperature increases above the optimum (Figure 4.6B; Alvarado and Bradford, 2002). At and below the optimal temperature, $\psi_b(g)$ values increased less markedly with temperature (Figure 4.6B). This behavior of $\psi_b(g)$ distributions with temperature explains the observation that T_b values are often similar among seeds in a (nondormant) population, while T_c values vary with germination fraction (Figure 4.4A). Since ψ_b values vary in the seed population, as the mean of the distribution increases with temperature, those seeds having higher ψ_b values than the mean will be prevented from germinating (i.e., $\psi_b \geq 0$ MPa) at a lower temperature, while those having a lower ψ_b value than the mean will be able to germinate at a higher temperature than the mean (Figure 4.6). The range of T_c values is therefore a consequence of the variation in $\psi_b(g)$ values. On the other hand, at suboptimal temperatures where the $\psi_b(g)$ distribution is much lower than 0 MPa (cf. Figure 4.5F), the ψ_b threshold will not be a factor and all seeds will have a similar minimum germination temperature.

4.5.2 *Bromus tectorum*

Bromus tectorum (cheatgrass, Junegrass, or downy brome) is a winter annual grass native to Eurasia that has become established on millions of hectares of wildlands in the western United States. It is considered by some to be the most significant plant invasion in North America (D'Antonio and Vitousek, 1992). *B. tectorum* seed populations exhibit varying levels of dormancy at maturity, which has been meaningfully correlated with ecological habitat at the site of collection (Meyer *et al.*, 1997). The obligately self-pollinating behavior of *B. tectorum* facilitates intensive greenhouse studies that contribute to its desirability as a model system to study genetic and environmental contributions to seed dormancy. These studies have revealed that dormancy at maturity is primarily due to genetic variation among and within populations (Meyer and Allen, 1999a,b; Ramakrishnan *et al.*, 2004), although environmental conditions present during seed production also contribute to dormancy levels. In particular, both high water availability and low temperatures during seed

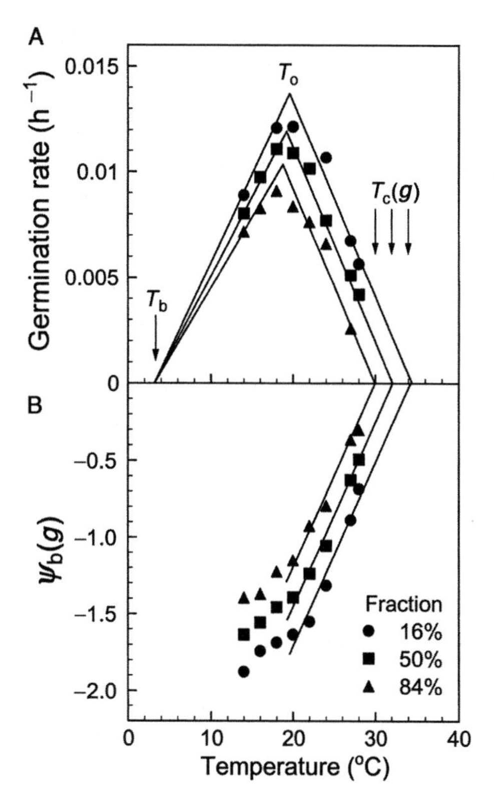

Figure 4.6 (A) The cardinal temperatures for germination of true potato seeds. Germination rates at different temperatures show the minimum or base temperature (T_b), optimal temperature (T_o), and the maximum or ceiling temperatures (T_c). For this seed lot, T_b is 3.2°C, T_o is 19.3°C, and T_c varies with the seed fraction, being 34.3, 32.1, and 29.9°C for the 16th, 50th, and 84th percentiles, respectively. The symbols are the actual data derived from germination time courses. The lines are based upon the parameters found after fitting the hydrotime model at sub- and supraoptimal T. (B) $\psi_b(g)$ increases linearly as T increases in the supraoptimal range. The $\psi_b(g)$ values calculated from the hydrotime model at supraoptimal temperatures are plotted for the 16th, 50th, and 84th percentiles (symbols), with the linear increases (lines) predicted by a model developed in Alvarado and Bradford (2002). The projected lines for different seed fractions intercept the $\psi_b(g) = 0$ MPa axis at the T_c values for these fractions. (From Alvarado and Bradford, 2002, with permission from Blackwell Publishing.)

maturation increased seed dormancy levels in multiple populations from contrasting habitats (Meyer and Allen, 1999a,b; Allen and Meyer, 2002).

B. tectorum seeds ripen in early to mid-summer. Dormancy loss through after-ripening is characterized by a suite of changes in germination behavior of seed populations (Allen *et al.*, 1995). Seeds first acquire the ability to germinate, and then gradually germinate more rapidly and uniformly (see similar examples in Figures 4.3 and 4.5). The optimum temperature for germination progressively increases as after-ripening progresses.

Hydrothermal time concepts can effectively be applied to the study of seed dormancy loss in *B. tectorum*. The $\psi_b(50)$ decreases during after-ripening, which accounts for several observations associated with dormancy loss: progressively faster and more complete germination as after-ripening progresses, variation in the time required for seed populations to after-ripen completely, and the upward shift in optimum incubation temperature as seeds lose dormancy (Christensen *et al.*, 1996). As with true potato seeds, the value of $\psi_b(50)$ is therefore a meaningful index of dormancy status. Other parameters of hydrothermal time (θ_{HT}, T_b, and $\sigma_{\psi b}$) can all be held constant for purposes of modeling dormancy loss in this species.

In order to predict dormancy loss in the field, after-ripening of *B. tectorum* has further been characterized using a thermal after-ripening (TAR) time model (Bauer *et al.*, 1998). This can account for the widely fluctuating temperatures that occur in the field by assuming that the rate of change in $\psi_b(50)$ is a linear function of temperature above a base temperature and that the successive thermal history seeds experience can be integrated as the sum of hourly temperature values in the seed zone. The TAR equation is:

$$\theta_{AT} = (T_s - T_{sl})t_{ar}, \tag{4.8}$$

where θ_{AT} is the thermal time required for after-ripening, T_s is the storage temperature, T_{sl} is the lower limit storage temperature below which after-ripening does not occur, and t_{ar} is the actual time in storage required for completion of after-ripening (the time required for $\psi_b(50)$ to change from its initial to its final value). As θ_{AT} is a constant, the time required for after-ripening decreases proportionately as T_s increases above T_{sl}. While the assumption that changes in $\psi_b(50)$ at a particular temperature are linear through time is not completely valid (e.g., the rate of after-ripening is more rapid earlier during storage) (Bair *et al.*, 2006), using the linear initial slope yields sufficiently accurate predictions and simplifies model development.

The TAR model describes changes in after-ripening in terms of the decrease in $\psi_b(50)$, which is predicted based on hourly measurements of field seed zone temperature. The model was validated by comparing predicted $\psi_b(50)$ values with those observed in field retrieval studies conducted at the seed zone monitoring site. Predicted values for $\psi_b(50)$ were generally close to observed values during a year when seed zone water potential generally remained above −150 MPa (Bauer *et al.*, 1998), but consistently overestimated the rate of after-ripening during a very dry year (seed zone water potentials frequently below −300 MPa) (Bair *et al.*, 2006).

Reports that after-ripening can be slowed or prevented at very low seed water potentials (e.g., Leopold *et al.*, 1988; Foley, 1994) led to efforts to expand the TAR model to account for the effects of seed ψ on dormancy loss, i.e., to develop a hydrothermal after-ripening (HTAR) model. Results of studies conducted with *B. tectorum* and other wildland grasses over the ψ range of −40 to −400 MPa revealed that at very negative water potentials after-ripening is delayed or completely prevented, and field predictions of dormancy loss are improved by the HTAR model (Figure 4.7). Results of HTAR modeling have led to a proposed conceptual framework to describe the influence of ψ on after-ripening (Figure 4.8). At water potentials

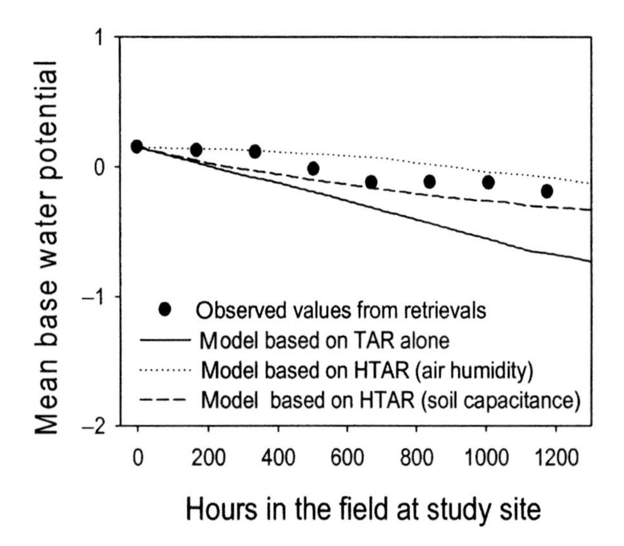

Figure 4.7 Predicted and observed changes in $\psi_b(50)$ during field after-ripening for *Elymus elymoides* seeds. Results shown are for a year characterized by very dry soil conditions (soil water potential generally below −150 MPa). The model based on thermal after-ripening (TAR) time alone consistently predicted a rate of dormancy loss that was too rapid. Two estimates of soil water potential were used for the hydrothermal after-ripening (HTAR) time model: (1) estimated soil water potential based on atmospheric humidity; and (2) soil capacitance readings calibrated to a soil moisture release curve. For similar results with *Bromus tectorum* seeds, see Bair *et al.* (2006). (Previously unpublished data of P.S. Allen and S.E. Meyer.)

above approximately −40 MPa or below approximately −400 MPa, seeds do not after-ripen, although the reasons for this differ. At storage above −40 MPa, seeds deteriorate, slowly progress toward germination, or remain imbibed but dormant. In contrast, seeds at water potentials below −150 MPa show a progressively decreased rate of after-ripening until at about −400 MPa after-ripening completely ceases. Thus, seeds stored from about −40 to −150 MPa after-ripen as a linear function of temperature according to the TAR model, while the HTAR model applies over the range of water potentials from −150 to −400 MPa.

Differences in seed dormancy that are evident in recently harvested seeds largely disappear during after-ripening. Collections of nondormant seeds germinate rapidly and completely, regardless of genotype or environmental factors likely to be encountered. This suggests that dormancy in *B. tectorum* functions largely to regulate germination timing primarily during the summer and autumn of maturation. Recently harvested *B. tectorum* seed populations, for example, initially have high ψ_b values. This means that they will have a lengthy hydration requirement if they can germinate at all, greatly reducing the potential for summer field germination (i.e., because soils do not remain moist for extended periods during this season in semi-arid habitats) (Bair *et al.*, 2006). As after-ripening progresses, decreasing ψ_b values increase the probability of germination following a given precipitation event and permit more rapid and higher percentages of germination when water is available.

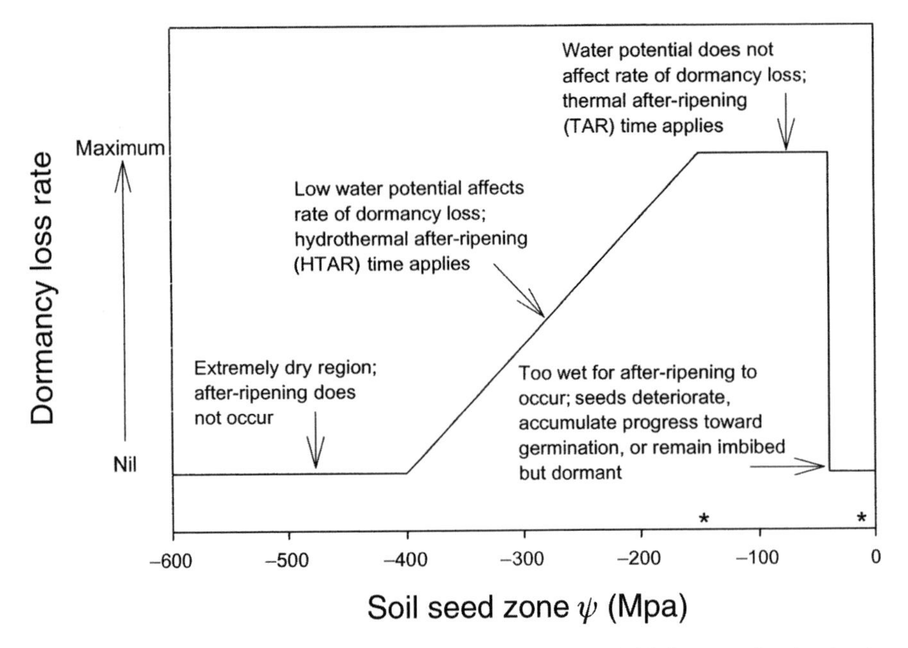

Figure 4.8 Conceptual diagram showing how storage water potential influences after-ripening in *Bromus tectorum*. Threshold water potentials are approximate. Asterisks above the *x*-axis correspond to critical hydration thresholds proposed by Vertucci and Farrant (1995).

Thus, a particular rainfall event will elicit a different germination response depending on whether it occurs in summer (dormant seeds) or autumn (after-ripened seeds). However, germination rate and percentage always remain closely linked to the current water potential of the surrounding environment. For seeds that require stratification and/or undergo dormancy cycles, shifts in ψ_b values are closely tied to annual temperature cycles. Release from dormancy is timed to coincide with periods most favorable for seedling establishment and/or to assure that a fraction of the seed population will persist across years in the soil seed bank.

4.5.3 *Polygonum aviculare*

Polygonum aviculare (prostrate knotweed, hogweed, or wireweed) is a cosmopolitan summer annual weed, usually invading cereal crops (Holm *et al.*, 1997). The seeds are dispersed near the end of the summer and are usually fully dormant at maturity. *P. aviculare* seeds are released from dormancy through stratification (Figure 4.1). Under field conditions, 'natural' stratification of the seed bank takes place during winter, resulting in weed emergence when soil temperature increases during early spring. Consequently, the possibility of predicting *P. aviculare* emergence to assist the planning of effective weed control strategies depends on understanding and modeling the stratification-induced dormancy release process in this species. To illustrate potential approaches to modeling seed stratification-dependent dormancy loss, a series of predictive models developed for *P. aviculare* seeds will be presented (see Batlla and Benech-Arnold, 2003, 2005, 2006; Batlla *et al.*, 2003, for details).

4.5.3.1 Modeling seed germination responses to temperature

In *P. aviculare*, as in many other summer annual species, seed dormancy loss through stratification is characterized by a widening of the thermal range permissive for germination as a consequence of a decrease in the lower limit temperature for seed germination (T_l) (i.e., seeds can germinate at progressively lower temperatures as stratification-dependent dormancy loss progresses). In contrast, other parameters associated with germination thermal responses of the seed population (T_b, T_o, T_c, θ_T, and T_h) show only minor changes during stratification (Kruk and Benech-Arnold, 1998; Batlla and Benech-Arnold, 2003). Therefore, the mean value of T_l ($T_l(50)$) can be used as a meaningful index of the mean dormancy status of the seed population. $T_l(50)$ values linearly decrease throughout stratification, at a rate that is inversely related to stratification temperature (Figure 4.9A). The effect of different stratification times and temperatures on dormancy loss (evaluated via changes in $T_l(50)$) can be quantified through a thermal time index (Figure 4.9B) calculated using the following equation:

$$S_{tt} = (T_{dc} - T_s)t_s \qquad (4.9)$$

where S_{tt} is accumulated stratification thermal time units (°C days), T_{dc} is the dormancy release 'ceiling' temperature (°C) (the temperature at, or above, which dormancy release does not occur), T_s is the daily mean storage temperature (°C), and t_s is the storage time (days).

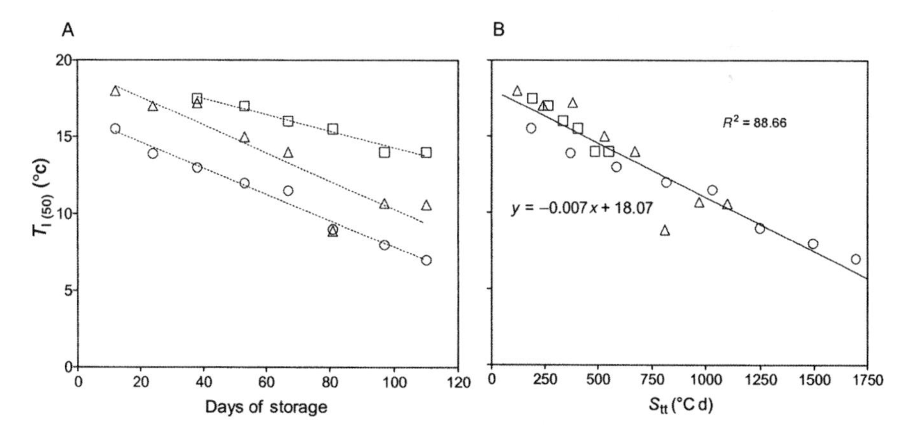

Figure 4.9 Estimated values of mean lower limit temperature ($T_l(50)$) for *Polygonum aviculare* seeds stored at 1.6 (○), 7 (Δ), and 12°C (□), plotted against days of storage (A) and against stratification thermal time (S_{tt}) (B). The dotted lines in (A) were fitted linear equations for each storage temperature, with R^2 values of 0.98, 0.84, and 0.96, respectively. The fitted line in (B) is the result of repeated regression analysis to obtain the threshold 'ceiling' temperature (T_c), with the best fit ($T_c = 17$°C) according to Equation 4.9. (From Batlla and Benech-Arnold, 2003, with permission of CABI Publishing International.)

This thermal time approach is similar to that previously presented for *B. tectorum* seeds. However, in contrast to the TAR time model in which degree-days are accumulated above a base storage temperature, this stratification thermal time index accumulates degree-days relative to a ceiling threshold temperature below which dormancy loss occurs. Thus, using this thermal time index, seeds stratified at lower temperatures would accumulate more thermal time units than seeds stratified at higher stratification temperatures for equal time periods. The ability to quantify temperature effects using a thermal time approach allows prediction of the dormancy level of a seed population exposed to a variable thermal environment similar to that prevailing under field conditions. Therefore, $T_l(50)$ for seeds stratified under different temperatures for different time periods can be predicted with a linear function that depends on S_{tt} accumulation (Figure. 4.9B).

Under field conditions, seeds germinate when current temperature overlaps the range of temperatures permissive for seed germination (Karssen, 1982; Benech-Arnold *et al.*, 2000). Consequently, germination of a given fraction of the seed bank population occurs when increasing spring soil temperature exceeds $T_l(g)$ for that fraction. The proportion of the seed bank that emerges at a given time can be predicted if the T_l distribution within the seed population and how this distribution changes with seed dormancy level are known. Results obtained by Batlla and Benech-Arnold (2003) showed that the distribution of T_l within the seed population, quantified through its standard deviation σ_{T_l}, also varied during *P. aviculare* seed stratification. However, variation in σ_{T_l} depended not only on the amount of S_{tt} units accumulated during the stratification period, but also on the stratification temperature at which S_{tt} units were accumulated. Based on these findings, a new empirical index was developed which permitted the prediction of σ_{T_l} in relation to the number of accumulated S_{tt} units and the daily mean stratification temperature (Batlla and Benech-Arnold, 2003).

This stratification thermal time model, as with other models described in this chapter, is based on a threshold-population approach. It accounts not only for the mean dormancy status of the seed population quantified through $T_l(50)$, but also for the distribution of dormancy levels within the population quantified through σ_{T_l}. It further allows the prediction of the fraction of the seed-bank population that will germinate under a given thermal environment when exhumed after different periods of burial in the field (Figure 4.10). Changes in germination behavior as seeds were progressively released from dormancy during winter were generally predicted with good accuracy by the model. Moreover, the model predicted changes in T_l of the seed population (i.e., the minimum temperature for germination) as dormancy loss progressed during 'natural' stratification in the soil.

The *P. aviculare* model is based on the assumption that dormancy loss due to the effect of low temperatures in summer annuals can be related to changes in a single variable (T_l), while all other thermal parameters associated with the germination response of the seed population remain constant or change little during burial (Kruk and Benech-Arnold, 1998; Batlla and Benech-Arnold, 2003). For species in which dormancy loss is also accompanied by significant changes in germination rate, changes in other germination parameters (e.g., θ_T) should also be considered.

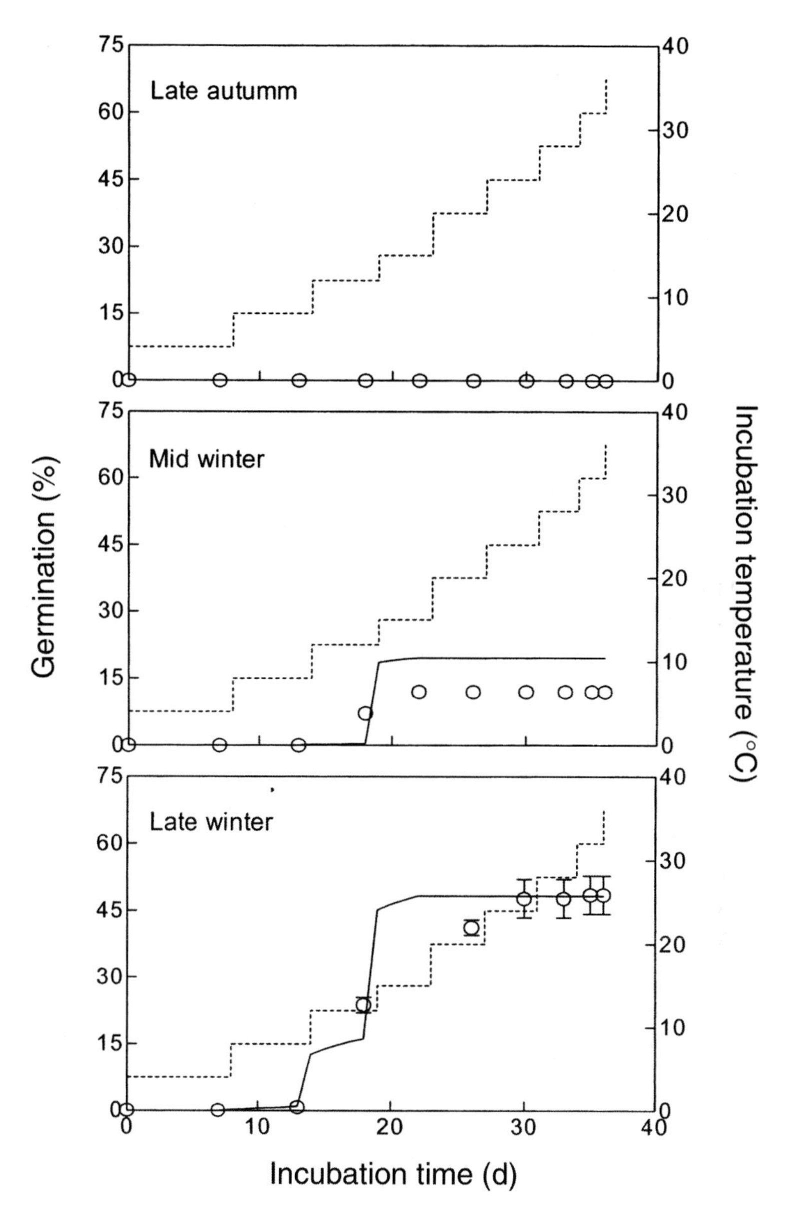

Figure 4.10 Predicted and observed germination time course curves for seeds of *Polygonum aviculare* exhumed from the field during different emergence seasons and incubated at an increasing temperature regime. Symbols represent observed germination percentages for seeds exhumed from the field, while solid lines represent values predicted from simulation modeling. Dotted lines represent the course of temperature change during incubation of seeds in the increasing temperature regime. Vertical bars indicate standard error. (Extracted from Batlla and Benech-Arnold, 2003, with permission of CABI Publishing International.)

4.5.3.2 Modeling seed responses to germination-stimulating factors

Light and fluctuating temperatures can be important in stimulating germination of many species under field situations (Figure 4.1), especially agricultural weeds. Therefore, the inclusion of seed sensitivity to germination-stimulating factors in simulation models may be important for accurately predicting emergence in the field. For example, the proportion of *P. aviculare* seeds that can respond to light in both the VFLR and LFR ranges depends on the dormancy status of the seed population, which decreases during stratification (Batlla *et al.*, 2003; Batlla and Benech-Arnold, 2005). The increasing sensitivity to light during stratification at different temperatures can be represented on a common thermal time scale using the stratification thermal time index (S_{tt}) (Figure 4.11). Similarly, *P. aviculare* seeds exhibited an increasing response to fluctuating temperature cycles of 10°C (14 h)/24°C (10 h) following exposure to light as they accumulated stratification thermal time (Figure 4.12). These increasing sensitivities to light and fluctuating temperatures were characterized by an increasing capacity of seeds to germinate in response to lower Pfr/P values as well as a decrease in the number of fluctuating temperature cycles required to saturate the germination response as stratification progressed. Quantifying these relationships permits the development of models to predict the germination responses of seeds stratified at different temperatures for different time periods and exposed to various conditions in the field (Batlla *et al.*, 2004). For example, the fraction of the seed bank population exhibiting VLFR can be predicted on the basis of soil temperature records. Because acquisition of VLFR by seeds is a prerequisite for germination in response to the light flash perceived during soil disturbance (Casal and Sánchez, 1998), this model can be used to predict the proportion of the seed bank that would germinate in response to tillage operations after the accumulation of a certain amount of S_{tt} during winter burial, contributing to weed control decisions (Batlla and Benech-Arnold, 2005).

The examples presented for *P. aviculare* show potential approaches for the development of models to predict dormancy loss (characterized through seed germination responses to different environmental factors) of summer annual species in relation to stratification temperature. Quantitative relationships between the dormancy status of the seed population (assessed through changes in T_l or through the sensitivity of seeds to light and fluctuating temperatures) and stratification temperature were established using a common thermal time approach. This approach allows the prediction of how seeds will respond to different environmental factors for a given accumulation of S_{tt} units through stratification during the dormancy release season (i.e., winter).

4.6 Population-based threshold models of seed dormancy

The examples and models described above share several common features that make them useful for characterizing seed dormancy and changes in dormancy levels due to after-ripening or stratification. First, the models explicitly recognize and incorporate variation in dormancy levels among individual seeds. Characterizing

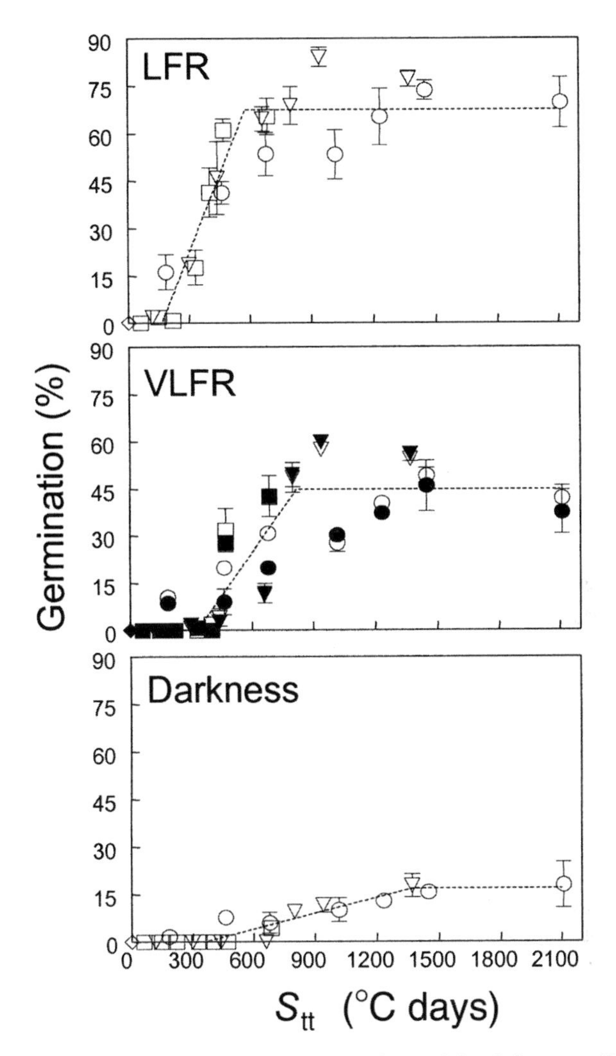

Figure 4.11 Germination percentages of *Polygonum aviculare* seeds in relation to stratification thermal time (S_{tt}) for seeds stored at 1.6 (○, •), 7 (Δ, ▲), and 12°C (□, ■) in response to different light treatments in the LFR range (Pfr/P = 0.76), the VLFR range (Pfr/P = 0.03 [closed symbols], or 7.6×10^{-6} [open symbols]) and for seeds incubated in darkness. Dotted lines correspond to the adjustment of trilineal models for different phases of the germination response. Vertical bars represent standard error. (From Batlla and Benech-Arnold, 2005, with permission of the New Phytologist Trust.)

and quantifying that variation is of considerable value in modeling and predicting overall germination behavior of seed populations. In most cases, the variation in dormancy characteristics is normally distributed within a given seed population, but this is not critical to the models. Second, the models are based upon threshold-type responses. That is, minimum or maximum thresholds are present that set limits on the range of environmental parameters (e.g., for temperature, water potential or

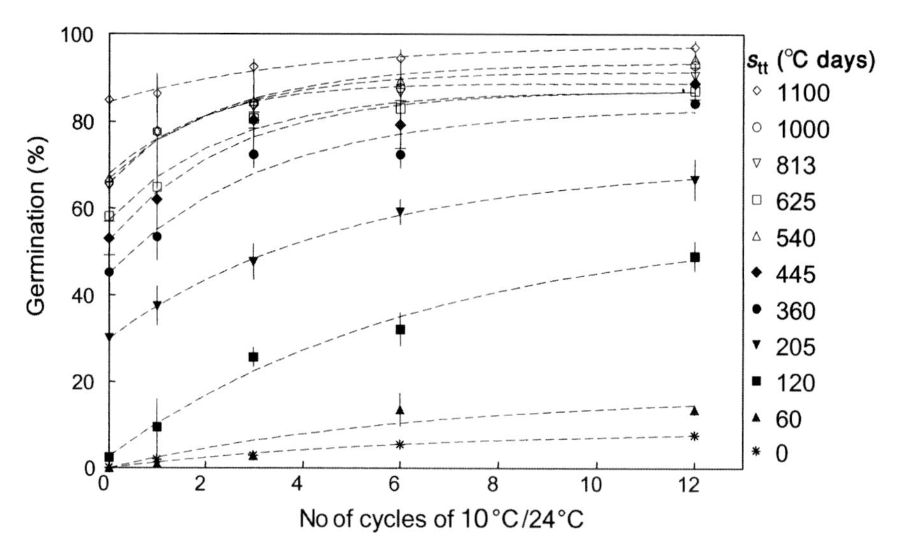

Figure 4.12 Germination responses of *Polygonum aviculare* seeds to cycles of a fluctuating temperature regime of 10°C (14 h)/24°C (10 h) determined for different values of stratification thermal time (S_{tt}, °C days). Germination data used for fitting curves correspond to seeds exhumed after different days of storage at 1.6, 7, and 12°C that had accumulated similar values of S_{tt}. Dashed lines were fitted with an exponential equation: $G_n = C\ [1 - \exp(-KN)] + G_0$, where G_n is the germination percentage obtained after N cycles of fluctuating temperature, C is the difference between the germination percentage obtained with 0 (i.e., constant 15°C) and N cycles, K is the effect of the number of cycles on seed germination given by the slope of the response curve, and G_0 is the germination percentage obtained for seeds incubated at constant temperature (0 cycles). The derived parameters C and K can be used to model and predict the response to fluctuating temperatures of seeds previously receiving different amounts of stratification (S_{tt}). Vertical bars indicate standard error. (From Batlla *et al.*, 2003, with permission of CABI Publishing International.)

light) permitting germination. These thresholds set the boundary sensitivity limits for seed responses to the various environmental factors controlling dormancy and germination. For the thermal and hydrotime models, the rates at which dormancy-breaking or germinative processes occur are also proportional to the difference between the current level of an environmental parameter and its threshold limit. Thus, as the environmental signal (i.e., difference from the threshold) increases, the rate of either dormancy release or progress toward germination also increases. Since thresholds can vary among individual seeds, the rates of dormancy release or progress toward germination also vary among seeds, resulting in the characteristic germination response curves to dormancy-releasing factors (e.g., Figures 4.3 and 4.5). Third, the sensitivity threshold distributions for a seed population shift to higher or lower mean values in response to dormancy-releasing or germination-stimulating factors. Shifting response threshold distributions underlie the widening or narrowing of the ranges of environmental factors over which germination will occur as dormancy is released or induced. Together, these features of population-based threshold models provide considerable flexibility and explanatory power for describing and predicting germination behavior.

Some examples of how threshold distributions for different dormancy-related factors are affected by environmental conditions are shown in Figure 4.13. As described above, dormancy levels can often be characterized by the temperature ranges permissive for germination. The lower limit distribution ($T_l(g)$) determines the range of minimum temperatures at which germination will occur, while the upper limit distribution ($T_h(g)$) specifies the maximum temperature limit range (Washitani, 1987; Kruk and Benech-Arnold, 2000). Seeds whose T_l or T_h values are either above or below, respectively, the current temperature are prevented from germinating. Under dormancy-releasing conditions (e.g., after-ripening or stratification), these distributions shift to lower and higher mean values, respectively, widening the range of temperatures permissive of germination (Figures 4.13A and 4.13B). Similarly, $\psi_b(g)$ distributions shift to more negative values as dormancy is released and to more positive values as dormancy is induced (Figure 4.13C). Thus, the potential for rapid germination when water is available increases as dormancy is released, but the rate and extent of germination remain sensitive to the current water potential. Water potential thresholds have also been connected to the hormonal regulation of dormancy, as high mean ψ_b values of GA-deficient tomato seeds decreased following GA_{4+7} applications that stimulated germination; in contrast, mean ψ_b values increased following ABA applications that prevented radicle emergence (Ni and Bradford, 1993). Ethylene, which expanded the high temperature range for germination of lettuce seeds, prevented the increase in mean ψ_b values that occurs as T approaches T_h (Dutta and Bradford, 1994). Thus, shifts in the distributions of T_l, T_h, and ψ_b are likely to be interconnected, although the specific physiological and molecular mechanisms underlying these connections remain to be definitively identified (see Chapter 11).

Changes in dormancy level are also associated with modifications in the sensitivity of the seed population to germination-stimulating factors (e.g., light and fluctuating temperature). For example, changes in seed sensitivity to light can be characterized through the Pfr/P values required for germination, both in the LFR and VLFR range. Whenever the Pfr/P value established by a certain light environment is above that required by a certain fraction of the population, all the fractions requiring lower Pfr/P values will advance toward germination. Dormancy release shifts the light sensitivity distribution toward lower Pfr/P values, allowing seeds to respond to a wider range of light environments (Figures 4.13D and 4.13E; Bradford, 1996, 2005). A similar approach can be used to characterize changes in seed sensitivity to fluctuating temperatures due to dormancy alleviation. In this case, the number of cycles with a specific amplitude range can be used as the stimulating factor required by the different fractions of the population. As dormancy release occurs over time, seeds first become responsive to temperature cycling, and then a lower number of cycles is required by an increasing fraction of the population until finally no response to cycling temperatures remains in nondormant seeds (Figure 4.13F).

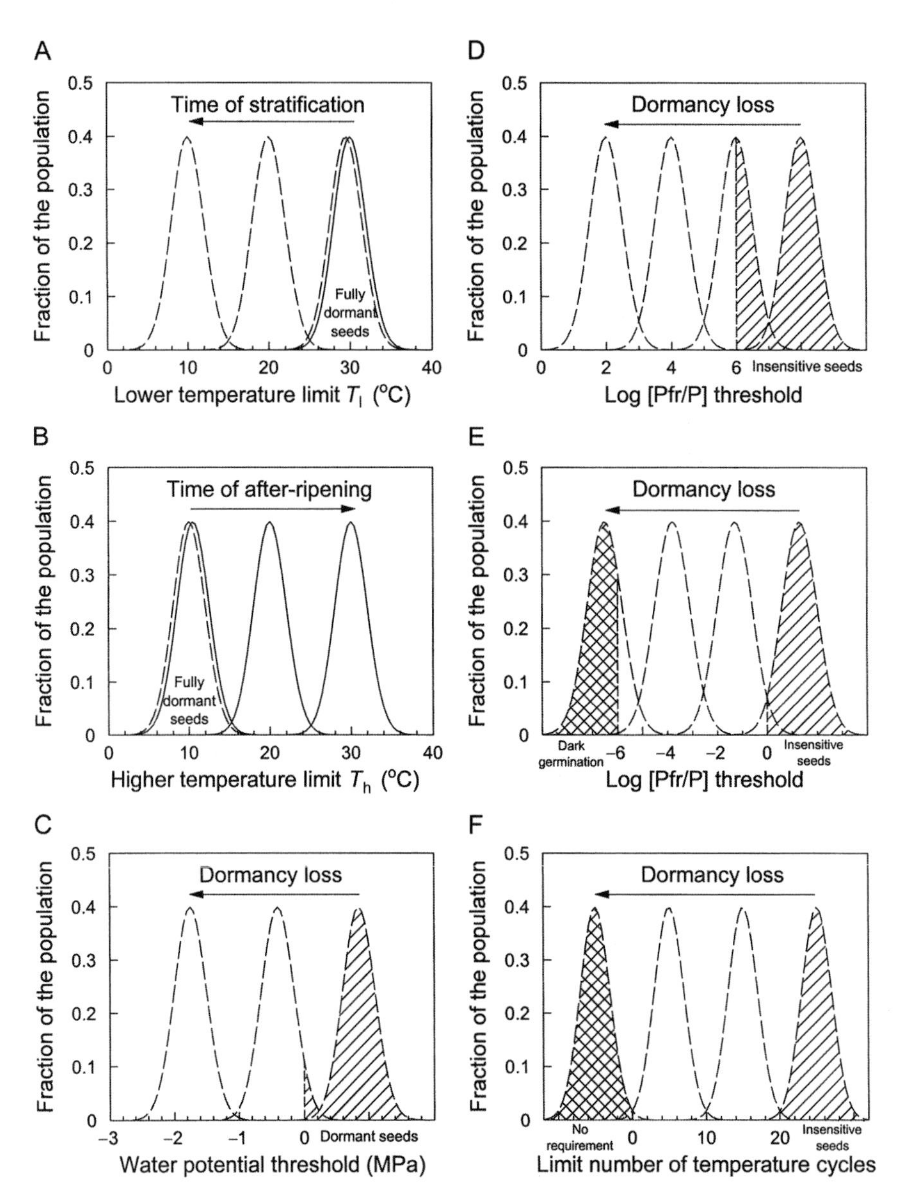

Figure 4.13 Population-based representation of changes in dormancy level through displacement of environmental thresholds for dormancy expression (i.e., temperature and water availability) (left panels) and of sensitivity thresholds to dormancy-terminating factors (i.e., light and fluctuating temperatures) (right panels). (A) The distribution of the lower limit temperature (T_l) permissive for germination (dashed curve) shifts toward lower temperatures with time of stratification while that of the higher limit temperature (T_h) (solid curve) remains constant (i.e., summer annual species). (B) The distribution of the higher limit temperature (T_h) permissive for germination (solid curve) shifts toward higher temperatures with time of after-ripening while that of the lower limit temperature (T_l) (dashed curve) remains constant (i.e., winter annual species). *(Continued over)*

4.7 Conclusions and future directions

Mathematical models are powerful tools to evaluate changes in seed dormancy and the influences of environmental factors that modulate these changes or stimulate germination. The primary focus of this chapter has been to present insights and approaches to describing seed dormancy based on threshold models that account for behavior of all viable seeds within a population. Most of the assumptions on which these models are constructed are, or lead to, hypotheses that aim to give explanations for aspects of seed functioning from either a physiological or an ecological perspective; model validation constitutes a way of testing those hypotheses. In addition, these models may have very important practical applications. In the case of weeds, for instance, the possibility of predicting changes in dormancy allows the identification of periods of maximum probability for seedling emergence according to the environmental conditions experienced by the soil seed bank, permitting the design of more efficient or effective control schemes.

Although the examples presented demonstrate that considerable progress has been made in modeling seed dormancy, a number of challenges remain. An important goal must be to integrate the influences of different factors that affect dormancy into a combined model. As illustrated in Figure 4.13, changes in threshold distributions for multiple factors may be occurring simultaneously during dormancy release, and better methods are needed to integrate these into a comprehensive model. While temperature and water potential have been integrated via the hydrothermal time model, this model has not been fully integrated with the upper and lower temperature limits approach. However, changes in both the upper and lower temperature limits appear to be associated with shifts in ψ_b distributions to more negative values (Figure 4.6; Batlla and Benech-Arnold, 2004, 2006). Efforts to extend the threshold modeling approach to account for multiple interacting factors have been attempted, but on a relatively limited scale thus far. In most cases, extension of models to incorporate additional factors has been accomplished by using primary parameters derived from threshold models in secondary models that empirically describe how the primary parameters change over time in response to other factors influencing

---

Figure 4.13 (*Continued*) In (A) and (B), overlapping of the distribution of T_l with that of T_h indicates full dormancy. (C) The distribution of the base water potential threshold (ψ_b) within the population shifts toward more negative values with dormancy loss. The shaded area indicates the fraction of the population that will not germinate even if incubated in pure water (because the seeds are fully dormant, they are incubated at a temperature outside their permissive range, or they require dormancy terminating factors). (D and E) The distribution of sensitivity thresholds to Pfr shifts toward lower values of Log [Pfr/P] in both the low fluence response (LFR) (D) and the very low fluence response (VLFR) ranges (E). The diagonally shaded areas in D and E represent seeds insensitive to light in the LFR and the VLFR ranges, respectively. The cross-hatched area in (E) represents seeds that are able to germinate in the dark. (F) The distribution of sensitivity thresholds to number of cycles of fluctuating temperatures shifts toward lower values with dormancy loss. The diagonally shaded area represents seeds insensitive to fluctuating temperatures while the cross-hatched area represents seeds that do not require fluctuating temperatures.

dormancy (e.g., Ni and Bradford, 1993; Alvarado and Bradford, 2002; Batlla *et al.*, 2003, 2004; Batlla and Benech-Arnold, 2004; Finch-Savage, 2004; Bair *et al.*, 2006). While these approaches are practically useful and provide new insights, we should continue to seek broader conceptual models based on physiological mechanisms rather than on empirical curve fitting, as the former have greater potential for generality. For example, knowledge of the multiple phytochrome genes and their diverse and unique roles in light reception and signaling should inform models attempting to account for different light response modes. Such specific targets for the physiological basis of ψ_b remain elusive as yet (see Chapter 11). Nonetheless, the overall goal should continue to be the development of models that describe and predict seed behavior based upon knowledge of the physiological and molecular bases of that behavior. Such an approach provides mutually reinforcing hypotheses for both physiologists and ecologists.

References

J.M. Adamoli, A.D. Goldberg and A. Soriano (1973) El desbloqueo de las semillas de chamico (*Datura ferox* L.) enterradas en el suelo: análisis de los factores causales. *Revista de Investigaciones Agropecuarias* **10**, 209–222.

P.S. Allen and S.E. Meyer (1998) Ecological aspects of seed dormancy loss. *Seed Science Research* **8**, 183–191.

P.S. Allen and S.E. Meyer (2002) Ecology and ecological genetics of seed dormancy in downy brome. *Weed Science* **50**, 241–247.

P.S. Allen, S.E. Meyer and J. Beckstead (1995) Patterns of seed after-ripening in *Bromus tectorum* L. *Journal of Experimental Botany* **279**, 1737–1744.

P.S. Allen, S.E. Meyer and M.A. Khan (2000) Hydrothermal time as a tool in comparative germination studies. In: *Seed Biology: Advances and Applications* (eds M. Black, K.J. Bradford and J. Vazquez-Ramos), pp. 401–410. CAB International, Wallingford, UK.

V. Alvarado and K.J. Bradford (2002) A hydrothermal time model explains the cardinal temperatures for seed germination. *Plant Cell and Environment* **25**, 1061–1069.

V. Alvarado and K.J. Bradford (2005) Hydrothermal time analysis of seed dormancy in true (botanical) potato seeds. *Seed Science Research* **15**, 77–88.

N.B. Bair, S.E. Meyer and P.S. Allen (2006) A hydrothermal after-ripening time model for seed dormancy loss in *Bromus tectorum* L. *Seed Science Research* **16**, 17–28.

C.C. Baskin and J.M. Baskin (1998) *Seeds: Ecology, Biogeography, and Evolution of Dormancy and Germination*. Academic Press, New York.

J.M. Baskin and C.C. Baskin (1976) High temperature requirement for after-ripening in seeds of winter annuals. *New Phytologist* **77**, 619–624.

J.M. Baskin and C.C. Baskin (1977) Role of temperature in the germination ecology of three summer annual weeds. *Oecologia* **30**, 377–382.

J.M. Baskin and C.C. Baskin (1980) Ecophysiology of secondary dormancy in seeds of *Ambrosia artemisiifolia*. *Ecology* **61**, 475–480.

J.M. Baskin and C.C. Baskin (1984) Role of temperature in regulating timing of germination in soil seed reserves of *Lamium purpureum* (L.). *Weed Research* **30**, 341–349.

J.M. Baskin and C.C. Baskin (1985) The annual dormancy cycle in buried weed seeds: a continuum. *BioScience* **35**, 392–398.

J.M. Baskin and C.C. Baskin (2004) A classification system for seed dormancy. *Seed Science Research* **14**, 1–16.

D. Batlla and R.L. Benech-Arnold (2003) A quantitative analysis of dormancy loss dynamics in *Polygonum aviculare* L. seeds: development of a thermal time model based on changes in seed population thermal parameters. *Seed Science Research* **13**, 55–68.

D. Batlla and R.L. Benech-Arnold (2004) A predictive model for dormancy loss in *Polygonum aviculare* L. seeds based on changes in population hydrotime parameters. *Seed Science Research* **14**, 277–286.

D. Batlla and R.L. Benech-Arnold (2005) Changes in the light sensitivity of buried *Polygonum aviculare* seeds in relation to cold-induced dormancy loss: development of a predictive model. *New Phytologist* **165**, 445–452.

D. Batlla and R.L. Benech-Arnold (2006) The role of fluctuations in soil water content on the regulation of dormancy changes in buried seeds of *Polygonum aviculare* L. *Seed Science Research* **16**, 47–59.

D. Batlla, B.C. Kruk and R.L. Benech-Arnold (2000) Very early detection of canopy presence by seeds through perception of subtle modifications in red:far red signals. *Functional Ecology* **14**, 195–202.

D. Batlla, B.C. Kruk and R.L. Benech-Arnold (2004) Modeling changes in dormancy in weed soil seed banks: implications for the prediction of weed emergence. In: *Handbook of Seed Physiology: Applications to Agriculture* (eds R.L. Benech-Arnold and R.A. Sánchez), pp. 245–270. Haworth Press, New York.

D. Batlla, V. Verges and R.L. Benech-Arnold (2003) A quantitative analysis of seed responses to cycle-doses of fluctuating temperatures in relation to dormancy: development of a thermal time model for *Polygonum aviculare* L. seeds. *Seed Science Research* **13**, 197–207.

M.C. Bauer, S.E. Meyer and P.S. Allen (1998) A simulation model to predict seed dormancy loss in the field for *Bromus tectorum* L. *Journal of Experimental Botany* **49**, 1235–1244.

R.L. Benech-Arnold, C.M. Ghersa, R.A. Sánchez and P. Insausti (1990a) A mathematical model to predict *Sorghum halepense* germination in relation to soil temperature. *Weed Research* **30**, 81–89.

R.L. Benech-Arnold, C.M. Ghersa, R.A. Sánchez and P. Insausti (1990b) Temperature effects on dormancy release and germination rate in *Sorghum halepense* (L.) Pers. seeds: a quantitative analysis. *Weed Research* **30**, 91–99.

R.L. Benech-Arnold and R.A. Sánchez (1995) Modeling weed seed germination. In: *Seed Development and Germination* (eds J. Kigel and G. Galili), pp. 545–566. Marcel Dekker, New York.

R.L. Benech-Arnold, R.A. Sánchez, F. Forcella, B.C. Kruk and C.M. Ghersa (2000) Environmental control of dormancy in weed seed banks in soil. *Field Crops Research* **67**, 105–122.

J.D. Bewley and M. Black (1982) *Physiology and Biochemistry of Seeds in Relation to Germination*. Springer-Verlag, Berlin.

J.F. Bierhuizen and W.A. Wagenvoort (1974) Some aspects of seed germination in vegetables. I: The determination and application of heat sums and minimum temperature for germination. *Scientia Horticulturae* **2**, 213–219.

H.J. Bouwmeester and C.M. Karssen (1992) The dual role of temperature in the regulation of the seasonal changes in dormancy and germination of seeds of *Polygonum persicaria* L. *Oecologia* **90**, 88–94.

H.J. Bouwmeester and C.M. Karssen (1993) Annual changes in dormancy and germination in seeds of *Sisymbrium officinale* (L.) Scop. *New Phytologist* **124**, 179–191.

K.J. Bradford (1990) A water relations analysis of seed germination rates. *Plant Physiology* **94**, 840–849.

K.J. Bradford (1995) Water relations in seed germination. In: *Seed Development and Germination* (eds J. Kigel and G. Galili), pp. 351–396. Marcel Dekker, New York.

K.J. Bradford (1996) Population-based models describing seed dormancy behaviour: implications for experimental design and interpretation. In: *Plant Dormancy: Physiology, Biochemistry and Molecular Biology* (ed. A. Lang), pp. 313–339. CAB International, Wallingford, UK.

K.J. Bradford (1997) The hydrotime concept in seed germination and dormancy. In: *Basic and Applied Aspects of Seed Biology* (eds R.H. Ellis, M. Black, A.J. Murdoch and T.D. Hong), pp. 349–360. Kluwer Academic Publishers, Boston.

K.J. Bradford (2002) Applications of hydrothermal time to quantifying and modeling seed germination and dormancy. *Weed Science* **50**, 248–260.

K.J. Bradford (2005) Threshold models applied to seed germination ecology. *New Phytologist* **165**, 338–341.

K.J. Bradford and A.M. Haigh (1994) Relationship between accumulated hydrothermal time during seed priming and subsequent seed germination rates. *Seed Science Research* **4**, 1–10.

K.J. Bradford and O.A. Somasco (1994) Water relations of lettuce seed thermoinhibition. I: Priming and endosperm effects on base water potential. *Seed Science Research* **4**, 1–10.

J.J. Casal and R.A. Sánchez (1998) Phytochromes and seed germination. *Seed Science Research* **8**, 317–329.

Z.Y. Cheng and K.J. Bradford (1999) Hydrothermal time analysis of tomato seed germination responses to priming treatments. *Journal of Experimental Botany* **50**, 89–99.

M. Christensen, S.E. Meyer and P.S. Allen (1996) A hydrothermal time model of seed after-ripening in *Bromus tectorum* L. *Seed Science Research* **6**, 1–9.

F. Corbineau and D. Côme (1995) Control of seed germination and dormancy by the gaseous environment. In: *Seed Development and Germination* (eds J. Kigel and G. Galili), pp. 397–424. Marcel Dekker, New York.

S. Covell, R.H. Ellis, E.H. Roberts and R.J. Summerfield (1986) The influence of temperature on seed germination rate in grain legumes. 1: A comparison of chickpea, lentil, soybean and cowpea at constant temperatures. *Journal of Experimental Botany* **37**, 705–715.

C.M. D'Antonio and P.M. Vitousek (1992) Biological invasions by exotic grasses, the grass/fire cycle, and global change. *Annual Review of Ecological Systems* **23**, 63–87.

P. Dahal and K.J. Bradford (1990) Effects of priming and endosperm integrity on seed germination rates of tomato genotypes. II: Germination at reduced water potential. *Journal of Experimental Botany* **41**, 1441–1453.

P. Dahal and K.J. Bradford (1994) Hydrothermal time analysis of tomato seed germination at suboptimal temperature and reduced water potential. *Seed Science Research* **4**, 71–80.

P. Dahal, K.J. Bradford and R.A. Jones (1990) Effects of priming and endosperm integrity on seed germination rates of tomato genotypes. I: Germination at suboptimal temperature. *Journal of Experimental Botany* **41**, 1431–1439.

V.A. Deregibus, J.J. Casal, E.J. Jacobo, D. Gibson, M. Kauffman and A.M. Rodriguez (1994) Evidence that heavy grazing may promote the germination of *Lolium multiflorum* seeds via phytochrome-mediated perception of high red/far-red ratios. *Functional Ecology* **8**, 536–542.

M.P.M. Derkx and C.M. Karssen (1993) Changing sensitivity to light and nitrate but not to gibberellins regulates seasonal dormancy patterns in *Sisymbrium officinale* seeds. *Plant, Cell and Environment* **16**, 469–479.

K. Donohue, L. Dorn, C. Griffith, *et al.* (2005) The evolutionary ecology of seed germination of *Arabidopsis thaliana*: variable natural selection on germination timing. *Evolution* **59**, 758–770.

S. Dutta and K.J. Bradford (1994) Water relations of lettuce seed thermoinhibition. II: Ethylene and endosperm effects on base water potential. *Seed Science Research* **4**, 11–18.

G.H. Egley (1986) Stimulation of weed seed germination in soil. *Review of Weed Science* **2**, 67–89.

R.H. Ellis and P.D. Butcher (1988) The effects of priming and natural differences in quality amongst onion seed lots on the response of the rate of germination to temperature and the identification of the characteristics under genotypic control. *Journal of Experimental Botany* **39**, 935–950.

R.H. Ellis, S. Covell, E.H. Roberts and R.J. Summerfield (1986) The influence of temperature on seed germination rate in grain legumes. II: Intraspecific variation in chickpea (*Cicer arietinum* L.) at constant temperatures. *Journal of Experimental Botany* **37**, 1503–1515.

R.H. Ellis, G. Simon and S. Covell (1987) The influence of temperature on seed germination rate in grain legumes. III: A comparison of five faba bean genotypes at constant temperatures using a new screening method. *Journal of Experimental Botany* **38**, 1033–1043.

Y. Esashi, M. Ogasawara, R. Gorecki and A.C. Leopold (1993) Possible mechanisms of afterripening in *Xanthium* seeds. *Physiologia Plantarum* **87**, 359–364.

J. Evelyn (1664) *A Discourse of Forest Trees and the Propagation of Timber in his Majesties Dominions* (eds J. Marryn and J. Allestry). Printer to the Royal Society, London.

M. Fenner (2000) *Seeds: The Ecology of Regeneration in Plant Communities*. CABI Publishing, Oxon, UK.

S.A. Fennimore and M.E. Foley (1998) Genetic and physiological evidence for the role of gibberellic acid in the germination of dormant *Avena fatua* seeds. *Journal of Experimental Botany* **49**, 89–94.

W.E. Finch-Savage (2004) The use of population-based threshold models to describe and predict the effects of seedbed environment on germination and seedling emergence of crops. In: *Handbook of Seed Physiology: Applications to Agriculture* (eds R. Benech-Arnold and R.A. Sánchez), pp. 51–95. Food Products Press, New York.

W.E. Finch-Savage, H.R. Rowse and K.C. Dent (2005) Development of combined imbibition and hydrothermal threshold models to simulate maize (*Zea mays*) and chickpea (*Cicer arietinum*) seed germination in variable environments. *New Phytologist* **165**, 825–838.

M.E. Foley (1994) Temperature and water status of seed affect afterripening in wild oat (*Avena fatua*). *Weed Science* **42**, 200–204.

F. Forcella (1998) Real-time assessment of seed dormancy and seedling growth for weed management. *Seed Science Research* **8**, 201–209.

F. Forcella, R.L. Benech-Arnold, R.A. Sánchez and C.M. Ghersa (2000) Modeling seedling emergence. *Field Crops Research* **67**, 123–139.

T.P. Fyfield and P.J. Gregory (1989) Effects of temperature and water potential on germination, radicle elongation and emergence of mungbean. *Journal of Experimental Botany* **40**, 667–674.

J. Garcia-Huidobro, J.L. Monteith and G.R. Squire (1982) Time, temperature and germination of pearl millet (*Pennisetum typhoides* S & H). I: Constant temperature. *Journal of Experimental Botany* **33**, 288–296.

S.C. Garvin and S.E. Meyer (2003) Multiple mechanisms for seed dormancy regulation in shadscale (*Atriplex confertifolia*: Chenopodiaceae). *Canadian Journal of Botany* **81**, 601–610.

C.M. Ghersa, R.L. Benech-Arnold and M.A. Martinez Ghersa (1992) The role of fluctuating temperatures in germination and establishment of *Sorghum halepense* (L.) Pers. Regulation of germination at increasing depths. *Functional Ecology* **6**, 460–468.

A.C. Grundy (2003) Predicting weed emergence: a review of approaches and future challenges. *Weed Research* **43**, 1–11.

A.C. Grundy, K. Phelps, R.J. Reader and S. Burston (2000) Modelling the germination of *Stellaria media* using the concept of hydrothermal time. *New Phytologist* **148**, 433–444.

R.J. Gummerson (1986) The effect of constant temperatures and osmotic potential on the germination of sugar beet. *Journal of Experimental Botany* **37**, 729–741.

S.P. Hardegree, G.N. Flerchinger and S.S. Van Vactor (2003) Hydrothermal germination response and the development of probabilistic germination profiles. *Ecological Modelling* **167**, 305–322.

T.W. Hegarty (1978) The physiology of seed hydration and dehydration, and the relation between water stress and the control of germination: a review. *Plant, Cell and Environment* **1**, 101–119.

B. Heim and E. Schäfer (1982) Light-controlled inhibition of hypocotyl growth in *Sinapis alba* L. seedlings. *Planta* **154**, 150–155.

B. Heim and E. Schäfer (1984) The effect of red and far-red light in the high irradiance reaction of phytochrome (hypocotyl growth in dark-grown *Sinapis alba* L.). *Plant, Cell and Environment* **7**, 39–44.

H.W.M. Hilhorst (1990) Dose–response analysis of factors involved in germination and secondary dormancy of seeds of *Sisymbrium officinale*. II. Nitrate. *Plant Physiology* **94**, 1096–1102.

H.W.M. Hilhorst (1998) The regulation of secondary dormancy. The membrane hypothesis revisited. *Seed Science Research* **8**, 77–90.

L. Holm, J. Doll, E. Holm, J. Pancho and J. Herberger (1997) *World Weeds: Natural Histories and Distribution*. John Wiley & Sons, New York.

P. Insausti, A. Soriano and R.A. Sánchez (1995) Effects of flood-influenced factors on seed germination of *Ambrosia tenuifolia*. *Oecologia* **103**, 127–132.

International Seed Testing Association (2004) *International Rules for Seed Testing*. ISTA, Bassersdorf, Switzerland.

C.M. Karssen (1980/81) Patterns of change in dormancy during burial of seeds in soil. *Israel Journal of Botany* **29**, 65–73.

C.M. Karssen (1982) Seasonal patterns of dormancy in weed seeds. In: *The Physiology and Biochemistry of Seed Development, Dormancy and Germination* (ed. A.A. Khan), pp. 243–270. Elsevier, Amsterdam.

C.M. Karssen and H.W.M. Hilhorst (1992) Effect of chemical environment on seed germination. In: *Seeds: The Ecology of Regeneration in Plant Communities* (ed. M. Fenner), pp. 327–348. CAB International, Wallingford, UK.

E. Kebreab and A.J. Murdoch (1999a) A quantitative model for loss of primary dormancy and induction of secondary dormancy in imbibed seeds of *Orobanche* spp. *Journal of Experimental Botany* **50**, 211–219.

E. Kebreab and A.J. Murdoch (1999b) Modelling the effects of water stress and temperature on germination rate of *Orobanche aegyptiaca* seeds. *Journal of Experimental Botany* **50**, 655–664.

E. Kebreab and A.J. Murdoch (2000) The effect of water stress on the temperature range for germination of *Orobanche aegyptiaca* seeds. *Seed Science Research* **10**, 127–133.

G.H.M. Kronenberg and R.E. Kendrik (1986) The physiology of action. In: *Photomorphogenesis in Plants* (eds R.E. Kendrik and G.H.M. Kronenberg), pp. 99–114. Marthinus Nijhoff/Dr. W. Junk Publishers, Dordrecht.

B.C. Kruk and R.L. Benech-Arnold (1998) Functional and quantitative analysis of seed thermal responses in prostrate knotweed (*Polygonum aviculare*) and common purslane (*Portulaca oleracea*). *Weed Science* **46**, 83–90.

B.C. Kruk and R.L. Benech-Arnold (2000) Evaluation of dormancy and germination responses to temperature in *Carduus acanthoides* and *Anagallis arvensis* using a screening system, and relationship with field-observed emergence patterns. *Seed Science Research* **10**, 77–88.

L.G. Labouriau (1970) On the physiology of seed germination in *Vicia graminea* Sm. I. *Annals Academia Brasilia Ciencia* **42**, 235–262.

L.G. Labouriau and J.H. Osborn (1984) Temperature dependence of the germination of tomato seeds. *Journal of Thermal Biology* **9**, 285–295.

A.C. Leopold, R. Glenister and M.A. Cohn (1988) Relationship between water content and afterripening in red rice. *Physiologia Plantarum* **74**, 659–662.

G. Leubner-Metzger (2005) $\beta-1$,3-Glucanase gene expression in low-hydrated seeds as a mechanism for dormancy release during tobacco after-ripening. *Plant Journal* **41**, 133–145.

R.M. Lynch and J.S. Clegg (1986) A study of metabolism in dry seeds of *Avena fatua* L. evaluated by incubation with ethanol-1-^{14}C. In: *Membranes, Metabolism, and Dry Organisms* (ed. A.C. Leopold), pp. 50–58. Comstock Publishing, Ithaca, NY.

S.E. Meyer and P.S. Allen (1999a) Ecological genetics of seed germination regulation in *Bromus tectorum* L. I. Phenotypic variance among and within populations. *Oecologia* **118**, 27–34.

S.E. Meyer and P.S. Allen (1999b) Ecological genetics of seed germination regulation in *Bromus tectorum* L. II. Reaction norms in response to a water stress gradient imposed during seed maturation. *Oecologia* **118**, 35–43.

S.E. Meyer, P.S. Allen and J. Beckstead (1997) Seed germination regulation in *Bromus tectorum* L. (Poaceae) and its ecological significance. *Oikos* **78**, 475–485.

S.E. Meyer, S.B. Debaene-Gill and P.S. Allen (2000) Using hydrothermal time concepts to model seed germination response to temperature, dormancy loss, and priming effects in *Elymus elymoides*. *Seed Science Research* **10**, 213–223.

S.E. Meyer, S.G. Kitchen and S.L. Carlson (1995) Seed germination patterns in Intermountain Penstemon (Scrophulariaceae). *American Journal of Botany* **82**, 377–389.

A.J. Murdoch, E.H. Roberts and C.O. Goedert (1988) A model for germination responses to alternating temperatures. *Annals of Botany* **63**, 97–111.

W.M. Nascimento, D.J. Cantliffe and D.J. Huber (2000) Thermotolerance in lettuce seeds: association with ethylene and endo-β-mannanase. *Journal of the American Society for Horticultural Science* **125**, 518–524.

B.R. Ni and K.J. Bradford (1992) Quantitative models characterizing seed germination responses to abscisic acid and osmoticum. *Plant Physiology* **98**, 1057–1068.

B.R. Ni and K.J. Bradford (1993) Germination and dormancy of abscisic acid-deficient and gibberellin-deficient mutant tomato seeds. Sensitivity of germination to abscisic acid, gibberellin, and water potential. *Plant Physiology* **101**, 607–617.

A. Orozco-Segovia, L. González-Zertuche, A. Mendoza and S. Orozco (1996) A mathematical model that uses Gaussian distribution to analyze the germination of *Manfreda brachystachya* (Agavaceae) in a thermogradient. *Physiologia Plantarum* **98**, 431–438.

N. Pallais (1995) High temperature and low moisture reduce the storage requirement of freshly harvested true potato seeds. *Journal of the American Society for Horticultural Science* **120**, 699–702.

K. Phelps and W.E. Finch-Savage (1997) A statistical perspective on threshold-type germination models. In: *Basic and Applied Aspects of Seed Biology* (eds R.H. Ellis, M. Black, A.J. Murdoch and T.D. Hong), pp. 361–368. Kluwer Academic Publishers, Boston.

T.L. Pons (1992) Seed response to light. In: *Seeds: The Ecology of Regeneration in Plant Communities* (ed. M. Fenner), pp. 259–284. CAB International, Melksham.

R.J. Probert (1992) The role of temperature in germination ecophysiology. In: *Seeds: The Ecology of Regeneration in Plant Communities* (ed. M. Fenner), pp. 285–325. CAB International, Melksham.

A.P. Ramakrishnan, S.E. Meyer, J. Waters, M.R. Stevens, C.E. Coleman and D.J. Fairbanks (2004) Correlation between molecular markers and adaptively significant genetic variation in *Bromus tectorum* (Poaceae), an inbreeding annual grass. *American Journal of Botany* **91**, 797–803.

O. Reisman-Berman, J. Kigel and B. Rubin (1991) Dormancy pattern in buried seeds of *Datura ferox* L. *Canadian Journal of Botany* **69**, 173–179.

E.H. Roberts and R.D. Smith (1977) Dormancy and the pentose phosphate pathway. In: *The Physiology and Biochemistry of Seed Dormancy and Germination* (ed. A.A. Khan), pp. 385–411. Elsevier/North-Holland Biomedical Press, Amsterdam.

E.H. Roberts and S. Totterdell (1981) Seed dormancy in *Rumex* species in response to environmental factors. *Plant, Cell and Environment* **4**, 97–106.

H.R. Rowse and W.E. Finch-Savage (2003) Hydrothermal threshold models can describe the germination response of carrot (*Daucus carota* L.) and onion (*Allium cepa* L.) seed populations across both sub- and supra-optimal temperatures. *New Phytologist* **158**, 101–108.

H.R. Rowse, J.M.T. McKee and E.C. Higgs (1999) A model of the effects of water stress on seed advancement and germination. *New Phytologist* **143**, 273–279.

A.L. Scopel, C.L. Ballaré and R.A. Sánchez (1991) Induction of extreme light sensitivity in buried weed seeds and its role in the perception of soil cultivations. *Plant, Cell and Environment* **14**, 501–508.

A. Shrestha, E.S. Roman, A.G. Thomas and C.J. Swanton (1999) Modelling germination and shoot-radicle elongation of *Ambrosia artemisiifolia*. *Weed Science* **47**, 557–562.

H. Smith (1982) Light quality, photoperception, and plant strategy. *Annual Review of Plant Physiology* **33**, 481–518.

K.J. Steadman and H.W. Pritchard (2004) Germination of *Aesculus hippocastanum* seeds following cold-induced dormancy loss can be described in relation to a temperature-dependent reduction in base temperature (T_b) and thermal time. *New Phytologist* **161**, 415–425.

S.J. Steinmaus, T.S. Prather and J.S. Holt (2000) Estimation of base temperatures for nine weed species. *Journal of Experimental Botany* **51**, 275–286.

P. Stokes (1965) Temperature and seed dormancy. In: *Encyclopedia of Plant Physiology* (ed. W. Ruhland), pp. 746–803. Springer, Berlin.

A.M. Tarquis and K.J. Bradford (1992) Prehydration and priming treatments that advance germination also increase the rate of deterioration of lettuce seeds. *Journal of Experimental Botany* **43**, 307–317.

S. Totterdell and E.H. Roberts (1979) Effects of low temperatures on the loss of innate dormancy and the development of induced dormancy in seeds of *Rumex obtusifolius* L. and *Rumex crispus* L. *Plant, Cell and Environment* **2**, 131–137.

A. Vegis (1964) Dormancy in higher plants. *Annual Review of Plant Physiology* **115**, 185–224.

C.W. Vertucci and J.M. Farrant (1995) Acquisition and loss of desiccation tolerance. In: *Seed Development and Germination* (eds J. Kigel and G. Galili), pp. 237–271. Marcel Dekker, New York.

C.W. Vertucci and A.C. Leopold (1986) Physiological activities associated with hydration level in seeds. In: *Membranes, Metabolism and Dry Organisms* (ed. A.C. Leopold), pp. 35–49. Comstock Publishing Associates, Ithaca, NY.

L.M. Vleeshouwers and H.J. Bouwmeester (2001) A simulation model for seasonal changes in dormancy and germination of seeds. *Seed Science Research* **11**, 77–92.

L.M. Vleeshouwers and M.J. Kropff (2000) Modeling field emergence patterns in arable weeds. *New Phytologist* **148**, 445–457.

C. Walters, N.W. Pammenter, P. Berjak and J. Crane (2001) Desiccation damage, accelerated ageing and respiration in desiccation tolerant and sensitive seeds. *Seed Science Research* **11**, 135–148.

I. Washitani (1987) A convenient screening test system and a model for thermal germination responses of wild plant seeds: behaviour of model and real seed in the system. *Plant, Cell and Environment* **10**, 587–598.

5 Genetic aspects of seed dormancy

Leónie Bentsink, Wim Soppe and Maarten Koornneef

5.1 Introduction

Every new plant generation starts with a seed, which usually contains a fully developed embryo that can survive the period between seed maturation and germination. Dormancy, defined as the failure of an intact viable seed to complete germination under conditions that are favorable for germination, is an adaptive trait optimizing germination to the best suitable time that enables the seed to complete its life cycle. In temperate climates, delaying germination until spring can prevent winter mortality. However, autumn germination may have a selective advantage when the risk of winter mortality is low by enabling the plants to flower earlier or at a larger size the following year (Donohue, 2002).

Changes in seed dormancy were selected during the domestication of crop plants from wild-plant species, because some features of dormancy that provide ecological advantages presented agronomic disadvantages within a farmed system. At one extreme, many weed species show very high levels of dormancy when shed from the plant. This causes infestation of land that subsequently requires long-term treatment to remove succeeding generations (Basu *et al.*, 2004). Crop plants might have only low levels of dormancy that provide a benefit for early and uniform seedling establishment. However, this has the disadvantage of possible preharvest sprouting (PHS), germination before harvest that reduces the quality of the seed. This is a problem especially for cereals when cool and damp conditions occur prior to harvest (Holdsworth *et al.*, 2006). For specific uses, such as barley (*Hordeum vulgare*) for malting, low dormancy is desired although this increases the risk of PHS. Therefore, a moderate dormancy just sufficient to prevent PHS is usually desired for this trait in crop plants (Ullrich *et al.*, 1997; Li *et al.*, 2003; Gubler *et al.*, 2005).

Seed dormancy varies widely among seed batches, even for seeds of the same genotype, indicating large environmental influences. This, together with other factors such as polygenic inheritance, makes dormancy a typical quantitative trait, which is often described as deep or shallow, or strong or weak, dormancy. Two dormancy types can be distinguished by the time of their induction. Primary dormancy is induced during seed development and secondary dormancy during imbibition of mature seeds. Furthermore, there is a distinction between embryo and coat (testa, hull, and/or endosperm)-imposed dormancy. The germination characteristics of seeds also depend on their physiological quality or vigor (germination potential) in addition to their dormancy status.

For a better understanding of seed dormancy and germination it is important (but difficult) to distinguish between dormancy, which is assumed to be induced during seed maturation on the mother plant, and germination. These are two different physiological states or processes that depend on each other. Germination can occur only after dormancy is released, and the release of dormancy is visible only when germination takes place. In genetic terminology, dormancy is epistatic to germination. Until we have clear markers that distinguish dormancy (either embryo- or coat-imposed) and germination potential *per se*, researchers must mainly rely on germination tests for their analyses of dormancy.

Germination *sensu stricto* is defined as the events following imbibition until visible embryo protrusion. The completion of germination depends on embryo expansion mainly due to cell elongation driven by water uptake. After radicle protrusion, seedling establishment takes place, which can be considered as a separate process from germination.

Germination assays are a measure of the integration of many events that happened in the history of the seed on the mother plant and the various environmental factors encountered during seed storage and germination. Dormancy is a property of the seed and the degree of dormancy defines which conditions should be met to allow the seeds to germinate. Therefore, a more precise method of defining the dormancy status of a seed batch requires a description of the environmental necessities for germination (temperature range or time of after-ripening required to overcome dormancy). For genetic studies, it is essential to obtain such quantitative data on seed material without the effects of variation in environmental conditions, including the maternal growing and seed storage conditions. In general, dormancy is assayed by a seed germination test under well-defined conditions. These conditions refer to the incubation medium, light, and temperature conditions during the assay. It is essential to find those conditions that discriminate maximally between the parents. This implies, for example, that when comparing nondormant with weakly dormant genotypes one often uses freshly harvested seeds and germination conditions that may not be optimal for germination, such as low or high temperatures. The comparison of weakly with strongly dormant genotypes will need the use of stored seeds and more promotive temperature conditions.

This chapter begins with a description of the genetic approach to understand seed dormancy using mutants in *Arabidopsis* and other plant species. Thereafter it continues with an overview of studies that analyze natural variation for seed dormancy in both *Arabidopsis* and other species with well-developed genetics and linkage maps (mainly grasses). Finally, an overview is given of the insights into seed dormancy and germination that resulted from these genetic approaches.

5.2 Mutant approaches in *Arabidopsis*

Mutants with defects in seed dormancy and germination provide important tools for understanding these processes because they directly identify the genes involved, nowadays often down to the molecular level. Most genetic variants that have been studied were those with dormancy or germination effects that were part of a pleiotropic phenotype, whereby other traits (e.g., dwarfness in gibberellin [GA]-deficient mutants or reduced stomatal closure for abscisic acid [ABA]-deficient mutants) were originally used for genetic analysis. This pleiotropism can be expected as these hormones have additional functions besides their roles in seed dormancy and germination. However, several mutants have been isolated that seem to be specific for seed germination, for instance the *dog1* (*delay of germination 1*) mutant (Bentsink, 2002).

Nondormant mutants have been selected directly by identifying germinating seeds within nongerminating dormant batches of wild-type seeds (Léon Kloosterziel *et al.*, 1996b; Peeters *et al.*, 2002; L. Bentsink, unpublished results). Another approach would be to select for mutants with increased dormancy levels. Screens for altered dormancy phenotypes are hampered by the variability of the germination trait, which may lead to genetic misclassification of individual seeds and therefore to false positives in mutant screens. The *Arabidopsis* genotypes Landsberg *erecta* (L*er*) and Columbia (Col), which are mostly used in *Arabidopsis* research, show only a low level of dormancy. This dormancy disappears after approximately 1 month of after-ripening (van der Schaar *et al.*, 1997). Despite this low level of dormancy, a mutant screen in L*er* for reduced dormancy yielded several mutants (*reduced dormancy 1-4*; *rdo1-4*) that also showed mild pleiotropic phenotypes (Léon-Kloosterziel *et al.*, 1996b; Peeters *et al.*, 2002). Because of the relatively low level of dormancy in L*er* and Col, it is impossible to saturate the mutations in dormancy genes in these accessions. However, this problem can be overcome by the use of more dormant accessions such as Cape Verde Islands (Cvi) (Koornneef *et al.*, 2000).

The difference between mutants with enhanced germination and wild type can also be increased by the selection of mutants under conditions that are less favorable for germination or that extend the germination period. This procedure was applied by Salaita *et al.* (2005), who selected for faster germination at 10°C. Whether the mutants identified specifically affect some component of the germination process or dormancy/germination regulation in general is not clear as yet.

Selection of mutants with reduced dormancy or increased germination in the presence of germination-inhibiting compounds such as ABA is very effective. A number of *ABA-insensitive* (*abi*) mutants were isolated using this procedure (Koornneef *et al.*, 1989; Finkelstein, 1994). On the other hand, the selection of nongerminating seeds in the presence of low, normally not inhibitory, concentrations of ABA resulted in so-called *enhanced response to ABA* (*era*) mutants (Cutler *et al.*, 1996) or *ABA-hypersensitive germination* (*ahg*) mutants (Nishimura *et al.*, 2004). The use of GA biosynthesis inhibitors such as paclobutrazol identified mutants in ABA biosynthesis (e.g., *aba2* and *aba3*; Léon-Kloosterziel *et al.*, 1996a) and ABA signal transduction (e.g., *abi3*; Nambara *et al.*, 1992) in addition to GA signal transduction mutants such as *spindly* (*spy*) (Jacobsen *et al.*, 1996). The same ABA biosynthesis mutants were found during the selection of revertants of nongerminating GA biosynthesis mutants (Koornneef *et al.*, 1982). Sugars, such as sucrose and various hexoses, inhibit seed germination independently of their osmotic effects (Pego *et al.*, 2000). Mutants that were insensitive to the inhibitory effect of glucose and sucrose were isolated by several groups and some appeared to be defective in ABA biosynthesis or appeared to belong to the ABA-insensitive mutants. The *sugar-insensitive sis4* and *sis5* mutants are allelic to respectively *aba2* and *abi4* (Laby *et al.*, 2000) and *sucrose uncoupled 6* (*sun6*) is an *abi4* allele (Huijser *et al.*, 2000). These results indicate that germination is controlled in an ABA- and sugar-dependent way (see Chapter 12). Garciarrubio *et al.* (1997) showed that the addition of sugars and amino acids allowed the seeds to germinate in otherwise inhibitory concentrations of ABA and suggested that ABA inhibits the mobilization of food reserves. However, it cannot be excluded that these sugar effects are mediated by sugar-signaling effects (Smeekens, 1998; Gibson and Graham, 1999). Penfield *et al.* (2004) showed that this inhibition of reserve mobilization by ABA occurs only in the embryo but not in the endosperm, which fuels hypocotyl elongation in darkness.

Mutant screens can also be performed by making use of reporter genes driven by promoters that are known to be involved in a pathway affecting germination. Mutants that affect the reporter genes can thereafter be checked for their germination phenotype. This approach has been used in genetic screens based on seed maturation and ABA-regulated reporter genes. Carles *et al.* (2002) used the seed-specific *Em* (*EARLY METHIONINE-LABELLED*) promoter and identified an additional allele of *ABI5*. Screens with ABA-induced promoters in vegetative tissues revealed two additional ABA-related mutants. The *ade1* (*ABA-deregulated gene expression*) mutation enhances gene expression in response to ABA (Foster and Chua, 1999) and *hos5* (*high expression of osmotically responsive gene 5*) displays an increased sensitivity of gene expression to ABA and osmotic stress (Xiong *et al.*, 1999). The *ade1* and *hos5* mutations, though, have little effect on seed germination. In addition, such screens have also identified ABA biosynthesis mutants such as *los6* (*low expression of osmotically responsive gene 6/aba1*) and *los5* (*aba3*) (Xiong *et al.*, 2001, 2002), which have reduced dormancy.

A different and more unbiased strategy, aiming specifically at seed germination mutants, makes use of enhancer or promoter trap transgenic lines where the randomly inserted reporter gene can be driven by a neighboring enhancer or promoter sequence that can have a germination-specific expression pattern. Using this procedure, Dubreucq *et al.* (2000) identified an extensin-like gene that is expressed in the endosperm during germination at the site of radicle protrusion. A similar approach identified the *mediator of ABA-regulated dormancy* (*MARD1*) gene as being involved in dormancy/germination (He and Gan, 2004). A T-DNA insertion in the promoter of this gene resulted in faster germination, germination in darkness, and a slight increase in ABA resistance. Liu *et al.* (2005) recently described a large set of such reporter inserts and emphasized that this approach is feasible for the GUS reporter gene only after rupture of the testa to allow penetration of the GUS substrate (see Chapter 11). Knockout mutants of genes with germination- or dormancy-related expression obtained from enhancer or gene trap screens do not always show phenotypes related to dormancy or germination. This can imply that the disrupted genes are not required for germination or that other redundant genes can provide a similar function.

When redundancy is a cause for the absence of mutant phenotypes in a dysfunctional gene, one needs to study double mutants. This approach is well illustrated in the identification of the GA signal transduction proteins of the DELLA family that control germination (Tyler *et al.*, 2004; Cao *et al.*, 2005). The study of single, double, and triple mutants combined with the GA-deficient *ga1-3* mutant and germination experiments in light and darkness allowed the conclusion that RGL2 (RGA [REPRESSOR OF *gal-3*]-LIKE2) is a predominant repressor of seed germination, which is degraded by the ubiquitin-26S proteasome pathway activated by GA (see Chapter 10). However, additional DELLA proteins (GAI [GA-INSENSITIVE], RGA, RGL1) enhance the function of RGL2 and their stability may be affected by light, implying that GA synthesis is induced not only by light but also by GA sensitivity (Cao *et al.*, 2005).

For *Arabidopsis*, knockouts for nearly every gene can be ordered from the stock centers and an increasing number of studies are published in which knockouts were tested for a germination phenotype. T-DNA knockouts of two DOF transcription factors (*DOF AFFECTING GERMINATION1 [DAG1]* and *DAG2*) revealed a seed germination phenotype that is determined by the maternal genotype. This is in agreement with expression of the gene in the vascular tissue of the funiculus, which is a maternal tissue connected to developing seeds (Papi *et al.*, 2000). *DAG1* and *DAG2* are the first reported genes to be specifically involved in the maternal control of seed germination, other than by affecting the maternal seed structure. Surprisingly, these two related, very similar genes have opposite effects on germination; *DAG1* increases dormancy and *DAG2* reduces it (Papi *et al.*, 2000; Gualberti *et al.*, 2002). Another example of a knockout mutant with a dormancy phenotype is *gpa1* (*G protein in Arabidopsis*), which encodes a heterotrimeric G protein (Ullah *et al.*, 2002). This is also an example of germination phenotypes that are expressed only under very specific conditions. Probably heterotrimeric G proteins affect GA signal transduction, as overexpression of *GCR1*, a putative G protein-coupled receptor gene, which is suggested to be involved in the GA signaling pathway, also shows reduced dormancy (Colucci *et al.*, 2002). A similar small effect was observed due to mutation in an arabinoglactan protein where faster germination than wild type was observed in seeds immediately after harvest or stratified in an ABA-containing medium (van Hengel and Roberts, 2003). While these analyses suggest a role in germination, in view of the mild phenotype these are not essential players, although functional redundancy can be an alternative explanation for small-effect phenotypes.

Since a comparison of a limited number of genotypes in several environments allows the discovery of relatively small differences in seed germination behavior, the outcome of these studies implies that the genes are involved in germination. However, their effects on germination phenotype can be limited; a better quantification could be achieved by comparison to mutants exhibiting stronger seed germination phenotypes such as ABA and GA mutants in the same experiment.

An important additional source of mutants to understand dormancy comes from mutants that were originally isolated in screens that did not make use of germination tests. An example of testing known mutants for their effects on germination is the use of phytochrome mutants. The germination of *Arabidopsis* seeds is under phytochrome-mediated photocontrol. It was therefore expected that phytochrome-deficient mutants would be affected in seed germination. Detailed action spectra for seed germination performed in wild type and *phyA* and *phyB* mutants revealed a typical red/far-red-reversible response. The low fluence response was mediated by PHYB, whereas the germination response mediated by PHYA turned out to be a very low fluence response with a 10^4-fold higher sensitivity to light (Shinomura *et al.*, 1996). The observation that *phyA phyB* double mutants show some light-dependent germination indicates the involvement of additional photoreceptors (Yang *et al.*, 1995; Poppe and Schäfer, 1997).

Testa mutants in *Arabidopsis* were originally identified based on their altered seed color or shape. The analysis of these mutants provided strong evidence for the importance of the testa in the control of seed germination (Debeaujon *et al.*, 2000; reviewed in Chapter 2).

Another important factor that is known to promote germination is nitrate. With the help of nitrate reductase-deficient mutants, it was shown that nitrate acts via signaling and not via nutritional effects. This was concluded because these mutant plants, which show N-deficiency symptoms but accumulate nitrate, produced seeds that were less dormant than wild-type seeds (Alboresi *et al.*, 2005).

Brassinosteroids (BRs) are a group of over 40 naturally occurring plant steroid hormones found in a wide variety of plant species (Clouse and Sasse, 1998; Schumacher and Chory, 2000). BRs are also involved in the control of germination in *Arabidopsis*. It is suggested that the BR signal is required to reverse ABA-induced dormancy and stimulate germination (Steber and McCourt, 2001). BRs overcome the lack of germination in the *sleepy1* (*sly1*) mutant, probably by bypassing its GA requirement (see Chapter 10). Two BR mutants (*det2* [*deetiolated2*] and *bri1* [*BR-insensitive1*]) showed reduced germination rates (speed) but eventually germinated without BR, indicating that in contrast to GAs, BRs are not absolutely required for germination (Steber and McCourt, 2001).

Ethylene mimics the action of GAs, as the complete germination of seeds from the GA-deficient mutant (which germinate only in the presence of GA) occurs in the presence of ethylene (Karssen *et al.*, 1989). Ethylene-insensitive mutants such as *etr1* (*ethylene response1*) and *ein2* (*ethylene insensitive2*) are more dormant (Chiwocha *et al.*, 2005) or, when nondormant, require more time for germination than wild-type seeds (Siriwitayawan *et al.*, 2003). Ethylene mutants show phenotypes that resemble ABA- and sugar-signaling mutants and were identified in mutant screens that aimed at the discovery of ABA signal transduction mutants (Beaudoin *et al.*, 2000; Ghassemian *et al.*, 2000).

5.3 Mutant approaches in other species

Apart from *Arabidopsis*, mutants affecting dormancy and germination have also been isolated in other species. In maize, *viviparous* (*vp*) mutants have been known for many years (reviewed in McCarty, 1995). The best studied is the ABA-insensitive *vp1* mutant, which encodes a B3 domain seed-specific transcription factor, orthologous to the *Arabidopsis ABI3*. Several other *vp* mutants were found to have mutations in genes encoding ABA biosynthesis enzymes. Among these, the *vp14* mutant was instrumental to isolate the *NCED* (*9-CIS-EPOXYCAROTENOID DIOXYGENASE*) gene, which is responsible for the cleavage of *cis*-xanthophylls (Tan *et al.*, 1997; Nambara and Marion-Poll, 2005; see Chapter 9).

In rice (*Oryza sativa*), studies on dormancy mutants are limited (Kurata *et al.*, 2005). In a screen for vivipary in a population with tissue culture-activated *Tos17* transposons, a mutation was found in the gene encoding the ABA biosynthesis enzyme zeaxanthin epoxidase, similar to the *Arabidopsis aba1* mutant (Agrawal *et al.*, 2001). In this screen, 1400 lines with a viviparous phenotype were reported out of 30 000 regenerated plants, suggesting that many more loci might be detected in a more systematic analysis of this mutant group. Although these studies suggest that many nondormant mutants can be found in rice, none of them have been characterized in detail. It is not known whether there are any nondormant mutants that affect genes not directly related to plant hormones. In barley, a nondormant mutant with no obvious pleiotropic effects was isolated from the cultivar Triumph and the locus was mapped in the centromere region of chromosome 6 (Romagosa *et al.*, 2001, Prada *et al.*, 2005).

In tomato (*Lycopersicon esculentum*), mutants affecting seed germination have been isolated and characterized extensively. These are various ABA and GA biosynthesis mutants whose phenotypes all correspond to the phenotypes of similar mutants in *Arabidopsis*. These mutants have been used extensively to investigate seed germination in tomato where weakening of the micropylar region of the endosperm is a requirement for germination (reviewed by Hilhorst *et al.*, 1998; see Chapter 11).

5.4 Genetic analyses of natural variation

The genetics of dormancy has been studied in many plant species and demonstrated the existence of heritable variation. However, only when specific loci are identified by mapping of mutants and/or quantitative trait loci (QTL) can a follow-up analysis using molecular biology identify the molecular basis of the genetic variation. This means that only species with well-developed genetic and linkage maps are accessible to this analysis, including *Arabidopsis* and several species that belong to the Gramineae.

Genetic studies on dormancy have benefited much from the development of quantitative genetics. The analysis of QTL allows the identification of the number of loci differing between two genotypes, their map positions, and the relative contribution of each QTL to the total variance for the trait. Germination is a variable trait that requires analysis of a high number of seeds. Therefore, the use of so-called permanent or immortal mapping populations is attractive, where a single (often homozygous) genotype can produce many seeds that can be tested in multiple environments (Alonso Blanco and Koornneef, 2000). The molecular identification (cloning) of the underlying genes that show allelic variation is still a major challenge, but the first results are now becoming available. The genes identified in the study of natural variation can be the same as those identified in mutant screens. However, there are several reasons why this is not always the case. Firstly, the parent lines used for mutation experiments can be mutated in specific genes, e.g., those promoting seed dormancy. This could be the case for L*er* and Col in *Arabidopsis*, where many natural accessions show much stronger seed dormancy than these commonly used laboratory accessions. Furthermore, mutants that show strong pleiotropic effects, such as most ABA mutants, will not survive in nature. Therefore, genes identified by analyzing natural variation are expected to be ecologically relevant.

5.4.1 Genetic analysis of natural variation in Arabidopsis

For *Arabidopsis*, QTL analysis for seed dormancy has been conducted using several recombinant inbred line (RIL) populations (van der Schaar *et al.*, 1997; Alonso-Blanco *et al.*, 2003; Clerkx *et al.*, 2004; L. Bentsink *et al.*, unpublished results). These RIL populations have been made by crossing accessions that have different levels of seed dormancy compared to the standard laboratory accession L*er*. Thus far these analyses have identified more than 12 regions in the *Arabidopsis* genome where QTL associated with dormancy are located (Alonso-Blanco *et al.*, 2003, Clerkx *et al.*, 2004; L. Bentsink *et al.*, unpublished results). These QTL have been called *DOG* (*Delay of Germination*) and can be divided into different groups. Some QTL are detected only in the low-dormancy accessions, whereas others are identified as major QTL in the moderate and strongly dormant accessions. Many of these do not colocate with known dormancy loci (Figure 5.1). In the analysis of van der Schaar *et al.* (1997), the lines were grown in three independent greenhouse experiments and tested in different germination conditions. Despite little difference in dormancy between the parental lines, altogether 14 QTL were identified, of which most had only small effects. Many of these were detected only in specific germination conditions or in seeds from one or two harvest dates, indicating the complexity of genetic control. The feasibility to clone the genes underlying major seed dormancy QTL has been demonstrated for the *DOG1* locus of which L*er* contains a weak allele and the dormant accession Cvi contains a strong allele, as was shown by analyses of near-isogenic lines (NILs) (Alonso-Blanco *et al.*, 2003). *DOG1* was fine-mapped, using the progeny of a cross between the NIL *DOG17-1* and the less dormant L*er* genotype. This indicated that the gene was in a region containing 22 candidate genes. The *DOG1* gene was isolated by the identification of nondormant loss-of-function mutants in one of these 22 candidates. This gene, which is expressed during seed development and downregulated during imbibition, encodes a protein with unknown molecular function. In view of the strong dormant phenotypes observed, it is assumed to be a major and specific regulator of this process (L. Bentsink, unpublished data). The cloning of additional QTL and the study of lines containing strong dormancy alleles that have been introgressed from other accessions into a L*er* background are ongoing and should identify additional genes involved in dormancy induction, maintenance, and breakage.

5.4.2 Natural variation for dormancy in grasses

Gramineae species in which the genetics of dormancy have been studied are the important cereals rice, barley, wheat (*Triticum aestivum*), and sorghum (*Sorghum*

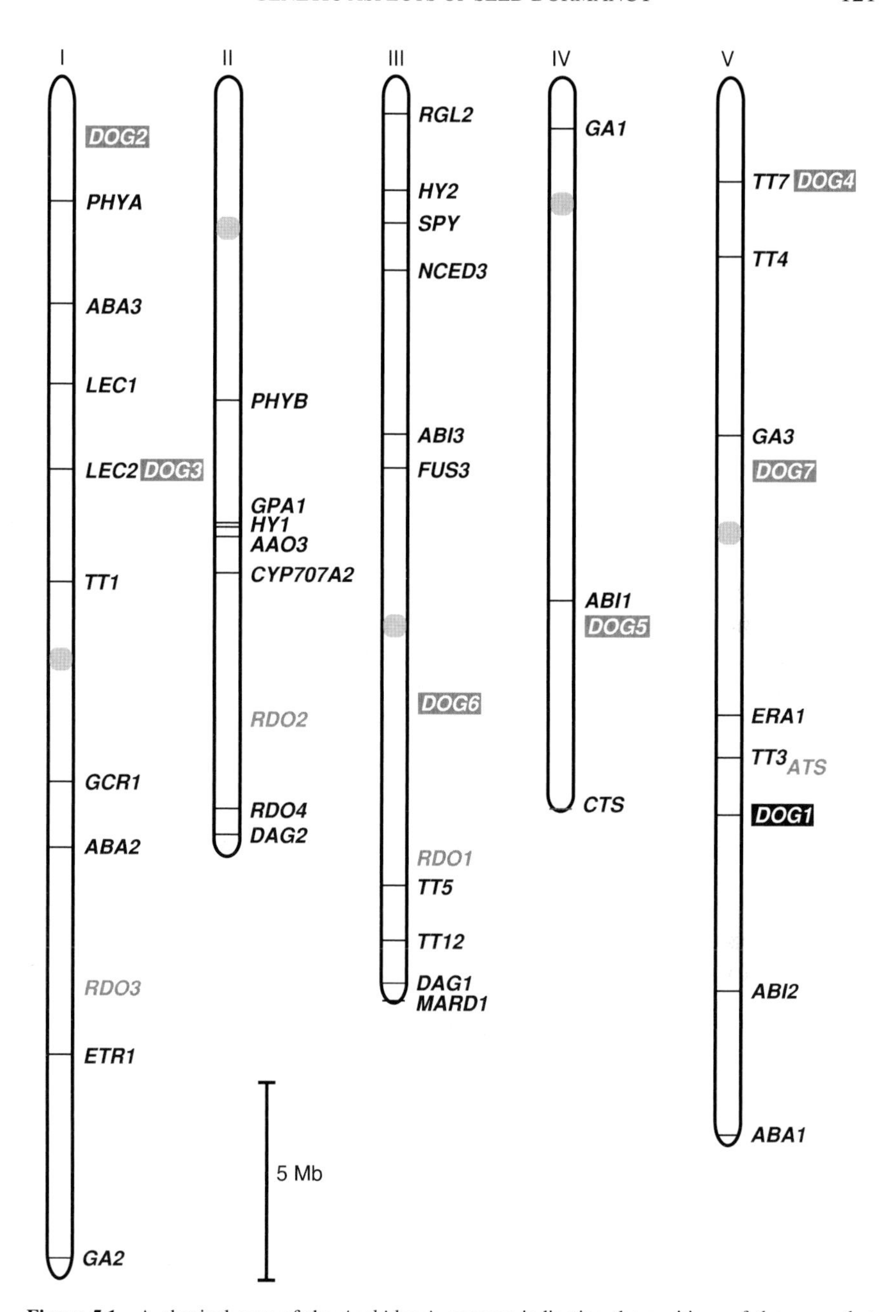

Figure 5.1 A physical map of the *Arabidopsis* genome indicating the positions of the genes that play roles in dormancy and germination. Cloned genes are shown in black and their position on the chromosome is indicated with a horizontal line; the exact positions of the gray colored genes are not yet known and QTL are boxed.

bicolor). The trait has also been analyzed in the weed species *Avena fatua* (wild oat). Some studies, especially in barley and rice, made use of combinations of wild and cultivated genotypes.

In the domesticated species, low levels of dormancy may result in PHS but could be required for efficient malting (Li *et al.*, 2003) and uniform germination. Nonetheless, despite selection for low dormancy, residual differences among cultivars are present and have been analyzed. However, due to the low dormancy and short after-ripening requirement of most cultivated species, many QTL are not observed in all environments and cannot be considered as confirmed QTL. Therefore, estimates of the total number of QTL are not always relevant. In this chapter we mainly focus on studies that identified major genes. Such loci can be expected to be amenable for cloning and could advance our knowledge of dormancy mechanisms. However, knowledge of the positions of minor QTL may also be useful for marker-assisted breeding.

Barley is of particular interest because the two major uses of this crop have different dormancy requirements. Barley production and end-use quality require a certain level of seed dormancy, and low or no seed dormancy can lead to PHS and low malting quality. High seed dormancy can result in nonuniformity of seed germination, causing field stand establishment problems and low malt extract yield (Ullrich *et al.*, 1997). Generally, a moderate level of seed dormancy is thought to be appropriate for barley cultivars (Han *et al.*, 1999; Romagosa *et al.*, 1999). A QTL analysis in a doubled haploid population derived from a cross between the dormant fodder variety Steptoe and the nondormant malting variety Morex was very informative (Gao *et al.*, 2003). In addition, several other cultivar combinations were analyzed, including the cross of Morex with the relatively dormant European malting variety Triumph. In both populations a QTL called *SD1*, located near the centromere of chromosome 5H, explained 50% or more of the phenotypic variation in all environments tested. This QTL region probably represents three closely linked loci in a region of 4.4 cM (Han *et al.*, 1999). In crosses involving the cultivar Harrington most of the variation was explained by a QTL near the telomere of 5HL (Li *et al.*, 2003). The latter locus was identified as *SD2* in the Steptoe/Morex (S/M) population and also in the Triumph/Morex (T/M) cross (Prada *et al.*, 2004). However, in the latter cross, the more dormant allele came from Morex, which provided the nondormant allele in the S/M cross, suggesting that this locus may not be identical to *SD2* (Ullrich *et al.*, 2002). Li *et al.* (2004) compared the syntenic regions of barley, wheat, and rice around the *SD2* locus, making use of the available EST sequence data, and suggested that a *GA20-oxidase* might be a candidate gene for this QTL. In addition, minor QTL were detected on chromosomes 7H (*SD3*), 4H (*SD4*), 1H, 2H, and 3H, but these were not detected in all environments and in all studied crosses.

Larger differences in dormancy exist between cultivars and wild-type accessions. In crosses of the cultivar Mono with a desert-derived dormant barley (*H. spontaneum*) accession, eight QTL regions were detected, of which five regions overlapped with the QTL regions identified in cultivar × cultivar crosses, including the *SD1* and *SD2* regions on 5H. In addition, three new regions were identified (Zhang *et al.*,

2005). Although more dormant, the wild barley accession also harbored alleles that result in reduced dormancy.

In rice, only low levels of dormancy variation are segregating in many cultivar crosses. Despite this, five QTL could be identified in the frequently studied Nipponbare × Kasalath population (Lin *et al.*, 1998). One locus called *sdr1*, located at the lower part of chromosome 3, explained almost 25% of the phenotypic variation. Its colocation with the *Hd8* flowering time locus appeared to be due to linkage, since the two loci could be separated by fine-mapping. The cloning of this locus is currently underway (M. Yano, personal communication). An analysis of the same population in the same experimental field, but in a different year, also yielded five QTL (Miura *et al.*, 2002). However, only two of these overlapped with the ones reported by Lin *et al.* (1998), indicating the strong interaction of dormancy QTL with the environment that the maternal plant experiences during seed development. The map position of *sdr1* seemed to be identical to those of the barley and wheat dormancy QTL on the barley 4HS (*SD4*) and the wheat 4AL chromosomes, respectively. However, this conclusion is somewhat ambiguous because of the complex rearrangements in that region of rice chromosomes 3 and 11 (Li *et al.*, 2004).

In an extensive analysis of a RIL population derived from a cross of an indica cultivar with a common wild rice accession (*O. rufipogon*), Cai and Morishima (2000) identified 17 QTL, with the alleles from the wild parent increasing dormancy. Several of these loci were also found in studies involving only cultivars. A study by Gu *et al.* (2004), involving the cross between an indica cultivar and weedy rice, identified five dormancy QTL that were reproducible across years (Gu *et al.*, 2004, 2005). The major QTL, explaining around 20% of the variation in a BC1 and up to 50% in a BC4, was found on chromosome 12 (*qSD12*) (Gu *et al.*, 2005). This map position did not colocate with a major QTL identified by Cai and Morishima (2000), indicating that variation in the genetic control is present among wild and weedy rice genotypes. It is of interest that this locus is the only one that does not colocate with other domestication-related QTL such as shattering, awn length, black hull color, or red pericarp color (Gu *et al.*, 2005). Such colocation may be due to very tightly linked genes or due to pleiotropism, as is suggested for the red pericarp locus (*Rc*), which is orthologous to the major loci for seed dormancy in wheat and may be related to the structure of the tissue surrounding the embryo (hull-imposed dormancy). This linkage of dormancy to morphological domestication characters might have enhanced the removal of genetic variation in the cultivated germplasm pool.

Bread wheat is an allohexaploid species consisting of three genomes (A, B, and D) that have homeologous colinear chromosomes. PHS has been extensively studied in wheat by the analysis of premature germination on ears kept at high humidity and by α-amylase activity measurements. The α-amylase activity is related to dormancy, but also to water uptake and drying, which explains why the presence of awns (controlled by the *Hd* loci) and epicuticular wax deposition loci affect these traits (King and Wettstein-Knowles, 2000). Although higher α-amylase activity is correlated with the progress of germination (indicating low dormancy), late maturity α-amylase activity appears independently of germination (Flintham, 2000).

A major component of dormancy (and PHS) in wheat is pericarp color, which is controlled by the Red grain color (*R*) loci (*R-A1*, *R-B1* and *R-D1*) located on group 5 chromosomes (Flintham, 2000; Groos *et al.*, 2002). Dominant alleles promote expression of red pigments (phlobaphenes) within the maternally derived pericarp that tightly surrounds the embryo. It is assumed that the *R* genes themselves provide increased dormancy (Flintham *et al.*, 2002). It has now been shown that these genes encode MYB-type transcription factors controlling flavonoid biosynthesis in a seed-specific manner. These *MYB* genes are similar to the *TT2* gene in *Arabidopsis*, where *tt2* mutants also affect seed dormancy (Debeaujon *et al.*, 2000). Even among the red varieties, additional dormancy loci can be found and used to reduce PHS. Although the *R* loci can be used to provide some resistance to PHS, they also provide negative characters that reduce the quality of the flour for certain markets. These include color (red wheat seeds produce colored flour, which is perceived negatively by the noodle-making industry), production energy costs in food processing, and taste perception. White wheat contains no active *R* loci, and is therefore in general more prone to PHS. Alternative loci conferring increased dormancy are needed for white-grained wheat.

Many QTL have been detected for PHS in various studies (reviewed by Flintham *et al.*, 2002) and their map positions have been compared among the cereals (Gale *et al.*, 2002). Although many of these have been reported only in specific crosses and environments, some QTL were confirmed across species. Mares *et al.* (2002) identified several dormancy QTL, including one on chromosome 3D which maps closely to the *VP1* homologue. A major QTL on chromosome 4AL, named *PHS*, was identified in several mapping populations (Kato *et al.*, 2001; Flintham *et al.*, 2002; Torada *et al.*, 2005) and could be homologous to the barley *SD4* locus. Other studies comparing dormant and nondormant seeds of varieties or induced mutants demonstrated large differences in responsiveness of embryos to ABA (Walker-Simmons, 1987; Kawakami *et al.*, 1997). Therefore, it is likely that sensitivity to ABA at specific developmental periods, together with the induction of desiccation, is a key process related to dormancy induction. The chromosomal locations of loci that influence wheat embryo sensitivity to ABA and dormancy have been investigated (Noda *et al.*, 2002). A cross between wheat cultivar Chinese Spring (nondormant and ABA-insensitive) and line Kitakei-1354 (dormant and ABA-sensitive) demonstrated the importance of loci located at 4AL and 2D for these characters.

Genetic studies in sorghum identified two major dormancy QTL (Lijavetzky *et al.*, 2000). The low dormancy in the PHS-sensitive parent is characterized by low embryonic ABA sensitivity and higher GA levels (Steinbach *et al.*, 1997; Perez-Flores *et al.*, 2003), which correlated with higher levels of a GA20-oxidase (Perez-Flores *et al.*, 2003). The higher level of *VP1* transcript that was maintained in imbibed seeds of the dormant line suggested a relationship between dormancy and *VP1* expression. However, no PHS QTL colocated with the *VP1* locus. This could mean that the difference in *VP1* expression is only a downstream effect of the dormancy QTL (Carrari *et al.*, 2003).

Wild oat is a noxious weed because dormant seeds may persist several years in agricultural soils. Genetic variation for seed dormancy has been reported in this species and at least three loci were identified (Fennimore *et al.*, 1999). Two RAPD markers were identified that showed linkage with the dormancy trait but these markers have not yet been integrated into the cereal linkage maps.

From the studies reviewed above, it is evident that wild relatives of cultivated plants are rich sources of dormant alleles, although residual variation is present even within cultivated species. Genetic synteny, especially in cereals, is expected to help identify dormancy loci in different species.

5.5 What do the genetics teach us about dormancy and germination?

The identification of genetic variation in seed dormancy can assist in understanding this important process. The number of genes that are known to be involved in dormancy or germination is rapidly increasing. Figure 5.1 shows the map positions of the genes that play a role in dormancy and germination in *Arabidopsis*, based on genetic evidence. The identification of the molecular and biochemical functions of these genes will increase our knowledge about the relevant processes. However, even before the respective genes were cloned, the study of plant hormone mutants in *Arabidopsis* and tomato showed in a convincing way the importance of ABA and GA for regulating seed germination. For mutants that show only a dormancy phenotype, cloning of the respective genes can reveal novel processes that are involved in dormancy.

Genetic analyses are more efficient in *Arabidopsis* and rice, compared to other plant species, because of the availability of many genetic resources, including mutations in almost every gene and access to the full genome sequence. In addition, the complete DNA sequence allows the application of '-omic' approaches, including full genome gene expression analysis and efficient proteomics. Therefore, it is not surprising that biological research is focusing on *Arabidopsis* and rice. The identification of mutants and QTL associated with dormancy will ultimately extend this knowledge to crop plants where genomics approaches are also becoming available. Until such genes are cloned, QTL analyses can identify genomic regions that can be followed in breeding programs using marker-assisted selection.

Studies using the available mutants that are involved in the GA biosynthetic pathway, especially in *Arabidopsis* and also in tomato, together with GA biosynthesis inhibitors, have shown that GAs play a major role as a promotive agent for germination. The role of GA during *Arabidopsis* seed development seems limited, since normal seeds are formed even on rather extreme dwarf mutants. Arguments that GAs play their major role only during seed germination come from a combination of proteomic and mutant analyses showing that imbibed *ga1* mutant seeds are not different from wild-type seeds until the moment of radicle protrusion when they show lack of germination (Gallardo *et al.*, 2002). This is in agreement with a careful analysis of the induction of GA biosynthesis and with gene expression studies (Ogawa *et al.*, 2003). GA biosynthesis occurs after imbibition because germination can be inhibited at that time by GA biosynthesis inhibitors. These studies all indicated that *de novo* synthesis of GA after seed maturation is required for germination.

Mutant approaches have shown that *RGL2*, one of the DELLA proteins that acts as a repressor of GA action and is degraded in a GA-inducible manner, is the major repressor acting during seed germination (Tyler *et al.*, 2004; Cao *et al.*, 2005; see Chapter 10). Possibly, other signal transduction pathway components involve trimeric G proteins and SPY.

ABA is a potent inhibitor of germination as seen in the strong phenotypes of many ABA-related mutants. ABA most likely plays roles in the induction of dormancy during seed maturation and the maintenance of dormancy during imbibition. Reciprocal crosses between wild type and the ABA-deficient *aba1* mutants showed that dormancy is controlled by the ABA genotype of the embryo and not by that of the mother plant. However, the mother plant is still responsible for the relatively high ABA levels found in seeds midway through seed development (Karssen *et al.*, 1983). At this time, ABA may prevent precocious germination, as shown by the maternal ABA effects in the extreme *aba abi3-1* double mutants (Koornneef *et al.*, 1989). Since ABA levels decrease in wild-type seeds at the end of maturation, it was proposed that endogenous ABA is not required for dormancy maintenance after its onset during seed development (Karssen *et al.*, 1983). In contrast, the observation that ABA biosynthesis inhibitors such as norfluorazon promote germination (Debeaujon and Koornneef, 2000) indicates that the maintenance of dormancy in imbibed seeds is an active process involving *de novo* ABA synthesis (Ali-Rachedi *et al.*, 2004). ABA catabolism is also involved in reducing ABA levels during imbibition, as is shown by the reduced germination of knockout mutants of the *CYP707A* gene, which encodes an ABA 8′-hydroxylase (Nambara and Marion-Poll, 2005; see Chapter 9).

Genetic approaches revealed interesting relationships among ethylene, ABA, and sugar signaling. *Ein* mutants show increased dormancy and an ethylene-hypersensitive *constitutive triple response* (*ctr1*) mutant resembles ABA- and sugar-signaling mutants (Beaudoin *et al.*, 2000; Ghassemian *et al.*, 2000). These observations, in combination with the nondormant phenotype of the *ein2 abi3-4* double mutant, indicated that ethylene negatively regulates seed dormancy by inhibiting ABA action (Beaudoin *et al.*, 2000). However, Ghassemian *et al.* (2000) suggested that in addition to signaling, *EIN2* may also affect ABA biosynthesis. Detailed analyses of plant hormone levels in the *etr1-2* mutant by Chiwocha *et al.* (2005) indicated that the higher ABA levels in the mutant are probably caused by an effect on ABA catabolism. Auxin, cytokinin, and GA levels were also affected, indicating the complexity of cross-talk among plant hormones (see Chapter 8). The presence of cross-talk between sugar signaling and ethylene was suggested by the sugar-insensitive phenotype of *ctr1*, which has a constitutive response to ethylene (Gibson *et al.*, 2001) and the sugar-hypersensitive phenotype of *etr* (Zhou *et al.*, 1998). Apparently, strong interactions among ABA, ethylene, and sugar signaling are present during germination and early seedling growth.

The involvement of these hormone biosynthesis and sensitivity genes in dormancy is not a surprise, since the application of different hormones and biosynthesis inhibitors already provided strong indications for the roles of these hormones. However, mutants have helped refine these roles and also identified several of the hormone signaling factors. It is expected that many discoveries will still be made by the identification of genes with clear effects on dormancy and germination that do not have specific hormone-related phenotypes. Examples of these are the many dormancy QTL that have been identified, although some might be hormone-related, as suggested in barley (Li *et al.*, 2004). Cloning of the first pleiotropic *RDO* genes indicated that these genes may be involved in cellular functions that are not specific to dormancy and germination (Y. Liu, personal communication).

Microarray and proteomic data on germination using GA-deficient mutants have been published (Gallardo *et al.*, 2002; Ogawa *et al.*, 2003). These studies, together with the use of dormancy-specific mutants and their complemention with wild-type genes under the control of tissue-specific or inducible promoters, may finally allow a better distinction of the various processes from dormancy induction to seedling establishment. The combination of mutant approaches with in-depth molecular analysis of the regulators and '-omic' approaches is expected to give many new insights into seed dormancy, which is still a relatively poorly understood process that has importance for both ecology and agriculture.

References

G.K. Agrawal, M. Yamazaki, M. Kobayashi, R. Hirochika, A. Miyao and H. Hirochika (2001) Screening of the rice viviparous mutants generated by endogenous retrotransposon *Tos17* insertion. Tagging of a zeaxanthin epoxidase gene and a novel *OsTATC* gene. *Plant Physiology* **125**, 1248–1257.

A. Alboresi, C. Gestin, M.T. Leydecker, M. Bedu, C. Meyer and H.N. Truong (2005) Nitrate, a signal relieving seed dormancy in *Arabidopsis*. *Plant Cell and Environment* **28**, 500–512.

S. Ali-Rachedi, D. Bouinot, M.H. Wagner, *et al.* (2004) Changes in endogenous abscisic acid levels during dormancy release and maintenance of mature seeds: studies with the Cape Verde Islands ecotype, the dormant model of *Arabidopsis thaliana*. *Planta* **219**, 479–488.

C. Alonso-Blanco, L. Bentsink, C.J. Hanhart, H. Blankestijn-De Vries and M. Koornneef (2003) Analysis of natural allelic variation at seed dormancy loci of *Arabidopsis thaliana*. *Genetics* **164**, 711–729.

C. Alonso-Blanco and M. Koornneef (2000) Naturally occurring variation in *Arabidopsis*: an underexploited resource for plant genetics. *Trends in Plant Science* **5**, 22–29.

C. Basu, M.D. Halfhill, T.C. Mueller and C.N. Stewart, Jr. (2004) Weed genomics: new tools to understand weed biology. *Trends in Plant Science* **9**, 391–398.

N. Beaudoin, C. Serizet, F. Gosti and J. Giraudat (2000) Interactions between abscisic acid and ethylene signaling cascades. *The Plant Cell* **12**, 1103–1115.

L. Bentsink (2002) *Genetic Analysis of Seed Dormancy and Seed Composition in* Arabidopsis thaliana *Using Natural Variation*. Ph.D. Dissertation, Department of Plant Sciences, Laboratory of Genetics, Wageningen University, 159 pp.

H.W. Cai and H. Morishima (2000) Genomic regions affecting seed shattering and seed dormancy in rice. *Theoretical and Applied Genetics* **100**, 840–846.

J.S. Cao, H.F. Yu, W.Z. Ye, *et al.* (2005) Identification and characterisation of a gibberellin-related dwarf mutant in pumpkin (*Cucurbita moschata*). *Journal of Horticultural Science and Biotechnology* **80**, 29–31.

C. Carles, N. Bies-Etheve, L. Aspart, *et al.* (2002) Regulation of *Arabidopsis thaliana Em* genes: role of ABI5. *The Plant Journal* **30**, 373–383.

F. Carrari, R. Benech-Arnold, R. Osuna-Fernandez, *et al.* (2003) Genetic mapping of the *Sorghum bicolor vp1* gene and its relationship with preharvest sprouting resistance. *Genome* **46**, 253–258.

S.D. Chiwocha, A.J. Cutler, S.R. Abrams, *et al.* (2005) The *etr1-2* mutation in *Arabidopsis thaliana* affects the abscisic acid, auxin, cytokinin and gibberellin metabolic pathways during maintenance of seed dormancy, moist-chilling and germination. *The Plant Journal* **42**, 35–48.

E.J.M. Clerkx, M.E. El Lithy, E. Vierling, *et al.* (2004) Analysis of natural allelic variation of *Arabidopsis* seed germination and seed longevity traits between the accessions Landsberg *erecta* and Shakdara, using a new recombinant inbred line population. *Plant Physiology* **135**, 432–443.

S.D. Clouse and J.M. Sasse (1998) Brassinosteroids: essential regulators of plant growth and development. *Annual Review of Plant Physiology and Plant Molecular Biology* **49**, 427–451.

G. Colucci, F. Apone, N. Alyeshmerni, D. Chalmers and M.J. Chrispeels (2002) *GCR1*, the putative *Arabidopsis* G protein-coupled receptor gene is cell cycle-regulated, and its overexpression abolishes seed dormancy and shortens time to flowering. *Proceedings of the National Academy of Sciences of the United States of America* **99**, 4736–4741.

S. Cutler, M. Ghassemian, D. Bonetta, S. Cooney and P. McCourt (1996) A protein farnesyl transferase involved in ABA signal transduction in *Arabidopsis*. *Science* **273**, 1239–1241.

I. Debeaujon and M. Koornneef (2000) Gibberellin requirement for *Arabidopsis thaliana* seed germination is determined both by testa characteristics and embryonic ABA. *Plant Physiology* **122**, 415–424.

I. Debeaujon, K.M. Léon-Kloosterziel and M. Koornneef (2000) Influence of the testa on seed dormancy, germination and longevity in *Arabidopsis thaliana*. *Plant Physiology* **122**, 403–414.

K. Donohue (2002) Germination timing influences natural selection on life-history characters in *Arabidopsis thaliana*. *Ecology* **83**, 1006–1016.

B. Dubreucq, N. Berger, E. Vincent, *et al.* (2000) The *Arabidopsis AtERP1* extensin-like gene is specifically expressed in endosperm during seed germination. *The Plant Journal* **23**, 643–652.

S.A. Fennimore, W.E. Nyquist, G.E. Shaner, R.W. Doerge and M.E. Foley (1999) A genetic model and molecular markers for wild oat (*Avena fatua* L.) seed dormancy. *Theoretical and Applied Genetics* **99**, 711–718.

R.R. Finkelstein (1994) Maternal effects govern variable dominance of two abscisic acid response mutations in *Arabidopsis thaliana*. *Plant Physiology* **105**, 1203–1208.

J. Flintham, R. Adlam, M. Bassoi, M. Holdsworth and M. Gale (2002) Mapping genes for resistance to sprouting damage in wheat. *Euphytica* **126**, 39–45.

J.E. Flintham (2000) Different genetic components control coat-imposed and embryo-imposed dormancy in wheat. *Seed Science Research* **10**, 43–50.

R. Foster and N.H. Chua (1999) An *Arabidopsis* mutant with deregulated ABA gene expression: implication for negative regulator function. *The Plant Journal* **17**, 363–372.

M.D. Gale, J. Flintham and K.M. Devos (2002) Cereal comparative genetics and preharvest sprouting. *Euphytica* **126**, 21–25.

K. Gallardo, C. Job, S.P.C. Groot, *et al.* (2002) Proteomics of *Arabidopsis* seed germination: A comparative study of wild-type and gibberellin-deficient seeds. *Plant Physiology* **129**, 823–837.

W. Gao, J.A. Clancy, F. Han, D. Prada, A. Kleinhofs and S.E. Ullrich (2003) Molecular dissection of a dormancy QTL region near the chromosome 7 (5H) L telomere in barley. *Theoretical and Applied Genetics* **107**, 552–559.

A. Garciarrubio, J.P. Legaria and A.A. Covarrubias (1997) Abscisic acid inhibits germination of mature *Arabidopsis* seeds by limiting the availability of energy and nutrients. *Planta* **203**, 182–187.

M. Ghassemian, E. Nambara, S. Cutler, H. Kawaide, Y. Kamiya and P. McCourt (2000) Regulation of abscisic acid signaling by the ethylene response pathway in *Arabidopsis*. *The Plant Cell* **12**, 1117–1126.

S.I. Gibson and I.A. Graham (1999) Another player joins the complex field of sugar-regulated gene expression in plants. *Proceedings of the National Academy of Sciences of the United States of America* **96**, 4746–4748.

S.I. Gibson, R.J. Laby and D. Kim (2001) The *sugar-insensitive1* (*sis1*) mutant of *Arabidopsis* is allelic to *ctr1*. *Biochemical and Biophysical Research Communications* **280**, 196–203.

C. Groos, G. Gay, M.R. Perretant, *et al.* (2002) Study of the relationship between pre-harvest sprouting and grain color by quantitative trait loci analysis in a white × red grain bread-wheat cross. *Theoretical and Applied Genetics* **104**, 39–47.

X.Y. Gu, S.F. Kianian and M.E. Foley (2004) Multiple loci and epistases control genetic variation for seed dormancy in weedy rice (*Oryza sativa*). *Genetics* **166**, 1503–1516.

X.Y. Gu, S.F. Kianian, G.A. Hareland, B.L. Hoffer and M.E. Foley (2005) Genetic analysis of adaptive syndromes interrelated with seed dormancy in weedy rice (*Oryza sativa*). *Theoretical and Applied Genetics* **110**, 1108–1118.

G. Gualberti, M. Papi, L. Bellucci, *et al.* (2002) Mutations in the Dof zinc finger genes *DAG2* and *DAG1* influence with opposite effects germination of *Arabidopsis* seeds. *The Plant Cell* **14**, 1253–1263.

F. Gubler, A.A. Millar and J.V. Jacobsen (2005) Dormancy release, ABA and pre-harvest sprouting. *Current Opinion in Plant Biology* **8**, 183–187.

F. Han, S.E. Ullrich, J.A. Clancy and I. Romagosa (1999) Inheritance and fine mapping of a major barley seed dormancy QTL. *Plant Science* **143**, 113–118.

Y. He and S. Gan (2004) A novel zinc-finger protein with a proline-rich domain mediates ABA-regulated seed dormancy in *Arabidopsis*. *Plant Molecular Biology* **54**, 1–9.

H.W.M. Hilhorst, S.P.C. Groot and R.J. Bino (1998) The tomato seed as a model system to study seed development and germination. *Acta Botanica Neerlandica* **47**, 169–183.

M. Holdsworth, L. Bentsink and M. Koornneef (2006) Conserved mechanisms of dormancy and germination as targets for manipulation of agricultural problems. In: *Model Plants, Crop Improvement* (eds R. Koebner and R. Varshney). CRC Press, Boca Raton, FL.

C. Huijser, A.J. Kortstee, J.V. Pego, P. Weisbeek, E. Wisman and S. Smeekens (2000) The *Arabidopsis sucrose uncoupled-6* gene is identical to *abscisic acid insensitive-4*: involvement of abscisic acid in sugar responses. *The Plant Journal* **23**, 577–585.

S.E. Jacobsen, K.A. Binkowski and N.E. Olszewski (1996) SPINDLY, a tetratricopeptide repeat protein involved in gibberellin signal transduction in *Arabidopsis*. *Proceedings of the National Academy of Sciences of the United States of America* **93**, 9292–9296.

C.M. Karssen, D.L.C. Brinkhorst-van der Swan, A.E. Breekland and M. Koornneef (1983) Induction of dormancy during seed development by endogenous abscisic acid: studies on abscisic acid deficient genotypes of *Arabidopsis thaliana* (L.) Heynh. *Planta* **157**, 158–165.

C.M. Karssen, S. Zaórski, J. Kepczynski and S.P.C. Groot (1989) Key role for endogenous gibberellins in the control of seed germination. *Annals of Botany* **63**, 71–80.

K. Kato, W. Nakamura, T. Tabiki, H. Miura and S. Sawada (2001) Detection of loci controlling seed dormancy on group 4 chromosomes of wheat and comparative mapping with rice and barley genomes. *Theoretical and Applied Genetics* **102**, 980–985.

N. Kawakami, Y. Miyake and K. Noda (1997) ABA insensitivity and low ABA levels during seed development of non-dormant wheat mutants. *Journal of Experimental Botany* **48**, 1415–1421.

R.W. King and P. Wettstein-Knowles (2000) Epicuticular waxes and regulation of ear wetting and pre-harvest sprouting in barley and wheat. *Euphytica* **112**, 157–166.

M. Koornneef, C. Alonso-Blanco, L. Bentsink, *et al.* (2000) The genetics of seed dormancy in *Arabidopsis thaliana*. In: *Dormancy in Plants* (eds J.D. Viémont and J. Crabbé), pp. 365–373. CAB International, Wallingford, UK.

M. Koornneef, C.J. Hanhart, H.W.M. Hilhorst and C.M. Karssen (1989) *In vivo* inhibition of seed development and reserve protein accumulation in recombinants of abscisic acid biosynthesis and responsiveness mutants in *Arabidopsis thaliana*. *Plant Physiology* **90**, 463–469.

M. Koornneef, M.L. Jorna, D.L.C. Brinkhorst-van der Swan and C.M. Karssen (1982) The isolation of abscisic acid (ABA) deficient mutants by selection of induced revertants in non-germinating gibberellin sensitive lines of *Arabidopsis thaliana* (L.) Heynh. *Theoretical and Applied Genetics* **61**, 385–393.

N. Kurata, K. Miyoshi, K. Nonomura, Y. Yamazaki and Y. Ito (2005) Rice mutants and genes related to organ development, morphogenesis and physiological traits. *Plant and Cell Physiology* **46**, 48–62.

R.J. Laby, S. Kincaid, D. Kim and S.I. Gibson (2000) The *Arabidopsis* sugar-insensitive mutants *sis4* and *sis5* are defective in abscisic acid synthesis and response. *The Plant Journal* **23**, 587–596.

K.M. Léon-Kloosterziel, M. Alvarez Gil, G.J. Ruijs, *et al.* (1996a) Isolation and characterization of abscisic acid-deficient *Arabidopsis* mutants at two new loci. *The Plant Journal* **10**, 655–661.

K.M. Léon-Kloosterziel, G.A. van de Bunt, J.A.D. Zeevaart and M. Koornneef (1996b) *Arabidopsis* mutants with a reduced seed dormancy. *Plant Physiology* **110**, 233–240.

C.D. Li, A. Tarr, R.C.M. Lance, *et al.* (2003) A major QTL controlling seed dormancy and pre-harvest sprouting/grain α-amylase in two-rowed barley (*Hordeum vulgare* L.). *Australian Journal of Agricultural Research* **54**, 1303–1313.

J.M. Li, J.H. Xiao, S. Grandillo, *et al.* (2004) QTL detection for rice grain quality traits using an interspecific backcross population derived from cultivated Asian (*O. sativa* L.) and African (*O. glaberrima* S.) rice. *Genome* **47**, 697–704.

D. Lijavetzky, M.C. Martinez, F. Carrari and H.E. Hopp (2000) QTL analysis and mapping of preharvest sprouting resistance in *Sorghum*. *Euphytica* **112**, 125–135.

S.Y. Lin, T. Sasaki and M. Yano (1998) Mapping quantitative trait loci controlling seed dormancy and heading date in rice, *Oryza sativa* L., using backcross inbred lines. *Theoretical and Applied Genetics* **96**, 997–1003.

P.-P. Liu, N. Koizuka, T.M. Homrichhausen, J.R. Hewitt, R.C. Martin and H. Nonogaki (2005) Large-scale screening of *Arabidopsis* enhancer-trap lines for seed germination-associated genes. *The Plant Journal* **41**, 936–944.

D. Mares, K. Mrva, M.K. Tan and P. Sharp (2002) Dormancy in white-grained wheat: progress towards identification of genes and molecular markers. *Euphytica* **126**, 47–53.

D.R. McCarty (1995) Genetic control and integration of maturation and germination pathways in seed development. *Annual Review of Plant Physiology and Plant Molecular Biology* **46**, 71–93.

K. Miura, S.Y. Lin, M. Yano and T. Nagamine (2002) Mapping quantitative trait loci controlling seed longevity in rice (*Oryza sativa* L.). *Theoretical and Applied Genetics* **104**, 981–986.

E. Nambara and A. Marion-Poll (2005) Abscisic acid biosynthesis and catabolism. *Annual Review of Plant Biology* **56**, 165–185.

E. Nambara, S. Naito and P. McCourt (1992) A mutant of *Arabidopsis* which is defective in seed development and storage protein accumulation is a new *abi3* allele. *The Plant Journal* **2**, 435–441.

N. Nishimura, T. Yoshida, M. Murayama, T. Asami, K. Shinozaki and T. Hirayama (2004) Isolation and characterization of novel mutants affecting the abscisic acid sensitivity of *Arabidopsis* germination and seedling growth. *Plant and Cell Physiology* **45**, 1485–1499.

K. Noda, T. Matsuura, M. Maekawa and S. Taketa (2002) Chromosomes responsible for sensitivity of embryo to abscisic acid and dormancy in wheat. *Euphytica* **123**, 203–209.

M. Ogawa, A. Hanada, Y. Yamauchi, A. Kuwahara, Y. Kamiya and S. Yamaguchi (2003) Gibberellin biosynthesis and response during *Arabidopsis* seed germination. *The Plant Cell* **15**, 1591–1604.

M. Papi, S. Sabatini, D. Bouchez, C. Camilleri, P. Costantino and P. Vittorioso (2000) Identification and disruption of an *Arabidopsis* zinc finger gene controlling seed germination. *Genes and Development* **14**, 28–33.

A.J.M. Peeters, H. Blankestijn-De Vries, C.J. Hanhart, K.M. Léon-Kloosterziel, J.A.D. Zeevaart and M. Koornneef (2002) Characterization of mutants with reduced seed dormancy at two novel *rdo* loci and a further characterization of *rdo1* and *rdo2* in *Arabidopsis*. *Physiologia Plantarum* **115**, 604–612.

J.V. Pego, A.J. Kortstee, C. Huijser and S. Smeekens (2000) Photosynthesis, sugars and the regulation of gene expression. *Journal of Experimental Botany* **51**, 407–416.

S. Penfield, E.L. Rylott, A.D. Gilday, S. Graham, T.R. Larson and I.A. Graham (2004) Reserve mobilization in the *Arabidopsis* endosperm fuels hypocotyl elongation in the dark, is independent of abscisic acid, and requires *PHOSPHOENOLPYRUVATE CARBOXYKINASE1*. *The Plant Cell* **16**, 2705–2718.

L. Perez-Flores, F. Carrari, R. Osuna-Fernandez, *et al.* (2003) Expression analysis of a GA 20-oxidase in embryos from two sorghum lines with contrasting dormancy: possible participation of this gene in the hormonal control of germination. *Journal of Experimental Botany* **54**, 2071–2079.

C. Poppe and E. Schäfer (1997) Seed germination of *Arabidopsis thaliana phyA/phyB* double mutants is under phytochrome control. *Plant Physiology* **114**, 1487–1492.

D. Prada, I. Romagosa, S.E. Ullrich and J.L. Molina-Cano (2005) A centromeric region on chromosome 6(6H) affects dormancy in an induced mutant in barley. *Journal of Experimental Botany* **56**, 47–54.

D. Prada, S.E. Ullrich, J.L. Molina-Cano, L. Cistue, J.A. Clancy and I. Romagosa (2004) Genetic control of dormancy in a Triumph/Morex cross in barley. *Theoretical and Applied Genetics* **109**, 62–70.

I. Romagosa, F. Han, J.A. Clancy and S.E. Ullrich (1999) Individual locus effects on dormancy during seed development and after ripening in barley. *Crop Science* **39**, 74–79.

I. Romagosa, D. Prada, M.A. Moralejo, *et al.* (2001) Dormancy, ABA content and sensitivity of a barley mutant to ABA application during seed development and after ripening. *Journal of Experimental Botany* **52**, 1499–1506.

L. Salaita, R.K. Kar, M. Majee and A.B. Downie (2005) Identification and characterization of mutants capable of rapid seed germination at 10°C from activation-tagged lines of *Arabidopsis thaliana*. *Journal of Experimental Botany* **56**, 2059–2069.

K. Schumacher and J. Chory (2000) Brassinosteroid signal transduction: still casting the actors. *Current Opinion in Plant Biology* **3**, 79–84.

T. Shinomura, A. Nagatani, H. Hanzawa, M. Kubota, M. Watanabe and M. Furuya (1996) Action spectra for phytochrome A- and B-specific photoinduction of seed germination in *Arabidopsis thaliana*. *Proceedings of the National Academy of Sciences of the United States of America* **93**, 8129–8133.

G. Siriwitayawan, R.L. Geneve and A.B. Downie (2003) Seed germination of ethylene perception mutants of tomato and *Arabidopsis*. *Seed Science Research* **13**, 303–314.

S. Smeekens (1998) Sugar regulation of gene expression in plants. *Current Opinion in Plant Biology* **1**, 230–234.

C.M. Steber and P. McCourt (2001) A role for brassinosteriods in germination in *Arabidopsis*. *Plant Physiology* **125**, 763–769.

H.S. Steinbach, R.L. Benech-Arnold and R.A. Sanchez (1997) Hormonal regulation of dormancy in developing sorghum seeds. *Plant Physiology* **113**, 149–154.

B.C. Tan, S.H. Schwartz, J.A.D. Zeevaart and D.R. McCarty (1997) Genetic control of abscisic acid biosynthesis in maize. *Proceedings of the National Academy of Sciences of the United States of America* **94**, 12235–12240.

A. Torada, S. Ikeguchi and M. Koike (2005) Mapping and validation of PCR-based markers associated with a major QTL for seed dormancy in wheat. *Euphytica* **143**, 251–255.

L. Tyler, S.G. Thomas, J.H. Hu, *et al.* (2004) DELLA proteins and gibberellin-regulated seed germination and floral development in *Arabidopsis*. *Plant Physiology* **135**, 1008–1019.

H. Ullah, J.G. Chen, S.C. Wang and A.M. Jones (2002) Role of a heterotrimeric G protein in regulation of *Arabidopsis* seed germination. *Plant Physiology* **129**, 897–907.

S.E. Ullrich, F. Han, W. Gao, *et al.* (2002) Summary of QTL analyses of the seed dormancy trait in barley. *Barley Newsletter* **45**, 39–41.

S.E. Ullrich, F. Han and B.L. Jones (1997) Genetic complexity of the malt extract trait in barley suggested by QTL analysis. *Journal of the American Society of Brewing Chemists* **55**, 1–4.

W. Van der Schaar, C. Alonso-Blanco, K.M. Léon-Kloosterziel, R.C. Jansen, J.W. Van Ooijen and M. Koornneef (1997) QTL analysis of seed dormancy in *Arabidopsis* using recombinant inbred lines and MQM mapping. *Heredity* **79**, 190–200.

A.J. van Hengel and K. Roberts (2003) AtAGP30, an arabinogalactan-protein in the cell walls of the primary root, plays a role in root regeneration and seed germination. *The Plant Journal* **36**, 256–270.

M.K. Walker-Simmons (1987) ABA levels and sensitivity in developing wheat embryos of sprouting resistant and susceptible cultivars. *Plant Physiology* **84**, 61–66.

L. Xiong, M. Ishitani, H. Lee and J.K. Zhu (1999) *HOS5* – a negative regulator of osmotic stress-induced gene expression in *Arabidopsis thaliana*. *The Plant Journal* **19**, 569–578.

L. Xiong, M. Ishitani, H. Lee and J.K. Zhu (2001) The *Arabidopsis LOS5/ABA3* locus encodes a molybdenum cofactor sulfurase and modulates cold stress- and osmotic stress-responsive gene expression. *The Plant Cell* **13**, 2063–2083.

L. Xiong, H. Lee, M. Ishitani and J.K. Zhu (2002) Regulation of osmotic stress-responsive gene expression by the *LOS6/ABA1* locus in *Arabidopsis*. *The Journal of Biological Chemistry* **277**, 8588–8596.

Y.Y. Yang, A. Nagatani, Y.J. Zhao, B.J. Kang, R.E. Kendrick and Y. Kamiya (1995) Effects of gibberellins on seed germination of phytochrome-deficient mutants of *Arabidopsis thaliana*. *Plant and Cell Physiology* **36**, 1205–1211.

F. Zhang, G. Chen, Q. Huang, *et al.* (2005) Genetic basis of barley caryopsis dormancy and seedling desiccation tolerance at the germination stage. *Theoretical and Applied Genetics* **110**, 445–453.

L. Zhou, J.C. Jang, T.L. Jones and J. Sheen (1998) Glucose and ethylene signal transduction crosstalk revealed by an *Arabidopsis* glucose-insensitive mutant. *Proceedings of the National Academy of Sciences of the United States of America* **95**, 10294–10299.

6 Lipid metabolism in seed dormancy

Steven Penfield, Helen Pinfield-Wells and Ian A. Graham

6.1 Introduction

One of the key features of seeds from both angiosperms and gymnosperms is the propensity to accumulate stores of nutrient compounds to fuel early seedling development. Stored reserves are used for respiration and growth in the period immediately following germination, but before photosynthesis and uptake from the roots begins in earnest. In this review we will focus on the biochemical pathways required for the catabolism of seed storage reserves and the role these might play in the control of seed dormancy and germination. In practice we will concentrate on the model plant *Arabidopsis thaliana*, although it is reasonable to expect the results to be broadly applicable to other species. Even seeds conventionally viewed as starch-containing, such as cereals, accumulate significant amounts of oil in both the embryo and the aleurone layer during seed development (Gram, 1982). Consequently, the activities of key enzymes in lipid catabolism have been demonstrated in cereals during early seedling growth (Eastmond and Jones, 2005). Perhaps more surprising is that the activities of many of the enzymes required for storage oil catabolism are present in maturing and dormant seeds prior to germination. In fact, although storage reserve mobilization has traditionally been viewed as a postgerminative event, it has become clear that significant breakdown of lipids occurs during the period of their synthesis in developing *Arabidopsis* and rape (*Brassica napus*) seeds. Furthermore, during the final stages of seed maturation, seed oil content actually falls (Baud *et al.*, 2002; Chia *et al.*, 2005). Hence, there is ample scope for these pathways to be regulating events prior to seed germination, such as the control of dormancy.

Fatty acids are stored in seeds as triacylglycerol (TAG). This consists of three fatty acids esterified to a glycerol backbone, and forms a highly efficient and compact energy source for germinating seeds. Fatty acids and glycerol are released from TAG and undergo further catabolism to provide energy and carbon skeletons for postgerminative growth. The overall mobilization of TAG involves several consecutive pathways in various cellular compartments (Figure 6.1). TAGs are hydrolyzed by lipases (Huang, 1992), and then the fatty acids are passed into single-membrane-bound organelles, termed glyoxysomes, where β-oxidation and the glyoxylate cycle occur (Cooper and Beevers, 1969). Glyoxysomes are identified as distinct from the more ubiquitous peroxisomes on the basis that glyoxysomes have an active glyoxylate cycle. β-Oxidation converts fatty acids to acetyl-CoA, which is subsequently condensed into four-carbon sugars via the glyoxylate cycle. These are then transported to the mitochondria from where they can be either converted to

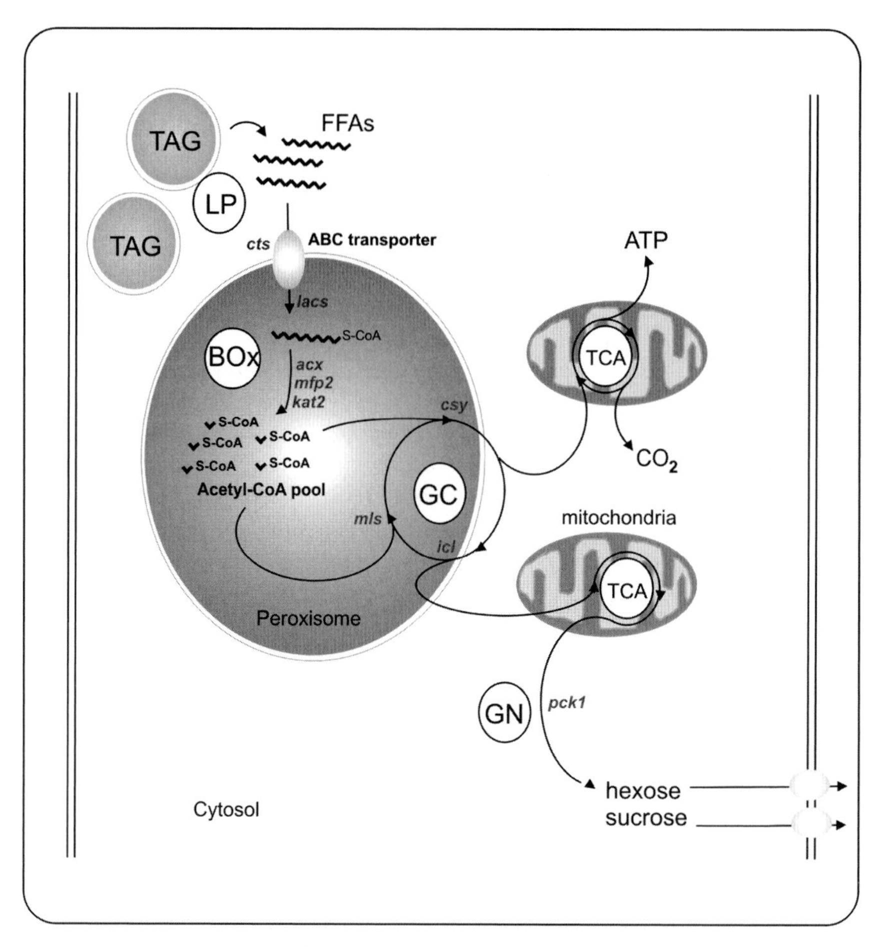

Figure 6.1 The catabolism of seed storage oil and conversion to carbohydrate by gluconeogenesis. Fatty acids are imported into the peroxisome where they are activated to acyl-CoA esters and catabolized by β-oxidation to acetyl-CoA. The acetyl-CoA is converted to citrate. This either enters the glyoxylate cycle, a partial TCA cycle, and gluconeogenesis, or is transported to the mitochondria and used for respiration. Abbreviations: BOx, β-oxidation; FFAs, free fatty acids; GC, glyoxylate cycle; GN, gluconeogenesis; LP, lipolysis; TAG, triacylglycerol; TCA, tricarboxylic acid cycle. Mutant symbols in gray show the key genetically characterized steps in storage reserve mobilization: *cts, comatose*; *lacs, long chain acyl-CoA synthetase*; *acx, acyl-CoA oxidase*; *mfp2, multifunctional protein2; kat2, α-ketoacyl-CoA thiolase2*; *csy, citrate synthase*; *icl, isocitrate lyase*; *mls, malate synthase*; *pck1, phosphoenolpyruvate carboxykinase1*.

malate and transported to the cytosol for gluconeogenesis or used as substrates for respiration.

In order to consider the potential role for the pathways required for storage reserve mobilization in the control of seed dormancy and germination we will first briefly review the current state of knowledge on lipid mobilization and then apply this to interpret the germination phenotypes of various mutants affected in lipid breakdown.

6.2 Metabolic pathways for TAG breakdown and conversion to sucrose

6.2.1 TAG hydrolysis and activation

TAG is stored in organelles called lipid bodies (spherosomes or oleosomes). Following germination, TAG is catabolized and the carbon yielded is used to support seedling growth and development. TAG lipase (*EC 3.1.1.3*) catalyzes the first step in this process, hydrolyzing TAG to yield fatty acids and glycerol in a 3:1 molar ratio (Bewley and Black, 1982). The Arabidopsis TAG lipase gene product has recently been characterised and shown to possess a patatin-like serine esterase domain with activity against long chain TAGs *in vitro* (Eastmond, 2006). The corresponding mutant *sdp1(sugar dependent 1)* mutant cannot break down seed storage lipid and shows poor hypocotyl elongation in the dark. Recovery of growth to wild type levels is achieved by applying an alternative carbon source such as sucrose. This phenotype is diagnostic for an impairment in storage reserve mobilisation (Eastmond *et al.*, 2000a; Carnah *et al.*, 2004; Penfield *et al.*, 2004). Various other lipases such as the acid lipase from castor bean (*Ricinus communis*) (Eastmond, 2004) and a TAG lipase from tomato (*Lycopersicon esculentum*) (Matsui *et al.*, 2004) have also been cloned but *SDP1* remains the only one with convincing genetic evidence for *in-vivo* function. Following lipolysis the free fatty acids that are released from TAG are then thought to be activated by the attachment of coenzyme A (CoA) to the carboxyl group of the fatty acids (Figure 6.1). This priming is necessary for oxidative attack at the C-3 position during β-oxidation.

6.2.2 Import of fatty acids into the peroxisome

The mechanism for fatty acid or acyl-CoA transport across the peroxisomal membrane in plants is still unknown. In *Saccharomyces cerevisiae* (yeast), in which β-oxidation occurs exclusively in the peroxisomes, two mechanisms have been identified for the import of fatty acids. Medium-chain fatty acids are imported as free fatty acids that are activated inside the peroxisome by acyl-CoA synthase Faa2p (Hettema *et al.*, 1996). Long-chain fatty acids (LCFAs) appear to be activated in the cytoplasm and transported as CoA esters across the peroxisomal membrane by two ABC (ATP-binding cassette) proteins, Pxa1p and Pxa2p (Hettema *et al.*, 1996; Verleur *et al.*, 1997). In mammals the orthologous transporter is called adrenoleukodystrophy protein (ALDP; Mosser *et al.*, 1993), named after its role in the aforementioned disease. The *Arabidopsis* homologue has been isolated in three independent genetic screens. The first two involve seedling resistance to two pro-herbicides, indole-3-butyric acid (IBA) and 2,4-dichlorophenoxybutyric acid (2,4-DB), which require the action of peroxisomal β-oxidation to convert them to the herbicides IAA (indole-3-acetic acid) and 2,4-D (2,4-dichlorophenoxyacetic acid), respectively (Zolman *et al.*, 2001; Hayashi *et al.*, 2002). The third screen involved the isolation of mutants with increased seed dormancy (Russell *et al.*, 2000; Footitt *et al.*, 2002). This screen identified the *comatose* (*cts*) mutant, named after its dormancy phenotype, also known as *pxa1* and *peroxisome defective3* (*ped3*). Hereafter we refer to this gene as *CTS* in order to emphasize the germination phenotype. In addition to the block on germination, the *cts* mutants exhibit impaired seedling establishment that can be restored by application of an alternative carbon source, such as sucrose. They cannot break down TAG and accumulate acyl-CoA esters, the substrates for β-oxidation. All these phenotypes suggest that β-oxidation is blocked in *cts* seeds. The germination phenotype of the *cts* mutants is considered in detail in a later section.

6.2.3 *Activation of fatty acids to acyl-CoA thioesters for β-oxidation*

Before free fatty acids can be degraded by β-oxidation they need to be activated to acyl-CoA thioesters. This reaction is catalyzed by long-chain acyl-CoA synthetases (LACS) (ACS; *EC 6.2.1.3*; Shockey *et al.*, 2002). Two LACS proteins (LACS6 and LACS7) that are strongly expressed during germination have been identified. Using green fluorescent protein fusions, it was shown that these are localized to the peroxisomes (Fulda *et al.*, 2004). Although neither the *lacs6* or *lacs7* mutants exhibit a seedling phenotype, *lacs6 lacs7* double mutant seedlings fail to grow and the cotyledons remain unexpanded, indicating that these two isoforms are functionally redundant. The double mutant phenotype can be rescued by sucrose application (Fulda *et al.*, 2004). The *lacs6 lacs7* double mutant is also deficient in TAG breakdown. Possible models showing explanations for the requirement for both a peroxisomal acyl-CoA transporter and a peroxisomal acyl-CoA synthetase activity are shown in Figure 6.2. Two alternative models were proposed by Fulda *et al.* (2004). In the first, the transport of acyl-CoA esters results in their cleavage and the peroxisomal LACS activity is required to reactivate them (Figure 6.2A). In the second, an

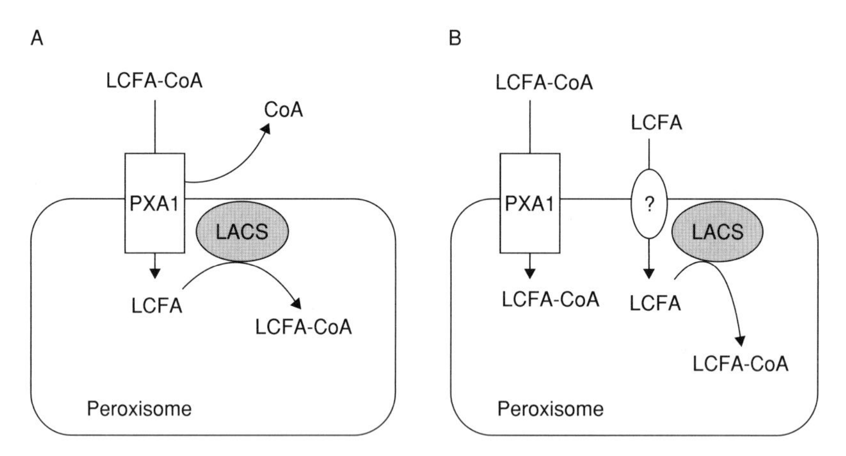

Figure 6.2 Two models proposed by Fulda *et al.* (2004) to explain the requirement for CTS/PXA1 putative acyl-CoA transporter and peroxisomal LACS for storage oil breakdown. In the first (A), long-chain acyl-CoAs synthesized in the cytosol are cleaved during peroxisomal import and subsequently reesterified in the peroxisome. In the second (B), two parallel import channels are envisaged, one for free fatty acids and one for acyl-CoAs. Acyl-CoA import could be important for coenzyme A import, which could then be used by peroxisomal LACS to provide further substrates for β-oxidation. (Copyright: American Society of Plant Biologists.)

alternative route for entry of free fatty acids into the peroxisome operates in addition to the transport of acyl-CoA esters by CTS. These free fatty acids would then require activation by the LACS in the peroxisomal lumen.

6.2.4 *β-Oxidation*

In each round of the *β*-oxidation spiral, acetyl-CoA (C_2) is cleaved from acyl-CoA (C_n). The remaining acyl-CoA (C_{n-2}) reenters the *β*-oxidation spiral and by repetitive passages through the pathway acetyl-CoA is produced. This is converted to sucrose via the glyoxylate cycle and gluconeogenesis (Graham and Eastmond, 2002). The breakdown of straight-chain saturated acyl-CoAs requires the enzymes acyl-CoA oxidase (ACX, *EC 1.3.3.6*), multifunctional protein, and 3-L-ketoacyl-CoA thiolase. These enzymes catalyze oxidation, hydration and dehydrogenation, and thiolytic cleavage, respectively, of acyl-CoA. Degradation of unsaturated fatty acids with double bonds at odd numbered carbons or *cis* double bonds requires additional auxiliary enzymes to convert these double bonds to the 2-*trans*-enoyl-CoA intermediate, which is the substrate for the 2-*trans*-enoyl-CoA hydratase (*EC 4.2.1.17*) activity of the multifunctional protein (Graham and Eastmond, 2002; Goepfert *et al.*, 2005). All three families of *β*-oxidation genes are upregulated co-ordinately during *Arabidopsis* seed germination and early postgerminative growth (Rylott *et al.*, 2001). This correlates with the period of most rapid fatty acid degradation in *Arabidopsis* (Germain *et al.*, 2001).

6.2.4.1 *Acyl-CoA oxidases*

Acyl-CoA oxidases (ACXs; *EC 1.3.3.6*) catalyze the first step of the peroxisomal *β*-oxidation sequence, where acyl-CoA is oxidized to Δ^2-*trans*-enoyl-CoA. The reaction requires a FAD as a cofactor to generate $FADH_2$, which is then oxidized by flavoprotein dehydrogenase to produce hydrogen peroxide. Studies have shown that *Arabidopsis* contains a family of six ACX isoenzymes with distinct but slightly overlapping acyl-CoA chain length specificities (Hooks *et al.*, 1999). Mutants disrupted in all six *ACX* genes have been described (Eastmond *et al.*, 2000b; Rylott *et al.*, 2003; Adham *et al.*, 2005). Of these, the *acx1*, *acx3*, and *acx4* mutants show weak resistance of root growth to IBA, and *acx3* and *acx4* show resistance to 2,4-DB, suggesting a defect in *β*-oxidation. However, only the *acx1 acx2* double mutant was found to be defective in seedling establishment, suggesting that the overlapping substrate specificities of these two isoforms are most important for germinating seeds (Adham *et al.*, 2005; Pinfield-Wells *et al.*, 2005). The double mutant also accumulates acyl-CoA esters and cannot break down TAG.

6.2.4.2 *Multifunctional protein*

Multifunctional protein (MFP) has a complex role in *β*-oxidation, exhibiting multiple activities required in different combinations for the *β*-oxidation of saturated and polyunsaturated fatty acids (reviewed in Graham and Eastmond, 2002). Two activities, 2-*trans*-enoyl-CoA hydratase and L-3-hydroxyacyl-CoA dehydrogenase (*EC 1.1.1.35*), are required for the *β*-oxidation of saturated fatty acids and those unsaturated fatty acids containing a *trans* double bond at the Δ^2 position. The dehydrogenase step requires NAD^+ and generates NADH. *Arabidopsis* contains two isoforms of MFP. The first (MFP1) has been characterized genetically as the *abnormal inflorescence meristem1* (*aim1*) locus (Richmond and Bleecker, 1999). This mutant shows aberrant vegetative and reproductive development, underlining the role of β-oxidation throughout plant development. Seedling establishment is normal in the *aim1* mutant, however. This is not surprising, as the second isoform (MFP2) is strongly and specifically expressed during postgerminative growth, suggesting that it plays the predominant role at this stage in development (Eastmond and Graham, 2000). In support of this, mutants in *MFP2* show a sugar-dependent short hypocotyl phenotype in the dark, accumulate acyl-CoAs during seedling establishment, and break down TAG only slowly (Rylott *et al.*, 2006). Although the *mfp2* mutant phenotype is weaker than other β-oxidation mutants, it is probable that the activity of AIM1 (MFP1) somewhat compensates for the loss of MFP2 during germination and seedling establishment. *aim1 mfp2* double mutants are embryo lethal (Rylott *et al.*, 2006).

6.2.4.3 3-L-Ketoacyl-CoA thiolase

The enzyme 3-L-ketoacyl-CoA thiolase (KAT; *EC 2.3.1.16*) catalyzes the last step of fatty acid β-oxidation involving thiolytic cleavage of 3-ketoacyl-CoA to acyl-CoA (C_{n-2}) and acetyl-CoA (C_2). The acyl-CoA (C_{n-2}) then reenters the β-oxidation spiral and the process is repeated until the entire molecule is broken down to acetyl-CoA units. There are four putative thiolase genes in *Arabidopsis* that contain putative peroxisomal targeting sequences (Hooks, 2002). Of these, *KAT2* is the most important in the seed. Allelic mutants in *KAT2* have been isolated by 2,4-DB resistance and reverse genetics (Hayashi *et al.*, 1998; Germain *et al.*, 2001). These exhibit the typical syndrome of compromised seedling establishment that can be rescued by sugar, accumulation of acyl-CoA esters, and inability to break down TAG.

6.2.4.4 Peroxisomal citrate synthase

Although traditionally viewed as a glyoxylate cycle enzyme, peroxisomal citrate synthase (pCSY) has recently been shown to be required for respiration of TAGs (Pracharoenwattana *et al.*, 2005). There are three putative *pCSY* genes in *Arabidopsis*, but one, *CSY1*, is only expressed in siliques. Consequently, the *csy2 csy3* double mutant has revealed the role of this enzyme during seed germination and vegetative growth. This double mutant fully phenocopies the β-oxidation mutants: TAG breakdown is blocked and seedling establishment is compromised without an alternative carbon source. Mutant seedlings are also 2,4-DB resistant. These phenotypes are not shared with other glyoxylate cycle mutants (see below; Eastmond *et al.*, 2000a; Cornah *et al.*, 2004). Pracharoenwattana *et al.* (2005) concluded that in the absence of CSY, acetyl-CoA cannot be transported from peroxisomes on further metabolized, and so β-oxidation is blocked. Thus, CSY activity is required for carbon export from the peroxisome.

6.2.5 Glyoxylate cycle and gluconeogenesis

The acetyl-CoA produced via the β-oxidation spiral is further metabolized by the glyoxylate cycle. This can be viewed as a modified form of the tricarboxylic acid

(TCA) cycle, which bypasses the decarboxylative steps, allowing the net production of carbon skeletons without the loss of carbon as CO_2 (Eastmond and Graham, 2001). Five enzymes are involved in this cycle, two that are specific to the glyoxylate cycle, isocitrate lyase (ICL; *EC 4.1.3.1*), and malate synthase (MLS; *EC 4.1.3.2*), and three that also appear in the TCA cycle, citrate synthase (CSY; *EC 4.1.3.7*), aconitase (ACO; *EC 4.2.1.3*) and malate dehydrogenase (MDH; *EC 1.1.1.37*). All these enzymes are localized in the peroxisome with the exception of aconitase, which is reported to be a cytosolic enzyme in castor bean, potato (*Solanum tuberosum*), and pumpkin (*Cucurbita* sp.) (Courtois-Verniquet and Douce, 1993; Hayashi *et al.*, 1995). Carbon from the glyoxylate cycle enters gluconeogenesis via the action of phosphoenolpyruvate carboxykinase.

Many of these glyoxylate enzymes have been investigated individually to determine their roles in the glyoxylate cycle and the role of this cycle in lipid utilization, gluconeogenesis, and seedling growth in *Arabidopsis*. Much of this has been achieved by the isolation and characterization of mutants.

6.2.5.1 Isocitrate lyase

Two allelic *Arabidopsis* mutants *isocitrate lyase* 1 and 2 (*icl-1* and *icl2*), both lacking ICL and therefore the glyoxylate cycle, have been characterized (Eastmond *et al.*, 2000b). Under favorable conditions the mutant seedlings establish at high frequency; however, when day-length is shortened or the light intensity decreased, seedling establishment is much reduced. In addition, hypocotyl elongation in the dark is markedly impaired in the *icl* mutants. Again, these phenotypes are restored by the provision of an alternative carbon source. The action of the glyoxylate cycle must be vital for buried seeds, for those in competition with other seedlings, and for those whose lipid reserves are contained within the endosperm. However, fatty acid breakdown still occurs in the mutant despite the block in the net conversion of acetate to carbon, suggesting that acetyl units are respired.

6.2.5.2 Malate synthase

Isolation of two *malate synthase* (*mls*) mutants (Cornah *et al.*, 2004) supports the findings of Eastmond *et al.* (2000b) showing that the glyoxylate cycle is not absolutely required for seedling growth or lipid breakdown. Comparison of *icl* and *mls* mutants showed that *mls* seedlings grew and used lipid faster than *icl* seedlings, yet still had a short hypocotyl phenotype in the dark. The much less severe phenotype of *mls* appears to be due to the seedling's ability to use an alternative gluconeogenic mechanism, potentially by hijacking some of the activities of the photorespiratory enzymes (Cornah *et al.*, 2004).

6.2.5.3 Phosphoenolpyruvate carboxykinase

The seedling phenotype of *phosphenolpyruvate carboxykinase* (*pck1*) mutants is essentially similar to that described for *mls* (Penfield *et al.*, 2004). Hence gluconeogenesis is not an absolute requirement for seedling establishment. It was also shown that this enzyme is required for the utilization of endospermic lipid reserves by the embryo. In castor bean seeds, carbon from endospermic lipid is transported to the embryo for utilization in the form of sucrose (Kornberg and Beevers, 1957).

6.3 Lipid metabolism and seed dormancy

6.3.1 Importance of the ABC transporter for the transition from dormancy to germination

The *cts* locus was originally isolated from a screen aimed at identifying novel factors affecting seed dormancy in *Arabidopsis* (Russell *et al.*, 2000). As discussed above, *cts* is allelic to *pxa1* and *ped3* and encodes a putative fatty acid or acyl-CoA transporter required for fatty acid β-oxidation in the peroxisome (Footitt *et al.*, 2002; Figure 6.2). Two alleles have been thoroughly characterized in relation to their seed germination characteristics: *cts-1* in the Landsberg *erecta* (L*er*) background and *cts-2* in the Wassilewskija (WS) background. In essence, *cts* mutant seeds exhibit a deep dormancy syndrome. The after-ripened seeds fail to germinate, and cold stratification only slightly increases the germination percentage but does not break *cts* dormancy (Russell *et al.*, 2000; Figure 6.3). Promotion of germination is observed when *cts* seeds are exposed to light, stratified, and placed on nitrogen-containing medium such as Murashige and Skoog (Pinfield-Wells *et al.*, 2005). However, even under these optimum conditions only a small percentage of *cts* mutant seeds will germinate, generally averaging around 10% using seeds 2 weeks after harvest (Figure 6.3). During the first experiments using *cts-1* seed, Russell *et al.* (2000) found that the best way to recover *cts-1* mutant seedlings was to excise the embryo from the endosperm and testa and grow in the presence of sucrose. This process bypassed the coat-imposed dormancy of *Arabidopsis* seeds (see Chapter 2) and allowed *cts-1* embryos to grow and establish seedlings. In this way the *cts-1* mutant was propagated and analyzed, and the responsible gene was eventually cloned using map-based techniques (Footitt *et al.*, 2002).

cts mutant seeds have been described as exhibiting a 'forever dormant' phenotype (Russell *et al.*, 2000). The primary evidence that CTS is required for normal seed dormancy comes from the analysis of the genetic interactions of the *cts-1* mutant with other loci known to be required for the onset of dormancy (Russell *et al.*, 2000; Footitt *et al.*, 2002). *Arabidopsis* mutants such as *leafy cotyledon1* (*lec1*), *fusca3* (*fus3*), and *abscisic acid insensitive3* (*abi3*) accumulate decreased amounts of storage reserves, are not desiccation tolerant, and do not enter seed dormancy (Baumlein *et al.*, 1994; West *et al.*, 1994; Nambara *et al.*, 1995; Chapter 1). However, the mutant embryos are still enclosed within the endosperm and testa and so germination must occur before seedling establishment. Therefore, if CTS is required for germination then the *cts-1* mutant should be epistatic to *lec1*, *abi3*, and *fus3*. Yet, when these three mutant loci are combined with *cts-1* all three in fact show epistasis to *cts-1* and the germination percentages of *cts-1 lec1*, *cts-1 fus3*, and *cts-1 abi3-4* seeds closely parallel those of the *fus3*, *lec1*, and *abi3* single mutants (Figure 6.4). These results

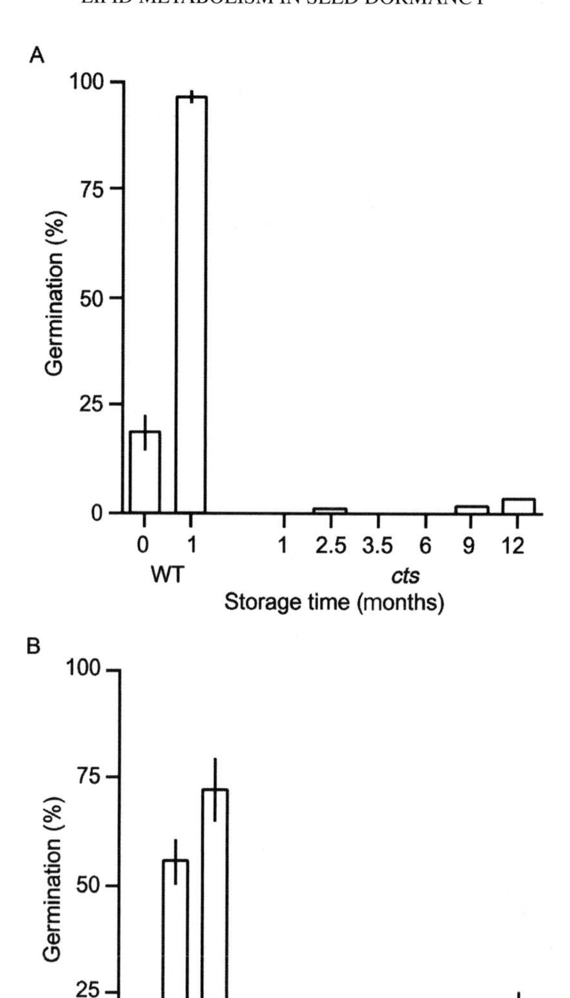

Figure 6.3 The germination phenotype of the *cts-1* mutant. (A) The effect of storage on the germination of wild type and the *cts-1* mutant. (B) *cts-1* does not germinate in response to after-ripening. Although cold stratification overcomes dormancy in wild-type seeds, even extended chilling times do not break the deep dormancy of *cts* mutants. (Reproduced from Russell *et al.*, 2000, with permission from The Company of Biologists Ltd.).

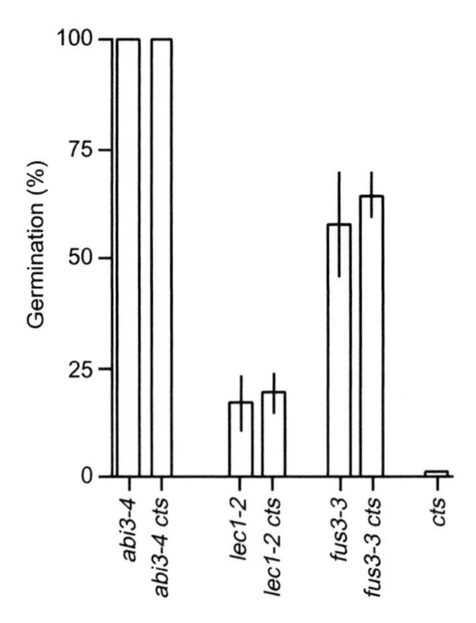

Figure 6.4 Although the *cts* single mutant exhibits strong dormancy, when combined with the nondormant mutants *abi3-4*, *lec1-2*, and *fus3-3*, *cts* mutants are capable of germination. Hence *CTS* is required only for the germination of dormant seeds. (Adapted from Russell *et al.*, 2000, with permission from the Company of Biologists Ltd.).

clearly show that CTS is required for the transition from the dormant to nondormant state but not for the seed germination process itself.

6.3.2 Defects in β-oxidation enzymes, but not in LACS, affect seed dormancy

The presumed biochemical role of the CTS protein is the transport of fatty acids or acyl-CoAs into the peroxisome, although this is yet to be proven experimentally in *Arabidopsis*. Once in the peroxisome, acyl-CoAs serve as substrates for β-oxidation, producing acetyl-CoA for respiration and gluconeogenesis. At the time of the identification of the *CTS* locus (Footitt *et al.*, 2002), few mutants disrupted in fatty acid β-oxidation during seed germination had been described. One exception to this was the *ped1/kat2* locus, which encodes the germination-specific isoform of 3-keto-thiolase. Two independent analyses of thiolase mutants both showed that exogenous carbohydrate application was required for mutant seedling establishment, but neither reported a seed dormancy phenotype (Hayashi *et al.*, 1998; Germain *et al.*, 2001). These observations led Footitt *et al.* (2002) to conclude that the role of CTS in seed dormancy was unique among genes required for peroxisomal fatty acid breakdown via β-oxidation.

However, subsequent observations suggest that other genes involved in lipid metabolism can affect seed germination. An analysis of mutants in long- and

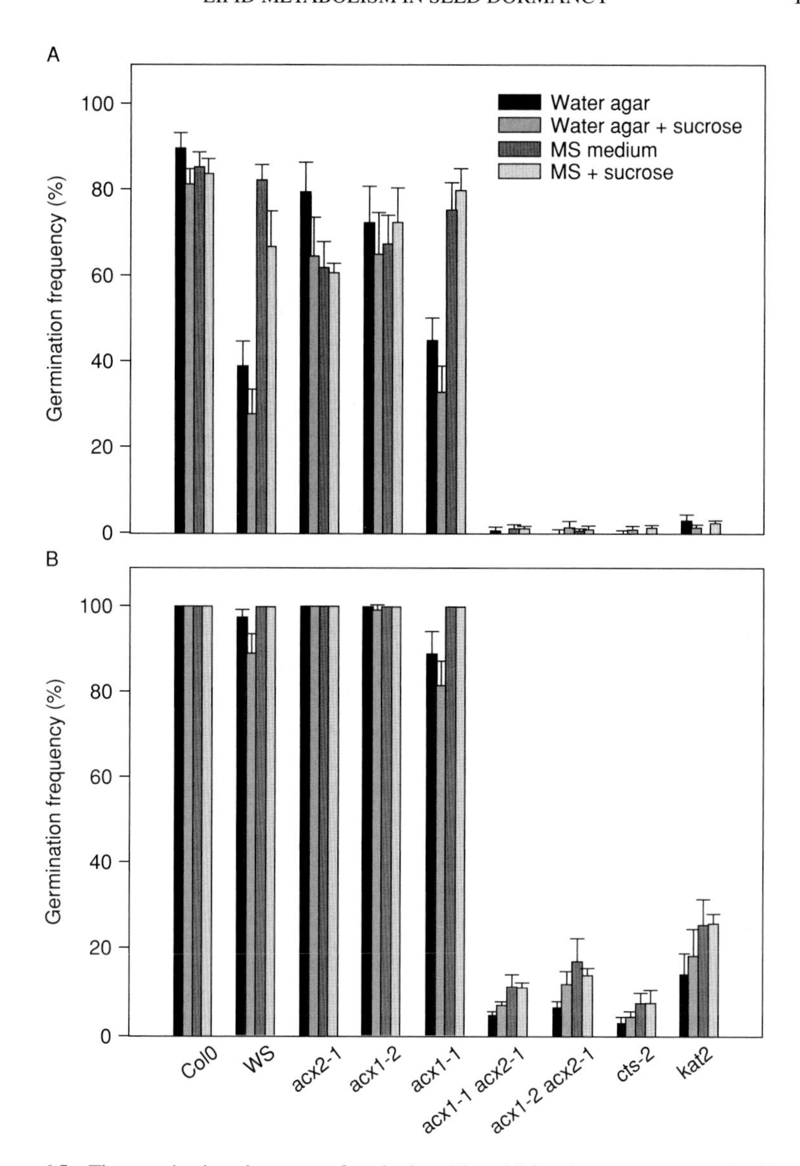

Figure 6.5 The germination phenotype of *cts* is shared by additional mutants compromised in peroxisomal β-oxidation during seed germination. Seed used for this experiment was stored for 2 weeks after harvest and sown without cold treatment (A) or after three nights stratification (B). (Adapted from Pinfield-Wells *et al.*, 2005, with permission from Blackwell Publishing.)

medium-chain ACX, the first step in peroxisomal β-oxidation, indicates that a general defect in β-oxidation can lead to the extreme dormancy phenotype exhibited by the *cts* mutants (Figure 6.5). *acx1 acx2* double mutants that cannot catabolize seed storage lipids also exhibit an increased dormancy phenotype (Pinfield-Wells *et al.*, 2005). We also re-examined the germination kinetics of the *kat2* (thiolase) mutant and found that it also exhibited a *cts*-like dormancy phenotype that had been overlooked in both previous mutant analyses. This was not surprising, as no dormancy phenotype had been described for two further alleles of the *CTS* locus, *ped3* and *pxa1*, prior to the discovery that these are allelic to *cts* (Zolman *et al.*, 2001; Hayashi *et al.*, 2002). In addition, Pracharoenwattana *et al.* (2005) noted that the *csy2 csy3* double mutant seeds that were defective in β-oxidation (Figure 6.1) also germinated very poorly and demonstrated a *cts*-like 'forever dormant' phenotype. Hence, the *csy2 csy3* double mutant represents a fourth lesion in lipid metabolism that leads to an increased dormancy phenotype. These examples underline the importance of including seed germination and dormancy among the phenotypes examined when characterizing mutants in lipid metabolism.

For storage lipid breakdown, fatty acids must be activated to the corresponding acyl-CoA ester to enter the β-oxidation pathway. In the case of long-chain fatty acids, of which the majority of TAG is comprised, this reaction is performed by peroxisomally localized LACS (Figure 6.2). In common with the β-oxidation mutants, *lacs6 lacs7* double mutants exhibit a sugar-dependent seedling establishment phenotype and impaired fatty acid breakdown. Nonetheless, seed germination is reported to be normal in the *lacs6 lacs7* double mutants (Fulda *et al.*, 2004; Pinfield-Wells *et al.*, 2005). In the presence of exogenously applied sucrose, lipid breakdown will take place in *lacs6 lacs7* double mutant seedlings, albeit at an extremely slow rate (Fulda *et al.*, 2004). This has not been observed in *kat2* mutants or in the *acx1 acx2* double mutant, but importantly has been reported in the *cts* mutants (Footitt *et al.*, 2002). The fact that this slow fatty acid breakdown phenotype is also exhibited by dormant *cts* seeds suggests that this cannot be responsible for the normal germination of *lacs6 lacs7* double mutants. The *lacs* mutants are therefore particularly instructive because they appear to show that the β-oxidation of LACS substrates is not required for germination, even though these substrates comprise the majority of TAG and are required for seedling establishment.

6.3.3 Storage lipid mobilization (glyoxylate cycle and gluconeogenesis) is not required for seed dormancy release

At this point it is instructive to stop and briefly consider pathways and enzymes that are required for normal storage lipid mobilization. The best characterized of these are the glyoxylate cycle and gluconeogenesis. These were shown in germinating castor bean seeds to be part of the mechanism for the metabolism and export of carbon from the endosperm to the germinating embryo (Kornberg and Beevers, 1957). Mutants in two glyoxylate cycle enzymes, ICL and MLS, have been described in *Arabidopsis*. Unlike pCSY, ICL and MLS have no role in β-oxidation (Eastmond *et al.*, 2000a; Cornah *et al.*, 2004) and mutants defective in these enzymes do not exhibit enhanced levels of seed dormancy. The same is true for the mutant alleles of the gluconeogenic enzyme PCK (Penfield *et al.*, 2004). *pck1* mutants show abnormal seedling establishment and defects in the mobilization of endospermic reserves and their transport to the embryo, but like the *icl* and *mls* mutants, catabolize fatty acids and germinate in a wild-type fashion. Therefore, it seems that altered seed dormancy is not a general result of compromising storage lipid conversion to soluble carbohydrate; rather it appears to be restricted to mutants affecting fatty acid β-oxidation in the peroxisome.

6.4 Mechanisms for the involvement of β-oxidation in dormancy release

6.4.1 *β-Oxidation does not fuel seed germination*

As we have seen previously, lesions in genes vital for the peroxisomal β-oxidation of fatty acids during seed germination lead to the failure of seedling establishment and the retention of oil bodies in the cotyledons. In all known cases to date, the provision of an alternative carbon source rescues affected seedlings and allows them to attain photosynthetic competence, albeit at a slower rate than the corresponding wild-type seedlings. This is consistent with the role of β-oxidation in providing acetyl-CoA from seed storage reserves as a substrate for respiration, the glyoxylate cycle, and gluconeogenesis. So can the dormancy phenotype of *cts* and related mutants also be explained by the same metabolic block imposed on fatty acid β-oxidation and the lack of carbon availability to fuel germination? So far the weight of results suggests that limited carbon availability is not the cause of the *cts* phenotype. Firstly, the application of exogenous sugars does not rescue the germination of *cts-1* single mutants, even following a period of after-ripening (Footitt *et al.*, 2002). The germination of *cts-1* seeds is improved by the weak acids propionate and butyrate, which could conceivably be used as a carbon source in the dormant seeds. However, this effect could not be demonstrated on dormant *csy2 csy3* (Pracharoenwattana *et al.*, 2005), *acx1 acx2*, *kat2*, or *cts-2* seeds (H. Pinfield-Wells and I.A. Graham, unpublished results). *csy2 csy3* seed germination also was unaffected by the presence of sugars. The observation that exogenous sugars do not promote the germination of mutants defective in peroxisomal β-oxidation was confirmed by the parallel analysis performed by Pinfield-Wells *et al.* (2005). The germination percentages of *cts2*, *kat2*, or *acx1 acx2* seeds were not improved by sugars, yet small promotive effects of nitrogen-containing media or cold stratification were observed. From these results it seems clear that the germination phenotype of β-oxidation mutants is separable from the seedling establishment phenotypes in terms of the effect of sugar. These results suggest that although seedling establishment requires carbon from β-oxidation, seed germination does not.

6.4.2 *β-Oxidation and hormonal signaling*

Gibberellin biosynthesis and perception are essential for seed germination in *Arabidopsis* (see Chapters 5, 8, and 9). This is clear because strong gibberellin-deficient mutants such as *ga1-3* display a complete nongerminating phenotype. This phenotype can be fully restored by an exogenous supply of gibberellins such as GA_3 (Koornneef and Van der Veen, 1980). In addition, the F-BOX protein

SLEEPY1 (SLY1) has been shown to be necessary for gibberellin perception in the seed (Steber *et al.*, 1998; McGinnis *et al.*, 2003; see Chapter 10). Consistent with a role in gibberellin signaling, germination of *sly1* mutant seeds is not rescued by gibberellin treatment (Steber *et al.*, 1998). Interestingly, the germination of both *ga1-3* and *sly1* mutants can be promoted by breaking the testa, indicating that the role of gibberellin lies in overcoming coat-imposed dormancy in *Arabidopsis*. This effect is paralleled in the β-oxidation mutants; mechanical damage to or complete removal of the testa reverts the increased dormancy phenotypes of *cts* mutants, as well as the *kat2*, *acx1 acx2*, and *csy2 csy3* mutants (Footitt *et al.*, 2002; Pinfield-Wells *et al.*, 2005; Pracharoenwattana *et al.*, 2005). The germination of *cts-1* mutant seeds is also partially restored by either the *aberrant testa shape* (*ats*; Léon-Kloosterziel *et al.*, 1994) or *transparent testa glabra* (*ttg*) mutants that affect the integrity of the testa (see Chapter 2). This genetic interaction is also shared with gibberellin-deficient mutants, as seeds of the *ga1-1 transparent testa4* and *ga1-1 ttg* double mutants germinate at low frequency, while those of the *ga1-1* single mutant do not (Debeaujon and Koornneef, 2000). In this way, the germination characteristics of *cts* and other β-oxidation mutants strongly resemble those of *ga1-3* and *sly1*, indicating that β-oxidation is required to overcome coat-imposed dormancy. One key difference is that the β-oxidation mutant germination phenotypes are generally leaky, whereas strong gibberellin biosynthesis and signaling mutants can show absolutely no germination. Thus it is important to precisely determine the relationship between gibberellin action and the increased dormancy exhibited by β-oxidation mutants.

It has been shown that exogenous gibberellin does not rescue germination of *cts-1* seeds, even at high concentrations (Russell *et al.*, 2000). As shown in Figure 6.6,

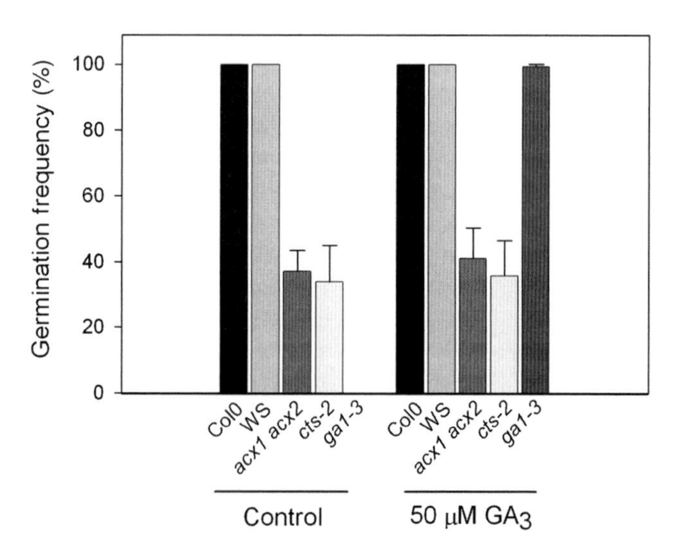

Figure 6.6 The germination of wild type and β-oxidation mutants on media with or without gibberellin. Note that while gibberellin restores the germination of the gibberellin-deficient *ga1-3* mutant, it is ineffective in promoting the germination of *acx1 acx2* or *cts-2*.

exogenous gibberellin does not rescue the germination of seeds of *cts-2* or *acx1 acx2* mutants (similar data for the *kat2* mutant not shown). Thus, the action of β-oxidation must lie genetically downstream of gibberellin biosynthesis, and must be either in the perception of or reaction to the gibberellin signal that overcomes coat-imposed dormancy. This is also consistent with the finding that signals that promote germination largely by upregulating gibberellin biosynthesis, such as light and cold, have only minor effects on the germination of β-oxidation mutants.

6.4.3 *Possible biosynthetic roles for β-oxidation in regulating germination*

Besides the well-known function of peroxisomal β-oxidation during reserve mobilization in germinating oilseeds, there are numerous other potential roles for β-oxidation in plants. Indeed, an analysis of *acx3 acx4* double mutants has shown them to be embryo lethal, demonstrating a constitutive function for β-oxidation during embryo development (Rylott *et al.*, 2003). *csy2 csy3* double mutants cannot complete the vegetative stage of development and senesce and die before the onset of flowering, while the *aim1* mutant exhibits a dwarf phenotype and abnormal flower formation (Richmond and Bleeker, 1999; Pracharoenwattana *et al.*, 2005). These phenotypes point to the multiple roles of fatty acid β-oxidation in plants.

Perhaps the best understood of these is the requirement of β-oxidation for the synthesis of the oxylipin signal molecule methyl jasmonate. Jasmonate synthesis begins with the peroxidation of linolenic acid in the chloroplast and the subsequent production of 12-oxo-phytodienoic acid (OPDA), which can be stored in the chloroplast membranes. OPDA is transported to the peroxisomes where reduction and two rounds of β-oxidation results in the production of jasmonate. It has recently been shown that the ACX1 isoform is specifically required for jasmonate biosynthesis in both *Arabidopsis* and tomato (Cruz Castillo *et al.*, 2004; Li *et al.*, 2005). It is also known that both KAT2 and CTS are required for normal jasmonate synthesis in *Arabidopsis* (Cruz Castillo *et al.*, 2004; Theodoulou *et al.*, 2005). The three mutants *acx1*, *cts-2*, and *kat2* all exhibit highly reduced levels of jasmonate in control and wounded plants, yet it is likely that some residual jasmonate synthesis remains as they exhibit none of the strong phenotypes evident in jasmonate signaling mutants. For instance, *coronatine insensitive1* (*coi1*) and jasmonate resistant1 (*jar1*) are male sterile, while *cts*, *kat2*, and *acx1* all produce viable pollen. So could jasmonate be the dormancy-breaking signal molecule? Germination experiments with the *acx1* and *acx1acx2* double mutants suggest that this is unlikely. Applied jasmonate does not restore the germination of β-oxidation mutants (H. Pinfield-Wells and I.A. Graham, unpublished results). Furthermore, while the *acx1* mutant is impaired in jasmonate biosynthesis, seed dormancy is unaffected (Pinfield-Wells *et al.*, 2005). Only after combining *acx1* with *acx2* is a seed dormancy phenotype observed. Furthermore, strong jasmonate-insensitive mutants such as *coi1* do not exhibit a *cts*-like 'forever dormant' phenotype. However, it is quite possible that an alternative unknown oxylipin signal is required for seed germination in *Arabidopsis*. For example, several intermediates in the jasmonate biosynthetic pathway have biological activity (Blechert *et al.*, 1995). Plants produce a wide range of volatile oxylipin-derived compounds and the functions of most of these remain unknown (Weber *et al.*, 1997).

It is also possible that β-oxidation metabolizes a non-oxylipin-derived signal. Here a precedent exists in the form of IBA. IBA is a naturally occurring auxin that requires the action of β-oxidation to convert it to the more biologically active IAA (Epsteinand Ludwig-Müller, 1993). This conversion to IAA requires two rounds of peroxisomal β-oxidation and is dependent on the action of CTS and KAT2 in young seedlings (Hayashi *et al.*, 1998; Zolman *et al.*, 2001). Indeed, this is the basis of the 2,4-DB resistance genetic screens that first isolated the *ped1/kat2* and *pxa1/ped3/cts* mutants. No systematic attempt to identify all coenzyme A esters in germinating seeds has ever been attempted, partly due to the technical difficulties in quantifying and identifying molecules present at very low abundances in seed samples. However, further characterization of coenzyme A esters in germinating seeds may lead to the identification of a signal molecule important for seed germination that requires β-oxidation for its production. Alternatively, a product of β-oxidation may be important for the synthesis of a signal molecule. One candidate would be acetyl-CoA, which is an essential substrate for many secondary metabolic pathways.

6.4.4 β-Oxidation, reactive oxygen species, and redox control

The process of fatty acid β-oxidation impacts greatly on the redox balance of the cell. The action of ACX produces reactive oxygen intermediates that are believed to be oxidized to molecular oxygen by catalase or by a peroxisomal ascorbate peroxidase system. It has been hypothesized that the peroxisome could be an important source of reactive oxygen signals, including nitric oxide (Corpas *et al.*, 2001; see Chapter 7). Reactive oxygen species (ROS) produced during germination have been shown to modulate seed dormancy in some species (Bailly, 2004). The precise extent of the oxidative load produced during germination and seedling establishment by fatty acid β-oxidation has been revealed by the characterization of the *Arabidopsis vitaminE2* (*vte2*) mutant (Sattler *et al.*, 2004). The *vte2* mutant is deficient in tocopherol biosynthesis in the seed and exhibits a sugar-dependent seedling establishment phenotype consistent with a role for seed tocopherols in peroxisomal fatty acid β-oxidation. Germinated *vte2* mutants accumulate lipid hydroperoxides and hydroxy fatty acids that cannot proceed through β-oxidation, leading to the block in seedling establishment. Oxidative damage to seed proteins during germination has also been documented (Job *et al.*, 2005). These observations suggest that sufficient reactive oxygen exists to cause extensive protein damage even in wild-type seedlings. Much of this could be a by-product of fatty acid β-oxidation, although other sources of ROS do exist in germinating seeds, notably the NADPH oxidases (Kwak *et al.*, 2003). Perhaps more interestingly, tocopherol-deficient seeds exhibit reduced longevity (Sattler *et al.*, 2004). This finding is of particular significance, for it suggests the presence of ROS in seeds prior to germination that could function in dormancy control. ROS generated through the action of the NADPH oxidases have been shown to modulate abscisic acid signaling in germinating seeds

(Kwak *et al.*, 2003). However, there is no evidence that this affects seed dormancy. Although there is ample scope for the production of ROS by fatty acid β-oxidation, there is as yet no direct evidence that this has any regulatory roles during seed dormancy and germination in *Arabidopsis*.

In addition to acetyl-CoA and hydrogen peroxide, the β-oxidation pathway also generates NADH through the action of the MFP. Shuttle mechanisms that involve transport of metabolites between the peroxisome, mitochondria, and cytosol to enable reoxidation of NADH have been proposed (Mettler and Beevers, 1980; Cornah and Smith, 2002). These ultimately result in reducing equivalents being transferred from the peroxisome and production of NADH from NAD in the cytosol and mitochondria. Germination studies in nondormant and dormant caryopses of oat (*Avena sativa*) have shown that there is an increase in NADH content and in the ratio NADH/(NADH + NAD^+) (catabolic redox charge [CRC]) during early germination of nondormant caryopses, which differed significantly from the NADH content and CRC of nongerminating dormant caryopses (Gallais *et al.*, 1998). These studies also showed that ethanol provokes increases in NADH and consequently in CRC, suggesting that germination is enhanced or accelerated when the NADH content increases. *Arabidopsis* mutants compromised in peroxisomal β-oxidation of fatty acids derived from storage lipids during germination could therefore be compromised in reducing equivalents such as NADH, which in turn could impact on germination. However, the fact that the *lacs6 lacs7* double mutant shows normal germination despite the fact that it is as compromised as the other mutants in fatty acid breakdown argues against CRC being the cause of the reduced germination in *cts*, *kat2*, and *acx1 acx2*.

6.5 Conclusions

Using molecular genetic approaches, it has been shown that peroxisomal β-oxidation is required for the termination of dormancy and initiation of germination in *Arabidopsis* seeds. All the evidence suggests that this effect is independent of the well-known role of β-oxidation in the catabolism of seed oil reserves. Further work to determine the precise function of β-oxidation in seeds will provide key insights into the control of seed dormancy and germination.

References

A.R. Adham, B.K. Zolman, A. Millius and B. Bartel (2005) Mutations in *Arabidopsis* acyl-CoA oxidase genes reveal distinct and overlapping roles in β-oxidation. *The Plant Journal* **41**, 859–874.

C. Bailly (2004) Active oxygen species and antioxidants in seed biology. *Seed Science Research* **14**, 93–107.

S. Baud, J.-P. Boutin, M. Miquel, L. Lepiniec and C. Rochat (2002) An integrated overview of seed development in *Arabidopsis thaliana* ecotype Ws. *Plant Physiology and Biochemistry* **40**, 151–160.

H. Baumlein, S. Misera, H. Luerssen, *et al.* (1994) The *fus3* gene of *Arabidopsis thaliana* is a regulator of gene expression during late embryogenesis. *The Plant Journal* **6**, 379–387.

J.D. Bewley and M. Black (1982) *Physiology and Biochemistry of Seeds in Relation to Germination, Volume 2: Viability, Dormancy and Environmental Control*, 375 pp. Springer-Verlag, Berlin.

S. Blechert, W. Brodschelm, S. Holder, *et al.* (1995) The octadecanoic pathway: signal molecules for the regulation of secondary pathways. *Proceedings of the National Academy of Sciences of the United States of America* **92**, 4099–4105.

T.Y. Chia, M.J. Pike and S. Rawsthorne (2005) Storage oil breakdown during embryo development of *Brassica napus* (L.). *Journal of Experimental Botany* **56**, 1285–1296.

T.G. Cooper and H. Beevers (1969) β-Oxidation in glyoxysomes from castor bean endosperm. *Journal of Biological Chemistry* **244**, 3514–3520.

J.E. Cornah, V. Germiane, J.L. Ward, M.H. Beale and S.M. Smith (2004) Lipid utilization, gluconeogenesis and seedling growth in *Arabidopsis* mutants lacking the glyoxylate cycle enzyme malate synthase. *Journal of Biological Chemistry* **279**, 42916–42923.

J.E. Cornah and S.M. Smith (2002) Synthesis and function of glyoxylate cycle enzymes. In: *Plant Peroxisomes: Biochemistry, Cell Biology and Biotechnological Applications* (eds A. Baker and I.A. Graham), pp. 57–101. Kluwer Academic Publishers, London.

F.J. Corpas, J.B. Barroso and L.A. del Rio (2001). Peroxisomes as a source of reactive oxygen species and nitric oxide signal molecules in plant cells. *Trends in Plant Science* **6**, 145–150.

F. Courtois-Verniquet and R. Douce (1993) Lack of aconitase in glyoxysomes and peroxisomes. *Biochemical Journal* **294**, 103–107.

M. Cruz Castillo, C. Martinez, A. Buchala, J.P. Metraux and J. Leon (2004) Gene-specific involvement of β-oxidation in wound-activated responses in *Arabidopsis*. *Plant Physiology* **135**, 85–94.

I. Debeaujon and M. Koornneef (2000) Gibberellin requirement for *Arabidopsis* seed germination is determined both by testa characteristics and embryonic abscisic acid. *Plant Physiology* **122**, 415–424.

P.J. Eastmond (2004) Cloning and characterization of the acid lipase from castor beans. *Journal of Biological Chemistry* **279**, 45540–45545.

P.J. Eastmond (2006) *SUGAR-DEPENDENT1* encodes a patatin domain triacylglycerol lipase that initiates storage oil breakdown in germinating *Arabidopsis* seeds. *The Plant Journal* **18**, 665–675.

P.J. Eastmond, V. Germain, P.R. Lange, J.H. Bryce, S.M. Smith and I.A. Graham (2000a) Postgerminative growth and lipid catabolism in oilseeds lacking the glyoxylate cycle. *Proceedings of the National Academy of Sciences of the United States of America* **97**, 5669–5674.

P.J. Eastmond and I.A. Graham (2000) The multifunctional protein AtMFP2 is co-ordinately expressed with other genes of fatty acid β-oxidation during seed germination in *Arabidopsis thaliana* (L.) Heynh. *Biochemical Society Transactions* **28**, 95–99.

P.J. Eastmond and I.A. Graham (2001) Re-examining the role of the glyoxylate cycle in oilseeds. *Trends in Plant Science* **6**, 72–77.

P.J. Eastmond, M.A. Hooks, D. Williams, *et al.* (2000b) Promoter trapping of a novel medium-chain acyl-CoA oxidase, which is induced transcriptionally during *Arabidopsis* seed germination. *Journal of Biological Chemistry* **275**, 34375–34381.

P.J. Eastmond and R.L. Jones (2005) Hormonal regulation of gluconeogenesis in cereal aleurone is strongly cultivar-dependent and gibberellin action involves SLENDER1 but not GAMYB. *The Plant Journal* **44**, 483–493.

E. Epstein and J. Ludwig-Müller (1993) Indole-3-butyric acid in plants: occurrence, synthesis, metabolism, and transport. *Physiologia Plantarum* **88**, 382–389.

S. Footitt, S.P. Slocombe, V. Larner, *et al.* (2002) Control of germination and lipid mobilization by COMATOSE, the *Arabidopsis* homologue of human ALDP. *EMBO Journal* **21**, 2912–2922.

M. Fulda, J. Schnurr, A. Abbadi, E. Heinz and J. Browse (2004) Peroxisomal acyl-CoA synthetase activity is essential for seedling development in *Arabidopsis thaliana*. *The Plant Cell* **16**, 394–405.

S. Gallais, M-A. Pou de Crescenzo and D.L. Laval-Martin (1998) Pyridine nucleotides and redox charges during germination of non-dormant and dormant caryopses of *Avena sativa* L. *Journal of Plant Physiology* **135**, 664–669.

V. Germain, E.L. Rylott, T.R. Larson, *et al.* (2001) Requirement for 3-ketoacyl-CoA thiolase-2 in peroxisome development, fatty acid β-oxidation and breakdown of triacylglycerol in lipid bodies of *Arabidopsis* seedlings. *The Plant Journal* **28**, 1–12.

S. Goepfert, C. Vidoudez, E. Rezzonico, J.K. Hiltunen and Y. Poirier (2005) Molecular identification and characterization of the *Arabidopsis* $\Delta^{3,5}$,$\Delta^{2,4}$-dienoyl-coenzyme A isomerase, a peroxisomal enzyme participating in the β-oxidation cycle of unsaturated fatty acids. *Plant Physiology* **138**, 1947–1956.

I.A. Graham and P.J. Eastmond (2002) Pathways of straight and branched chain fatty acid catabolism in higher plants. *Progress in Lipid Research* **41**, 156–181.

N.H. Gram (1982) The ultrastructure of germinating barley seeds. I: Changes in the scutellum and aleurone layer in Nordal barley. *Carlsberg Research Communications* **47**, 143–162.

M. Hayashi, L. De Bellis, A. Alpi and M. Nishimura (1995) Cytosolic aconitase participates in the glyoxylate cycle in etiolated pumpkin cotyledons. *Plant and Cell Physiology* **36**, 669–680.

M. Hayashi, K. Nito, R. Takei-Hoshi, *et al.* (2002) Ped3p is a peroxisomal ATP-binding cassette transporter that might supply substrates for fatty acid β-oxidation. *Plant and Cell Physiology* **43**, 1–11.

M. Hayashi, K. Toriyama, M. Kondo and M. Nishimura (1998) 2,4-Dichlorophenoxybutyric acid-resistant mutants of *Arabidopsis* have defects in glyoxysomal fatty acid β-oxidation. *The Plant Cell* **10**, 183–195.

E.H. Hettema, C.W. van Roermund, B. Distel, *et al.* (1996) The ABC transporter proteins Pat1 and Pat2 are required for import of long-chain fatty acids into peroxisomes of *Saccharomyces cerevisiae*. *The EMBO Journal* **15**, 3813–3822.

M.A. Hooks (2002) Molecular biology, enzymology and physiology of β-oxidation. In: *Plant Peroxisomes* (eds A. Baker and I.A. Graham), pp. 19–55. Kluwer Academic Publishers, London.

M.A. Hooks, F. Kellas and I.A. Graham (1999) Long-chain acyl-CoA oxidases of *Arabidopsis*. *The Plant Journal* **20**, 1–13.

A.H.C. Huang (1992) Oil bodies and oleosins in seeds. *Annual Review of Plant Physiology and Plant Molecular Biology* **43**, 177–200.

C. Job, L. Rajjou, Y. Lovigny, M. Belghazi and D. Job (2005) Patterns of protein oxidation in *Arabidopsis* seeds and during germination. *Plant Physiology* **138**, 790–802.

M. Koornneef and J.H. Van der Veen (1980) Induction and analysis of gibberellin-sensitive mutants in *Arabidopsis thaliana* (L.) Heynh. *Theoretical and Applied Genetics* **58**, 257–263.

H.L. Kornberg and H. Beevers (1957) A mechanism of conversion of fat to carbohydrate in castor beans. *Nature* **180**, 35–36.

J.M. Kwak, I.C. Mori, Z.M. Pei, *et al.* (2003) NADPH oxidase *AtrbohD* and *AtrbohF* genes function in ROS-dependent ABA signaling in *Arabidopsis*. *The EMBO Journal* **22**, 2623–2633.

K.M. Leon-Kloosterziel, C.J. Keijzer and M. Koornneef (1994) A seed shape mutant of *Arabidopsis* that is affected in integument development. *The Plant Cell* **6**, 385–392.

C. Li, A.L. Schilmiller, G. Liu, *et al.* (2005) Role of β-oxidation in jasmonate biosynthesis and systemic wound signaling in tomato. *The Plant Cell* **17**, 971–986.

K. Matsui, S. Fukutomi, M. Ishii and T. Kajiwara (2004) A tomato lipase homologous to DAD1 (LeLID1) is induced in post-germinative growing stage and encodes a triacylglycerol lipase. *FEBS Letters* **569**, 195–200.

K.M. McGinnis, S.G. Thomas, J.D. Soule, *et al.* (2003) The *Arabidopsis SLEEPY1* gene encodes a putative F-box subunit of an SCF E3 ubiquitin ligase. *The Plant Cell* **15**, 1120–1130.

I.J. Mettler and H. Beevers (1980) Oxidation of NADH in glyoxysomes by a malate-aspartate shuttle. *Plant Physiology* **66**, 555–560.

J. Mosser, A.-M. Douar, C.-O. Sarde, *et al.* (1993) Putative X-linked adrenoleukodystrophy gene shares unexpected homology with ABC transporters. *Nature* **361**, 726–730.

E. Nambara, K. Keith, P. McCourt and S. Naito (1995) A regulatory role for the *ABI3* gene in the establishment of embryo maturation in *Arabidopsis thaliana*. *Development* **121**, 629–636.

S. Penfield, E.L. Rylott, A.D. Gilday, S. Graham, T.R. Larson and I.A. Graham (2004) Reserve mobilization in the *Arabidopsis* endosperm fuels hypocotyl elongation in the dark, is independent of abscisic acid, and requires *PHOSPHOENOLPYRUVATE CARBOXYKINASE1*. *The Plant Cell* **16**, 2705–2718.

H. Pinfield-Wells, E.L. Rylott, A.D. Gilday, *et al.* (2005) Sucrose rescues seedling establishment but not germination of *Arabidopsis* mutants disrupted in peroxisomal fatty acid catabolism. *The Plant Journal* **43**, 861–872.

I. Pracharoenwattana, J.E. Cornah and S.M. Smith (2005) *Arabidopsis* peroxisomal citrate synthase is required for fatty acid respiration and seed germination. *The Plant Cell* **17**, 2037–2048.

T.A. Richmond and A.B. Bleecker (1999) A defect in β-oxidation causes abnormal inflorescence development in *Arabidopsis*. *The Plant Cell* **11**, 1911–1923.

L. Russell, V. Larner, S. Kurup, S. Bougourd and M. Holdsworth (2000) The *Arabidopsis COMATOSE* locus regulates germination potential. *Development* **127**, 3759–3767.

E.L. Rylott, P.J. Eastmond, A.D. Gilday, *et al.* (2006) The *Arabidopsis thaliana* multifunctional protein gene (*MFP2*) of peroxisomal β-oxidation is essential for seedling establishment. *The Plant Journal* **45**, 930–941.

E.L. Rylott, M.A. Hooks and I.A. Graham (2001) Co-ordinate regulation of genes involved in storage lipid mobilization in *Arabidopsis thaliana*. *Biochemical Society Transactions* **29**, 283–287.

E.L. Rylott, C.A. Rodgers, A.D. Gilday, T. Edgell, T.R. Larson and I.A. Graham (2003) *Arabidopsis* mutants in short- and medium-chain acyl-CoA oxidase activities accumulate acyl-CoAs and reveal that fatty acid β-oxidation is essential for embryo development. *Journal of Biological Chemistry* **278**, 21370–21377.

S.E. Sattler, L.U. Gilliland, M. Magallanes-Lundback, M. Pollard and D. DellaPenna (2004) Vitamin E is essential for seed longevity and for preventing lipid peroxidation during germination. *The Plant Cell* **16**, 1419–1432.

J.M. Shockey, M.S. Fulda and J.A. Browse (2002) *Arabidopsis* contains nine long-chain acyl-coenzyme A synthetase genes that participate in fatty acid and glycerolipid metabolism. *Plant Physiology* **129**, 1710–1722.

C.M. Steber, S.E. Cooney and P. McCourt (1998) Isolation of the GA-response mutant *sly1* as a suppressor of *ABI1-1* in *Arabidopsis thaliana*. *Genetics* **149**, 509–521.

F.L. Theodoulou, K. Job, S.P. Slocombe, *et al.* (2005) Jasmonic acid levels are reduced in COMATOSE ATP-binding cassette transporter mutants. Implications for transport of jasmonate precursors into peroxisomes. *Plant Physiology* **137**, 835–840.

N. Verleur, E.H. Hettema, C.W. van Roermund, H.F. Tabak and R.J. Wanders (1997) Transport of activated fatty acids by the peroxisomal ATP-binding-cassette transporter Pxa2 in a semi-intact yeast cell system. *European Journal of Biochemistry* **249**, 657–661.

H. Weber, B.A. Vick and E.E. Farmer (1997) Dinor-oxo-phytodienoic acid – a new hexadecanoid signal in the jasmonate family. *Proceedings of the National Academy of Sciences United States of America* **94**, 10473–10478.

M. West, K.M. Yee, J. Danao, *et al.* (1994) *LEAFY COTYLEDON1* is an essential regulator of late embryogenesis and cotyledon identity in *Arabidopsis*. *The Plant Cell* **6**, 1731–1745.

B.K. Zolman, I.D. Silva and B. Bartel (2001) The *Arabidopsis pxa1* mutant is defective in an ATP-binding cassette transporter-like protein required for peroxisomal fatty acid β-oxidation *Plant Physiology* **127**, 1266–1278.

7 Nitric oxide in seed dormancy and germination

Paul C. Bethke, Igor G.L. Libourel and Russell L. Jones

7.1 Nitric oxide in plant growth and development

Nitric oxide (NO) is a reactive, gaseous, free radical that functions as a potent signaling molecule in plants and animals. Numerous roles for NO in plant growth and development have been uncovered during the past decade (Lamattina *et al.*, 2003; Neill *et al.*, 2003; Wendehenne *et al.*, 2004; Delledonne, 2005; Shapiro, 2005). In early reports it was shown that NO acts during the hypersensitive response to protect plants from pathogens. NO has subsequently been shown to inhibit maturation and senescence, and to function as an antioxidant that slows the rate of programmed cell death. NO promotes vigorous seedling development, represses the transition to flowering, facilitates abscisic acid (ABA)-induced stomatal closure, and may be required for the adaptation of roots to low oxygen concentrations. The number of processes in which NO participates is increasing rapidly, and additional NO-dependent events in plants are likely to be discovered. In this chapter we review data that indicate that NO plays important roles in seed dormancy and germination. Some of the literature discussed here was written prior to the recognition that NO functions as a signaling molecule, and we discuss it with the benefit of hindsight. The rest of the data summarized here come from experiments devised to test directly for an involvement of NO in seed dormancy or germination. When taken together, these data support the hypothesis that NO promotes the germination of seeds, either by reducing seed dormancy or by minimizing the effects of environmental conditions that inhibit germination.

7.2 Challenges in NO chemistry and biology

NO is a challenging molecule to work with for both chemists and biologists. These challenges derive from the nature of the NO molecule itself. NO is a highly reactive, free radical gas that dissolves in aqueous solution but preferentially partitions into hydrophobic environments such as biomembranes. NO is not stable in aerobic environments. NO in the gas phase reacts with itself and molecular oxygen to produce nitrogen dioxide (NO_2), as indicated in Equation 7.1, and NO_2 that diffuses into an aqueous environment is converted to equal molar concentrations of nitrite (NO_2^-) and nitrate (NO_3^-), as indicated in Equations 7.2 and 7.3 (Ignarro *et al.*, 1993).

$$2NO + O_2 \rightarrow 2NO_2 \quad (7.1)$$

$$2NO_2 \leftrightarrow N_2O_4 \quad (7.2)$$

$$N_2O_4 + H_2O \rightarrow NO_2^- + NO_3^- + 2H^+ \quad (7.3)$$

This conversion of one oxide of nitrogen into another is a theme that runs throughout NO chemistry and biology, and is the most serious hurdle to understanding NO function. NO is also unstable in aerobic, aqueous solutions and undergoes a set of reactions (Equations 7.1, 7.4, and 7.5) that result in the preferential formation of nitrite (Feelisch, 1991).

$$2NO + O_2 \rightarrow 2NO_2 \quad (7.1)$$

$$NO + NO_2 \leftrightarrow N_2O_3 \quad (7.4)$$

$$N_2O_3 + H_2O \rightarrow 2NO_2^- + 2H^+ \quad (7.5)$$

Under suitably acidic conditions, the overall reaction of Equations 7.4 and 7.5 is reversible, as shown in Equations 7.6, 7.7, and 7.4′ (Yamasaki, 2000).

$$NO_2^- + H^+ \leftrightarrow HNO_2 \quad (pK_a \sim 3.2) \quad (7.6)$$

$$2HNO_2 \leftrightarrow N_2O_3 + H_2O \quad (7.7)$$

$$N_2O_3 \leftrightarrow NO + NO_2 \quad (7.4')$$

In biological systems the presence of oxidizing species such as oxyhemoglobin may alter the final products formed by these reactions and may result in the complete oxidation of NO and NO_2^- into NO_3^- (Ignarro *et al.*, 1993).

It is important to note that many reactions in Equations 7.1–7.7 are reversible. Hence, NO gas can be converted to nitrogen dioxide, nitrous acid, nitrite, or nitrate; and nitrous acid, nitrogen dioxide, and nitrite can be converted into NO. The rate and extent of these conversions depend on the concentration of NO, the concentration of oxygen, the presence of metals that can act as catalysts, the redox environment, and the pH of the aqueous phase, among other parameters. It is into this morass of chemical uncertainty that biologists must walk in order to investigate the biological effects of NO. Plant biologists working with nitrate, nitrite, other nitrogen oxides, and perhaps other N-containing compounds, may have trod there unknowingly in the past.

7.3 Tools used in NO research

A variety of tools are available for NO research, and these have greatly facilitated our understanding of NO action in biological systems, including seeds. The most useful tools include NO donors, scavengers, and detection systems. The most straightforward NO donor is NO gas, but purified NO gas has seldom been used as a NO donor by plant scientists because of its instability and the difficulties inherent in using controlled gas flows for experiments. An alternative NO donor is acidified nitrite, which produces NO through a relatively simple set of chemical reactions (Equations 7.6, 7.7, 7.4′), but requires a pH of less than approximately 5. Other NO donor compounds, such as SNP (sodium nitroprusside), SNAP (*S*-nitroso-*N*-acetylpenicillamine), SIN-1 (3-morpholinosydnonimine, hydrochloride), or DEANO (diethylamine nitric oxide), and the NONOates (diazeniumdiolates containing the $[N(O)NO]^-$ functional group) have been used more commonly by biologists to apply NO (Feelisch, 1998). These donor compounds decompose under the appropriate conditions and liberate free NO along with secondary products. Donor compounds have been developed with a range of half-lives, from a few seconds to many hours, and they are relatively simple to use. Controls for the secondary products formed by NO donors are warranted as some of these, such as NO^+, NO^-, spermine from spermine-NONOate and cyanide from SNP, may have biological effects.

The potential pitfalls of using SNP as a NO donor are highlighted here, as it has been one of the most widely used NO donors in germination experiments. SNP in an illuminated solution undergoes photolysis to produce NO as well as CN^- (Feelisch, 1998). Using a simple method to deliver gasses to seeds in a closed system (Figure 7.1A), we showed that far more hydrogen cyanide (HCN) than NO was volatilized from solutions of SNP (Bethke *et al.*, 2006b). HCN was released rapidly

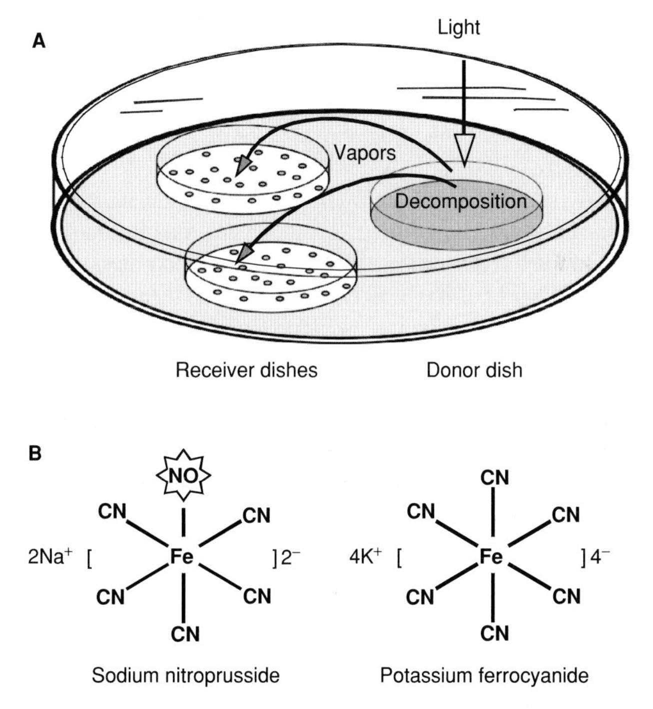

Figure 7.1 Schematics of the NO experimental system and chemicals. A simple means of applying volatile components to imbibed seeds is diagrammed in (A). The structures of sodium nitroprusside (SNP) and potassium ferrocyanide (K-FeCN) are indicated in (B). Both SNP and K-FeCN can decompose and release cyanide, but only SNP can release NO.

from solutions of SNP, whereas NO was very poorly volatilized. Salts of ferricyanide and ferrocyanide have been used as controls for experiments with SNP because they are structurally related to SNP but lack a NO-donating group (Figure 7.1B). Like SNP, these compounds decompose readily in the light and release CN^- (Bethke *et al.*, 2006b). We showed that volatile HCN released from SNP, potassium ferrocyanide, and potassium ferricyanide reduced dormancy of *Arabidopsis* seeds and stimulated germination (Bethke *et al.*, 2006b). Cyanide is well known to break seed dormancy (Roberts, 1964a; Taylorson and Hendricks, 1973). From these data it is clear that the observed effect on dormancy in response to treatment with SNP was inconclusive evidence for the involvement of NO in this process.

In contrast to the situation with NO donor compounds, where a broad range of reagents is available, relatively few NO scavengers have been developed. Among the most widely used are PTIO (2-phenyl-4,4,5,5,-tetramethylimidazoline-1-oxyl-3-oxide) and its derivatives, especially c-PTIO (2-[4-carboxyphenyl]-4,4,5,5-tetramethylimidazoline-1-oxyl-3-oxide). These compounds have been shown repeatedly to function in biological tissues, where they interact with NO and produce nitrite as a by-product (Akaike *et al.*, 1993; Goldstein *et al.*, 2003). Both PTIO and c-PTIO have been used successfully with seeds and seed tissues to scavenge NO. Although the PTIOs are relatively specific for NO, concerns have been raised about their ability to react with other compounds. Of those, the potential to react with reduced ascorbic acid seems most worrisome (Feelisch, 1998). Hemoglobin is another NO scavenger, but opportunities for its use with intact cells and tissues are limited by the relatively low concentrations that can be used, and by its exclusion from cells. To our knowledge, hemoglobin has not been used successfully as a NO scavenger with seeds.

Several NO detection systems are in use by plant biologists. No one system has yet been demonstrated to be superior to the others. The simplest method for measuring NO production is to convert the NO to nitrite and then measure the amount of nitrite with the Greiss reaction (Granger *et al.*, 1996). This method is not available in situations where nitrite is added to, or produced by, the experimental material. It is available when NO in a biological system is released into the gas phase since $NO_{(g)}$ can be chemically reduced to NO_2^- in the presence of appropriate catalysts. The detection limit for nitrite in the Greiss reaction is $\sim$100 nM. More elaborate methods for measuring NO include NO-specific electrodes, a chemiluminescent method based on the interaction between NO and ozone, and molecular probes that react with NO and produce a fluorescent product. The latter method may also provide data on the specific location of NO synthesis within a tissue. Commonly used NO-reactive fluorescent probes include DAF-2 (4,5-diaminofluorescein diacetate), DAF-FM (4-amino-5-methylamino-2′,7′-difluorofluorescein), and their cell-permeant acetoxymethyl esters. These NO-reactive probes are relatively simple to use and have been used with seeds and seed tissues. Nonspecific signals have been reported in some cases (Beligni *et al.*, 2002; Stohr and Stremlau, 2006), and rigorous controls are necessary to preclude misidentification of a nonspecific signal as an indication of NO production. All three of these NO detection systems are extremely sensitive and can detect NO in solution to 1–10 nM. NO in gas streams at $\sim$1 ppb can be detected with NO electrode systems and systems employing ozone. Comparisons have rarely been done across platforms, however, to determine the robustness of the measured data and to compare quantitative measurements made with one system to those made with another system.

7.4 Roles of NO and other N-containing compounds in seed dormancy and germination

The literature showing that various nitrogen-containing compounds stimulate seed germination is extensive and dates back to at least 1909 (Lehman, 1909). Table 7.1 lists N-containing compounds that stimulate germination of seeds from a wide range of species. What is striking about this list is that it contains a broad range of compounds, including simple inorganic salts of nitrate and nitrite, reduced forms of nitrogen such as ammonium salts, urea, hydroxylamine, and methylamine; metabolic inhibitors such as azide and cyanide; various classes of NO donors such as SNP and SNAP; and gaseous NO_2 and NO. As suggested by Equations 7.1 through 7.7, many N-containing compounds can undergo abiotic or enzymatic reactions that convert

Table 7.1 Nitrogen-containing compounds known to stimulate seed germination

Compound	Reference
Nitrate	Toole *et al.*, 1956; Roberts, 1969;
Nitrite	Toole *et al.*, 1956; Roberts, 1969
Acidified nitrite	Cohn *et al.*, 1983; Giba *et al.*, 1998
Nitric oxide	Libourel *et al.*, 2006
Nitrogen dioxide	Cohn and Castle, 1984; Keeley and Fotheringham, 1997
Azide	Roberts, 1969; Taylorson and Hendricks, 1973; Hendricks and Taylorson, 1974
Cyanide	Roberts, 1964a,b; Taylorson and Hendricks, 1973; Hendricks and Taylorson, 1974
Ammonium salts	Hendricks and Taylorson, 1974
Arsenate	Ballard and Lip, 1967
Hydroxylamine	Major and Roberts, 1968; Hendricks and Taylorson, 1974, 1975
Thiourea	Roberts, 1969; Hendricks and Taylorson, 1975
Urea	Crocker and Barton, 1953
Glycine	Poljakoff-Mayber and Mayer, 1963
Sodium nitroprusside	Giba *et al.*, 1998; Beligni and Lamatina 2000, etc.
SNAP	Giba *et al.*, 1998; Beligni and Lamatina 2000; Kopyra and Gwozdz, 2003, etc.
SIN-1	Giba *et al.*, 1998
Nitroglycerine	Batak *et al.*, 2002
Potassium ferrocyanide	Bethke *et al.*, 2006b
Potassium ferricyanide	Bethke *et al.*, 2006b
Isosorbide mononitrate	Grubisic *et al.*, 1992
Isosorbide dinitrate	Grubisic *et al.*, 1992
Pentaerythrityl tetranitrate	Grubisic *et al.*, 1992
Methylene blue	Roberts, 1964b

them into other N-containing compounds. The possibility exists, therefore, that when seeds were treated with N-containing compounds such as nitrite, nitrate, and others, they might have been treated indirectly with NO.

7.4.1 Nitrate, nitrite, and ammonium

Simple inorganic forms of nitrogen such as nitrate, nitrite, and ammonium are often potent promoters of germination, but the biochemical mechanism underlying this phenomenon remains unclear despite decades of research. Roberts (1964a,b) argued that because nitrate and nitrite could act as electron acceptors, their stimulatory effects on germination were likely to be on respiratory metabolism at the level of electron transport or by oxidation of NADPH. He argued that this would allow for greater carbon flow through the pentose phosphate pathway (Figure 7.2). This pathway supplies the pentose sugars required in early germination for nucleic acids (ribose), cell walls (xylulose), and the shikimic acid pathway (erythrose). One argument in favor of this proposal came from observations showing that in many seeds, nitrate, and especially nitrite, is effective in stimulating germination of dormant seeds, whereas ammonium, which is fully reduced and therefore cannot act as an electron acceptor, is generally much less effective (Roberts, 1969; Hendricks and Taylorson, 1974). Notable exceptions to this general conclusion are seeds of the genus *Barbarea*, such as *B. verna* (winter cress) and *B. vulgaris* (yellow rocket), whose germination is dramatically stimulated by ammonium (Hendricks and Taylorson, 1974). The stimulatory effect of ammonium salts in seeds from *Barbarea* and other species was ascribed to the requirement for reduced nitrogen in metabolism (Hendricks and Taylorson, 1974).

In a particularly insightful publication, Hendricks and Taylorson (1974) analyzed the effects of various inorganic forms of nitrogen on the germination of seeds from a

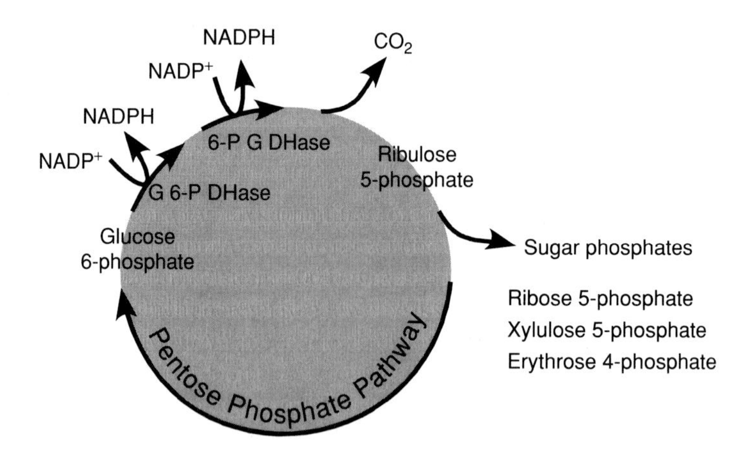

Figure 7.2 Production of sugar phosphates through the pentose phosphate pathway. The pentose phosphate pathway utilizes $NADP^+$ to produce sugar phosphates from glucose 6-phosphate. The enzymes that use $NADP^+$ and produce NADPH are glucose 6-phosphate dehydrogenase (G 6-P DHase) and 6-phosphogluconate dehydrogenase (6-P G DHase).

range of species and concluded that the effectiveness of nitrate on germination was via its reduction to nitrite, hydroxylamine, or NO, but not from further reduction to ammonium. Their model concluded that the final product resulting from treatment of seeds with nitrites was likely to be NO, and that NO might accelerate flux through the pentose phosphate pathway by indirectly increasing the oxidation of NADPH.

7.4.2 Cyanide and azide

Many metabolic inhibitors have been shown to stimulate germination (Roberts, 1964a). Azide and cyanide, in particular, are a paradox for plant physiologists. These two compounds are among the most effective in promoting a loss of dormancy and stimulating germination of seeds, yet they are potent metabolic inhibitors. In a series of papers in the 1960s, Roberts (1963, 1964b, 1969) proposed that the effects of cyanide and azide were likely to be mediated via effects on respiratory metabolism. Support for this proposal came from experiments showing that other inhibitors of cytochrome *c* oxidase such as carbon monoxide (CO), methylene blue, and hydroxylamine were also effective in stimulating germination of rice (*Oryza sativa*) and barley (*Hordeum vulgaris*) grains, as well as seeds of other species (Roberts, 1964b; Hendricks and Taylorson, 1974). Roberts' hypothesis posited that treatment with respiratory inhibitors would increase O_2 concentrations within dormant seeds. Elevated O_2 concentrations would promote the oxidation of NADPH and would facilitate the operation of the pentose phosphate pathway, which is limited by the availability of $NADP^+$ (Figure 7.2). Thus, O_2 would facilitate the synthesis of key intermediates required for DNA and RNA synthesis by making available more $NADP^+$.

Roberts' hypothesis was supported by early work by Hendricks and Taylorson (1972), but their later work suggested that the effects of cyanide and azide were unlikely to be mediated by an interaction with cytochromes and the respiratory chain. Taylorson and Hendricks (1973) showed that low concentrations of azide and cyanide that strongly stimulated germination did not affect O_2 consumption, as a measure of respiration, by seeds of *Amaranthus albus*, *Lepidium virginicum*, or *Lactuca sativa* (lettuce). Furthermore, there was limited uptake of cyanide by seeds, and much of that which was taken up was metabolized (Taylorson and Hendricks, 1973). Cyanide is known to react with cysteine in plant tissues to produce cyanoalanine and H_2S. Cyanoalanine in turn is converted to asparagine (Asn) and then to aspartic acid (Asp) (Castric *et al.*, 1972). When *A. albus* seeds were incubated in radiolabeled KCN, 9% of radioactivity was found in cyanoalanine, 20% in Asn, and 10% in Asp. Asn and Asp were both incorporated into proteins in *A. albus* seed, leading to the supposition that cyanide might exert its effect on germination via the proteins into which it was eventually incorporated (Taylorson and Hendricks, 1973).

Further research led Hendricks and Taylorson (1974) to conclude that NO might be a product resulting from treatment of seeds with azide or hydroxylamine. They based this model on the observation that catalase in the presence of H_2O_2 could oxidize azide. The intermediate could be a NO-ferrochelatase that would dissociate to form NO. Hendricks and Taylorson (1974) argued that lipid-storing seeds are uniquely poised to carry out the strong oxidation of N-containing compounds such

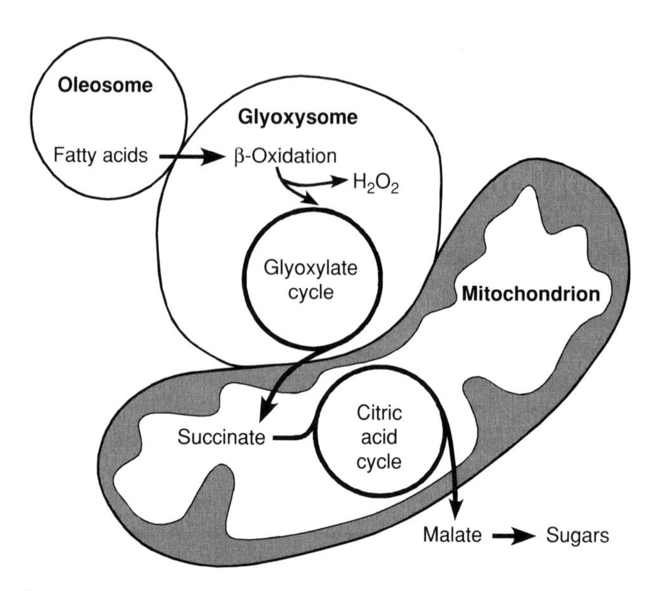

Figure 7.3 Generation of hydrogen peroxide by β-oxidation. Gluconeogenesis is used to produce sugars from stored lipids. Central to this process are β-oxidation and the glyoxylate cycle. Hydrogen peroxide is an obligate by-product of β-oxidation.

as azide, hydroxylamine, and nitrite because they carry out lipid catabolism and gluconeogenesis. As part of these processes, β-oxidation in the glyoxysome (peroxisome) produces H_2O_2 and the glyoxysome also contains catalase (Figure 7.3). Most seeds, including those of *Arabidopsis* and lettuce, and the grains of cereals store lipid and convert these reserves into sugars by β-oxidation (see Chapter 6 for β-oxidation in seeds). Conditions therefore exist in the glyoxysomes of most seeds to produce NO by oxidation of natural compounds such as hydroxylamine and applied compounds such as azide. The offshoot of this is that NO may be a common chemical intermediate in the response of seeds to many of the compounds listed in Table 7.1, and NO may be the molecule responsible for promoting germination.

7.4.3 NO donors and germination

Several lines of direct investigation indicate that NO is involved in one or more processes that promote the germination of seeds. Although only a few species have been examined in detail, Table 7.2 shows that they include representatives from distinct angiosperm lineages. In some cases NO was shown to reduce seed dormancy. In other cases, NO augmented the effects of light on germination, or increased the final germination percentage of largely nondormant seeds. Most of these experiments were done with NO-donor compounds, and the most reliable data are likely to be those where the effects of several donor compounds were similar, or where the effects of NO donor compounds were reversed by a NO scavenger such as c-PTIO. A particularly good example is shown in Figure 7.4. Giba *et al.* (1998) used a range

Table 7.2 Species whose seeds are stimulated to germinate by NO

Species (common name)	Reference
Arabidopsis thaliana (thale cress)	Batak *et al.*, 2002; Bethke *et al.*, 2006b
Brassica napus (kale, rape, turnip, etc.)	Zanardo *et al.*, 2005
Descurainia sophia (flixweed)	Li *et al.*, 2005
Emmenanthe penduliflora (whispering bells)	Keeley and Fotheringham, 1998
Hordeum vulgaris (barley)	Bethke *et al.*, 2004b
Lactuca sativa (lettuce)	Beligni and Lamattina, 2000
Lupinus luteus (lupin)	Kopyra and Gwozdz, 2003
Oryza sativa (rice)	Cohn and Castle, 1984
Panicum virgatum (switchgrass)	Sarath *et al.*, 2006
Paulownia tomentosa (Empress tree)	Giba *et al.*, 1998
Nicotiana plumbaginifolia (tobacco)	Grappin *et al.*, 2000
Pinus mugo (mugo pine)	Cited in Batak *et al.*, 2002
Stellaria media (chickweed)	Cited in Batak *et al.*, 2002
Suaeda salsa	Li *et al.*, 2005
Triticum aestivum (wheat)	Hua *et al.*, 2003
Vaccinium myrtillus (blueberry)	Giba *et al.*, 1995

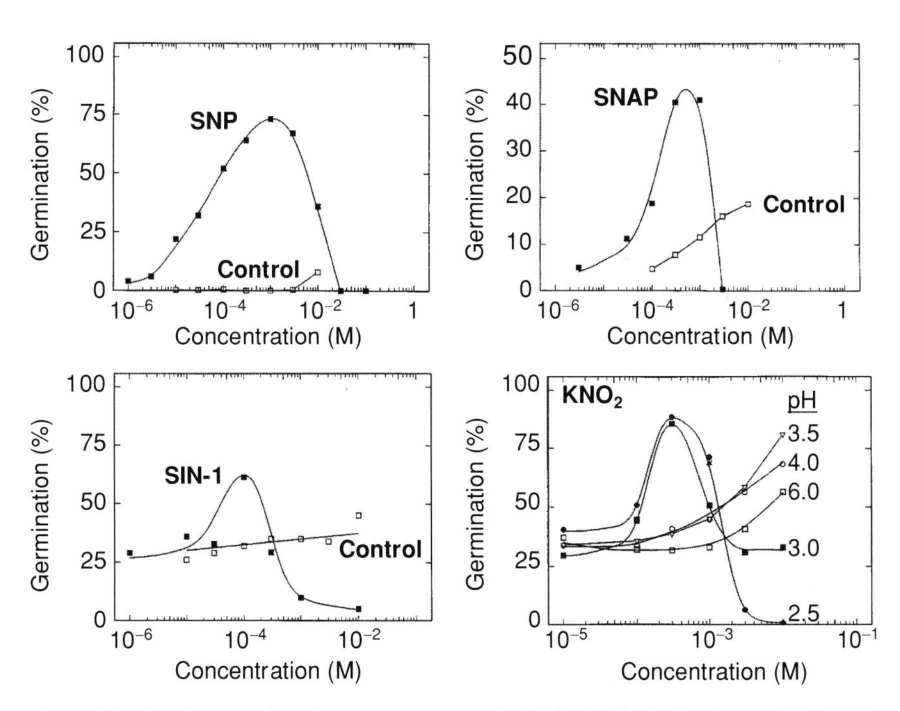

Figure 7.4 Germination of *Paulownia tomentosa* seeds imbibed with the NO donors SNP, SNAP, SIN-1, or nitrite. Seeds were imbibed in respective chemicals and then irradiated with a 5-min pulse of red light. Controls for SNP, SNAP, and SIN-1 were carbonyl prussiate, acetylpenicillamine, and molsidomine respectively. (Data are taken from Giba *et al.*, 1998.)

of NO donors including SNP, SNAP, SIN-1, and acidified nitrite as sources of NO. Controls for the decomposition products of these NO donors were also done, and the data on the whole show convincingly that NO increased red-light-stimulated germination of Empress tree (*Paulownia tomentosa*) seeds.

7.4.4 NO scavengers and germination

NO scavengers have proven to be extremely useful in probing the effects of N-containing compounds on dormancy and germination. As in the case of NO donors, there are important caveats that must be considered when synthetic chemicals are used to scavenge NO. Are these compounds specific, are they toxic to plant cells, and do they form inhibitory complexes when they bind NO? The most commonly used NO scavengers are PTIO and c-PTIO. Very little is published about the specificity of these compounds, especially in plants, but they apparently have very low toxicity in plants even at concentrations of 200 μM or higher.

NO scavengers have been useful in implicating NO as an endogenous regulator of dormancy in plants. For example, dormancy was enhanced when partially dormant lots of *Arabidopsis* and barley seeds were incubated with c-PTIO, but c-PTIO did not affect the germination of seeds whose dormancy had been released by after-ripening or stratification before c-PTIO treatment (Bethke *et al.*, 2004b, 2006b).

NO scavengers have also been used to implicate NO as the effective component when dormancy is broken by a range of N-containing compounds, including NO donors. For example, c-PTIO prevented germination of *Arabidopsis* seeds that would otherwise be induced by SNP, KCN, NaN_3, KNO_3, and KNO_2 (Bethke *et al.*, 2006b; R. Jones, unpublished results). The specificity of the effect of c-PTIO in blocking KCN-stimulated germination of dormant *Arabidopsis* seeds was tested by pretreating c-PTIO with NO, HCN, or air. Whereas pretreatment of c-PTIO with purified NO canceled its inhibitory effect on KCN-treated seeds, pretreatment of c-PTIO with KCN did not impair the ability of c-PTIO to inhibit cyanide-stimulated germination (Bethke *et al.*, 2006b). Hence, under these conditions, the reversal of KCN-stimulated dormancy loss resulting from c-PTIO treatment was not caused by an interaction between c-PTIO and CN^-, but was likely to be an interaction with NO.

7.5 Biochemical and molecular basis of NO action in seeds

Whereas there is a growing body of evidence that NO plays a role in seed dormancy and germination, there is little consensus concerning how or where it is synthesized or its mechanism of action at the cellular level.

7.5.1 Synthesis of NO by plants

NO is made by plants, as was first demonstrated in the 1980s. At that time it was widely known that NO was produced by nitrate reductase (NR) in denitrifying bacteria (Firestone, 1982). Dean and Harper (1986) extended these findings to show

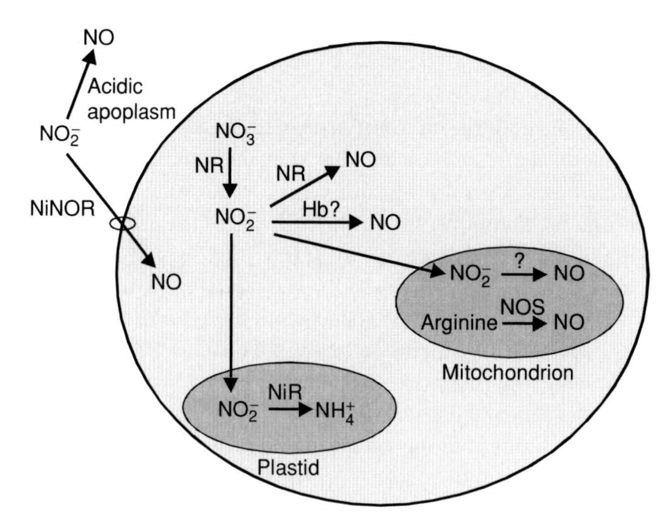

Figure 7.5 Known and proposed mechanisms for the production of NO from nitrite and arginine in plants. Nitrite is reduced to NO via nonenzymatic reactions in the apoplast, nitrite-nitric oxide reductase (NiNOR) at the plasma membrane, nitrate reductase (NR) in the cytosol, potentially a hemoglobin (Hb), and unknown mechanisms in the mitochondrion. Nitrite is also reduced to ammonium by nitrite reductase (NiR) in the plastid. The arginine-dependent production of NO involves nitric oxide synthase (NOS). (Modified from Crawford and Guo, 2005.)

that NO is a by-product of *in vivo* NR action when the enzyme reduces nitrite to NO (Figure 7.5). Further research has shown that NO produced from NO_3^- or NO_2^- in leaves (Rockel *et al.*, 2002), roots (Stohr and Ullrich, 2002), and algae (Sakihama *et al.*, 2002) is a result of NR activity.

Plants have enzymatic mechanisms for the production of NO, in addition to the NR pathway, which may be more important than NR for some NO-dependent processes. One of these enzymes is the product of the *Arabidopsis AtNOS1* gene (Guo and Crawford, 2005). This plant nitric oxide synthase (NOS) is different in sequence from mammalian neuronal, endothelial, and inducible NOS. Rather, it is more similar to a snail NO-synthesizing enzyme (Crawford and Guo, 2005). Although the plant NOS differs in sequence and size (62 kDa) from animal NOS (130–150 kDa), both enzymes use arginine as a substrate and generate NO and citrulline as products. Inhibitors of mammalian NOS, such as arginine analogs, also inhibit NO formation by AtNOS1. Localization studies strongly suggest that the AtNOS1 protein is found in mitochondria (Guo and Crawford, 2005). Other NO-synthesizing enzymes, including a plasma membrane-bound nitrite–NO reductase, have been reported (Stohr *et al.*, 2001). The activity of catalase, as described above, and some oxidases and peroxidases could also result in NO production (Corpas *et al.*, 2001; del Rio *et al.*, 2004). Nitrite binding to oxyhemoglobin can result in NO synthesis (discussed in detail below).

The nonenzymatic conversion of nitrite to NO has also been demonstrated and this source of NO may have special significance for seeds (Bethke *et al.*, 2004a). Nitrite concentrations in agricultural soils can exceed 50 μM, and at low pH, NO_2^-

can be reduced to NO. The pK_a of this reaction is 3.2 and it is dramatically accelerated by reducing agents. Seeds and grains often maintain a low pH in the apoplast. For example, the cereal aleurone layer secretes phosphoric and organic acids and acidifies the endosperm to between pH 3.5 and 4 (Drozdowicz and Jones, 1995). Furthermore, the testa/pericarp that surrounds the aleurone layer contains proanthocyanidins, which are strong antioxidants/reductants (Bethke *et al.*, 2004a). Proanthocyanidins are solubilized from the cell wall of the grain and can contribute to the reduction of NO_2^- in the cereal endosperm (see Chapter 2 for proanthocyanidin function in seed dormancy). When NO_2^- was added to the incubation medium in which barley aleurone layers were incubated, copious amounts of NO were produced (Bethke *et al.*, 2004a). Barley aleurone layers also have a nitrate-inducible NR activity. When nitrate is supplied to aleurone layers under anaerobic conditions, nitrite is released from the aleurone into the apoplast and starchy endosperm (Ferrari and Varner, 1969). The interior of the cereal grain is likely to have reduced O_2 concentrations during germination, allowing NO_3^- to be reduced to NO_2^- and released into the acidic apoplast where it can be readily converted to NO (Bethke *et al.*, 2004a).

There is increasing recognition of the importance that nonenzymatic formation of NO from NO_2^- has for human biology as well. In fact, NO_2^- 'may represent the largest intravascular and tissue storage form of nitric oxide', playing a central role in the 'human nitrogen cycle' where nitrite is converted to NO via reactions with hemoglobin, myoglobin, and xanthine oxidoreductase (Gladwin *et al.*, 2005). The presence of nitrate or nitrite in the soil, the ability of plants to synthesize nitrite from nitrate, and the existence of acidic compartments in plants that are accessible to nitrite are strong indications that nitrate or nitrite may be equally important storage forms of NO for plant growth and development.

7.5.2 NO binding to metal-containing proteins

NO has a high affinity for metal-containing proteins, and proteins containing hemes are attractive targets for NO action (Shapiro, 2005). In mammalian systems, much of NO action is transduced through the heme-containing enzyme guanalyl cyclase. Guanalyl cyclase may be a target for NO as part of plant responses to pathogens (Durner *et al.*, 1998), but an involvement of this enzyme in NO signaling in seeds is unclear.

Cytochrome *c* oxidase is a heme-binding protein that is part of the mitochondrial electron transport chain. It has been shown that NO prevents respiration through the cytochrome pathway, and that this prevents ATP synthesis when the alternate oxidase pathway is inhibited (Yamasaki *et al.*, 2001). NO is likely to be produced in mitochondria by AtNOS1 (Guo and Crawford, 2005), making cytochrome *c* oxidase a potential target for NO binding.

Catalase is a heme-containing protein that is inhibited by NO. Hendricks and Taylorson (1974) hypothesized that NO derived from N-containing compounds might function to stimulate germination by inhibiting catalase activity. They postulated that by inhibiting catalase, NO would increase intracellular H_2O_2 concentrations, and that H_2O_2 might act as a substrate for enzymes such as peroxidases.

Peroxidase can oxidize NADPH to $NADP^+$, and $NADP^+$ is one of the limiting electron acceptors for the pentose phosphate pathway. Increasing the concentration of $NADP^+$ might allow for the more rapid operation of this pathway.

Plants contain several classes of nonsymbiotic hemoglobins (nsHb), but their functions remain unclear. In mammals, chemical reactions involving Hb and NO have been well characterized and are central to some aspects of NO biology. It is perhaps not surprising that plant nsHb have been implicated in the synthesis and scavenging of NO (Igamberdiev *et al.*, 2005). Two classes of nsHb have been identified in dicots, but only members of the class-1 family are found in monocots. Class-1 Hb is expressed in seeds of many species including *Arabidopsis*, barley, and rice. In barley and rice seeds Hb is present in the outer living cells including the aleurone layer and embryo. Class-1 nsHb are induced by hypoxia and sucrose as well as NO_3^-, NO_2^-, and NO (Hunt *et al.*, 2002; Dordas *et al.*, 2004; Igamberdiev *et al.*, 2005; Ohwaki *et al.*, 2005). Despite being induced by hypoxia, there is general agreement that class-1 nsHb are not involved in facilitating O_2 transport or acting as O_2 sensors. It is believed that their induction during hypoxia is related to ATP availability or is a consequence of reduced ATP concentrations (Nie and Hill, 1997; Hunt *et al.*, 2002).

The affinity of class-1 nsHb for various ligands has been studied in detail to address their function. Hill and his colleagues (Igamberdiev *et al.*, 2005) have advanced a model for the action of this class of Hb under anoxia that envisages a Hb/NO cycle. In this model, NADH and NADPH are consumed and oxidized forms of these electron acceptors are produced via reactions that involve the oxidation and reduction of NO_3^-, NO_2^-, and NO on the one hand, and Hb/metHb on the other (Figure 7.6). This cycle provides an alternative mechanism for the oxidation of

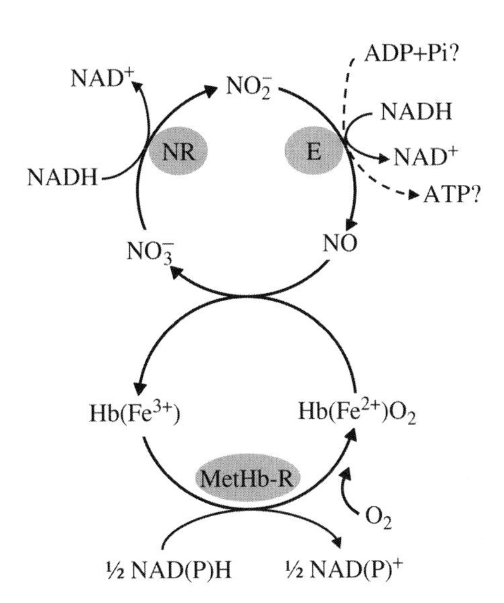

Figure 7.6 A schematic showing proposed reactions of the hemoglobin (Hb)/NO cycle. Abbreviations: NR, nitrate reductase; E, putative NO sources in hypoxia; MetHb-R, methemoglobin reductase. For details see Igamberdiev *et al.* (2005).

cofactors by molecular oxygen under low oxygen tensions and is consistent with lower expression of alcohol dehydrogenase in plants that overexpress Hb (Igamberdiev *et al.*, 2004).

The Hb/NO cycle bears remarkable similarity to the hypothesis advanced by Roberts (1969) to explain the effects of compounds such as azide, cyanide, NO_3^-, and NO_2^- on germination. Increased recycling of cofactor could result in an increase in the flux through metabolic pathways that rely on a constant supply of electron acceptors. In germinating seeds that catabolize lipids, this could result in making available a pool of carbon skeletons for synthesis of RNA, cell walls, and the shikimic acid pathway. Whereas Roberts (1969) and others would favor this interpretation, Hill and his colleagues favor a modification of this idea and suggest that increased oxidation of NADH resulting from the Hb/NO cycle leads to increased ATP synthesis under hypoxic conditions (Igamberdiev *et al.*, 2005).

A more prosaic hypothesis for the role of nsHb is that they protect plant tissues by acting as scavengers of NO. It is useful to note that soils are a major source of NO and evolution may have favored seeds that contain nsHbs because they are able to convert environmental NO to less toxic nitrite or nitrate. It should also be noted that overexpression of the *Arabidopsis* class-1 hemoglobin AtGLB1 protects *Arabidopsis* leaves from ROS (reactive oxygen species)-induced damage (Yang *et al.*, 2005). Sakamoto *et al.* (2004) have shown that all three *Arabidopsis* Hb proteins exhibit peroxidase activity and that in the presence of NO_2^- and H_2O_2, they could carry out tyrosine nitration.

7.5.3 NO as an antioxidant

The ability of NO to act as a strong antioxidant in cells has led to the notion that it plays a role in scavenging ROS, particularly lipid peroxides and superoxide (Lamattina *et al.*, 2003). Lipid peroxides are formed in membranes through exposure to ROS and result in a cascade of chain reactions that cause extensive membrane damage. NO can interact with lipid peroxides to limit this damage by acting as a chain breaker. NO reacts with superoxide to form peroxynitrite, and although this is also a highly reactive molecule, cells may benefit from the reaction between NO and superoxide under certain conditions. This is especially true when the production of NO exceeds the production of superoxide (Wink and Mitchell, 1998).

An illustration of the protective role of NO in seeds is seen in work on programmed cell death (PCD) in barley aleurone tissue (Beligni *et al.*, 2002). Cereal aleurone cells store abundant reserves of lipids that are broken down to sugar by β-oxidation and the glyoxylate cycle (Eastmond and Jones, 2005). This process results in concomitant production of ROS in the glyoxysome (Figure 7.3). In gibberellic acid (GA)-treated aleurone tissue, catalase, ascorbate peroxidase, and superoxide dismutase are dramatically downregulated (Fath *et al.*, 2001). As a result, ROS accumulate in GA-treated aleurone cells which subsequently undergo PCD. GA-induced PCD could be arrested by antioxidants such as butylated hydroxytoluene (Bethke *et al.*, 1999), as well as by the NO donors SNP and SNAP (Beligni *et al.*, 2002). The effects of NO donors on PCD in aleurone could be reversed by c-PTIO. In the absence of NO donors, PCD was even more pronounced in GA-treated aleurone layers that were treated with c-PTIO than in controls without c-PTIO (Bethke *et al.*, 2004b). These data indicate that endogenous NO reduced the rate of PCD in this tissue.

7.6 Interactions between NO and phytochrome or ABA

Light is a ubiquitous signal for plant growth and development that has profound effects on seed dormancy and germination. Many cases of interactions between light and N-containing compounds, especially nitrate and nitrite, have been reported (see Chapter 3). In general, the germination response of seeds to light is greater in the presence of N-containing compounds than in their absence. Phytochrome is the photoreceptor that influences germination. Accumulating data (described below) suggest that NO derived from N-containing compounds may be a signaling molecule that acts in conjunction with phytochrome.

Red light acting through phytochrome promotes germination of Grand Rapids lettuce seeds, while seeds imbibed in the dark do not germinate. Beligni and Lamattina (2000) showed that germination of Grand Rapids lettuce seeds was greater than 95% when seeds were imbibed in the dark with 100 μM SNP or 100 μM SNAP. SNP- or SNAP-induced germination in the dark was reversed ($<5\%$ germination) by co-incubation with c-PTIO. These data provide strong evidence for the involvement of NO in photo-induced seed germination. Light and Van Staden (2003), however, did not observe increased germination of Grand Rapids lettuce seeds imbibed in the dark with SNP or PBN (*N-tert*-butyl-α-phenylnitrone). The reason for this discrepancy is unclear, but the authors suggested that it could be attributed to differences in seed batches. Alternatively, there may have been differences in the uptake of the applied compounds. Germination of moderately dormant *Arabidopsis* seeds exposed to a pulse of red light is promoted by KNO_3 application. The involvement of NO in light-stimulated germination in *Arabidopsis* seeds is also supported by the observation that KNO_3, nitrogylcerine, and SNP all augmented light-stimulated germination (Batak *et al.*, 2002). This effect was greatest in imbibed seeds that contained primarily PhyA rather than PhyB. In this case, SNP and KNO_3 were equally effective, while nitroglycerine was approximately 10 times more effective than either of these compounds in promoting germination.

Seeds of *Paulownia tomentosa* require extended periods of illumination in order to germinate, but this light requirement can be dispensed with if seeds are imbibed with one of several N-containing compounds and given a brief period of red light. Grubisic (1992) showed that KNO_3, nitroglycerine, and isosorbidine dinitrate all increased red-light-stimulated germination of *P. tomentosa* seeds. The optimal concentration for KNO_3 was 1–10 mM, while the optimal concentrations for SNP and nitroglycerin were $\sim$100 μM. In a later report it was shown that nitroglycerine application resulted in the production of NO as measured by electron paramagnetic resonance, and that the NO donors SNAP and SIN-1 also promoted red-light-stimulated germination (Figure 7.4; Giba *et al.*, 1998). In the same report, the effectiveness of nitrite was found to be highly pH dependent. At pH 2.5 or 3, KNO_2 was as effective as SNP, SNAP, or SIN-1 in promoting germination. Under these conditions NO_2^- is readily converted to NO (Equations 7.6, 7.7, and 7.4′). The effect of KNO_2 was reduced as the pH increased from pH 3.5 to 6 (Figure 7.4).

While phytochrome and NO generally promote germination in the light, ABA always inhibits germination. The balance between the actions of promoters and of inhibitors often determines the net effect on germination. For example, SNP decreased the sensitivity of dormant and weakly dormant *Arabidopsis* seeds to ABA (Bethke *et al.*, 2006a). In this case, the relationship between SNP and NO was indirect, as SNP generated CN^-, which then promoted a NO-dependent loss of dormancy. More direct experiments that examine the relationships between ABA and NO will be required to confirm that NO can decrease the sensitivity of seeds to ABA.

A simple model to explain the effects of NO on dormancy and germination posits that NO reacts with enzymes that are involved in the synthesis or catabolism of ABA. Genetic and molecular evidence strongly implicates ABA in maintaining *Arabidopsis* seed dormancy, and the endogenous concentrations of ABA are maintained by a balance between its synthesis and catabolism (see Chapter 9). Inhibiting ABA synthesis in dormant *Arabidopsis* seeds was insufficient to remove dormancy under conditions where SNP readily removed dormancy (Bethke *et al.*, 2006a). These data suggest that ABA catabolism is likely to be an important aspect of dormancy loss. ABA is inactivated by ABA-8′-hydroxylase, a heme-containing cytochrome P450 enzyme. It will be important to determine whether an increase in the activity of this enzyme resulting from a NO signal leads to a decrease in ABA content. It is not expected that a direct interaction of NO with ABA-8′-hydroxylase would increase enzyme activity, as NO has been shown previously to inactivate cytochrome P450-containing enzymes (Wink and Mitchell, 1998).

7.7 Ecological significance of NO

7.7.1 Nitrogen and vegetation gap sensing

It has been proposed that soil NO_3^- or NO_2^- concentrations might function as a 'gap sensor', allowing seeds to germinate in an environment that is conducive to seedling growth (Pons, 1989; Giba *et al.*, 2003). The formation of an open space or gap in the local vegetation, such as occurs following a fire or a reduction in the forest canopy, is often associated with increased nutrient availability. An increase in soil nitrogen might result in increased NO production and this could be perceived by seeds as a germination cue. It is well established that soil is an important source of NO and a large proportion of atmospheric NO_x arises from reactions in soil (Ludwig *et al.*, 2001). Although chemical reduction of NO_2^- to NO occurs in acidic soils, the extent to which this source of NO contributes to overall NO production is thought to be small. By far the most important contributors to soil NO are the denitrifying and nitrifying bacteria. Both reduction and oxidation of soil nitrogen are known to generate NO, and the amount of NO produced varies with soil nitrogen, oxygen, and water content. The highest rates of NO production from soils are associated with addition of N fertilizers, and NO from heavily fertilized soils can contribute as much NO to the atmosphere as is generated by human activity in urban areas (Ludwig *et al.*, 2001). One impact of increased production of NO by soil would be the release of dormant seeds from the seed bank. It is known that germination of seeds from seed banks is seasonal and correlated with changes in temperature and/or rainfall. Since increased temperature and reduced soil water content increase NO production when nitrogen sources are present in the soil, it is easy to imagine that NO production serves as a gap detection mechanism, bringing about an increase in seed germination (Ludwig *et al.*, 2001).

7.7.2 Smoke and NO

Seeds of many plants that live in fire-prone areas are deeply dormant at maturity and dormancy persists until seeds are exposed to a local wildfire. In most cases, smoke, rather than heat, results in a loss of seed dormancy, and there has been much interest in identifying the compounds in smoke that cause this effect. It is clear that smoke contains a complex mixture of compounds, and that most of them have little or no effect on seed dormancy (Baldwin *et al.*, 1994). It is also clear that burning vegetation produces copious amounts of nitrogen oxides, especially NO_2.

Keeley and Fotheringham (1997) showed that NO_2 or NO_x (NO_2 + NO) was effective in breaking dormancy of the fire-chasing species *Emmenanthe penduliflora* (Californian chaparral). They estimated that NO_x production during a wildfire, and subsequent NO_x production resulting from microbial activity, was sufficient to bring about germination of *E. penduliflora* seeds. Additional experimentation extended this finding to other chaparral species (Keeley and Fotheringham, 1998) and to species whose germination is not usually triggered by fire such as lettuce and red rice (Brown and van Staden, 1997). Hence, it is clear that wildfires have the capacity to produce large amounts of nitrogen oxides, including NO_2 and NO, and that these compounds can effectively reduce dormancy of many species. It is also clear that other compounds in smoke can result in a loss of dormancy. One of the most effective dormancy-breaking compounds in smoke is a butenolide that results from the partial combustion of cellulose (Flematti *et al.*, 2004). The available data, therefore, suggest that smoke-induced dormancy release may be caused by NO_x, a butenolide, or both. Further experimentation will clarify the importance of N-containing compounds relative to other compounds in smoke that trigger a loss of dormancy. It would be particularly interesting to know if butenolide-stimulated germination requires endogenous NO and can be prevented by c-PTIO.

7.8 Unresolved questions and concluding remarks

Much has been learned about NO in seed dormancy and germination. It has been demonstrated convincingly that NO is produced by plants and by seeds. It is also clear that NO is involved in the process of dormancy loss and that NO acts synergistically with light to promote germination of seeds that are not strictly dormant. NO may have additional roles in seeds. For example, NO may function as an antimicrobial compound, a sensor for soil nutrient status, or a gap detector. The molecular interactions that allow NO to perform any of its roles, however, are completely unknown. For our understanding of NO in seed dormancy and germination to advance much farther it will be essential to establish where, when, and how NO is synthesized in seeds, the identity of the molecular targets for NO binding or covalent modification, and how NO signaling pathways interact with the phytochrome, GA, or ABA signaling pathways.

NO is reactive and volatile, and thus must be produced when it is needed. Definitive information on the details of NO production in seeds is lacking. NO can be synthesized both enzymatically and nonenzymatically in seeds. *Arabidopsis* mutants lacking *AtNOS1* are not defective in germination, making it clear that NO produced by *AtNOS1* is not required for germination. Until NO production by seeds is precisely quantified, it will be difficult to determine how its production is regulated by developmental or environmental cues. Likewise, it will be important to identify the sites of NO synthesis in seeds. All tissues in the embryo or endosperm may have the capacity to produce NO, but that does not preclude the possibility that some tissues make NO as a local or long-distance signal, and other tissues respond to NO but do not synthesize it themselves. In many seeds the activities of the embryo and the endosperm must be coordinated, and an attractive possibility is that NO plays a role in this coordination.

Molecular targets for NO in seeds have been proposed, but none have been shown to be required for NO activity in seed dormancy or germination. Among the most commonly mentioned are catalase, hemoglobin, cytochrome *c* oxidase, guanalyl cyclase, and other metal-containing enzymes including superoxide dismutase and peroxidases. Hemoglobin is an intriguing target for NO, if for no other reason than the observation that seeds contain nonsymbiotic hemoglobin, the function of which in seeds is unknown. Since some sites of NO production are in mitochondria, it is likely that NO binds to cytochrome *c* oxidase and prevents or reduces the electron transport capacity of this protein. Numerous proposals have been made over the years as to how redirecting electron transport through the alternative oxidase might promote seed germination. Perhaps some of this earlier literature should be revisited with NO binding to cytochrome *c* oxidase in mind.

Guanalyl cyclase is one of the principal targets for NO binding in mammals, and data from plants indicate that it is a target for NO produced during the hypersensitive response to pathogens. In seeds, NO-activated guanalyl cyclases have been proposed to be a node for cross-talk between NO signaling and phytochrome signaling (Giba *et al.*, 1998), but data supporting this hypothesis are lacking. NO works in concert with ROS in plant disease resistance and in some forms of plant PCD. Given this, and that NO has a very high affinity for hemes such as those found in superoxide dismutase and catalase, it is worth considering the possibility that NO potentiates its own activity by changing the rate of ROS production. Finally, NO can covalently modify proteins and thereby modify their activities. To date, nitrosylated and nitrosated proteins have not been reported to occur in seeds.

Research with seeds from lettuce, Empress tree, *Arabidopsis*, and other species indicates that there are interactions between NO and the phytochrome and ABA signaling pathways. This is expected given that ABA and phytochrome play significant roles in seed dormancy and germination. Importantly, it opens up a promising avenue for further exploration of NO signaling in seeds. Mutants in phytochrome and ABA biosynthesis and signaling exist, especially in *Arabidopsis*. These should be employed to full advantage in experiments designed to understand more about the molecular interactions between these signaling pathways. Mutant screens designed to identify components of the NO signaling network might add further depth to our understanding of NO action in seeds.

Finally, the relationships among plant hormones, NO, and simple nitrogen-containing compounds such as NO_2^-, NO_3^-, CN^-, and azide need further investigation. These nitrogen-containing compounds reduce seed dormancy across a wide range of species, but their mechanism of action remains unclear despite a century of research. Are they triggers for increased NO synthesis or NO availability, or are they substrates for NO production as we and others have suggested? Careful measurements of NO production by seeds following exposure to nitrate, nitrite, cyanide, and azide may at last reveal part of the picture as to how these compounds reduce dormancy and promote germination. Perhaps the central question at this time, however, is how NO signaling, ABA signaling, and the operation of key metabolic pathways work together to control dormancy and germination. For example, is there an interaction between NO or NO signaling, on the one hand, and carbon metabolism on the other hand? Numerous proposals have been made over the years suggesting that limitations to lipid metabolism or to the operation of the pentose phosphate pathway can inhibit germination. Yet much of the evidence for this is indirect or subject to alternative interpretations. Recent advances in metabolite profiling and the availability of mutants with defects in ABA signaling provide an entrée for detailed investigations in this area. Coupling these to *in vivo* measurements of NO production under experimental conditions that promote or prevent germination may help to explain more fully a century of experimentation with seeds.

References

T. Akaike, M. Yoshida, Y. Miyamoto, *et al.* (1993) Antagonistic action of imidazolineoxyl *N*-oxides against endothelium-derived relaxing factor NO (nitric oxide) through a radical reaction. *Biochemistry* **32**, 827–832.

I.T. Baldwin, L. Staszakkozinski and R. Davidson (1994) Up in smoke. 1: Smoke-derived germination cues for postfire annual, *Nicotiana attenuata* Torr Ex Watson. *Journal of Chemical Ecology* **20**, 2345–2371.

L.A.T. Ballard and A.E.G. Lipp (1967) Seed dormancy-breaking by uncouplers and inhibitors of oxidative phosphorylation. *Science* **156**, 398–400.

I. Batak, M. Devic, Z. Giba, D. Grubisic, K.L. Poff and R. Konjevic (2002) The effects of potassium nitrate and NO-donors on phytochrome A- and phytochrome B-specific induced germination of *Arabidopsis thaliana* seeds. *Seed Science Research* **12**, 253–259.

M. Beligni, A. Fath, P.C. Bethke, L. Lamattina and R.L. Jones (2002) Nitric oxide acts as an antioxidant and delays programmed cell death in barley aleurone layers. *Plant Physiology* **129**, 1642–1650.

M.V. Beligni and L. Lamattina (2000) Nitric oxide stimulates seed germination and de-etiolation, and inhibits hypocotyl elongation, three light-inducible responses in plants. *Planta* **210**, 215–221.

P.C. Bethke, M.R. Badger and R.L. Jones (2004a) Apoplastic synthesis of nitric oxide by plant tissues. *The Plant Cell* **16**, 332–341.

P.C. Bethke, F. Gubler, J.V. Jacobsen and R.L. Jones (2004b) Dormancy of *Arabidopsis* seeds and barley grains can be broken by nitric oxide. *Planta* **219**, 847–855.

P.C. Bethke, I. Libourel and R. Jones (2006a) Nitric oxide reduces seed dormancy in *Arabidopsis*. *Journal of Experimental Botany* **57**, 517–526.

P.C. Bethke, I.G.L. Libourel, V. Reinohl and R.L. Jones (2006b) Sodium nitroprusside, cyanide, nitrite, and nitrate break Arabidopsis seed dormancy in a nitric oxide-dependent manner. *Planta* **223**, 805–812.

P.C. Bethke, J.E. Lonsdale, A. Fath and R.L. Jones (1999) Hormonally regulated programmed cell death in barley aleurone cells. *The Plant Cell* **11**, 1033–1045.

N.A.C. Brown and J. van Staden (1997) Smoke as a germination cue: a review. *Plant Growth Regulation* **22**, 115–124.

P.A. Castric, K.J.F. Farnden and E.E. Conn (1972) Cyanide metaboism in higher plants. V: The formation of asparagine from β-cyanoline. *Archives of Biochemistry and Biophysics* **152**, 62–69.

M. Cohn, D. Butera and J. Hughes (1983) Seed dormancy in red rice. *Plant Physiology* **73**, 381–384.

M.A. Cohn and L. Castle (1984) Dormancy in red rice. IV: Response of unimbibed and imbibing seeds to nitrogen dioxide. *Physiologia Plantarum* **60**, 552–556.

F.J. Corpas, J.B. Barroso and L.A. del Rio (2001) Peroxisomes as a source of reactive oxygen species and nitric oxide signal molecules in plant cells. *Trends in Plant Science* **6**, 145–150.

N.M. Crawford and F.-Q. Guo (2005) New insights into nitric oxide metabolism and regulatory functions. *Trends in Plant Science* **10**, 195–200.

W. Crocker and L. Barton (eds) (1953) *Physiology of Seeds*. Chronica Botanica, Waltham, MA.

J. Dean and J.E. Harper (1986) Nitric oxide and nitrous oxide production by soybean and winged bean during the *in vivo* nitrate reductase assay. *Plant Physiology* **82**, 718–723.

L.A. del Rio, F.J. Corpas, A.M. Leon, *et al.* (2004) The nitric oxide synthase activity of plant peroxisomes. *Free Radical Biology and Medicine* **36**, S41–S42.

M. Delledonne (2005) NO news is good news for plants. *Current Opinion in Plant Biology* **8**, 390–396.

C. Dordas, B.B. Hasinoff, J. Rivoal and R.D. Hill (2004) Class-1 hemoglobins, nitrate and NO levels in anoxic maize cell-suspension cultures. *Planta* **219**, 66–72.

Y.M. Drozdowicz and R.L. Jones (1995) Hormonal regulation of organic and phosphoric acid release by barley aleurone layers and scutella. *Plant Physiology* **108**, 769–776.

J. Durner, D. Wendehenne and D.F. Klessig (1998) Defense gene induction in tobacco by nitric oxide, cyclic GMP, and cyclic ADP-ribose. *Proceedings of the National Academy of Sciences of the United States of America* **95**, 10328–10333.

P.J. Eastmond and R.L. Jones (2005) Hormonal regulation of gluconeogenesis in cereal aleurone is strongly cultivar-dependent and gibberellin action involves SLENDER1 but not GAMYB. *The Plant Journal* **44**, 483–493.

A. Fath, P.C. Bethke and R.L. Jones (2001) Enzymes that scavenge reactive oxygen species are down-regulated prior to gibberellic acid-induced programmed cell death in barley aleurone. *Plant Physiology* **126**, 156–166.

M. Feelisch (1991) The biochemical pathways of nitric oxide formation from nitrovasodilators: appropriate choice of exogenous NO donors and aspects of preparation and handling of aqueous NO solutions. *Journal of Cardiovascular Pharmacology* **17**(Suppl. 3), S25–S33.

M. Feelisch (1998) The use of nitric oxide donors in pharmacological studies. *Naunyn-Schmiedebergs Archives of Pharmacology* **358**, 113–122.

T.E. Ferrari and J.E. Varner (1969) Substrate induction of nitrate reductase in barley aleurone layers. *Plant Physiology* **44**, 85–88.

M.K. Firestone (1982) Biological dentrification. *Agronomy Journal* **22**, 289–336.

G.R. Flematti, E.L. Ghisalberti, K.W. Dixon and R.D. Trengove (2004) A compound from smoke that promotes seed germination. *Science* **305**, 977.

Z. Giba, D. Grubisic and R. Konjevic (1995) The involvement of phytochrome in light-induced germination of blueberry (*Vaccinium myrtillus* L.) seeds. *Seed Science and Technology* **23**, 11–19.

Z. Giba, D. Grubisic and R. Konjevic (2003) Nitrogen oxides as environmental sensors for seeds. *Seed Science Research* **13**, 187–196.

Z. Giba, D. Grubisic, S. Todorovic, L. Sajc, D. Stojakovic and R. Konjevic (1998) Effect of nitric oxide-releasing compounds on phytochrome-controlled germination of Empress tree seeds. *Plant Growth Regulation* **26**, 175–181.

M.T. Gladwin, A.N. Schechter, D.B. Kim-Shapiro, *et al.* (2005) The emerging biology of the nitrite anion. *Nature Chemical Biology* **1**, 308–314.

S. Goldstein, A. Russo and A. Samuni (2003) Reactions of PTIO and carboxy-PTIO with NO, NO_2, and O_2. *Journal of Biological Chemistry* **278**, 50949–50955.

D.L. Granger, R.R. Taintor, K.S. Boockvar and J.B. Hibbs (1996) Measurement of nitrate and nitrite in biological samples using nitrate reductase and Griess reaction. *Methods in Enzymology* **268**, 142–151.

P. Grappin, D. Bouinot, B. Sotta, E. Miginiac and M. Jullien (2000) Control of seed dormancy in *Nicotiana plumbaginifolia*: post-imbibition abscisic acid synthesis imposes dormancy maintenance. *Planta* **210**, 279–285.

D. Grubisic, Z. Giba and R. Konjevic (1992) The effect of organic nitrates in phytochrome-controlled germination of *Paulownia tomentosa* seeds. *Photochemistry and Photobiology* **56**, 629–632.

F.-G. Guo and N.M. Crawford (2005) The *Arabidopsis* nitric oxide synthase protein AtNOS1 is targeted to mitochondria and protects against oxidative damage and dark-induced senescence. *The Plant Cell* **17**, 3436–3450.

S.B. Hendricks and R.B. Taylorson (1972) Promotion of seed germination by nitrates and cyanide. *Nature* **237**, 169–170.

S. Hendricks and R. Taylorson (1974) Promotion of seed germination by nitrate, nitrite, hydroxylamine and ammonium salts. *Plant Physiology* **54**, 304–309.

S.B. Hendricks and R.B. Taylorson (1975) Breaking of seed dormancy by catalase inhibition. *Proceedings of the National Academy of Sciences of the United States of America* **72**, 306–309.

Z. Hua, W.-B. Shen and L.-L. Xu (2003) Effects of nitric oxide on the germination of wheat seeds and its reactive oxygen species metabolisms under osmotic stress. *Acta Botanica Sinica* **45**, 901–905.

P.W. Hunt, E.J. Klok, B. Trevaskis, *et al.* (2002) Increased level of hemoglobin 1 enhances survival of hypoxic stress and promotes early growth in *Arabidopsis thaliana*. *Proceedings of the National Academy of Sciences of the United States of America* **99**, 17197–17202.

A.U. Igamberdiev, K. Baron, N. Manac'h-Little, M. Stoimenova and R.D. Hill (2005) The haemoglobin/nitric oxide cycle: involvement in flooding stress and effects on hormone signaling. *Annals of Botany* **96**, 557–564.

A.U. Igamberdiev, C. Seregelyes, N. Manach and R.D. Hill (2004) NADH-dependent metabolism of nitric oxide in alfalfa root cultures expressing barley hemoglobin. *Planta* **219**, 95–102.

L.J. Ignarro, J.M. Fukuto, J.M. Griscavage, N.E. Rogers and R.E. Byrns (1993) Oxidation of nitric oxide in aqueous solution to nitrite but not nitrate – comparison with enzymatically formed nitric oxide from L-arginine. *Proceedings of the National Academy of Sciences of the United States of America* **90**, 8103–8107.

J.E. Keeley and C.J. Fotheringham (1997) Trace gas emissions and smoke-induced seed germination. *Science* **276**, 1248–1250.

J.E. Keeley and C.J. Fotheringham (1998) Smoke-induced seed germination in California chaparral. *Ecology* **79**, 2320–2336.

M. Kopyra and E.A. Gwozdz (2003) Nitric oxide stimulates seed germination and counteracts the inhibitory effect of heavy metals and salinity on root growth of *Lupinus luteus*. *Plant Physiology and Biochemistry* **41**, 1011–1017.

L. Lamattina, C. Garcia-Mata, M. Graziano and G. Pagnussat (2003) Nitric oxide: the versatility of an extensive signal molecule. *Annual Review of Plant Biology* **54**, 109–136.

E. Lehman (1909) Zur Keimungsphysiologie and biologie von *Ranunculus sclereatus* L. und einigen anderen Samen. *Berichte Deutschen Botanischen Gesellschaft* **27**, 476–494.

W.Q. Li, X.J. Liu, M.A. Khan, Y. Kamiya and S. Yamaguchi (2005) Hormonal and environmental regulation of seed germination in flixweed (*Descurainia sophia*). *Plant Growth Regulation* **45**, 199–207.

I.G.L. Libourel, P.C. Bethke, R. De Michele and R.L. Jones (2006) Nitric oxide gas stimulates germination of dormant *Arabidopsis* seeds: Use of a flow-through apparatus for delivery of nitric oxide. *Planta* **223**, 813–820.

M.E. Light and J. van Staden (2003) The nitric oxide-specific scavenger carboxy-PTIO does not inhibit smoke-stimulated germination of Grand Rapids lettuce seeds. *South African Journal of Botany* **69**, 217–219.

J. Ludwig, F.X. Meixner, B. Vogel and J. Forstner (2001) Soil–air exchange of nitric oxide: an overview of processes, environmental factors, and modeling studies. *Biogeochemistry* **52**, 225–257.

W. Major and E.H. Roberts (1968) Dormancy in cereal seeds. I: Effects of oxygen and respiratory inhibitors. *Journal of Experimental Botany* **19**, 77–89.

S. Neill, R. Desikan and J.T. Hancock (2003) Nitric oxide signaling in plants. *New Phytologist* **159**, 11–35.

X. Nie and R.D. Hill (1997) Mitochondrial respiration and hemoglobin gene expression in barley aleurone tissue. *Plant Physiology* **114**, 835–840.

Y. Ohwaki, M. Kawagishi-Kobayashi, K. Wakasa, S. Fujihara and T. Yoneyama (2005) Induction of class-1 non-symbiotic hemoglobin genes by nitrate, nitrite and nitric oxide in cultured rice cells. *Plant and Cell Physiology* **46**, 324–331.

A. Poljakoff-Mayber and A.M. Mayer (eds) (1963) *The Germination of Seeds*. Pergammon, Oxford.

T.L. Pons (1989) Breaking of seed dormancy by nitrate as a gap detection mechanism. *Annals of Botany* **63**, 139–143.

E.H. Roberts (1963) The effects of inorganic ions on dormancy in rice seed. *Physiologia Plantarum* **16**, 732–744.

E.H. Roberts (1964a) A survey of effects of chemical treatments on dormancy in rice seed. *Physiologia Plantarum* **17**, 30–43.

E.H. Roberts (1964b) Distribution of oxidation–reduction enzymes: effects of respiratory inhibitors, oxidising agents on dormancy in rice seed. *Physiologia Plantarum* **17**, 14–29.

E.H. Roberts (1969) Seed dormancy and oxidation processes. *Symposia of the Society of Experimental Biology* **23**, 161–192.

P. Rockel, F. Strube, A. Rockel, J. Wildt and W.M. Kaiser (2002) Regulation of nitric oxide (NO) production by plant nitrate reductase *in vivo* and *in vitro*. *Journal of Experimental Botany* **53**, 103–110.

A. Sakamoto, S. Sakurao, K. Fukunaga, *et al.* (2004) Three distinct *Arabidopsis* hemoglobins exhibit peroxidase-like activity and differentially mediate nitrite-dependent protein nitration. *FEBS Letters* **572**, 27–32.

Y. Sakihama, S. Nakamura and H. Yamasaki (2002) Nitric oxide production mediated by nitrate reductase in the green alga *Chlamydomonas reinhardtii*: an alternative NO production pathway in photosynthetic organisms. *Plant and Cell Physiology* **43**, 290–297.

G. Sarath, P.C. Bethke, R.L. Jones, L.M. Baird, G. Hou and R.B. Mitchell (2006) Nitric oxide accelerates seed germination in warm-season grasses *Planta* **223**, 1154–1164.

A.D. Shapiro (2005) Nitric oxide signaling in plants. *Advances in Research and Applications: Plant Hormones and Vitamins* **72**, 339–398.

C. Stohr and S. Stremlau (2006) Formation and possible roles of nitric oxide in plant roots. *Journal of Experimental Botany* **57**, 463–470.

C. Stohr, F. Strube, G. Marx, W.R. Ullrich and P. Rockel (2001) A plasma membrane-bound enzyme of tobacco roots catalyses the formation of nitric oxide from nitrite. *Planta* **212**, 835–841.

C. Stohr and W.R. Ullrich (2002) Generation and possible roles of NO in plant roots and their apoplastic space. *Journal of Experimental Botany* **53**, 2293–2303.

R.B. Taylorson and S.B. Hendricks (1973) Promotion of seed germination by cyanide. *Plant Physiology* **52**, 23–27.

E.H. Toole, S.B. Hendricks, H.A. Borthwick and V.K. Toole (1956) Physiology of seed germination. *Annual Review of Plant Physiology* **7**, 299–324.

D. Wendehenne, J. Durner and D.F. Klessig (2004) Nitric oxide: a new player in plant signalling and defence responses. *Current Opinion in Plant Biology* **7**, 449–455.

D.A. Wink and J.B. Mitchell (1998) Chemical biology of nitric oxide: insights into regulatory, cytotoxic, and cytoprotective mechanisms of nitric oxide. *Free Radical Biology and Medicine* **25**, 434–456.

H. Yamasaki (2000) Nitrite-dependent nitric oxide production pathway: implications for involvement of active nitrogen species in photoinhibition *in vivo*. *Philosophical Transactions of the Royal Society London. Series B: Biological Sciences* **355**, 1477–1488.

H. Yamasaki, H. Shimoji, Y. Ohshiro and Y. Sakihama (2001) Inhibitory effects of nitric oxide on oxidative phosphorylation in plant mitochondria. *Nitric Oxide* **5**, 261–270.

L.X. Yang, R.Y. Wang, F. Ren, J. Liu, J. Cheng and Y.T. Lu (2005) AtGLB1 enhances the tolerance of *Arabidopsis* to hydrogen peroxide stress. *Plant and Cell Physiology* **46**, 1309–1316.

D. Zanardo, F. Zanardo, M. Ferrarese, J. Magalhaes and O. Ferrarese-Filho (2005) Nitric oxide affecting seed germination and peroxidase activity in canola (*Brassica napus* L.). *Physiology and Molecular Biology of Plants* **11**, 81–86.

8 A merging of paths: abscisic acid and hormonal cross-talk in the control of seed dormancy maintenance and alleviation

J. Allan Feurtado and Allison R. Kermode

8.1 Introduction

The sesquiterpene phytohormone abscisic acid (ABA) regulates key events during seed formation, such as the deposition of storage reserves, prevention of precocious germination, acquisition of desiccation tolerance, and induction of primary dormancy (Finkelstein and Rock, 2002; Kermode, 2005). Dynamic changes in ABA biosynthesis and catabolism elicit hormone-signaling changes that affect downstream gene expression, and thereby regulate critical checkpoints during the plant life cycle, including the transitions from dormancy to germination and from germination to growth (Nambara and Marion-Poll, 2005). However, the regulatory role of ABA is in part achieved through interactions with other hormones and their associated signaling networks, by mechanisms that remain largely unknown. Thus, in many instances, a linear representation of hormone signaling pathways is no longer a viable model to explain downstream physiological events. Instead, cross-talk among hormones such as ABA, gibberellic acid (GA), ethylene, brassinosteroids (BRs), cytokinin, and auxin forms a complex web whereby key regulatory factors become nodes or hubs under combinatorial control. There is further intricacy when we consider that the hormone levels themselves are under regulation in this complex web and that environmental cues, in turn, modulate hormonal responses.

Quantitative genetics and functional genomics approaches have and will continue to contribute to the identification of genes and proteins that control seed dormancy and germination, including components of the ABA signal transduction pathway and elements that link this pathway to other hormonal networks. Interactions between ABA and other hormones during the control of key processes, from dormancy induction to seedling growth, include:

1. dormancy inception, in which GA and ethylene play antagonistic roles to ABA during seed development;
2. the maintenance and termination of primary dormancy, in which other hormones or hormone-signaling components, such as GA, may alter ABA homeostasis (e.g., *de novo* ABA synthesis or ABA catabolism) and/or the stability of ABA signal transduction components;

3. the initiation of germination in which ethylene and other hormones such as BRs may alter ABA or GA homeostasis and associated signal transduction components;
4. the developmental arrest during the early post-germinative phase, a state induced by conditions that are not optimal for a transition from germination to seedling growth. Here seeds may respond by increasing ABA biosynthesis and the expression of genes that impose a transient 'quiescence' until those conditions become more favorable.

Recent evidence for interactions among the different hormonal networks during these dormancy/growth-related events of the plant life cycle is discussed, and where possible, the potential mechanisms that form the basis of these interactions are speculated upon. We first outline the role of ABA in controlling these processes, followed by a discussion of the roles of GA, ethylene, auxin, cytokinin, and BRs in dormancy induction, maintenance, and termination. Emphasis will also be drawn to cross-talk among the hormones and interactions with environmental signals such as light and cold.

8.2 Abscisic acid

8.2.1 ABA in seed maturation and the induction of primary dormancy

ABA is involved in several specific processes during seed development and, in general, is thought to maintain seeds in a developmental program while on the parent plant (for reviews see Kermode, 1995, 2005; Holdsworth *et al.*, 1999; Finkelstein and Rock, 2002). Evidence for the role of ABA in such processes as the deposition of storage reserves, prevention of precocious germination, acquisition of desiccation tolerance, and dormancy induction has come from work with ABA-deficiency or -response mutants of *Arabidopsis thaliana* or *Zea mays* (maize) (Finkelstein and Rock, 2002). For example, many of the *viviparous* (*vp*) mutants of maize (which exhibit precocious germination of kernels on the ear) are linked to deficiencies in ABA synthesis or signaling (McCarty, 1995; Suzuki *et al.*, 2006; see Chapter 1). Recently, McCourt *et al.* (2005) have proposed a model in which ABA functions as a 'status quo' hormone, preserving or maintaining the present state and preventing life cycle transitions (e.g., the dormancy-to-germination transition). In addition to this ascribed role, ABA also seems to prevent seed abortion and promote embryo growth during embryogenesis (Cheng *et al.*, 2002; Frey *et al.*, 2004).

Seeds of different species show differences in the timing of dormancy inception during development; the role of ABA in this process is evident from genetic and physiological studies (reviewed in Bewley, 1997; Koornneef *et al.*, 2002). Typically, the ABA content in seeds is low during early development (i.e., during histodifferentiation and early pattern formation) and thereafter increases, usually peaking around mid-maturation. ABA levels usually decline precipitously during late development, particularly during the maturation drying phase (Bewley and

Black, 1994; Kermode, 1995; Meinke, 1995; Bewley, 1997). These developmental changes in endogenous ABA correlate well with the pattern of expression of the gene encoding the ABA biosynthesis enzyme zeaxanthin epoxidase (ZEP) (Audran *et al.*, 1998; Seo and Koshiba, 2002; Chapter 9). The *ZEP* gene is ubiquitously expressed during the maturation phase of *Arabidopsis* seed development but expression becomes restricted to the embryo and endosperm during the desiccation phase (Audran *et al.*, 2001). Less information is available for another ABA biosynthetic enzyme, 9-*cis*-epoxycarotenoid dioxygenase (NCED), which is present as a gene family in many species. Several *NCED* genes are expressed during development of *Arabidopsis* seeds, and their mRNAs are present in the dry seed (Tan *et al.*, 2003; Seo *et al.*, 2004). The *Arabidopsis NCED5* and *NCED6* genes are active during seed development until late maturation (Tan *et al.*, 2003). This is also true for the *Arabidopsis NCED3* gene, the major stress-induced *NCED* gene, although at earlier stages of development, expression is restricted to the basal region of the seed (Tan *et al.*, 2003). *AtNCED6* and *AtNCED9* encode the most abundant NCEDs in seed tissues, and Lefebvre *et al.* (2006) demonstrated that ABA synthesized within both the endosperm and embryo contributes to dormancy induction in *Arabidopsis*. In accordance with the expected redundancy in the *NCED* gene family, only seeds from double mutants (*Atnced6*/*Atnced9*) show increased germination percentages as compared to wild-type seeds; however, the germination percentage of fresh double mutant seeds is intermediate between that of wild-type and *aba3-1* (ABA-deficient) mutant seeds (Lefebvre *et al.*, 2006). The *AtABA3-1* gene encodes a molybdenum cofactor sulfurase required for abscisic aldehyde oxidase 3 (AAO3 or AB-AO) activity, which converts ABA aldehyde to ABA (Bittner *et al.*, 2001; Seo and Koshiba, 2002). The gene encoding this enzyme exhibits low expression in *Arabidopsis* seeds 10 days after flowering and transcripts are present in dry seeds (González-Guzmán *et al.*, 2004; Seo *et al.*, 2004). Similarly, the gene encoding SHORT CHAIN DEHYDROGENASE/REDUCTASE 1 (SDR1), which catalyzes the conversion of xanthoxin to abscisic aldehyde, is expressed at low levels in seeds and embryos and may contribute to maternally derived ABA synthesis (Cheng *et al.*, 2002).

In *Arabidopsis*, reciprocal crosses between wild-type and ABA-deficient mutants reveal two origins of ABA during seed development: the first peak of ABA is derived from maternal tissues and immediately precedes the maturation phase and a later peak occurs in the embryo prior to desiccation (Karssen *et al.*, 1983). Similar results were recently reported for *Nicotiana plumbaginifolia* (Frey *et al.*, 2004). However, *de novo* synthesis of ABA in developing embryos may also contribute to the first peak, with maternal ABA serving as a signal to trigger embryo/endosperm ABA synthesis. In support of this contention, ABA biosynthesis genes are actively expressed during mid-development (10 DAP [days after pollination]) in *Arabidopsis* embryo and endosperm tissues (Xiong and Zhu, 2003; Lefebvre *et al.*, 2006). The first peak of ABA may prevent precocious germination, perhaps in concert with proteins encoded by the *FUS3* (*FUSCA3*) and *LEC* (*LEAFY COTYLEDON*) genes, transcription factors that control seed developmental events (Raz *et al.*, 2001). The second 'lesser' peak in ABA produced by the embryo itself is important for dormancy imposition (Karssen *et al.*, 1983; Frey *et al.*, 2004). However, the surrounding seed tissues and parent plant may play an important role in maintaining or encouraging

ABA biosynthesis in developing embryos during later seed development (i.e., during the second ABA peak). Several factors add to the complexity of the pattern of ABA increase and decline, including developmental changes in the sensitivities of embryo and other seed tissues to ABA and the differential thresholds for the initiation and maintenance of developmental events (e.g., the expression of certain developmental genes and the induction and maintenance of dormancy) (Xu and Bewley, 1991; Kermode, 1995; Jiang *et al.*, 1996). Moreover, the environment during seed development (e.g., light, temperature, and water availability) can strongly influence the ABA content and sensitivity of the mature seed. In wheat (*Triticum aestivum*), low temperatures can affect ABA content or sensitivity, which in turn influence the degree of dormancy during development and in the mature grain (Walker-Simmons, 1990; Garello and LePage-Degivry, 1999). Likewise, water stress imposed during the development of *Sorghum bicolor* seeds decreases both their ABA content and sensitivity, and the seeds have an increased capacity for germination during development (Benech-Arnold *et al.*, 1991). In the barley (*Hordeum vulgare*) cultivar Triumph and its mutant line TL43, endogenous ABA contents during grain development alone do not adequately account for genetic differences in germination percentages between two environments, one conducive to low dormancy and the other promotive of high dormancy (Spain and Scotland, respectively) (Romagosa *et al.*, 2001).

Often there is no clear relationship between the ABA content of the mature dry seed or grain and the degree of dormancy. Thus, although ABA is important for dormancy inception in developing seeds, high levels of ABA need not be present in order to maintain the state of dormancy (e.g., during late maturation and desiccation). However, during imbibition, ABA synthesis and catabolism in dormant versus nondormant seeds can be clearly differentiated. Both dormant and nondormant seeds continue to synthesize ABA during imbibition (often after an initial lag), but in nondormant seeds the capacity for ABA catabolism is increased relative to that for biosynthesis (see Chapter 9 and further discussion below).

Various ABA mutants and transgenic studies have been important in providing more direct links between ABA and key developmental processes, including the prevention of vivipary and the induction of primary dormancy. Species or mutants whose seeds exhibit vivipary are often characterized by ABA deficiency or by a relative insensitivity to ABA. Relative ABA insensitivity occurs in seeds and seedlings of the mangrove (*Rhizophora mangle*), in which there is germination of the embryo within the developing fruit on the parent plant with no intervening period of quiescence, and in most cases, little or no dehydration (Sussex, 1975). The ABA-deficient *sitiens* mutant of tomato (*Lycopersicon esculentum*) has approximately 10% of the ABA content of its wild-type counterpart and exhibits reduced dormancy and vivipary in overripe fruits (Groot and Karssen, 1987; Groot *et al.*, 1991). However, low endogenous ABA may indirectly contribute to vivipary in this case, as the embryos also exhibit a reduced sensitivity to the highly negative osmotic potential of surrounding seed tissues. ABA-deficient mutants are typically capable of some ABA production and redundant genes/pathways often exist to compensate for defective genes. ABA contents of the *vp14* mutant of maize are about 30% of that found in wild-type seeds (Tan *et al.*, 1997); hence, the amount of ABA in these mutant seeds is sufficient to allow normal development to proceed (with respect to reserve deposition), and only the later maturation stages are altered (e.g., *vp14* seeds exhibit an enhanced capacity for precocious germination).

Transgenic studies also have provided evidence regarding the contribution of ABA biosynthesis and catabolism to dormancy status. In *N. plumbaginifolia*, overexpression of the gene encoding the ABA biosynthesis enzyme ZEP (i.e., *ABA2* gene overexpression) produces transgenic seeds with more pronounced dormancy than that exhibited by wild-type seeds; conversely, the downregulation of this gene by an antisense approach results in reduced levels of ABA and in reduced dormancy (Frey *et al.*, 1999). Similarly, constitutive expression of a gene encoding a major enzyme of ABA catabolism (an 8′-hydroxylase gene, cytochrome P450 *CYP707A* subfamily) in transgenic *Arabidopsis* seeds (ecotype C24) leads to reduced amounts of ABA at seed maturity and results in a lowered requirement for after-ripening of the seeds as compared to wild-type (i.e., nontransgenic C24) seeds. Conversely, a loss-of-function mutation in the *CYP707A* gene results in seeds requiring a longer after-ripening period to terminate dormancy (Millar *et al.*, 2006). Sequestration of ABA within developing transgenic tobacco (*Nicotiana tabacum*) seeds (via expression of an anti-ABA single chain variable fragment antibody) leads to a marked disruption of storage reserve deposition. Reserve accumulation is reduced to the extent that the storage parenchyma cells of the seed more closely resemble plant vegetative cells (Phillips *et al.*, 1997). The transgenic seeds exhibit other 'seedling' features, including green chloroplast-containing cotyledons, desiccation intolerance, and premature activation of the shoot apical meristem.

8.2.2 Transcription factors and combinatorial control of seed development and maturation

Elucidating the role of ABA in controlling seed and embryo maturation processes, including the inception of dormancy, has been facilitated by the generation and analysis of mutants (e.g., of tomato, maize, and *Arabidopsis*) that exhibit a relative insensitivity to ABA (so-called response mutants). These mutants have defects in ABA signaling pathways and are characterized by reduced dormancy, which is generally accompanied by disruption of seed maturation and precocious expression of germinative/postgerminative genes. For example, similar to the tobacco plants expressing an anti-ABA antibody (Phillips *et al.*, 1997), null alleles of the *Arabidopsis abi3* mutant (e.g., *abi3-4* and *abi3-6*) fail to complete seed maturation and produce green, desiccation-intolerant seeds that have reduced storage protein accumulation as well as embryos that resemble seedlings rather than their desiccated mature wild-type counterparts (Nambara *et al.*, 1992; Ooms *et al.*, 1993). Screens to identify suppressors of ABA signaling mutants have helped define some of the components of ABA signaling that interact with other signaling pathways and thus form the basis of hormonal networks that control the dormancy-to-germination transition (Figure 8.1).

Transcription factors essential for ABA- or 'seed-specific' gene expression include ABI3/VP1, ABI4, ABI5, LEC1, LEC2, and FUS3 (reviewed in Finkelstein *et al.*, 2002). Lesions in the genes encoding some of these factors underlie some

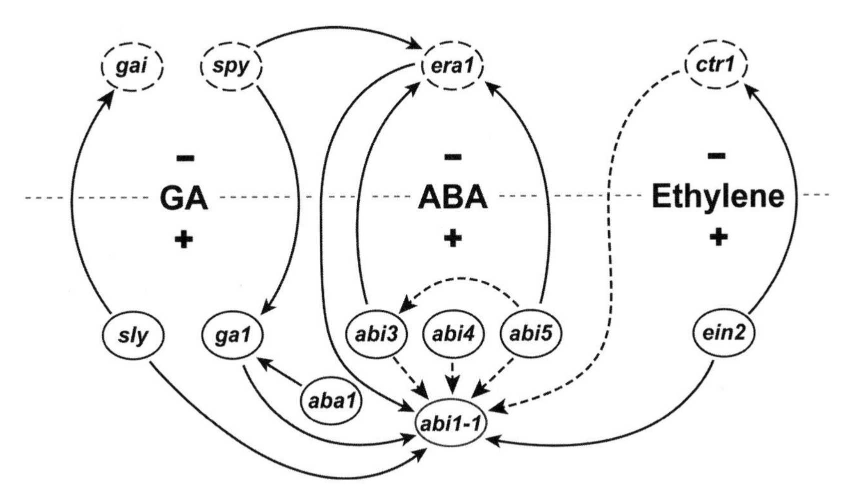

Figure 8.1 Genetic interactions between ABA, GA, and ethylene pathways. Interactions are based on (i) enhancer or suppressor screens (e.g., of *abi1-1*) or (ii) epistasis in double mutant analysis (e.g., *ein2 ctr1*) (Koornneef *et al.*, 1982; Kieber *et al.*, 1993; Jacobsen and Olszewski, 1993; Finkelstein, 1994; Steber *et al.*, 1998; Beaudoin *et al.*, 2000; Brady *et al.*, 2003; Fu *et al.*, 2004). Loci positively affecting each hormone response are shown in solid circles; those with a negative effect are shown in dashed circles. Suppressor interactions are shown by black lines and enhancer interactions by dashed lines. The *abi1-1* mutant phenotype is irregular; it is ABA insensitive but is actually a negative regulator of ABA signaling (Yoshida *et al.*, 2006). Abbreviations: *aba, abscisic acid deficient*; *abi, abscisic acid insensitive*; *ctr, constitutive triple response*; *ein, ethylene insensitive*; *era, enhanced response to abscisic acid*; *ga, gibberellic acid requiring*; *gai, gibberellic acid insensitive*; *sly, sleepy*; *spy, spindly.*

of the phenotypic characteristics of mutants that are disrupted in seed maturation, including dormancy inception. For example, the various proposed functions of some of the transcription factors belonging to the B3 domain family (e.g., VP1 and ABI3) include (a) regulation of the expression of ABA-responsive genes during seed development (e.g., storage-protein genes and genes encoding putative desiccation protectants) (McCarty *et al.*, 1991; Giraudat *et al.*, 1992; Parcy *et al.*, 1997; Zeng *et al.*, 2003; Zeng and Kermode, 2004; Kagaya *et al.*, 2005); (b) repression of postgerminative genes (Nambara *et al.*, 2000; Gazzarrini *et al.*, 2004); and (c) participation in dormancy inception during seed development and/or the maintenance of a dormant state in mature dispersed seeds (Jones *et al.*, 1997; Fukuhara *et al.*, 1999; McKibbin *et al.*, 2002; Zeng *et al.*, 2003).

Combinatorial control involving several transcription factors is a key theme in the regulation of seed and embryo developmental events, including dormancy inception. Superimposed on this is the capacity for other components of hormonal signal transduction to effect posttranslational modifications to proteins, thus modifying the actions of transcription factors and/or promoting their degradation. For example, there is strong evidence that the various functions of ABI3/VP1 are achieved through interactions with other proteins (Finkelstein *et al.*, 2002; see Chapter 1). In concert, these may activate or repress target genes, including those important for maintaining/terminating the dormant state.

ABI3 together with ABI4 and ABI5 participate in combinatorial control of gene expression, possibly by forming a regulatory complex mediating seed-specific and/or ABA-inducible expression (Finkelstein *et al.*, 2002). There appears to be extensive cross-regulation of expression among *ABI3*, *ABI4*, and *ABI5* genes (Söderman *et al.*, 2000). Mutations in the *ABI4* and *ABI5* gene loci have similar effects on seed development and ABA sensitivity to those associated with the *abi3* mutant; however, null mutations in the *ABI3* gene locus result in more severe phenotypes than those associated with defects in *ABI4* or *ABI5* (Finkelstein *et al.*, 2002). The ABI4 and ABI5 proteins contain putative DNA-binding and protein-interaction domains. ABI4 contains the AP2 domain characteristic of the APETALA2 family of transcription factors (Finkelstein *et al.*, 2002). The ABI5 family of proteins all share three conserved charged domains in the N-terminal half of the protein that contain potential phosphorylation sites and a bZIP domain at the C-terminus. Like VP1/ABI3, ABI5 can either activate or repress gene expression. ABI3 and ABI5 interact physically (Nakamura *et al.*, 2001) and may have antagonistic or synergistic effects on gene expression, depending on the targeted gene (Finkelstein and Lynch, 2000; Delseny *et al.*, 2001).

ABA (and the signal it transduces) is not the only regulator of the middle and late stages of seed development. For example, the *lec* class mutants (*lec1*, *lec2*, and *fus3*) act in a concerted fashion to promote events that are critical to the later stages of seed development, including the acquisition of desiccation tolerance and induction of dormancy, and to repress germinative and postgerminative functions (Keith *et al.*, 1994; Meinke *et al.*, 1994; Holdsworth *et al.*, 1999). Similar to *ABI3*, these three loci encode transcription factors regulating genes expressed during seed development and are thought to exert their effects by protein–DNA and/or protein–protein interactions. FUS3 and LEC2 belong to the B3-domain protein family of transcription factors that includes ABI3 (Luerßen *et al.*, 1998, Stone *et al.*, 2001); the *LEC1* gene encodes a protein with sequence similarity to the HAP3 subunit of CCAAT binding factors (Lotan *et al.*, 1998). In mammalian cells, these heterotrimeric binding factors enhance the transcription of a large number of genes (reviewed in Harada, 2001; see Chapter 1). FUS3 and LEC1 can modulate the abundance of ABI3 protein in seeds. Synergistic interactions among the three proteins (ABI3, FUS3, and LEC1) are thought to control the accumulation of chlorophyll and anthocyanins, sensitivity to ABA and expression of individual members of the 12S storage protein gene family (Parcy *et al.*, 1997). FUS3 and LEC2 also appear to control ABA and GA homeostasis (see below).

8.2.3 ABA in dormancy maintenance and termination

8.2.3.1 ABA synthesis and homeostasis during dormancy maintenance and termination

De novo synthesis of ABA appears to be necessary for dormancy maintenance following imbibition in seeds of sunflower (*Helianthus annuus*), barley, beechnut (*Fagus sylvatica*), lettuce (*Lactica sativa*), tobacco, and Douglas fir (*Pseudotsuga menziesii*) (Le Page-Degivry and Garello, 1992; Wang *et al.*, 1995; Bianco *et al.*,

1997; Le Page-Degivry *et al.*, 1997; Yoshioka *et al.*, 1998; Grappin *et al.*, 2000). For instance, when tobacco seeds are imbibed, there is an accumulation of ABA in dormant seeds but not in seeds that have been after-ripened. In some species, there is a lag before ABA begins to accumulate in the dormant-imbibed seed. The carotenoid- and ABA-biosynthesis inhibitor fluridone, when used in conjunction with gibberellic acid (GA_3), is effective in releasing dormancy; exogenous application of both chemicals to seeds inhibits accumulation of ABA during imbibition (Grappin *et al.*, 2000). Similar to tobacco, GA_3 and fluridone are effective in releasing dormancy of yellow-cedar (*Chamaecyparis nootkatensis*) seeds in the absence of any additional treatment (Schmitz *et al.*, 2001). These seeds normally require a 3-month dormancy-releasing treatment consisting of 1 month of warm, moist conditions followed by 2 months of moist chilling (Ren and Kermode, 1999). However, fluridone alone (i.e., with no GA_3) is much less effective in eliciting germination, indicating that a decline in ABA amount alone is not sufficient to release dormancy and other changes (e.g., synthesis of gibberellins) may also be necessary.

The termination of dormancy (through after-ripening, moist chilling, or smoke) is strongly associated with a decline in ABA content upon transfer to germination conditions, and the decline in ABA corresponds well with an increase in germination capacity. After-ripening itself has little effect on ABA levels within embryos of dry barley grains (Jacobsen *et al.*, 2002). Upon transfer of the after-ripened grain to germination conditions, the ABA content of the embryo declines markedly over the first 12 h, and low ABA concentrations are maintained up to 30 h, by which time 90% of the grains have germinated. Dormant grains placed immediately in germination conditions with no previous after-ripening period show some reduction of embryo ABA content over the first 12 h, but ABA content subsequently increases between 12 and 30 h of imbibtion. A similar transient reduction of ABA, followed by its accumulation, occurs in imbibed dormant *Arabidopsis* seeds of the Cape Verde Islands (Cvi) ecotype. This is in marked contrast to imbibed seeds that have received a previous dormancy-releasing treatment (moist chilling, nitrate or fluridone), in which ABA levels continue to decline during imbibition (Ali-Rachedi *et al.*, 2004). Seeds of the commonly used accessions of *Arabidopsis* such as Columbia (Col) (which are far less dormant than seeds of the Cvi ecotype) show a decline in ABA levels upon imbibition in the absence of a previous dormancy-releasing treatment; however, the seeds do not show the large increase in ABA glucosylester, an ABA metabolite which is characteristic of their nondormant (moist-chilled and imbibed) counterparts (Chiwocha *et al.*, 2005).

In Scots pine (*Pinus sylvestris*) seed, dormancy-releasing treatments that include either white or red light decrease ABA levels prior to radicle protrusion; seeds subjected to a far-red light pulse after red light do not exhibit as great a decline in ABA, nor is dormancy relieved (Tillberg, 1992). During dormancy release of yellow-cedar seeds, there is about a twofold reduction of ABA in the embryo, but ABA content does not change in the megagametophyte; however, the embryos exhibit a change both in ABA turnover and in their sensitivity to ABA (Schmitz *et al.*, 2000, 2002). ABA decreases in both the megagametophyte and embryo during moist chilling of Douglas-fir seeds; in the former, ABA declines fourfold during 7 weeks of moist chilling (Corbineau *et al.*, 2002). In addition, the longer the duration of moist chilling, the faster the rate of ABA decline during subsequent germination (Corbineau *et al.*, 2002).

In postfire environments, smoke can promote the germination of many species that otherwise remain dormant for long periods (reviewed in Gubler *et al.*, 2005; see Chapter 7). Dormancy release of *Nicotiana attenuata* seeds induced by smoke-water, in which the active stimulator is likely the butenolide 3-methyl-2*H*-furo [2,3-*c*]pyran-2-one (Flematti *et al.*, 2004), is accompanied by an eightfold decrease in seed ABA content. This is in marked contrast to seeds subjected to a control (water) treatment for an equivalent period (22 h), in which ABA contents increase shortly after imbibition and remain stable thereafter (Schwachtje and Baldwin, 2004).

Changes in ABA content are only part of the story, as the capacity for ABA biosynthesis versus catabolism (metabolic flux) is a superior indicator of whether a seed will or will not terminate dormancy and germinate. Increased ABA catabolism is associated with dormancy termination of seeds of yellow-cedar, beechnut, Douglas-fir, and barley (Le Page-Degivry *et al.*, 1997; Schmitz *et al.*, 2000, 2002; Corbineau *et al.*, 2002; Jacobsen *et al.*, 2002). The degradation or inactivation of ABA occurs via oxidation and conjugation (Zaharia *et al.*, 2005; see Chapter 9). The major pathway by which ABA is catabolized is through hydroxylation to form 8′-hydroxy-ABA, which reversibly cyclizes to phaseic acid (PA). Further reduction of PA can take place at the 4′ position to form dihydrophaseic acid (DPA). ABA and ABA metabolites (PA and DPA) can also become conjugated with glucose to form esters or glucosides. Other minor pathways include formation of 7′-hydroxy-ABA (7′OH-ABA) and ABA 1′,4′-diols (Cutler and Krochko, 1999; Zeevaart, 1999).

Effective dormancy release likely requires ABA catabolism, but at the same time may require a reduction in ABA biosynthesis as well. Moist chilling of western white pine (*Pinus monticola*) seeds is accompanied by a significant decrease in ABA in both the embryo and the megagametophyte (Feurtado *et al.*, 2004). More importantly, the decline in ABA contents after different durations of moist chilling correlate well with an increased capacity of seeds to germinate following their transfer to germination conditions. The decline of ABA (due to enhanced catabolism and/or reduced synthesis) continues during germination. In conditions that are not effective for releasing dormancy, ABA increases in western white pine seeds but there is a decrease of PA, 7’OH ABA, and DPA (or maintenance at steady-state amounts), suggesting a slower rate of ABA catabolism. Thus, moist chilling not only causes a decline in ABA by stimulating ABA catabolism as chilling proceeds, but, perhaps equally important, increases the capacity (or competence) for ABA catabolism, when the seeds are subsequently placed in germination conditions.

When barley grains are subjected to after-ripening and then imbibed, ABA in the embryo is rapidly metabolized to PA, which is not released into the incubation medium or into the endosperm (Jacobsen *et al.*, 2002). The products of 8′-hydroxylation of ABA (PA and DPA) also accumulate in after-ripened *Arabidopsis* (Col) seeds over the first 24 h of imbibition (Kushiro *et al.*, 2004). In previously moist-chilled Col seeds, the major product of ABA metabolism is ABA glucose ester (ABA-GE); PA and DPA remain at low levels over the first 48 h of imbibition (Chiwocha *et al.*, 2005).

Thermodormancy of lettuce seeds, induced by imbibing seeds at a supraoptimal temperature for germination (33°C instead of 23°C), is associated with surprisingly active hormone fluxes (Chiwocha *et al.*, 2003). Moreover, the hormone and hormone metabolite profiles of germinating and thermodormant lettuce seeds are distinct. This is particularly true for ABA and its metabolites; thermodormant seeds accumulate high levels of DPA, while germinating seeds accumulate high amounts of ABA-GE. Thermodormant seeds are further distinguished from germinating seeds by exhibiting accumulations of indole-3-acetic acid (IAA) and zeatin that are not accompanied by any significant increases in the levels of their conjugates indole-3-aspartate and zeatin riboside, respectively. The most striking changes potentially reflective of hormonal cross-talk include a marked accumulation of IAA levels in thermodormant seeds coincident with a major increase in the level of DPA; the significance of the rise in IAA and DPA is unknown as this occurs well after the induction of thermodormancy (Chiwocha *et al.*, 2003). Whether there is a mechanistic connection between auxin biosynthesis and ABA catabolism remains to be determined.

The cytochrome P450 CYP707A subfamily functions as ABA 8′-hydroxylases in *Arabidopsis* (Kushiro *et al.*, 2004; Saito *et al.*, 2004; Nambara and Marion-Poll, 2005; see Chapter 9). There are four *CYP707A* genes in *Arabidopsis* that exhibit tissue-specific regulation (Kushiro *et al.*, 2004; Saito *et al.*, 2004). *CYP707A2* is the major 8′-hydroxylase gene expressed during seed imbibition, with its transcripts reaching a peak at approximately 6 h after the start of imbibition in seeds receiving a previous after-ripening period. This timing of expression coincides with the decline in seed ABA and increase in PA contents. Expression of the *CYP707A2* gene is also induced by 30 μM ABA after 12 h of imbibition (Kushiro *et al.*, 2004). In contrast, the expression of the *CYP707A1* and*A3* genes remains low during early imbibition and increases only after 12 h of imbibition. T-DNA insertional mutants of *CYP707A2* have higher seed ABA content and exhibit increased dormancy as compared to their wild-type counterparts. Thus, CYP707A2 contributes to the reduction of ABA content in imbibed after-ripened seeds, a process important for dormancy release. However, these analyses were conducted in Col seeds that show weak primary dormancy and it will be important to extend these analyses to a more dormant ecotype or species. In this respect, the gene encoding a barley 8′-hydroxylase (*CYP707A2*) is induced to a much higher level in embryos of imbibed nondormant grains (previously subjected to after-ripening) than in embryos of dormant imbibed grains and the transcripts (and hence, possibly 8′- hydroxylase action) are localized to the coleorhiza in the region of primary root tip (Millar *et al.*, 2006).

Mechanisms that underlie the fine-tuned balance between ABA biosynthesis and catabolism and its potential transport within and between seed tissues need to be elucidated. Changes in ABA sensitivity (and sensitivity to certain metabolites) undoubtedly also play a role in the transition of seeds from a dormant to a nondormant state.

8.2.3.2 ABA signaling factors and the control of dormancy maintenance and termination

The role of ABA in controlling the transition from dormancy to germination likely involves actions at several levels, including effects on transcription, RNA processing, posttranslational modifications of proteins, and the metabolism of secondary messengers. Thus, as well as transcription factors, many of the participants in ABA signal transduction are protein-modifying enzymes, RNA-binding/processing proteins, GTP-binding proteins, enzymes of phospholipid or phosphoinositide metabolism, and proteins regulating vesicle trafficking or subcellular localization of proteins. For brevity, only selected examples of these are discussed; Table 8.1 summarizes the various ABA-regulatory mechanisms and indicates potential control points that involve cross-talk with other hormones and environmental factors.

Several individual genes have been identified in seeds of beechnut that may function in ABA signaling within the dormant-imbibed seed. These encode a GTP-binding protein (FsGTP1), protein kinases (FsPK1 and FsPK2), and type 2C protein phosphatases (FsPP2C1 and FsPP2C2); all are ABA-inducible in seeds and are expressed in a dormancy-related manner (Nicolás *et al.*, 1998; Lorenzo *et al.*, 2001, 2002, 2003; González-García *et al.*, 2003). For example, expression of the *FsPP2C1* gene is detected in dormant seeds and increases after ABA treatment. Downregulated expression of the gene occurs during dormancy release elicited by a moist chilling or GA treatment (Lorenzo *et al.*, 2001). Further, stable overexpression of the *FsPP2C1* gene in *Arabidopsis* results in plants that exhibit relative ABA-insensitivity and seeds having reduced dormancy as compared to the wild-type. Root growth is resistant or less sensitive to ABA, mannitol, NaCl, and paclobutrazol, a GA synthesis inhibitor (González-García *et al.*, 2003). Thus, FsPP2C1 is an important regulator modulating the ABA signal in dormant beechnut seeds that operates through a negative feedback loop (González-García *et al.*, 2003).

Another example of a putative regulator of dormancy in mature seeds is the ABI3 (VP1) family of transcription factors. In both wild oat (*Avena fatua*) and the common ice plant (*Mesembryanthemum crystallinum*), the expression of *VP1* genes is positively correlated with the degree of dormancy of the imbibed mature seeds (Jones *et al.*, 1997; Fukuhara *et al.*, 1999). In wild oat, if dormancy is reintroduced by exposure of seeds to nonoptimal germination conditions (secondary dormancy), the synthesis of *AfVP1* transcripts is reinduced (Jones *et al.*, 1997; Holdsworth *et al.*, 1999). Further, *VP1* genes in some wheat cultivars have splicing defects and this contributes to their increased susceptibility to preharvest sprouting (PHS) or precocious germination (McKibbin *et al.*, 2002). Interestingly, the expression pattern of the sorghum *VP1* gene is different in two cultivars/genotypes exhibiting differential resistance to PHS at physiological maturity. In embryos of the susceptible cultivar, transcripts encoding VP1 peak at a relatively early stage of grain development (20 DAP), while in the resistant cultivar transcript accumulation occurs at much later developmental stages when seed maturation is almost complete (Carrari *et al.*, 2001).

Because ABI3 has both repressor and activator functions during seed development and seedling growth, it is interesting to postulate that ABI3 has similar

Table 8.1 Selected ABA synthesis/catabolism and response genes involved in modulating seed dormancy

Species	Gene/locus	Protein	Mutant/ transgenic line	Effect on dormancy and ABA sensitivity	Hormone/ environment interaction (alleles)	References
			Synthesis/Catabolism			
Arabidopsis	*ABA1*	Zeaxanthin epoxidase (ZEP)	*aba1*	Reduced	Decreased induction of RD29::LUC during OS (*los6*); sugar insensitive; upregulated in *ein2* leaves	Karssen *et al.* (1983); Arenas-Huertero *et al.* (2000); Ghassemian et al. (2000); Xiong *et al.* (2002)
	ABA2	Short-chain dehydrogenase-reductase (AB-SDR)	*aba2*	Reduced	Sugar insensitive (*gin1*, *isi4*); salt insensitive (*san3*)	Rook *et al.* (2001); Cheng *et al.* (2002); González-Guzmán *et al.* (2002)
	ABA3	Molybdenum cofactor sulfurase (MCS)	*aba3*	Reduced	Sugar insensitive (*gin5*); decreased induction of RD29::LUC during OS (*los5*)	Arenas-Huertero *et al.* (2000); Xiong *et al.* (2001b)
	AtNCED6 *AtNCED9*	9-*cis*-Epoxycarotenoid dioxygenase (NCED)	*Atnced6/Atnced9* double mutant	Reduced		Tan *et al.*, 2003; Lefebvre *et al.*, 2006
	CYP707A2	ABA 8′-hydoxylase	*cyp707a2-1* *cyp707a2-2*	Enhanced		Kushiro *et al.* (2004)
Zea mays	*VP14*	9-*cis*-Epoxycarotenoid dioxygenase (NCED)	*vp14*	Vivipary; reduced		Tan *et al.* (1997)
Nicotiana plumbaginifolia	*NpABA2*	Zeaxanthin epoxidase (ZEP)	*Npaba2*	Reduced		Marin *et al.* (1996)

(*Continued*)

Table 8.1 Selected ABA synthesis/catabolism and response genes involved in modulating seed dormancy (*Continued*)

Species	Gene/locus	Protein	Mutant/ transgenic line	Effect on dormancy and ABA sensitivity	Hormone/ environment interaction (alleles)	References
				Response		
Arabidopsis	*ABI1*	PP2C Ser/Thr protein phosphatase	*abi1-1*	Reduced; ABA insensitive	Genetically, ET mutants *ctr1* enhance and *ein2* suppress *abi1-1*	Leung *et al.* (1994); Beaudoin *et al.* (2000)
	ABI3	B3 domain TF	*abi3*	Reduced; ABA insensitive	ABA–auxin root interaction; phyB and light	Brady *et al.* (2003); Mazzella *et al.* (2005)
	ABI4	DREB subfamily A-3 of ERF/APETALA2 TF	*abi4*	Normal; ABA insensitive	Sugar insensitive (*sis5*, *gin6*); negative regulator of light-induced genes (with sugar interaction); salt insensitive (*san5*)	Finkelstein (1994); Arenas-Huertero *et al.* (2000); Acevedo-Hernández *et al.* (2005)
	ABI5	bZIP TF	*abi5*	Normal; ABA insensitive	Sugar insensitive	Finkelstein (1994); Finkelstein and Lynch (2000); Laby *et al.* (2000)
	ERA1	β-Subunit farnesyl transferase	*era1*	Enhanced; ABA hypersensitive		Cutler *et al.* (1996)
	AHG2	PARN (Poly(A)-specific ribonuclease)	*ahg2-1*	Enhanced; ABA hypersensitive	SA hypersensitive	Nishimura *et al.* (2005)
	AHG3 (*AtPP2CA*)	PP2C Ser/Thr protein phosphatase	*ahg3-1*	Enhanced; ABA hypersensitive		Sheen (1998); Yoshida *et al.* (2006)
	MARD1	Zinc-finger protein (TF?)	*mard1*	Reduced; ABA insensitive		He and Gan (2004)

	RGS1	Regulator of G-protein signaling	*rgs1-2*	Reduced; ABA insensitive	Positive regulator of *AtNCED3*, *AtABA2* (esp. during sugar stress)	Chen *et al.* (2003); Chen *et al.* (2006)
	RPN10	26S Proteasome subunit	*rpn10-1*	Enhanced (?); ABA hypersensitive	Seedlings hypersensitive to NaCl, sugar; CK insensitive; Aux insensitive roots	Smalle *et al.* (2003)
	ROP2	Rop GTPase	*CA-rop2* (constitutively active GTP-bound)	Reduced; slightly reduced ABA sensitivity	hypocotyl elongation promoted by BR; increased LR formation by Aux compared to WT	Li *et al.* (2001)
			DN-rop2 (dominant negative GDP-bound)	Enhanced; ABA hypersensitive	Decreased LR formation by Aux compared to WT	
	SAD1	Sm-like snRNP protein	*sad1*	Enhanced; ABA hypersensitive		Xiong *et al.* (2001a)
Z. mays	*VP1*	B3 domain TF	*vp1*	Vivipary or reduced dormancy; ABA insensitive	ABA–auxin root interaction; increased ethylene production during seed development	McCarty *et al.* (1991); Young and Gallie (2000); Suzuki *et al.* (2001)

Abbreviations: Aux, auxin; CK, cytokinin; BL, brassinolide (brassinosteroid); ET, ethylene; LR, lateral root; OS, osmotic stress; SA, salicylic acid; TF, transcription factor.

functions during dormancy maintenance in the mature seed to regulate postimbibition dormancy-related genes/processes. However, conclusive evidence of a direct role for ABI3 awaits and will be difficult to demonstrate experimentally because of the more global role of this transcription factor in seed developmental processes. Stable overexpression in a transgenic system influences several processes during seed development (Zeng and Kermode, 2004); an inducible expression system (e.g., via use of a glucocorticoid receptor/DEX [dexamethasone]-inducible system) may pose a more viable alternative to test for a direct role of ABI3/VP1 specifically in dormancy maintenance. Maintenance of dormancy in yellow-cedar seeds appears to involve ABI3. A decline in *CnABI3* transcripts and in CnABI3 protein abundance is correlated with dormancy termination, and their synthesis is also downregulated during germination (Zeng *et al.*, 2003). Further, it has been suggested that regulation of *CnABI3* is not restricted to the transcriptional level and may also involve protein stability (Zeng *et al.*, 2003). Indeed, a key control over the dormancy-to-germination transition may occur at the level of regulation of proteasome components (Bove *et al.*, 2005; see Chapter 10). Interestingly, *AIP2* (*ABI3-INTERACTING PROTEIN 2*), which was originally uncovered in a yeast 2-hybrid screen for proteins that interact with ABI3, encodes an E3-ubiquitin ligase (Kurup *et al.*, 2000; Zhang *et al.*, 2005), which targets proteins for degradation via the 26S proteasome pathway (Vierstra, 2003). AIP2 can polyubiquitinate ABI3 *in vitro* and, in transgenic plants, induced *AIP2* gene expression leads to decreased ABI3 protein levels. Further, *Arabidopsis aip2-1* null mutants show higher ABI3 protein levels compared to the wild-type seeds after moist chilling, are hypersensitive to ABA, and mimic an ABI3-overexpression phenotype. Conversely, *AIP2* overexpressors contain lower levels of ABI3 protein compared to wild type and are more resistant to ABA, phenocopying the *abi3* mutant (Zhang *et al.*, 2005). Thus, AIP2 is a negative regulator of ABA responses via its modulation of ABI3 protein levels.

8.3 Gibberellin

8.3.1 GA is antagonistic to ABA during seed development

During development, GA plays a multitude of roles, participating in fertilization, embryo growth, nutrient uptake, and the prevention of seed abortion (Hays *et al.*, 2002; Singh *et al.*, 2002; Finkelstein, 2004). GA levels are usually high during embryo development, in which there may be two peaks of GA content; as seeds approach maturity, most active GAs are deactivated, for example via 2-β-hydroxylation or conjugate formation (Hedden, 1999; see Chapter 9). An accumulation of active GAs during later seed development would likely promote premature germination. Indeed, GA appears to act as an antagonist to ABA during seed development, especially with respect to dormancy induction (White *et al.*, 2000). When an ABA-deficient mutant of maize (*vp5*) is manipulated either genetically or via biosynthesis inhibitors to induce GA deficiency during early seed development, vivipary is suppressed in developing kernels and the seeds acquire desiccation tolerance and storage longevity. In cultured immature maize embryos, GA deficiency induced by inhibiting biosynthesis enhances many ABA-responsive developmental events including the accumulation of anthocyanins and transcripts encoding storage and LEA (Late Embryogenesis Abundant) proteins (White and Riven, 2000). *In planta*, the major accumulation of GA_1 and GA_3 occurs in wild-type maize kernels just prior to a peak in ABA content during development. It is speculated that these GAs induce a developmental program that leads to vivipary in the absence of normal amounts of ABA, and that a reduction of GAs reestablishes an ABA/GA ratio appropriate for suppression of germination and induction of maturation. Induction of GA deficiency does not suppress vivipary in *vp1* mutant kernels, suggesting that VP1 acts downstream of both GA and ABA in programming seed development. Thus, in contrast to the proposed role of ABA as a 'status quo' hormone, GA can be described as a 'transitionary' hormone, promoting growth and changes in developmental status (McCourt *et al.*, 2005). For example, GA is needed for the initiation of germination and can also influence the timing of leaf emergence and decrease the time to flowering (Evans and Poethig, 1995; Bentsink and Koornneef, 2002).

The embryonic proteins FUS3 and LEC2 that regulate multiple processes during seed development (e.g., repression of postgerminative programs and activation of seed storage protein genes) may be a nexus of hormone action during *Arabidopsis* embryogenesis controlling GA and ABA biosynthesis (Gazzarrini *et al.*, 2004). In the *Arabidopsis* mutants with defective expression of these genes (*fus3* and *lec2*), the GA biosynthesis pathway is misregulated, and this in turn contributes to some of the altered phenotypes (e.g., trichome cell formation) (Curaba *et al.*, 2004). Indeed, FUS3 negatively modulates GA synthesis and positively regulates ABA synthesis. In *fus3* mutants, ABA levels are decreased while GA levels increase; conversely, *FUS3* misexpressors show the opposite trend: an increase in ABA and decrease in GA levels (Gazzarrini *et al.*, 2004). In wild-type seeds, the FUS3 protein represses expression of the gene encoding the GA biosynthetic enzyme AtGA3ox2 by physically interacting with two RY elements (GATGCATG) of the *AtGA3ox2* promoter. Repression of the *AtGA3ox2* gene by FUS3 occurs primarily in epidermal cells of the embryonic axis, which is distinct from the pattern of expression of this gene during germination (Curaba *et al.*, 2004). Similarly, the expression of *AtGA3ox1* and *AtGA20ox1* genes is downregulated by inducible FUS3 misexpression in seedlings (Gazzarrini *et al.*, 2004). *FUS3* expression itself is modulated by the hormone auxin (see below). Gazzarrini *et al.* (2004) suggest that it is this modulation of the ABA/GA ratio by FUS3, and a positive or negative feedback regulation of FUS3 by ABA and GA, respectively, that helps regulate maturation events during seed development. Interestingly, maternally derived ABA inhibits viviparous germination of *fus3* mutants (Raz *et al.*, 2001), supporting the contention that FUS3 maintains embryo growth arrest indirectly by impinging on the ABA pathway.

8.3.2 GA promotes the transition to germination

Classically, the action of ABA and GA during seed development and germination was thought to be temporally distinct. GA was hypothesized to play a key role in the promotion of germination, while ABA induced dormancy during seed maturation (Karssen and Laçka, 1986; Bewley, 1997). However, as described above, we now know that ABA is also involved in dormancy maintenance during imbibition and that GA acts antagonistically to ABA during seed development.

GA is thought to specifically act by promoting the growth potential of the embryo and by mediating the weakening of tissues that enclose the embryo (Bewley, 1997; Koornneef *et al.*, 2002; see Chapter 11). Consistent with this, the major increase in GA levels during germination usually occurs just prior to radicle protrusion, while ABA content usually decreases during early imbibition (Jacobsen *et al.*, 2002; Poljakoff-Mayber *et al.*, 2002; Ogawa *et al.*, 2003; Pérez-Flores *et al.*, 2003; see Chapter 9). The temporal and spatial expression of GA biosynthesis genes during germination is well characterized in *Arabidopsis* (Yamaguchi *et al.*, 2001; Ogawa *et al.*, 2003). Expression of the early biosynthesis enzyme *ent*-COPALYL DIPHOSPHATE SYNTHASE (encoded by *AtCPS1*), which catalyzes the first step of geranylgeranyl diphosphate (GGDP) cyclization, occurs in embryo provasculature in germinating seeds. Later steps, including the second step of the cyclization reaction forming *ent*-kaurene, catalyzed by *ent*-KAURENE OXIDASE (*AtKO1*) and the production of GA_4 from GA_9, catalyzed by the GA-3-OXIDASES (*AtGA3ox1* and *AtGA3ox2*), seem to be restricted to the cortex and endodermis of the embryonic axis. Thus, during germination an intercellular transport mechanism has been proposed to account for formation of bioactive GAs (Yamaguchi *et al.*, 2001). Furthermore, the expansion of the cortical cells of the embryonic axis (radicle) during germination correlates well with the predicted sites for GA production. These findings support the contention that GA may help increase radicle growth potential and are also consistent with the hypothesis that embryonic GA is released to trigger weakening of the tissues surrounding the radicle. Movement of GA (or a GA signal) is evident in the expression of GA-regulated genes in non-GA-producing cell types such as the endosperm (Ogawa *et al.*, 2003). Of course, the classic model of the cereal endosperm provides the best example of release of an embryo GA signal, which induces the synthesis of hydrolytic enzymes such as α-amylase in the aleurone layer during postgerminative reserve mobilization (Woodger *et al.*, 2004).

The widely accepted notion that GA promotes germination has been supported through the analysis of mutants having defects in loci that control GA biosynthesis and signal transduction. Germination of GA biosynthesis mutants such as those of *Arabidopsis* (e.g., *ga1-3*) and tomato (e.g., *gib-1*) depends on the application of exogenous GA (Groot and Karssen, 1987; Karssen *et al.*, 1989). In a *ga1* background, the *aba1* mutation leading to ABA deficiency is able to counteract the GA requirement for germination (Koornneef *et al.*, 1982). However, *ga1-3 aba1-1* double mutants still require light and moist chilling to efficiently release dormancy and to maximize germination to 100% (Debeaujon and Koornneef, 2000; see Chapter 5).

GA/ABA antagonism during germination has also been revealed through application of exogenous ABA. However experiments using applied ABA do not necessarily reflect the actions of the endogenous hormone. Classic experiments demonstrate that applied ABA inhibits cell wall loosening and elasticity in *Brassica napus* embryos (Schopfer and Plachy, 1985); thus, in contrast to GA, ABA is thought to reduce the growth potential of the embryo (see Chapter 11 for a more detailed review). More recently, an arabinogalactan-protein (AGP) in the cell walls of *Arabidopsis* roots has been implicated in the ABA inhibition of seed germination (van Hengel and Roberts, 2003). Seeds of the *agp30* mutant germinate faster in the presence of 10 μM ABA, as compared to the wild-type seeds. Furthermore, the effect of this mutation appears to be the result of an enhanced force exerted by the radicle, rather than due to changes in the testa. The authors suggest that AGP30 plays a role in ABA perception, perhaps by interacting or modifying properties of cell surface receptors or other partner molecules associated with ABA perception. However, AGPs have also been implicated in cell elongation and it is this process that is likely affected rather than any direct effect on ABA response-mediators (van Hengel and Roberts, 2003). It is currently unknown if AGP30 is regulated by hormones such as GA or ethylene.

Also consistent with the role of GA in the germination process, application of GA can release seed dormancy in numerous species including some *Arabidopsis* accessions such as Landsberg *erecta* (L*er*), but not the more dormant Cvi accession (Derkx and Karssen, 1993; Bewley and Black, 1994; Ali-Rachedi *et al.*, 2004). However, a point for debate concerns the specific role of GA in dormancy termination elicited by moist chilling. Early results suggested that cold treatments do not stimulate GA biosynthesis in *Arabidopsis* seeds but, rather, promote enhanced sensitivity to GA (Derkx and Karssen, 1993). However, recent evidence from *Arabidopsis* suggests that treatments that terminate dormancy (e.g., moist chilling, with or without light) stimulate the expression of GA biosynthesis genes and subsequent accumulation of bioactive GAs (Yamauchi *et al.*, 2004). Consistent with these findings, factors that regulate the cold- and light-mediated increase in GA, via increased expression of the *AtGA3ox* gene, have been identified. These include the positive regulator BME3 (BLUE MICROPYLAR END 3), a GATA zinc finger transcription factor, and negative regulators SPT (SPATUALA) and PIL5 (PHYTOCHROME INTERACTING FACTOR-LIKE 5), basic helix-loop-helix transcription factors (Oh *et al.*, 2004; Liu *et al.*, 2005; Penfield *et al.*, 2005). The regulation of expression of *BME3*, *SPT*, or *PIL5* genes by ABA or other hormonal pathways is unknown; however, the involvement of PIL5 (and SPT) in germination is associated with phytochrome action (see below).

Thus, the mechanisms by which GA may participate in dormancy release have been partially elucidated by examining changes elicited by moist chilling of seeds. Exposure of seeds to moist chilling reverses the regulatory actions of ABA and GA in dormant seeds; ABA events are downregulated and GA events are upregulated. Changes in ABA flux during moist chilling support a catabolic mode leading to a decline in ABA levels during moist chilling itself, but especially after placement of seeds in germination conditions. The *CYP707A2* gene is upregulated during moist chilling and during early germination (Kushiro *et al.*, 2004; Yamauchi *et al.*, 2004). ABA signaling factors such as ABI2 and ABI3 are also downregulated during moist chilling (Yamauchi *et al.*, 2004). Accompanying the general trend toward ABA catabolism and the attenuation of ABA signaling related to dormancy, synthesis of bioactive GAs is upregulated during moist chilling. As such, selected GA-inducible genes are upregulated while selected GA-repressible genes are downregulated during chilling (Yamauchi *et al.*, 2004). One example of a gene downregulated by GA and moist chilling is the *ABI5*-homologue *EEL* (*ENHANCED Em LEVEL*; Bensmihen *et al.*, 2002; Ogawa *et al.*, 2003; Yamauchi *et al.*, 2004). Interestingly, the expression of the *EEL* gene is dormancy-related, being upregulated in dormant-imbibed Cvi seeds, but downregulated during after-ripening of Cvi seeds (Finch-Savage, 2004). However, the participation of *EEL* in dormancy and germination remains speculative, especially since *eel* mutants do not display ABA-insensitive phenotypes during germination (Bensmihen *et al.*, 2002).

Additional mechanisms of GA-signaling in seed dormancy release and germination are emerging and have been linked to hormone cross-talk. One of the most prominent examples of repressors of GA action during germination is the DELLA (motif) transcription factors, GAI (GA INSENSITIVE), RGA (REPRESSOR OF *ga1-3*), RGL1 (RGA-LIKE 1), RGL2, and RGL3 – a subfamily of GRAS (for GAI, RGA, and SCARECROW) regulatory proteins (Cao *et al.*, 2005; see Chapter 10). In addition to modulating GA response, DELLA proteins appear also to mediate ethylene and auxin responses and are suggested to play an integrative role in the phytohormone signal response network (Achard *et al.*, 2003). Further, the 'growth-repressing' DELLA proteins have been linked to ABA signaling as well. The roots of a quadruple-DELLA mutant (*gai*, *rga*, *rgl1*, *rgl2*) are relatively insensitive to ABA growth inhibition; in addition, in transgenic wild-type and *abi1-1* mutant plants expressing a reporter (green fluorescent protein)-RGA fusion driven by the native RGA promoter (*pRGA::GFP-RGA*), GFP-RGA accumulation in the roots is responsive to ABA only in the former (Achard *et al.*, 2006).

An additional factor that seems to interplay with the GA pathway is the *PICKLE* (*PKL*) locus. The *PKL* gene encodes an SWI (switch)/SNF (Sucrose Non-Fermenting) class chromatin remodeling factor that belongs to the CHD3 (Chromodomain Helicase DNA-binding 3) group that generally are repressors of transcription (Ogas *et al.*, 1999). PKL helps modulate the phase transition between seed development and dormancy/germination and is necessary for the repression of embryonic traits from germination onward (Ogas *et al.*, 1997; Li *et al.*, 2005). In mutant *pkl* plants, the embryonic identity genes, *LEC1* and *LEC2*, are de-repressed during germination and *LEC1*, *LEC2*, and *FUS3* are upregulated more than 100-fold in *pkl* roots (Ogas *et al.*, 1999; Rider *et al.*, 2003). Expression of these *LEC* family genes in *pkl* seedlings allows expression of seed developmental traits after germination, such as accumulation of seed storage reserves and the ability to undergo somatic embryogenesis (Ogas *et al.*, 1999; Stone *et al.*, 2001; Henderson *et al.*, 2004; Rider *et al.*, 2004). GA interacts with *pkl* to repress embryonic identity; decreasing the level of GA in germinating seeds increases the penetrance of the *pkl* root phenotype (swollen green appearance) from ~1–5% to 80% (Ogas *et al.*, 1997). Further, *pkl* plants are reminiscent of GA-response mutants; they are dwarfed, dark-green, and exhibit increased time to flowering. Exogenous GA can partially rescue *pkl* phenotypes but plants are not deficient in GA (Henderson *et al.*, 2004). Henderson *et al.* (2004) suggested that PKL may mediate a subset of GA-dependent responses during plant development, including its novel function of repressing embryonic traits. Recently, using a steroid-inducible *PKL*, it was demonstrated that only expression of *PKL* during germination (rather than during the postgerminative stage) repressed embryonic trait formation in the primary roots of *pkl* seedlings (Li *et al.*, 2005). Thus, in contrast to the *LEC* genes that help maintain the seed development state, *PKL* acts to repress this state and, in turn, promotes the transition to germination and seedling growth. It is currently unknown how *PKL* is itself regulated, as ABA or the GA-signaling factor *SPINDLY* does not affect it (Henderson *et al.*, 2004). However, *CYTOKININ-HYPERSENSITIVE 2* (*CKH2*) is allelic to the *PKL* locus, suggesting that *PKL* may integrate multiple hormonal pathways (Furuta *et al.*, 2005).

8.4 Light interactions

8.4.1 GA synthesis and signaling are promoted by light through the action of phytochrome

One of the more important environmental factors that regulate germination in many seeds is light. Light can stimulate germination of seeds of many species; key model systems are *Arabidopsis* and lettuce (Bewley and Black, 1994; Yamaguchi and Kamiya, 2002). The predominant pathway through which light is perceived by the seed is via the phytochrome group of photoreceptors that act through red and far-red light absorption (Sullivan and Deng, 2003; Chen *et al.*, 2004; Franklin and Whitelam, 2004). The connection between phytochrome and GA, particularly the activation of GA biosynthesis by light, has been known for some time. Red light upregulates GA biosynthesis genes (e.g., *AtGA3ox1* and *AtGA3ox2* genes) and leads to the subsequent production of bioactive GAs in lettuce and *Arabidopsis* seeds (Toyomasu *et al.*, 1993, 1998; Yamaguchi *et al.*, 1998, 2001; Ogawa *et al.*, 2003). In lettuce, red light suppresses the *LsGA2ox2* gene, which encodes a GA catabolic enzyme responsible for converting bioactive GAs to inactive forms. Thus several events may cause the changes in GA homeostasis in response to light (Nakaminami *et al.*, 2003).

In *Arabidopsis*, there are five phytochrome (Phy) apoprotein genes, *PhyA* to *PhyE*, that have distinct but overlapping functions (Sullivan and Deng, 2003). *PhyA*, *PhyB*, and *PhyE* have been implicated in the regulation of germination by light (Shinomura *et al.*, 1994, 1996; Hennig *et al.*, 2002). Phytochromes are cytoplasmic in the dark; light elicits the translocation of a significant proportion of Phy proteins to the nucleus, triggering subsequent changes in light-responsive gene expression (Nagatani, 2004; Mazzella *et al.*, 2005). Several related basic helix-loop-helix transcription factors (a subfamily of up to 15 proteins) appear to be involved in the negative regulation of phytochrome signaling (Duek and Fankhauser, 2005). As noted above, SPT and PIL5 modulate responses to cold and light to promote germination in *Arabidopsis* (Penfield *et al.*, 2005). PIL5 physically interacts with both PhyA and PhyB proteins to negatively regulate seed germination (Oh *et al.*, 2004).

Further, light triggers the degradation of PIL5 through an ubiquitin-mediated proteasome pathway (Shen *et al.*, 2005). In contrast, SPT is light stable and does not interact with PhyA or PhyB, but still modulates responses to red light (Penfield *et al.*, 2005). As mentioned above, both SPT and PIL5 modulate GA biosynthesis genes (either directly or indirectly). Further, another member of the basic helix-loop-helix transcription factor subfamily is a negative regulator of PhyB action. The gene encoding PHYTOCHORME INTERACTING FACTOR 4 (*PIF4*) is regulated at the level of transcription by GA (Ogawa *et al.*, 2003). Thus, interactions between light, phytochrome, and GA biosynthesis are connected through the actions of the various basic helix-loop-helix transcription factors. It would be interesting to determine whether the stability of these proteins is affected by temperature (i.e., moist chilling) as one would presume.

8.4.2 ABA-associated signaling processes are opposed by light signaling

During development, seeds of the phytochrome-deficient *pew1* (*partly etiolated in white light 1*) mutant of *N. plumbaginifolia* accumulate higher levels of ABA at maturity, suggesting that ABA metabolism is controlled, at least in part, by a phytochrome-mediated light signal (Kraepiel *et al.*, 1994). In addition, *pew1* seeds are hyperdormant and the plants are drought-resistant. Analysis of the phytochrome- and ABA-deficient double mutant (*Nppew1 Npaba1*, deficient in AAO due to a defect in the molybdenum cofactor) suggests that the effect of the *Nppew1* mutation is not on ABA biosynthesis (at least it does not affect steps in the ABA biosynthetic pathway earlier than AAO). The double mutant produces seeds that are nondormant due to ABA deficiency; however, the seeds accumulate no more *trans*-ABA-alcohol glucoside (the accumulation product in *Npaba1* mutants) than seeds of the single *Npaba1* mutant. The endogenous function of the NpPEW1 protein may be to modulate ABA levels by upregulating ABA catabolism rather than by downregulating ABA biosynthesis (Kraepiel *et al.*, 1994).

Connections between ABA signaling and phytochrome action have also been demonstrated during seedling growth. IMB1 (IMBIBITION-INDUCIBLE 1), a Bromodomain Extra Terminal (BET) protein induced during seed imbibition, modulates the ABA response factor ABI5. BET proteins are a class of poorly characterized transcriptional regulators whose functions have been linked to chromatin modification/remodeling (Duque and Chua, 2003). Seeds of the *imb1* mutant show impaired cotyledon greening in the presence of exogenous ABA and have higher levels of the ABI5 protein compared to wild-type seeds. The link to light signaling is apparent from the finding that *imb1* mutants are impaired in the PhyA-mediated promotion of germination. However, IMB1 does not appear to be linked to the GA pathway; for example, sensitivity to GA and GA-biosynthesis inhibition is similar in *imb1* mutants and wild-type seeds (Duque and Chua, 2003).

Light negatively modulates *ABI3* gene expression, mainly through changes in PhyB (Mazzella *et al.*, 2005). Genes bearing the RY *cis* motif (a target motif for interaction with ABI3) are upregulated in *phyB* mutants as is *ABI3* gene expression. Further, *abi3* mutant seedlings show enhanced responses to far-red light, whereas overexpressors display the opposite trend. Although PhyB is not the direct signal, light perceived by PhyB in the seed regulates the later developmental decision that leads to either seedling vegetative quiescence or photomorphogenesis, a transition in part modulated by the ABI3 protein (Mazzella *et al.*, 2005).

8.5 Ethylene

8.5.1 Ethylene counteracts ABA during seed development

The role of ethylene in seed development is less apparent than that of ABA, GA, auxin, or cytokinin, although ethylene has been suggested to modulate cotyledon expansion during embryo development (Hays *et al.*, 2000). Ethylene has also been linked to programmed cell death during maize endosperm development (Young *et al.*, 1997). Further, ethylene evolution is significantly higher (by two- to four-fold) in developing endosperms of ABA-insensitive *vp1* and ABA-deficient *vp9* mutants. As such, programmed cell death is accelerated in the mutants and a balance between ABA and ethylene may establish the appropriate timing for programmed cell death during endosperm development (Young and Gallie, 2000). Similarly, an opposing role for ethylene in dormancy induction by ABA has been revealed through recovery of ethylene mutants as modulators of ABA germination phenotypes.

Seeds of *era* (*enhanced response to ABA*) mutants 1 to 3 of *Arabidopsis* are characterized by their inability to germinate in concentrations of ABA that do not inhibit the germination of wild-type seeds (i.e., they are hypersensitive to ABA) (Cutler *et al.*, 1996). The most prominent example of the *era* mutations is *era1* in which there is a defect in the gene encoding a β-subunit farnesyltransferase (Cutler *et al.*, 1996; Finkelstein and Rock, 2002). The *era3* locus has been shown to be allelic to the *ethylene insensitive 2* (*ein2*) locus (Ghassemian *et al.*, 2000). The *EIN2* gene encodes a novel integral membrane protein with homology to the Nramp (natural resistance-associated macrophage protein) metal-ion transporter family (Alonso *et al.*, 1999). Mutations in this pathway lead to an overaccumulation of ABA in leaves, suggesting that *ERA3* (*EIN2*) is a negative regulator of ABA synthesis; indeed, *ZEP* transcripts are increased in leaves of *ein2* mutants (Ghassemian *et al.*, 2000). Similarly, a mutation in *ETR1* (*ETHYLENE RECEPTOR 1*), an *Arabidopsis* ethylene receptor mutant, results in an eightfold higher level of ABA in the mature dry seeds than in wild-type seeds (Chiwocha *et al.*, 2005). This is consistent with the observations that ethylene acts antagonistically to ABA during seed development and may be a main reason why *etr1-2* seeds display a greater degree of dormancy than wild-type Col seeds. That is, during dormancy induction, the absence of a fully functional ethylene signaling pathway to counteract ABA allows ABA to exert its effects to a greater extent.

In addition to modulating ABA levels, mutations that decrease ethylene sensitivity (such as *ein2*) increase the sensitivity of seeds to ABA. Conversely, seeds subjected to the ethylene precursor ACC (1-aminocyclopropane-1-carboxylic acid) exhibit decreased ABA sensitivity, as do seeds of constitutive ethylene response (*ctr1*, *constitutive triple response 1*) mutants. CTR1, a Raf-like serine/threonine protein kinase suggested to be a mitogen-activated protein kinase kinase kinase, is a negative regulator of ethylene signaling (Ouaked *et al.*, 2003). In a similar study, Beaudoin *et al.* (2000) screened for mutations that either enhanced or suppressed the ABA-resistant seed germination phenotype of *Arabidopsis abi1-1* seeds. Alleles of *ctr1* and *ein2* were recovered as enhancer and suppressor mutations, respectively (Beaudoin *et al.*, 2000).

One particular point of contention is that we do not know whether ethylene exerts its antagonistic actions both during seed development (dormancy induction) and during dormancy termination and germination. From the discussion above, it seems clear that ethylene counteracts ABA during seed development and therefore modulates dormancy induction (i.e., the degree of dormancy in the mature seed). Reciprocal crosses between wild-type tomato plants and *Nr* (*Never-ripe*) mutants (an ethylene receptor mutant) show that ethylene insensitivity during seed development plays a role in determining the subsequent time to complete germination (Siriwitayawan *et al.*, 2003). However, ethylene is involved in dormancy release and germination in many species, and the relative contributions of ethylene to dormancy induction and release are still debatable. The existence of multiple ethylene receptors (e.g., in *Arabidopsis*) allowing for functional redundancies only compounds the problem.

8.5.2 Ethylene promotes the transition from dormancy to germination

Ethylene acts antagonistically to ABA during seed development and during dormancy termination and germination and acts in concert with GA to promote these transitional changes. Ethylene promotes germination of many species and in others, such as peanut (*Arachis hypogaea*), apple (*Malus domestica*), cocklebur (*Xanthium strumarium*) and sunflower, participates in the release from seed dormancy (Kępczyński and Kępczyński, 1997; Matilla, 2000). In beechnut, ethephon, an ethylene-releasing compound, can accelerate dormancy release (as can GA_3) and the GA and ethylene pathways likely interact (Calvo *et al.*, 2003, 2004a,b). *FsGA20ox1* gene expression is stimulated by moist chilling, GA or ethephon treatment (all effective in releasing dormancy) (Calvo *et al.*, 2004b). Similarly, *FsACO1* (*ACC OXIDASE 1*) gene expression, ACC content, and ethylene evolution all increase when dormant seeds are treated with GA_3 (ACO catalyzes the last step in ethylene biosynthesis, the conversion of ACC to ethylene) (Calvo *et al.*, 2004a). Indeed, ethylene can replace the requirement for GA and restore germination of *Arabidopsis ga1* seeds in the light or promote germination of wild-type seeds in the dark (Karssen *et al.*, 1989). Conversely, GA_3 can stimulate germination of seeds of the ethylene receptor mutant *etr1* (Bleecker *et al.*, 1988).

In many species, a burst of ethylene evolution occurs around the time of radicle protrusion coincident with the major increase in bioactive GA in seeds (Kępczyński and Kępczyński, 1997; Lashbrook *et al.*, 1998; Kucera *et al.*, 2005). However, the role of ethylene in germination is still under debate; ethylene production may be a requirement for germination or it may simply be a consequence of the process (Matilla, 2000). Studies with ethylene inhibitors suggest that, in many species, ethylene is needed for germination (Matilla, 2000; Siriwitayawan *et al.*, 2003). However, ethylene may act to enhance germination especially under unfavorable conditions (Siriwitayawan *et al.*, 2003). In lettuce, treatments with GA, ethylene, or cytokinin (kinetin) alone are not able to overcome thermodormancy at 32°C in the dark; in addition to ethylene, at least one other hormone, or light, is needed. Experiments using ethylene inhibitors demonstrated that ethylene is an essential participant in the release of thermoinhibition elicited by GA or cytokinin in the dark or in light-imbibed seeds (Saini *et al.*, 1986, 1989). It is interesting to speculate that the combination of hormones (or light) is needed to overcome the inhibitory action of ABA; fluridone, an ABA biosynthesis inhibitor, and GA_3 can overcome thermoinhibition at 33°C (Gonai *et al.*, 2004). Detailed information regarding the precise mechanism of action of ethylene during dormancy termination and germination is lacking. The primary action of ethylene may be in the promotion of radial cell expansion in the embryonic axis, increased respiration, or increased water potential which potentially contributes to an increased embryo growth potential (Kucera *et al.*, 2005). Similar to GA, induction of β-1,3-glucanase in tobacco is regulated by endogenous ethylene, which subsequently promotes endosperm rupture and radicle protrusion (Leubner-Metzger *et al.*, 1998). Thus, GA and ethylene may go hand-in-hand to promote embryo growth extension and weakening of the tissues surrounding the radicle, although evidence for the action of ethylene in these processes is tenuous at best.

Genetic studies with *Arabidopsis* have contributed to some clarification of the role of ethylene in seed dormancy and germination and the links to other hormones. During GA_4 treatment of *ga1-3* seeds, which stimulates germination, expression of *AtACO* and of *AtERS1* (*ETHYLENE RESPONSE SENSOR* encoding a member of the ethylene receptor family) is increased (Ogawa *et al.*, 2003). However, since GA treatment promotes the transition to germination in *ga1-3* seeds, we cannot conclude with certainty which genes are regulated directly by GA and which genes change in expression as a result of the progression to germination. Nonetheless, ethylene is intimately involved in the promotion of germination as revealed through ethylene mutants.

Ethylene signal transduction involves the binding of ethylene to a series of receptors which subsequently 'turn-off' the negative regulation of the pathway. In brief, ethylene binds to the two-component receptor ETR1, which subsequently stops the receptor from activating CTR1. Lack of negative regulation by CTR1 allows activation of downstream positive regulators such as EIN2 and EIN3 and subseqent ethylene responses (Chen *et al.*, 2005). Consistent with their roles in ethylene signaling and with ethylene's positive influence during seed development, dormancy termination, and germination, seeds of *etr1* and *ein2* mutants display enhanced dormancy (ABA hypersensitivity) while *ctr* seeds have slightly reduced dormancy (relative ABA insensitivity) compared to the wild-type seeds (Beaudoin *et al.*, 2000; Ghassemian *et al.*, 2000; Chiwocha *et al.*, 2005). Hence the ability of *ein2* and *ctr* to suppress and enhance, respectively, the seed germination phenotype of the ABA-insensitive *abi1-1* mutant is not surprising (Beaudoin *et al.*, 2000). An interaction with the ABA pathway is further demonstrated by the epistasis of *ein2* and loss-of-function *abi3* alleles (i.e., *abi3-4*) (Beaudoin *et al.*, 2000). The nondormant phenotype of an *abi3-4/ein2* double mutant further supports the notion that ethylene acts by counteracting ABA action.

Further convergence of signaling pathways is revealed through the two-component nature of the ethylene family of receptors. Two component His-to-Asp phospho-relay signaling systems are composed of 'hybrid' histidine kinases (HKs, e.g., AHK2-4, *ARABIDOPSIS* HISTIDINE KINASE 2-4, that functions as a cytokinin receptor, and ETR1 and ERS1 that function as ethylene receptors), histidine-containing phosphotransfer domain proteins, and response regulators (e.g., ARR, *ARABIDOPSIS* RESPONSE REGULATOR) (Grefen and Harter, 2004). In the simplest scenario, HKs function to pass a phosphate group from a conserved His residue present within the HK (transmitter domain) to an Asp residue present in the receiver domain of a response regulator (RR). For example, this would occur when a hormone binds to the HK, and thereby activates the signaling system (Grefen and Harter, 2004). In addition to functioning as ethylene and cytokinin receptors, these signaling systems have also been linked to light and osmotic sensing (Urao *et al.*, 1999; Sweere *et al.*, 2001; see cytokinin section below). Recently, a direct link between ethylene and cytokinin signaling was revealed through the response regulator ARR2. ARR2 modulates both cytokinin and ethylene responses through the receptors AHK3 and ETR1, respectively (Hass *et al.*, 2004; Kim *et al.*, 2006), although a role for ARR2 in ethylene responses has been disputed (Mason *et al.*, 2005). As mentioned, ethylene signaling, through ETR1 for example, has always been presumed to act through the negative regulator CTR1 (Alonso and Stepanova, 2004). Although no seed phenotypes were reported in the previous examples, we hypothesize that ethylene and cytokinin may cross-talk during seed development, dormancy, and germination and these interactions may be mediated via different ARRs.

EIN2 is also another focal point for interactions among various hormones. Mutants of *ein2* have also been recovered in root phenotype screens using auxin transport inhibitors and for resistance to cytokinin application (Su and Howell, 1992; Fujita and Syono, 1996). Further, Alonso *et al.* (1999) suggested that EIN2 also modulates responses to jasmonic acid. Finally, EIN2 and several other ethylene loci have been linked to sugar signaling (León and Sheen, 2003). EIN2 is hypersensitive to glucose (sugar) application during seedling growth – a phenotype that is dependent on endogenous ABA. Other loci such as *etr1* and *ein3* are also sugar hypersensitive; however, *ctr1* is sugar insensitive (León and Sheen, 2003). The *ctr1* mutant has been recovered in screens for both sucrose and glucose insensitivity (*sis1* and *gin4*, respectively) (Gibson *et al.*, 2001; Cheng *et al.*, 2002; see Chapter 12). Thus, the sugar phenotypes (i.e., ethylene insensitive = sugar hypersensitive) are in stark contrast to that of the ABA-biosynthesis or -insensitive mutants, which are insensitive to sugars, and further suggest integration of signaling pathways (in this case for sugar, ethylene, and ABA).

8.6 Auxin and cytokinin

8.6.1 Auxin and cytokinin establish the embryo body plan during seed development

Within the past decade, the role of auxin in embryonic development has been further clarified and it is now known that auxin plays a fundamental role in establishing the basic body plan of the embryo. Auxin contributes to development of the apical–basal (shoot–root) axis, formation of the shoot and root apical meristems, and the cotyledons (Jenik and Barton, 2005). Consistent with this, levels of IAA are usually highest during embryogenesis and fall to lower levels during the middle to late stages of seed development when ABA levels are highest (Bewley and Black, 1994; Ficher-Iglesias and Neuhaus, 2001). Little information exists to suggest that auxin is involved in dormancy imposition during development per se. However, auxin has been implicated in the regulation of the transcriptional factor FUS3. The embryonic expression patterns of the *FUS3* gene correlate well with those reported for an auxin-responsive promoter *DR5*, and exogenous IAA induces FUS3 promoter-reporter fusion (*pFUS3::GFP* and *pFUS3::GUS* [glucuronidase]) constructs in isolated embryos and seedlings, respectively (Gazzarrini *et al.*, 2004). Thus, in addition to promoting embryogenesis, auxin may also regulate the embryo growth arrest that maintains the expression of embryonic traits and is, in part, modulated by FUS3 (as well as other factors such as LEC1 and LEC2, although it is unknown whether these proteins are regulated by auxin as well). FUS3, in turn, modulates the ABA/GA ratio, further maintaining the developmental arrest that occurs during seed development and allowing key developmental processes to proceed, such as reserve deposition, acquisition of desiccation tolerance, and induction of primary dormancy (Raz *et al.*, 2001; Gazzarrini *et al.*, 2004).

Cytokinins promote cell division and may be involved in this process during embryogenesis (Bewley and Black, 1994); these hormones also participate in suspensor function, in embryo pattern formation during embryogenesis, and in endosperm growth and grain filling (Bewley and Black, 1994; Mähönen *et al.*, 2000, Mok and Mok, 2001). For example, the cytokinin receptor WOODEN LEG (WOL), whose encoding gene is allelic to the *AHK4* gene, is involved in the radial patterning of vascular tissue during embryogenesis (Mähönen *et al.*, 2000; Yamada *et al.*, 2001). Intriguing results were obtained recently for triple mutants having defective genes encoding three cytokinin receptors, AHK2, AHK3, and AHK4. Seeds of the triple mutants are ~30% bigger through increases in both cell size and number, and analyses of crosses between wild-type plants and triple mutant plants revealed that maternal tissues play a major role in determining the increased seed size (Riefler *et al.*, 2006). It is unknown why seed size was increased through downregulation of cytokinin signaling; one would expect perhaps the opposite to be true, i.e., that mutations in cytokinin signaling would result in fewer cell divisions and smaller seeds. One explanation is that the triple receptor mutant contains significantly higher levels of cytokinins and its metabolites. Thus, the increased cytokinins may 'overcompensate' for the loss of the three receptors, and alternative cytokinin (receptor) signaling paths may be invoked. Alternatively, complex interactions with other hormonal pathways could account for the increase in seed size. As with auxin, no specific role for cytokinins in dormancy induction has been ascribed, although seeds of the triple cytokinin receptor mutants germinate faster than wild-type (Riefler *et al.*, 2006).

8.6.2 Auxin and cytokinins have not been intimately linked to dormancy maintenance or termination

Recently, IAA was suggested to participate in the dormancy of wheat caryopses (Ramaih *et al.*, 2003). IAA and its precursors tryptophan and indole-acetaldehyde inhibit germination of excised embryos from a dormant cultivar but not from a nondormant one. Furthermore, embryos from dormant caryopses gradually lose sensitivity to IAA throughout after-ripening (Ramaih *et al.*, 2003). Thus, a link has been drawn between exogenous auxin and embryo germinability but there is still no compelling evidence to suggest that auxin is involved in dormancy maintenance or termination per se. Auxin is linked to the resumption of growth; however, this role is especially prominent in the germinated seedling. IAA is released from conjugates (with amino acids or peptides) stored in seeds of Scots pine and *Arabidopsis* during germination (Ljung *et al.*, 2001; Rampey *et al.*, 2004). This IAA may be needed for germination and/or subsequent seedling growth. Postgerminative, shoot-derived auxin may be essential for the initiation of lateral roots in *Arabidopsis* seedlings (Bhalerao *et al.*, 2002). Interestingly, the first observable phenotypes resulting from deficiencies in auxin amount or response are often the presence of fewer lateral roots and shorter hypocotyls in the light. An auxin-conjugate hydrolase triple mutant (*ilr1* [*IAA-leucine resistant 1*] *iar3* [*IAA-alanine resistant 3*]*ill2* [*ilr-like2*]) displays these low-auxin phenotypes (Rampey *et al.*, 2004). In addition, these phenotypes are associated with the auxin response mutants *axr1* (*auxin resistant 1*), *axr4*, and *ibr5* (*IBA* [*indole-3-butyric acid*] *response 5*) (Estelle and Somerville, 1987; Hobbie and Estelle, 1995; Monroe-Augustus *et al.*, 2003). No significant seed germination phenotypes have been reported for auxin-related mutants, so whether auxin plays a role in germination is unclear. However, germination of seeds of *axr1* and *ibr5* mutants is slightly more insensitive to low ABA concentrations (0.5 to 1 μM) than that exhibited by wild-type seeds (Tiryaki and Staswick, 2002; Monroe-Augustus *et al.*, 2003). Indirect evidence indicating the involvement of auxin in the germination process comes from a microarray analysis of germinating *Arabidopsis* seeds. A number of auxin biosynthesis genes and genes encoding putative auxin carrier proteins are upregulated during germination in wild-type seeds and in *ga1-3* mutant seeds treated with exogenous GA_4 (Ogawa *et al.*, 2003).

Cytokinins can release the dormancy of seeds of numerous species such as selected species of maple (*Acer*), apple, beech, and peanut (Cohn and Butera, 1982). Kinetin together with at least one other hormone or light can release the thermodormancy of lettuce seeds (Saini *et al.*, 1986). Given the recent advances in hormone metabolite profiling, it would be interesting to revisit the ability of kinetin and light, for example, to stimulate the germination of thermodormant seeds, examining the effects of this treatment on the profiles of hormones and hormone metabolites (e.g., those of the ABA, GA, ethylene [ACC], and auxin pathways). Mutants of *Arabidopsis* have revealed hormonal interactions among cytokinin, ethylene, and GA. The ethylene-insensitive *ein2* mutant is allelic to the cytokinin-resistant mutant *ckr1* (Su and Howell, 1992). More recently, although not specifically in seeds, the GA negative regulator SPINDLY (SPY) was shown to integrate cytokinin responses (Greenboim-Wainberg *et al.*, 2005). The *spy-4* and *spy-3* mutants are more resistant to exogenously applied cytokinin (benzyladenine) than the wild-type, for example with respect to root growth inhibition and anthocyanin accumulation. Further, both GA_3 and the *spy* mutation suppress cytokinin induction of the two-component response regulator *ARR5* but not *ARR7*. The authors suggest that SPY positively affects a subset of cytokinin-dependent responses; when GA levels are high, SPY is downregulated and this diminishes cytokinin responses. Conversely, when GA levels are low, SPY downregulates GA responses and upregulates cytokinin responses. Whether these processes are occurring during germination is unknown. For example, it is not known whether cytokinin responses are downregulated at a time during germination when GA levels increase. A significant role for cytokinins in germination seemed unlikely, as the cytokinin benzyladenine (5 μM) has no effect on seed germination of wild-type or *spy* mutant seeds regardless of whether GA biosynthesis is inhibited (Greenboim-Wainberg *et al.*, 2005).

In contrast, recent analysis of triple cytokinin receptor mutants (*ahk2 ahk3 ahk4*) revealed a possible role for cytokinin in germination. Seeds of both the double- and triple-mutants germinated faster than wild-type seeds, although the effect was more pronounced in the triple mutant seeds. By 24 h, seeds of the triple mutant display a similar germination percentage to that exhibited by wild-type seeds at 48 h. The triple mutants also reveal an interaction with light. Seeds of triple mutants are able to germinate in the dark and are less inhibited by far-red light than are wild-type seeds (Riefler *et al.*, 2006). Thus, cytokinin appears to be a negative regulator of *Arabidopsis* germination. Interestingly, microarray analysis of cytokinin-treated seedlings reveals that cytokinin negatively affects GA synthesis/signaling (Brenner *et al.*, 2005). This perhaps provides a reason why the triple cytokinin receptor mutants germinate faster, as inhibited cytokinin signaling may have increased GA responses. In addition, seedlings of the triple mutants have significantly increased levels of zeatin (by 15-fold) as compared to wild-type seedlings, which may impact cytokinin signaling or cross-talk with other signaling pathways. The interaction of cytokinin signaling with light during germination should be more extensively investigated. For example, the cytokinin-associated two-component response regulator, ARR4, seems to physically interact with and stabilize the active form of PhyB (PfrB), promoting PhyB light-associated responses (Sweere *et al.*, 2001; Fankhauser, 2002). Is this interaction a critical part of the control of seed germination?

8.7 Brassinosteroids

Brassinosteroids are involved in a variety of developmental processes such as the promotion of cell elongation and cell division, pollen development and fertility, leaf morphogenesis and epinasty, induction of ethylene biosynthesis, and photomorphogenesis (Clouse and Sasse, 1998; Clouse, 2004). While BRs are generally not considered primary regulators of seed dormancy and germination, they promote the germination of seeds of rice (*Oryza sativa*), tobacco, tomato, and the parasitic species *Orobanche minor*, but not cress (*Lepidium sativum*) seed (Yamaguchi *et al.*, 1987; Takeuchi *et al.*, 1995; Jones-Held *et al.*, 1996; Leubner-Metzger, 2001; Vardhini and Rao, 2001).

In *Arabidopsis*, the BR compounds 24-*epi*-brassinolide (BL) and castasterone have been detected in mature dry seeds (Schmidt *et al.*, 1997); however, their levels have not been profiled in seeds during dormancy release and germination. BL and castasterone are both bioactive BRs, capable of activating BR signaling, although BL is much more active (Reid *et al.*, 2004; Kinoshita *et al.*, 2005). *Arabidopsis* mutants deficient in BR synthesis or response have revealed a possible role for BR in germination. Germination of seeds of two BR mutants is more sensitive to ABA as compared to wild-type seeds. This includes *det2-1* (*de-etiolated 2-1*), a BR-deficient mutant with a defect in the gene encoding a 5α-reducase, and *bri1-1* (*brassinosteroid insensitive 1*), a BR receptor mutant with a defect in the gene encoding a plasma-membrane-localized leucine-rich repeat receptor kinase (Steber and McCourt, 2001). In addition, BL treatment rescues the germination of seeds of the GA-biosynthesis mutants *ga1-3*, *ga2-1*, and *ga3-1*. Further, BL treatment also increases germination of the GA-insensitive mutant *sly1-2* (from 20 to 81%) and a *sly1* mutant (*sly1-10*) was recovered in a screen for a BR-dependent germination phenotype (Steber and McCourt, 2001). The G protein signaling mutants, *gcr1* (*g protein-coupled receptor 1*), *gpa1* (*g protein α-subunit 1*), and *agb1* (*arabidopsis g protein β-subunit 1*), are also less sensitive to BR in the promotion of germination (when GA synthesis is inhibited) (Ullah *et al.*, 2002; see below).

The role of BRs in tobacco seed germination has also been investigated. Similar to *Arabidopsis*, BR and GA seem to promote tobacco seed germination through overlapping yet distinct pathways. Unlike GA, BL cannot release photodormancy in dark-imbibed seeds and does not induce β-1,3-glucanase activity in the micropylar endosperm in the dark (Leubner-Metzger, 2001). BL, but not GA, accelerates endosperm rupture in the light and, subsequently, promotes germination in the light. In the dark, both BL and GA can overcome the inhibitory effects of ABA on germination, while in the light only BL is effective (Leubner-Metzger, 2001). Thus, since the effect of BRs on germination in light is additive, and GA promotes photodormancy release and β-1,3-glucanase activity in the dark, Leubner-Metzger (2001) suggested that BR acts in a GA-independent manner. Perhaps consistent with the role of BRs in cell expansion, xyloglucan endotransglycosylase activity in the embryos and endoperms of germinating tobacco seeds seems to be partially under BR control (Leubner-Metzger, 2003). Whether endogenous BRs participate in the germination of tobacco seeds is unknown.

Current evidence intimates that BRs are important for the promotion of germination and that these hormones act synergistically with GA to promote germination and counter the inhibitory actions of ABA. The effects of BRs on GA or ABA homeostasis or on ABA signaling during germination are unknown. However, while BR does not induce GA-stimulated β-1,3-glucanase activity in tobacco seeds (Leubner-Metzger, 2001), this hormone upregulates the expression of the GA biosynthesis gene *AtGA20ox1* in 2-week-old *Arabidopsis* seedlings (Bouquin *et al.*, 2001). Whether BRs act through part of the GA signaling pathway (i.e., not involving *SLY1*) and whether they can 'bypass' parts of the GA pathway to regulate common downstream genes are also unknown. One model is that the BR pathway enhances the growth potential of the embryo by enhancing cell elongation; such an action is consistent with the ability of BRs to act synergistically with GA, thus counteracting ABA, although possibly in an indirect manner. Supporting this contention, but by no means proving it, is the ability of BL (24-*epi*-brassinolide and 28-*homo*-brassinolide) to overcome the inhibitory effects of osmotic stress on the germination of three sorghum (*S. vulgare*) varieties (Vardhini and Rao, 2003).

8.8 G-protein signaling reveals integration of GA, BR, ABA, and sugar responses

Heterotrimeric G proteins and G-protein-coupled receptors are early components of signal transduction pathways that control growth and differentiation in eukaryotes. Activation of G-protein-coupled receptors triggered by the binding of an extracellular ligand sets off a transduction cascade that is mediated by GDP to GTP activation of the Gα subunit protein, causing a conformational change that allows the α subunit to dissociate from the G$\beta\gamma$ subunit. Dissociation of the Gα protein in this manner allows activation of downstream components and a cellular response (e.g., the opening of ion channels and activation of adenylate cyclase, phospholipase C, and phospholipase D) that can lead to increases in other intracellular mediators (e.g., Ca^{2+}, cAMP, IP_3, phosphatidic acid). Ultimately, changes in cellular functions including the activation/repression of genes are effected. In addition, regulators of G-protein signaling (RGS) have been discovered that act to accelerate the intrinsic activity of the Gα subunit and thus return the GTP-active state to its basal GDP-bound state. The G-protein signaling pathway has not diversified over the course of plant evolution nearly to the extent that has occurred in animals (Chen *et al.*, 2004; Assmann, 2005; Chen *et al.*, 2006). In *Arabidopsis* there is only one G-PROTEIN-COUPLED RECEPTOR (AtGCR1), one Gα (AtGPA1, G-PROTEIN ALPHA 1), one Gβ (AtAGB1, G-PROTEIN BETA 1), two Gγ (AtAGG1, AtAGG2, G-PROTEIN GAMMA), and one RGS protein (AtRGS1, REGULATOR OF G PROTEIN SIGNALING 1) (Assmann, 2005).

AtGCR1, AtGPA1, and AtAGB2 have been implicated in the promotion of germination via BR and GA. Mutants of *gcr1*, *gpa1*, and *agb1* are less sensitive to GA and BR in the promotion of seed germination (after GA synthesis has been inhibited) (Chen *et al.*, 2004). In addition, *gpa1* mutants also display increased sensitivity to glucose, sucrose, and ABA (Ullah *et al.*, 2002; Chen *et al.*, 2006). Transgenic plants overexpressing the *GPA1* gene are more sensitive to GA, but show relative insensitivity to ABA, and produce seeds that lack dormancy (Colucci *et al.*, 2002; Ullah *et al.*, 2002). Furthermore, there is an induction of the expression of a *MYB65* gene, an *Arabidopsis* homologue of the *GAMYB* gene of barley, whose encoded product regulates GA-mediated α-amylase induction in aleurone layer cells (Colucci *et al.*, 2002). Ullah *et al.* (2002) suggested that the promotion of germination mediated by G-protein signaling and GPA1 is through BR action; however, the evidence for this is currently circumstantial.

More recently, AtRGS1 has been identified as a regulator of sugar and ABA responses during germination (Chen *et al.*, 2006). In contrast to *gpa1*, mutant seeds of *rgs1* are less sensitive to ABA, glucose, and sucrose during germination. Overexpressors of *AtRGS1* are hypersensitive to ABA and sugar. Thus, AtRGS1 may operate to oppose the actions of AtGCR1, AtGPA1, and AtAGB2. It is unknown whether the action is a direct one, although the RGS domain of AtRGS1 can quench AtGPA1 activity *in vitro* (Chen *et al.*, 2003). AtRGS1 also has a membrane ligand-binding domain (similar to a GPCR [G-Protein Coupled Receptor] domain) and may be able to direct G-protein signaling itself, rather than simply opposing it (Chen *et al.*, 2003). Supporting this, ABA levels and the expression of the genes *AtNCED3* and *AtABA2* are increased in wild-type seeds but not in *rgs1* mutant seeds following glucose application (Chen *et al.*, 2006). It is unknown whether sugars are the ligands that bind to the GPCR domain of AtRGS1, but sugars do bind to GPCRs in yeast and mammalian systems (Assmann, 2005). Thus, sugar signaling, directly or upstream of AtRGS1, seems to modulate ABA biosynthesis, which may ultimately be the reason for induced quiescence during germination (see Chapter 12). Chen *et al.* (2006) provided the first evidence that this quiescence may operate through G-protein signaling. Determination of the true nature of G-protein signaling in the realm of germination promotion and opposition is awaited with interest.

8.9 Profiling of hormone metabolic pathways in *Arabidopsis* mutants reveals cross-talk

The ability of a specific mutation in a hormonal pathway to affect other hormone networks was recently demonstrated in the *Arabidopsis etr1-2* mutant (Figure 8.2). The *etr1-2* mutation confers dominant ethylene-insensitivity and as a consequence results in mature seed populations that exhibit more pronounced primary dormancy (Hall *et al.*, 1999; Chiwocha *et al.*, 2005). As discussed above, the inability of seeds to perceive ethylene results in seeds with higher ABA contents at maturity (~10-fold higher than in mature dry wild-type Col seeds), consistent with ethylene's antagonism of ABA action during seed development (Beaudoin *et al.*, 2000; Ghassemian *et al.*, 2000). The higher ABA levels in the mature *etr1-2* seeds suggest that ABA response may also be upregulated during dormancy induction; indeed, ABA sensitivity is greater in mature *etr1* seeds (Ghassemian *et al.*, 2000). Thus, the lack of ethylene signaling (due to the *etr1-2* mutation) likely results in a failure of ethylene to counteract both ABA synthesis and signaling. During imbibition of dormant and nondormant seeds, there is a striking difference in ABA metabolism in *etr1-2* mutant versus wild-type seeds; in particular, ABA-GE is decreased in the mutant. This is especially evident in germinating seeds, for example, where a

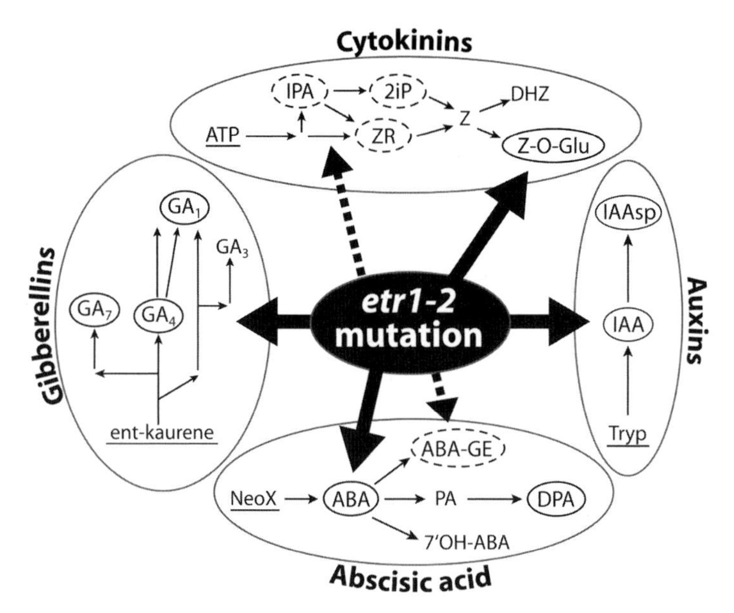

Figure 8.2 Schematic representation of the impact of the *etr1-2* mutation during germination in *Arabidopsis thaliana*. Plant hormones and hormone metabolites that increased in the *etr1-2* mutant are circled. Compounds that were present at higher levels in wild-type seeds are shown in hatched circles. Underlined compounds are precursors, which were not quantified in this study. *Note*: Only hormones and hormonal metabolites that were investigated in this study are shown in the auxin, abscisic acid, gibberellin and cytokinin pathways. In addition, although not shown in the above model, it should be pointed out that alternative biosynthetic pathways for cytokinins (Astot *et al.*, 2000) and for IAA (Normanly, 1997) exist in Arabidopsis. Abbreviations: 2iP, isopentenyladenine; 7′OH-ABA, 7′-hydroxy-abscisic acid; ABA-GE, ABA glucose ester; DHZ, dihydrozeatin; DPA, dihydrophaseic acid; GA, gibberellic acid; IAA, indole-3-acetic acid; IAA-asp, indole-3-aspartate; IPA, isopentenyladenosine; NeoX, neoxanthin; PA, phaseic acid; Tryp, tryptophan; Z-O-Glu, zeatin-O-glucoside; Z, zeatin; ZR, zeatin riboside. (Redrawn from Chiwocha *et al.*, 2005, with permission from Blackwell Publishing.)

substantial increase in ABA-GE is observed in wild-type seeds, but not in *etr1-2* seeds. In the *etr1-2* mutant seeds, ABA, together with the ABA metabolite DPA, increases as germination proceeds. This suggests that *etr1-2* mutants have a reduced capacity to glucosylate ABA, in this case *de novo* synthesized ABA, resulting in ABA accumulation and partial compensation by increased metabolism through the 8′-hydroxylation pathway to DPA (see Chapter 9).

In addition to a disruption in ABA homeostasis, auxin, cytokinin, and GA pathways are all affected in *etr1-2* mutant seeds (Chiwocha *et al.*, 2005). Auxin in the form of IAA, along with its amino acid conjugate indole-3-aspartate, is generally higher in *etr1-2* mature seeds, and is also higher during dormancy maintenance, termination, and germination. This suggests that the *ETR1* gene may normally function to promote auxin catabolism. Biologically active cytokinins are reduced in the *etr1-2* mutant, seemingly in favor of increased metabolism to the glucose conjugate, zeatin-O-glucoside; this is particularly evident during germination when increased levels of cytokinins are detected in wild-type seeds, but not in *etr1-2* seeds. Thus, ethylene signaling through ETR1 may help to maintain bioactive cytokinins by shifting the flux away from cytokinin catabolism toward biosynthesis. The ethylene and cytokinin pathways may converge – the *ein2* mutant is insensitive to both hormones. Levels of GAs (GA_1, GA_4, and GA_7) are significantly increased in mature seeds of *etr1-2* and increased levels are detected through dormancy maintenance, termination, and germination as compared to those found in wild-type seeds. The changes in the various hormonal pathways due to lack of ETR1 can be attributed to two possibilities: (1) ethylene signaling normally regulates these hormonal pathways in a positive or negative way, either directly or indirectly (e.g., ethylene signaling antagonizes ABA response); or (2) the lack of a functional ethylene signaling system causes compensatory responses in other hormonal pathways (e.g., ethylene insensitivity causes an increase in GA to further counter increased ABA response).

We have recently compared the hormonal metabolic pathways in seeds of the *Arabidopsis* ABA-insensitive *abi1-1* mutant to those in L*er* wild-type seeds. The *abi1-1* mutant produces seed populations that are less dormant than that of the wild-type (S.D.S. Chiwocha, A.R. Kermode *et al.*, unpublished results), even though the ABI1 protein seems to be a negative regulator of ABA-signaling (Gosti *et al.*, 1999; Yoshida *et al.*, 2006). In contrast to the *etr1-2* mutation described above, the *abi1-1* mutation did not significantly affect any hormonal pathway – hormones and hormone metabolites belonging to the ABA, GA, auxin, cytokinin, jasmonic acid, and salicylic acid classes were all similar in abundance in the mutant and wild-type seeds. Subtle differences in the *abi1-1* mutant included slightly higher levels of ABA and IAA and of the ABA metabolite DPA. Most of the differences between *abi1-1* and wild-type seeds occurred during imbibition of seeds that had not undergone a dormancy-releasing moist-chilling period and could be attributable to differences in germination capacity (i.e., *abi1-1* mutant seeds exhibited 98% germination by 2 days, while wild-type seeds exhibited only 13%). For example, the increase in GA_4 in nonchilled *abi1-1* seeds was virtually identical to that of moist-chilled wild-type seeds (S.D.S. Chiwocha, A.R. Kermode *et al.*, unpublished results).

The above examples illustrate the importance of integrating genomic (microarray) and proteomic data with hormone profiling to give a more comprehensive picture of hormone response and signaling. In the *etr1-2* mutant, for example, we cannot ascribe all of the effects of the mutation on gene expression and protein synthesis solely to the ETR1 protein or ethylene; results must take into account the effects of the mutation on all hormone pathways. Similarly, in experiments using hormone treatments, we must be cautious in assigning a regulatory function to a gene based on one hormone (i.e., gene X is GA-inducible); changes in developmental state and other hormones may also be involved.

8.10 Summary and future directions

The ability of mature desiccated seeds to withstand extremes in the environment and remain viable is a remarkable phenomenon. Even more so is the capacity for the imbibed seed to modulate its developmental schedule to germinate only when environment signals predict that growth and seedling establishment will be successful. Dormancy has evolved in most lineages and serves to distribute germination over the dimensions of time and space, ensuring that plants maximize their fitness through successful establishment of subsequent generations. In the past, dormancy and germination were viewed as 'black-box' processes; however, in recent years significant progress has been made such that factors and processes involved in controlling the transition from dormancy to germination are emerging at an ever-increasing rate. It has also become apparent in recent years that extensive hormonal interactions occur to modulate downstream physiological events. Some of the interactions were discussed in this review; however, we are only beginning to understand the complexities of 'integrated signaling'.

ABA is a well-known participant in the induction and maintenance of dormancy (Figure 8.3; Gubler *et al.*, 2005; Kermode, 2005; Nambara and Marion-Poll, 2005). Conversely, GA is required for germination and also promotes dormancy release, overcoming the inhibitory actions of ABA (Bentsink and Koornneef, 2002; Kucera *et al.*, 2005). During treatments that terminate dormancy such as moist chilling, ABA homeostasis shifts toward a catabolic state and ABA signals are downregulated; GA homeostasis shifts to favor biosynthesis and GA signaling is activated. However, it is unknown whether GA signaling components directly downregulate ABA-related genes during dormancy termination. Further, it is unknown whether negative regulators of GA signaling, such as DELLA proteins and basic helix-loop-helix proteins, are regulated by ABA signaling in seeds. However, DELLA proteins do seem to be regulated by ABA in *Arabidopsis* roots, and so this may indeed be

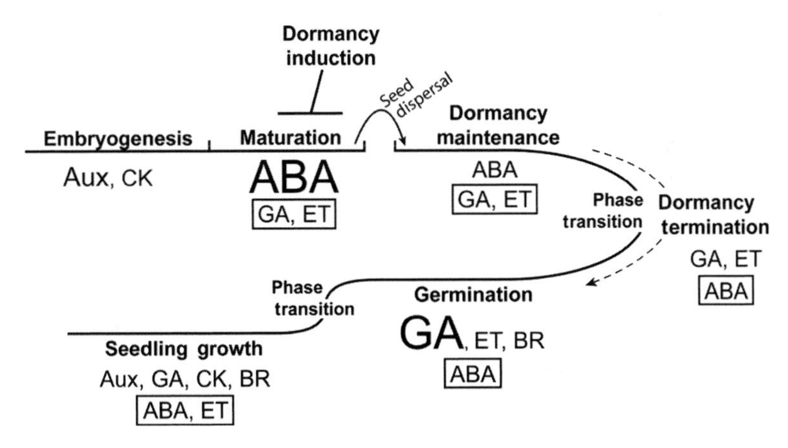

Figure 8.3 The progression from seed development to seedling growth and the major hormones regulating these processes. Positive hormones are shown below each process, and negative hormones are boxed. Effects are shown as generalities and hormones may switch from being positive to negative regulators depending on the context (developmental process) and concentration of the hormone. For example, ABA can maintain seedling root growth under water stress (Spollen *et al.*, 2000) and applied cytokinin can inhibit seedling root growth (Greenboim-Wainberg *et al.*, 2005). Two major phase transitions are suggested: (i) from the dormant imbibed state to germinative events, and (ii) from germination to seedling growth.

the case (Achard *et al.*, 2006). The role for ethylene and BRs in dormancy and germination is less obvious than that of ABA and GA. Ethylene, together with GA, antagonizes ABA actions during dormancy induction/termination and germination. Moreover, ethylene and GA can each activate the synthesis of the other. BRs seem to play a synergistic role with GA in germination, perhaps through their role in promoting cell elongation in the embryo. However, no specific BR mutants have been recovered in germination screens, leading us to speculate upon the true function of BRs during germination. Furthermore, identification of downstream 'effector' genes for both ethylene and BR signaling in seeds is lacking.

Genetic redundancy within a plant's genome appears to be commonplace. *Arabidopsis* is a key example: there are multiple receptors for hormones like ethylene, cytokinin, and auxin and there are large gene families that encode enzymes of hormone biosynthesis (e.g., the NCED enzymes involved in ABA synthesis). Likewise, signaling factors and transcription factors show redundancy (e.g., the protein phosphatase PP2C subfamily A involved in ABA signaling and subfamily 15 of the basic helix-loop-helix transcription factors involved in light/GA signaling). The list seems never-ending (Schwartz *et al.*, 2003; Grefen and Harter, 2004; Schweighofer *et al.*, 2004; Dharmasiri *et al.*, 2005; Duek and Fankhauser, 2005). With this high degree of redundancy, one may question why hormonal cross-talk/signal integration is needed. But when one considers the complexity of the environment surrounding the seed (and the parent plant), we can perhaps rationalize the need for an integrated hormonal web. Hormonal webs, akin to those of spiders, are stronger and more robust and give the seed the flexibility to relay complex information within the embryo itself and between the seed and its surroundings.

Additional complexity is added to the role of hormones and their interactions in dormancy and germination when we consider that signaling factors are not necessarily directly 'homologous' or have identical roles between species. For example, the ABA-binding protein, ABAP1, from barley was isolated from ABA-treated aleurone layers (Razem *et al.*, 2004). Its closest homologue in *Arabidopsis* is FCA, a nuclear RNA-binding protein that promotes flowering, and it too binds ABA. ABA binding to FCA disrupts FCA binding to FY (an mRNA 3′ end processing factor); together FCA–FY inhibit the accumulation of mRNAs encoding FLOWERING LOCUS C (FLC), a potent inhibitor of the transition to flowering. Thus, ABA binding to FCA leads to an increase in FLC activity and a delay in flowering. However, no seed germination phenotypes are present in *fca-1* mutants and seeds do not differ in ABA sensitivity as compared to wild-type (Razem *et al.*, 2006). Thus, the roles of AtFCA and ABAP1 appear to be distinct; however, the precise role of ABAP1 in barley seeds merits further investigation.

Other intricacies of the hormonal signaling web include the fact that signaling factors may behave differently in different tissues or at different developmental stages and may be dependent on hormone concentration (i.e., they are context dependent). The *gpa1-1* and *gpa1-2* mutants of *Arabidopsis*, encoding the G-protein Gα subunit, exhibit insensitivity to ABA, inhibition of stomatalopening, and altered ABA responsiveness of guard-cell inward K^+ channels and slow anion channels (Wang *et al.*, 2001). However, during seed germination, *gpa1-1* and *gpa1-2* mutants exhibit hypersensitivity to ABA and sugars and insensitivity to GA and BR (Ullah *et al.*, 2002; Chen *et al.*, 2006). The actions of ethylene are also context-dependent. In contrast to its role in seeds, where it counteracts ABA and promotes the transition to germination, ethylene in roots of *Arabidopsis* inhibits growth together with ABA (Beaudoin *et al.*, 2000; Ghassemian *et al.*, 2000).

We are perhaps at a 'cross-roads' in seed dormancy and germination research (and in plant research in general), in which it is becoming ever more important to integrate the data sets from microarrays, proteomics, and metabolic profiling as well as data from protein–protein interaction studies and phenotypic mutant analysis. Although well-planned seed physiological experiments are still essential, the importance of bioinformatics approaches to integrate the complexities of the processes is ever increasing. It is conceivable that, in the not too distant future, a comprehensive network map will be created showing the nodes (e.g., transcripts, proteins, hormones, environmental factors) and connections (interactions and regulatory points) among various hormonal factors regulating the seed's decision to terminate dormancy and germinate.

References

G.J. Acevedo-Hernández, P. León and L.R. Herrera-Estrella (2005) Sugar and ABA responsiveness of a minimal *RBCS* light-responsive unit is mediated by direct binding of ABI4. *The Plant Journal* **43**, 506–519.

P. Achard, H. Cheng, L. De Grauwe, *et al.* (2006) Integration of plant responses to environmentally activated phytohormonal signals. *Science* **311**, 91–94.

P. Achard, W.H. Vriezen, D. Van Der Straeten and N.P. Harberd (2003) Ethylene regulates *Arabidopsis* development via the modulation of DELLA protein growth repressor function. *The Plant Cell* **15**, 2816–2825.

S. Ali-Rachedi, D. Bouinot, M-H. Wagner, *et al.* (2004) Changes in endogenous abscisic acid levels during dormancy release and maintenance of mature seeds: studies with the Cape Verde Islands ecotype, the dormant model of *Arabidopsis thaliana*. *Planta* **219**, 479–488.

J.M. Alonso, G. Hirayama, S. Nourizadeh and J.R. Ecker (1999) EIN2, a bifunctional transducer of ethylene and stress responses in *Arabidopsis*. *Science* **284**, 2148–2152.

J.M. Alonso and A.N. Stepanova (2004) The ethylene signaling pathway. *Science* **306**, 1513–1515.

F. Arenas-Huertero, A. Arroyo, L. Zhou, J. Sheen and P. Leon (2000) Analysis of *Arabidopsis* glucose insensitive mutants, *gin5* and *gin6*, reveals a central role of the plant hormone ABA in the regulation of plant vegetative development by sugar. *Genes and Development* **14**, 2085–2096.

S.M. Assmann (2005) G proteins go green: a plant G protein signaling FAQ sheet. *Science* **310**, 71–73.

C. Astot, K. Dolezal, A. Nordstrom, *et al.* (2000) An alternative cytokinin biosynthesis pathway. *Proceedings of the National Academy of Sciences of the United States of America* **97**, 14778–14783.

C. Audran, C. Borel, A. Frey, *et al.* (1998) Expression studies of the zeaxanthin epoxidase gene from *Nicotiana plumbaginifolia*. *Plant Physiology* **118**, 1021–1028.

C. Audran, S. Liotenberg, M. Gonneau, *et al.* (2001) Localisation and expression of zeaxanthin epoxidase mRNA in *Arabidopsis* in response to drought stress and during seed development. *Australian Journal of Plant Physiology* **28**, 1161–1173.

N. Beaudoin, C. Serizet, F. Gosti and J. Giraudat (2000) Interactions between abscisic acid and ethylene signaling cascades. *The Plant Cell* **12**, 1103–1115.

R.L. Benech-Arnold, M. Fenner and P.J. Edwards (1991) Changes in germinability, ABA content and embryonic sensitivity in developing seeds of *Sorghum bicolor* (L.) Moench. induced by water stress during grain filling. *New Phytologist* **118**, 339–347.

S. Bensmihen, S. Rippa, G. Lambert, *et al.* (2002) The homologous ABI5 and EEL transcription factors function antagonistically to fine-tune gene expression during late embryogenesis. *The Plant Cell* **14**, 1391–1403.

L. Bentsink and M. Koornneef (2002) Seed dormancy and germination. In: *The Arabidopsis Book* (eds C.R. Somerville and E.M. Meyerowitz), pp. 1–18. American Society of Plant Biologists, Rockville, MD. http://www.aspb.org/publications/arabidopsis/.

J.D. Bewley (1997) Seed germination and dormancy. *The Plant Cell* **9**, 1055–1066.

J.D. Bewley and M. Black (1994) *Seeds: Physiology of Development and Germination*, 2nd edn, 445 pp. Plenum Publishing Corporation, New York.

R.P. Bhalerao, S. Eklöf, K. Ljung, A. Marchant, M. Bennett and G. Sandberg (2002) Shoot-derived auxin is essential for early lateral root emergence in *Arabidopsis* seedlings. *The Plant Journal* **29**, 325–332.

J. Bianco, G. Garello and M.T. Le Page-Degivry (1997) *De novo* ABA synthesis and expression of seed dormancy in a gymnosperm: *Pseudotsuga menziesii*. *Plant Growth Regulation* **21**, 115–119.

F. Bittner, M. Oreb and R.R. Mendel (2001) ABA3 is a molybdenum cofactor sulfurase required for activation of aldehyde oxidase and xanthine dehydrogenase in *Arabidopsis thaliana*. *Journal of Biological Chemistry* **276**, 40381–40384.

A.B. Bleecker, M.A. Estelle, C. Somerville and H. Kende (1988) Insensitivity to ethylene conferred by a dominant mutation in *Arabidopsis thaliana*. *Science* **241**, 1086–1089.

T. Bouquin, C. Meier, R. Foster, M.E. Nielsen and J. Mundy (2001) Control of specific gene expression by gibberellin and brassinosteroid. *Plant Physiology* **127**, 450–458.

J. Bove, P. Lucas, B. Godin, Ogé. Laurent, M. Jullien and P. Grappin (2005) Gene expression analysis by cDNA-AFLP highlights a set of new signaling networks and translational control during seed dormancy breaking in *Nicotiana plumbaginifolia*. *Plant Molecular Biology* **57**, 593–612.

S.M. Brady, S.F. Sarkar, D. Bonetta and P. McCourt (2003) The *ABSCISIC ACID INSENSITIVE 3* (*ABI3*) gene is modulated by farnesylation and is involved in auxin signaling and lateral root development in *Arabidopsis*. *The Plant Journal* **34**, 67–75.

W.G. Brenner, G.A. Romanov, I. Kollmer, L. Burkle and T. Schmulling (2005) Immediate-early and delayed cytokinin response genes of *Arabidopsis thaliana* identified by genome-wide expression profiling reveal novel cytokinin-sensitive processes and suggest cytokinin action through transcriptional cascades. *The Plant Journal* **44**, 314–333.

A.P. Calvo, J.A. Jiménez, C. Nicolás, G. Nicolás and D. Rodriguez (2003) Isolation and characterization of genes related with the breaking of beechnuts dormancy and putatively involved in ethylene signal perception and transduction. In: *The Biology of Seeds: Recent Research Advances* (eds G. Nicolás, K.J. Bradford, D. Come and H. Pritchard), pp. 141–149. CAB International, Wallingford, UK.

A.P. Calvo, C. Nicolás, O. Lorenzo, G. Nicolás and D. Rodriguez (2004a) Evidence for positive regulation by gibberellins and ethylene of ACC oxidase expression and activity during transition from dormancy to germination in *Fagus sylvatica* L. seeds. *Journal of Plant Growth Regulation* **23**, 44–53.

A.P. Calvo, C. Nicolás, G. Nicolás and D. Rodriguez (2004b) Evidence of a cross-talk regulation of a GA 20-oxidase (*FsGA20ox1*) by gibberellins and ethylene during the breaking of dormancy in *Fagus sylvatica* seeds. *Physiologia Plantarum* **120**, 623–630.

D. Cao, A. Hussain, H. Cheng and J. Peng (2005) Loss of fuction of four DELLA genes leads to light- and gibberellin-independent seed germination in *Arabidopsis*. *Planta* **223**, 105–113.

F. Carrari, L. Perez-Flores, D. Lijavetzky, *et al.* (2001) Cloning and expression of a sorghum gene with homology to maize *VP1*. Its potential involvement in pre-harvest sprouting resistance. *Plant Molecular Biology* **45**, 631–640.

M. Chen, J. Chory and C. Fankhauser (2004) Light signal transduction in higher plants. *Annual Review of Genetics* **38**, 87–117.

Y.F. Chen, N. Etheridge and G.E. Schaller (2005) Ethylene signal transduction. *Annals of Botany* **95**, 901–915.

Y. Chen, F. Ji, H. Xie, J. Liang and J. Zhang (2006) The regulator of G-protein signaling proteins involved in sugar and abscisic acid signaling in *Arabidopsis* seed germination. *Plant Physiology* **140**, 302–310.

J.G. Chen, S. Pandey, J. Huang, *et al.* (2004) GCR1 can act independently of heterotrimeric G-protein in response to brassinosteroids and gibberellins in *Arabidopsis* seed germination. *Plant Physiology* **135**, 907–915.

J.G. Chen, F.S. Willard, J. Huang, *et al.* (2003) A seven-transmembrane RGS protein that modulates plant cell proliferation. *Science* **301**, 1728–1731.

W.H. Cheng, A. Endo, L. Zhou, *et al.* (2002) A unique short chain dehydrogenase/reductase in *Arabidopsis* glucose signaling and abscisic acid biosynthesis and functions. *The Plant Cell* **14**, 2723–2743.

S.D.S. Chiwocha, S.R. Abrams, S.J. Ambrose, *et al.* (2003) A method for profiling classes of plant hormones and their metabolites using liquid chromatography–electrospray ionization tandem mass spectrometry: an analysis of hormone regulation of thermodormancy in lettuce (*Lactuca sativa* L.) seeds. *The Plant Journal* **35**, 405–417.

S.D.S. Chiwocha, A.J. Cutler, S.R. Abrams, S.J. Ambrose, J. Yang and A.R. Kermode (2005) The *etr1-2* mutation in *Arabidopsis thaliana* affects the abscisic acid, auxin, cytokinin and gibberellin metabolic pathways during maintenance of seed dormancy, moist-chilling and germination. *The Plant Journal* **42**, 35–48.

S.D. Clouse (2004) Brassinosteroid signal transduction and action. In: *Plant Hormones: Biosynthesis, Signal Transduction, Action!* (ed. P.J. Davies), pp. 413–436. Kluwer Academic Publishers, Dordrecht.

S.D. Clouse and J.M. Sasse (1998) Brassinosteroids: essential regulators of plant growth and development. *Annual Review of Plant Physiology and Plant Molecular Biology* **49**, 427–451.

M.A. Cohn and D.L. Butera (1982) Seed dormancy in red rice (*Oryza sativa*). II. Response to cytokinins. *Weed Science* **30**, 200–205.

G. Colucci, F. Apone, N. Alyeshmerni, D. Chalmers and M.J. Chrispeels (2002) GCR1, the putative *Arabidopsis* G protein-coupled receptor gene is cell cycle-regulated, and its overexpression abolishes seed dormancy and shortens time to flowering. *Proceedings of the National Academy of Sciences of the United States of America* **99**, 4736–4741.

F. Corbineau, J. Bianco, G. Garello and D. Côme (2002) Breakage of *Pseudotsuga menziesii* seed dormancy by cold treatment as related to changes in seed ABA sensitivity and ABA levels. *Physiologia Plantarum* **114**, 313–319.

J. Curaba, T. Moritz, R. Blervaque, *et al.* (2004) *AtGA3ox2*, a gene responsible for bioactive gibberellin biosynthesis, is regulated during embryogenesis by *LEAFY COTYLEDON2* and *FUSCA3* in *Arabidopsis*. *Plant Physiology* **136**, 3660–3669.

S. Cutler, M. Ghassesmian, D. Bonetta, S. Cooney and P. McCourt (1996) A protein farnesyl transferase involved in abscisic acid signal transduction. *Science* **273**, 1239–1241.

A.J. Cutler and J.E. Krochko (1999) Formation and breakdown of ABA. *Trends in Plant Science* **4**, 472–478.

I. Debeaujon and M. Koornneef (2000) Gibberellin requirement for *Arabidopsis* seed germination is determined both by testa characteristics and embryonic abscisic acid. *Plant Physiology* **122**, 415–424.

M. Delseny, N. Bies-Etheve, C.H. Carles, *et al.* (2001) Late Embryogenesis Abundant (LEA) protein gene regulation during *Arabidopsis* seed maturation. *Journal of Plant Physiology* **158**, 419–427.

M.P.M. Derkx and C.M. Karssen (1993) Effects of light and temperature on seed dormancy and gibberellin-stimulated germination in *Arabidopsis thaliana*: studies with gibberellin-deficient and -insensitive mutants. *Physiologia Plantarum* **89**, 360–368.

N. Dharmasiri, S. Dharmasiri, D. Weijers, *et al.* (2005) Plant development is regulated by a family of auxin receptor F box proteins. *Developmental Cell* **9**, 109–119.

P.D. Duek and C. Fankhauser (2005) bHLH class transcription factors take centre stage in phytochrome signaling. *Trends in Plant Science* **10**, 51–54.

P. Duque and N.H. Chua (2003) IMB1, a bromodomain protein induced during seed imbibition, regulates ABA- and phyA-mediated responses of germination in *Arabidopsis*. *The Plant Journal* **35**, 787–799.

M.A. Estelle and C. Somerville (1987) Auxin-resistant mutants of *Arabidopsis thaliana* with an altered morphology. *Molecular and General Genetics* **206**, 200–206.

M.M. Evans and R.S. Poethig (1995) Gibberellins promote vegetative phase change and reproductive maturity in maize. *Plant Physiology* **108**, 475–487.

C. Fankhauser (2002) Light perception in plants: cytokinins and red light join forces to keep phytochrome B active. *Trends in Plant Science* **7**, 143–145.

J.A. Feurtado, S.J. Ambrose, A.J. Cutler, A.R.S. Ross, S.R. Abrams and A.R. Kermode (2004) Dormancy termination of western white pine (*Pinus monticola* Dougl. Ex D. Don) seeds is associated with changes in abscisic acid metabolism. *Planta* **218**, 630–639.

C. Ficher-Iglesias and G. Neuhaus (2001) Zygotic embryogenesis: hormonal control of embryo development. In: *Current Trends in the Embryology of Angiosperms* (ed. S.S. Bhojwani), pp. 223–247. Kluwer Academic Publishers, Dordrecht.

W. Finch-Savage (2004) A genomic approach to understanding seed dormancy. Nottingham Arabidopsis Stock Centre Microarray Database, NASCARRAY-69, http://affymetrix.arabidopsis.info/narrays/experimentbrowse.pl.

R.R. Finkelstein (1994) Mutations at two new *Arabidopsis* ABA response loci are similar to the *abi3* mutations. *The Plant Journal* **5**, 765–771.

R.R. Finkelstein (2004) The role of hormones during seed development and germination. In: *Plant Hormones: Biosynthesis, Signal Transduction, Action!* (ed. P.J. Davies), pp. 513–537. Kluwer Academic Publishers, Dordrecht.

R.R. Finkelstein, S.S. Gampala and C.D. Rock (2002) Abscisic acid signaling in seeds and seedlings. *The Plant Cell* **14**(Suppl), S15–S45.

R.R. Finkelstein and T.J. Lynch (2000) The *Arabidopsis* abscisic acid response gene *ABI5* encodes a basic leucine zipper transcription factor. *The Plant Cell* **12**, 599–609.

R.R. Finkelstein and C.D. Rock (2002) Abscisic acid biosynthesis and response. In: *The Arabidopsis Book* (eds C.R. Somerville and E.M. Meyerowitz), pp. 1–52. American Society of Plant Biologists, Rockville, MD. http://www.aspb.org/publications/arabidopsis/.

G.R. Flematti, E.L. Ghisalberti, K.W. Dixon and R.D. Trengove (2004) A compound from smoke that promotes seed germination. *Science* **305**, 977.

K.A. Franklin and G.C. Whitelam (2004) Light signals, phytochromes and cross-talk with other environmental cues. *Journal of Experimental Botany* **395**, 271–276.

A. Frey, C. Audran, E. Marin, B. Sotta and A. Marion-Poll (1999) Engineering seed dormancy by modification of zeaxanthin epoxidase gene expression. *Plant Molecular Biology* **39**, 1267–1274.

A. Frey, B. Godin, M. Bonnet, B. Sotta and A. Marion-Poll (2004) Maternal synthesis of abscisic acid controls seed development and yield in *Nicotiana plumbaginifolia*. *Planta* **218**, 958–964.

X. Fu, D.E. Richards, B. Fleck, D. Xie, N. Burton and N.P. Harberd (2004) The *Arabidopsis* mutant sleepy1^{gar2-1} protein promotes plant growth by increasing the affinity of the SCF^{SLY1} E3 ubiquitin ligase for DELLA protein substrates. *The Plant Cell* **16**, 1406–1418.

H. Fujita and K. Syono (1996) Genetic analysis of the effects of polar auxin transport inhibitors on root growth in *Arabidopsis thaliana*. *Plant and Cell Physiology* **37**, 1094–1101.

T. Fukuhara, H.-H. Kirch and H.J. Bohnert (1999) Expression of *Vp1* and water channel proteins during seed germination. *Plant, Cell and Environment* **22**, 417–424.

K. Furuta, K. Minoru, Y-G. Liu, D. Shibata and T. Kakimoto (2005) CKH2/PICKLE negatively regulates a set of cytokinin responses. In : *Abstract, Arabidopsis 2005*, Madison, WI.

G. Garello and M.T. LePage-Degivry (1999) Evidence for the role of abscisic acid in the genetic and environmental control of dormancy in wheat (*Triticum aestivum* L.). *Seed Science Research* **9**, 219–226.

S. Gazzarrini, Y. Tsuchiya, S. Lumba, M. Okamoto and P. McCourt (2004) The transcription factor *FUSCA3* controls developmental timing in *Arabidopsis* through the action of the hormones gibberellins and abscisic acid. *Developmental Cell* **7**, 73–85.

M. Ghassemian, E. Nambara, S. Cutler, H. Kawaide, Y. Kamiya and P. McCourt (2000) Regulation of abscisic acid signaling by the ethylene response pathway in *Arabidopsis. The Plant Cell* **12**, 1117–1126.

S.I. Gibson, R.J. Laby and D. Kim (2001) The *sugar-insensitive1* (*sis1*) mutant of *Arabidopsis* is allelic to *ctr1*. *Biochemical and Biophysical Research Communications* **280**, 196–203.

J. Giraudat, B.M. Hauge, C. Valon, J. Smalle, F. Parcy and H.M. Goodman (1992) Isolation of the *Arabidopsis ABI3* gene by positional cloning. *The Plant Cell* **4**, 1251–1261.

T. Gonai, S. Kawahara, M. Tougou, *et al.* (2004) Abscisic acid in the thermoinhibition of lettuce seed germination and enhancement of its catabolism by gibberellin. *Journal of Experimental Botany* **55**, 111–118.

M.P. González-García, D. Rodríguez, C. Nicolás, P.L. Rodríguez, G. Nicolás and O. Lorenzo (2003) Negative regulation of abscisic acid signaling by the *Fagus sylvatica* FsPP2C1 plays a role in seed dormancy regulation and promotion of seed germination. *Plant Physiology* **133**, 135–144.

M. González-Guzmán, D. Abia, J. Salinas, R. Serrano and P.L. Rodríguez (2004) Two new alleles of the *abscisic aldehyde oxidase 3* gene reveal its role in abscisic acid biosynthesis in seeds. *Plant Physiology* **135**, 325–333.

M. González-Guzmán, N. Apostolova, J.M. Bellés, *et al.* (2002) The short-chain alcohol dehydrogenase ABA2 catalyzes the conversion of xanthoxin to abscisic aldehyde. *The Plant Cell* **14**, 1833–1846.

F. Gosti, N. Beaudoin, C. Serizet, A.A. Webb, N. Vartanian and J. Giraudat (1999) ABI1 protein phosphatase 2C is a negative regulator of abscisic acid signaling. *The Plant Cell* **11**, 1897–1910.

P. Grappin, D. Bouinot, B. Sotta, E. Miginiac and M. Jullien (2000) Control of seed dormancy in *Nicotiana plumbaginifolia*: post-imbibition abscisic acid synthesis imposes dormancy maintenance. *Planta* **210**, 279–285.

Y. Greenboim-Wainberg, I. Maymon, R. Borochov, *et al.* (2005) Cross talk between gibberellin and cytokinin: the *Arabidopsis* GA response inhibitor SPINDLY plays a positive role in cytokinin signaling. *The Plant Cell* **17**, 92–102.

Grefen, C. and Harter, K. (2004) Plant two-component systems: principles, functions, complexity and cross talk. *Planta* **219**, 733-742.

S.P.C. Groot and C.M. Karssen (1987) Gibberellins regulate seed germination in tomato by endosperm weakening: a study with GA-deficient mutants. *Planta* **171**, 525–531.

S.P.C. Groot, I.I. Van Yperen and C.M. Karssen (1991) Strongly reduced levels of endogenous abscisic acid in developing seeds of tomato mutant *sitiens* do not influence *in vivo* accumulation of dry matter and storage proteins. *Physiologia Plantarum* **81**, 83–87.

F. Gubler, A.A. Millar and J.V. Jacobsen (2005) Dormancy release, ABA, and pre-harvest sprouting. *Current Opinion in Plant Biology* **8**, 183–187.

A.E. Hall, Q.G. Chen, J.L. Findell, G.E. Schaller and A.B. Bleecker (1999) The relationship between ethylene binding and dominant insensitivity conferred by mutant forms of the ETR1 ethylene receptor. *Plant Physiology* **121**, 291–299.

J.J. Harada (2001) Role of *Arabidopsis LEAFY COTYLEDON* genes in seed development. *Journal of Plant Physiology* **158**, 405–409.

C. Hass, J. Lohrmann, V. Albrecht, *et al.* (2004) The response regulator 2 mediates ethylene signalling and hormone signal integration in *Arabidopsis*. *The EMBO Journal* **23**, 3290–3302.

D.B. Hays, D.M. Reid, E.C. Yeung and R.P. Pharis (2000) Role of ethylene in cotyledon development of microspore-derived embryos of *Brassica napus*. *Journal of Experimental Botany* **51**, 1851–1859.

D.B. Hays, E.C. Yeung and R.P. Pharis (2002) The role of gibberellins in embryo axis development. *Journal of Experimental Botany* **53**, 1747–1751.

Y. He and S. Gan (2004) A novel zinc-finger protein with a proline-rich domain mediates ABA-regulated seed dormancy in *Arabidopsis*. *Plant Molecular Biology* **54**, 1–9.

P. Hedden (1999) Regulation of gibberellin biosynthesis. In: *Biochemistry and Molecular Biology of Plant Hormones* (eds P.J.J. Hooykaas, M.A. Hall and K.R. Libbenga), pp. 161–188. Elsevier Science BV, Amsterdam.

J.T. Henderson, H.C. Li, S.D. Rider, *et al.* (2004) *PICKLE* acts throughout the plant to repress expression of embryonic traits and may play a role in gibberellin dependent responses. *Plant Physiology* **134**, 995–1005.

L. Hennig, W.M. Stoddart, M. Dieterle, G.C. Whitelam and E. Schäfer (2002) Phytochrome E controls light-induced germination of *Arabidopsis*. *Plant Physiology* **128**, 194–200.

L. Hobbie and M. Estelle (1995) The *axr4* auxin-resistant mutants of *Arabidopsis thaliana* define a gene important for root gravitropism and lateral root initiation. *The Plant Journal* **7**, 211–220.

M. Holdsworth, S. Kurup and R. McKibbin (1999) Molecular and genetic mechanisms regulating the transition from embryo development to germination. *Trends in Plant Science* **4**, 275–280.

S.E. Jacobsen and N.E. Olszewski (1993) Mutations at the *SPINDLY* locus of *Arabidopsis* alter gibberellin signal transduction. *The Plant Cell* **5**, 887–896.

J.V. Jacobsen, D.W. Pearce, A.T. Poole, R.P. Pharis and L.N. Mander (2002) Abscisic acid, phaseic acid and gibberellin contents associated with dormancy and germination in barley. *Physiologia Plantarum* **115**, 428–441.

P.D. Jenik and M.K. Barton (2005) Surge and destroy: the role of auxin in plant embryogenesis. *Development* **132**, 3577–3585.

L. Jiang, S. Abrams and A.R. Kermode (1996) Vicilin and napin storage protein gene promoters are responsive to abscisic acid in developing transgenic tobacco seed but lose sensitivity following premature desiccation. *Plant Physiology* **110**, 1135–1144.

H.D. Jones, N.C. Peters and M.J. Holdsworth (1997) Genotype and environment interact to control dormancy and differential expression of the *VIVIPAROUS 1* homologue in embryos of *Avena fatua*. *The Plant Journal* **12**, 911–920.

S. Jones-Held, M. Vandoren and T. Lockwood (1996) Brassinolide application to *Lepidium sativum* seeds and the effects on seedling growth. *Journal of Plant Growth Regulation* **15**, 63–67.

Y. Kagaya, R. Toyoshima, R. Okuda, H. Usui, A. Yamamoto and T. Hattori (2005) LEAFY COTYLEDON1 controls seed storage protein genes through its regulation of *FUSCA3* and *ABSCISIC ACID INSENSITIVE3*. *Plant and Cell Physiology* **46**, 399–406.

C.M. Karssen, D.L.C. Brinkhorst-van der Swan, A.E. Breekland and M. Koornneef (1983) Induction of dormancy during seed development by endogenous abscisic acid: studies on abscisic acid-deficient genotypes of *Arabidopsis thaliana* (L.) Heynh. *Planta* **157**, 158–165.

C.M. Karssen and E. Laçka (1986) A revision of the hormone balance theory of seed dormancy: studies on gibberellin and/or abscisic acid deficient mutants of *Arabidopsis thaliana*. In: *Plant Growth Substances* (ed. M. Bopp), pp. 315–323. Springer-Verlag, Berlin.

C.M. Karssen, S. Zagórsky, J. Kepczynski and S.P.C. Groot (1989) Key role for endogenous gibberellins in the control of seed germination. *Annals of Botany* **63**, 71–80.

K. Keith, M. Kraml, N.G. Dengler and P. McCourt (1994) *fusca 3*: a heterochronic mutation affecting late embryo development in *Arabidopsis*. *The Plant Cell* **6**, 589–600.

J. Kępczyński and E. Kępczyński (1997) Ethylene in seed dormancy and germination. *Physiologia Plantarum* **101**, 720–726.

A.R. Kermode (1995) Regulatory mechanisms in the transition from seed development to germination: interactions between the embryo and the seed environment. In: *Seed Development and Germination* (eds J. Kigel and G. Galili), pp. 273–332. Marcel Dekker, New York.

A.R. Kermode (2005) Role of abscisic acid in seed dormancy. *Journal of Plant Growth Regulation* **24**, 319–344.

J.J. Kieber, M. Rothenberg, G. Roman, K.A. Feldmann and J.R. Ecker (1993) CTR1, a negative regulator of the ethylene response pathway in *Arabidopsis*, encodes a member of the raf family of protein kinases. *Cell* **72**, 427–441.

H.J. Kim, H. Ryu, S.H. Hong, *et al.* (2006) Cytokinin-mediated control of leaf longevity by AHK3 through phosphorylation of ARR2 in *Arabidopsis*. *Proceedings of the National Academy of Sciences of the United States of America* **103**, 814–819.

T. Kinoshita, A. Cano-Delgado, H. Seto, *et al.* (2005) Binding of brassinosteroids to the extracellular domain of plant receptor kinase BRI1. *Nature* **433**, 167–171.

M. Koornneef, L. Bentsink and H. Hilhorst (2002) Seed dormancy and germination. *Current Opinion in Plant Biology* **5**, 33–36.

M. Koornneef, M.L. Jorna, D.L.C. Brinkhorst-van der Swan and C.M. Karssen (1982) The isolation of abscisic acid (ABA) deficient mutants by selection of induced revertants in non-germinating gibberellin sensitive lines of *Arabidopsis thaliana* (L.) Heynh. *Theoretical and Applied Genetics* **61**, 385–393.

Y. Kraepiel, P. Rousselin, B. Sotta, *et al.* (1994) Analysis of phytochrome- and ABA-deficient mutants suggests that ABA degradation is controlled by light in *Nicotiana plumbaginifolia. The Plant Journal* **6**, 665–672.

B. Kucera, M.A. Cohn and G. Leubner-Metzger (2005) Plant hormone interactions during seed dormancy release and germination. *Seed Science Research* **15**, 281–307.

S. Kurup, H.D. Jones and M.J. Holdsworth (2000) Interactions of the developmental regulator ABI3 with proteins identified from developing *Arabidopsis* seeds. *The Plant Journal* **21**, 143–155.

T. Kushiro, M. Okamoto, K. Nakabayashi, *et al.* (2004) The *Arabidopsis* cytochrome P450 *CYP707A* encodes ABA 8′-hydroxylases: key enzymes in ABA catabolism. *The EMBO Journal* **23**, 1647–1656.

R.J. Laby, M.S. Kincaid, D. Kim and S.I. Gibson (2000) The *Arabidopsis* sugar-insensitive mutants *sis4* and *sis5* are defective in abscisic acid synthesis and response. *The Plant Journal* **23**, 587–596.

C.C. Lashbrook, D.M. Tieman and H.J. Klee (1998) Differential regulation of the tomato *ETR* gene family throughout plant development. *The Plant Journal* **15**, 243–252.

V. Lefebvre, H. North, A. Frey, *et al.* (2006) Functional analysis of *Arabidopsis NCED6* and *NCED9* genes indicates that ABA synthesised in the endosperm is involved in the induction of seed dormancy. *The Plant Journal* **45**, 309–319.

P. León and J. Sheen (2003) Sugar and hormone connections. *Trends in Plant Science* **8**, 110–116.

M.T. Le Page-Degivry and G. Garello (1992) *In situ* abscisic acid synthesis. A requirement for induction of embryo dormancy in *Helianthus annuus*. *Plant Physiology* **98**, 1386–1390.

M.T. Le Page-Degivry, G. Garello and P. Barthe (1997) Changes in abscisic acid biosynthesis and catabolism during dormancy breaking in *Fagus sylvatica* embryo. *Journal of Plant Growth Regulation* **16**, 57–61.

G. Leubner-Metzger (2001) Brassinosteroids and gibberellins promote tobacco seed germination by distinct pathways. *Planta* **213**, 758–763.

G. Leubner-Metzger (2003) Brassinosteroids promote seed germination. In: *Brassinosteroids: Bioactivity and Crop Productivity* (eds S. Hayat and A. Ahmad), pp. 119–128. Kluwer Academic Publishers, Dordrecht.

G. Leubner-Metzger, L. Petruzzelli, R. Waldvogel, R. Vogeli-Lange and F. Meins Jr. (1998) Ethylene-responsive element binding protein (EREBP) expression and the transcriptional regulation of class I β-1,3-glucanase during tobacco seed germination. *Plant Molecular Biology* **38**, 785–795.

J. Leung, M. Bouvier-Durand, P.-C. Morris, D. Guerrier, F. Chefdor and J. Giraudat (1994) *Arabidopsis* ABA-response gene *ABI1*: features of a calcium-modulated protein phosphatase. *Science* **264**, 1448–1452.

H. Li, J.J. Shen, Z.L. Zheng, Y. Lin and Z. Yang (2001) The Rop GTPase switch controls multiple developmental processes in *Arabidopsis*. *Plant Physiology* **126**, 670–684.

H.C. Li, K. Chuang, J.T. Henderson, et al. (2005) PICKLE acts during germination to repress expression of embryonic traits. *The Plant Journal* **44**, 1010–1022.

P.P. Liu, N. Koizuka, R.C. Martin and H. Nonogaki (2005) The *BME3* (*Blue Micropylar End 3*) GATA zinc finger transcription factor is a positive regulator of *Arabidopsis* seed germination. *The Plant Journal* **44**, 960–971.

K. Ljung, A. Ostin, L. Lioussanne and G. Sandberg (2001) Developmental regulation of indole-3-acetic acid turnover in Scots pine seedlings. *Plant Physiology* **125**, 464–475.

O. Lorenzo, C. Nicolás, G. Nicolás and D. Rodríguez (2002) Molecular cloning of a functional protein phosphatase 2C (FsPP2C2) with unusual features and synergistically upregulated by ABA and calcium in dormant seeds of *Fagus sylvatica*. *Physiologia Plantarum* **114**, 482–490.

O. Lorenzo, C. Nicolás, G. Nicolás and D. Rodríguez (2003) Characterization of a dual plant protein kinase (FsPK1) up-regulated by abscisic acid and calcium and specifically expressed in dormant seeds of *Fagus sylvatica* L. *Seed Science Research* **13**, 261–271.

O. Lorenzo, D. Rodríguez, G. Nicolás, P.L. Rodríguez and C. Nicolás (2001) A new protein phosphatase 2C (FsPP2C1) induced by abscisic acid is specifically expressed in dormant beechnut seeds. *Plant Physiology* **125**, 1949–1956.

T. Lotan, M. Ohto, K.M. Yee, *et al.* (1998) *Arabidopsis LEAFY COTYLEDON1* is sufficient to induce embryo development in vegetative cells. *Cell* **93**, 1195–1205.

H. Luerßen, V. Kirik, P. Herrmann and S. Miséra (1998) *FUSCA3* encodes a protein with a conserved *VP1/ABI3*-like B3 domain which is of functional importance for the regulation of seed maturation in *Arabidopsis thaliana*. *The Plant Journal* **15**, 755–764.

A.P. Mähönen, M. Bonke, L. Kauppinen, M. RiiKonen, P.N. Benfey and Y. Helariutta (2000) A novel two-component hybrid molecule regulates vascular morphogenesis of the *Arabidopsis* root. *Genes and Develoment* **14**, 2938–2943.

M.G. Mason, D.E. Mathews, D.A. Argyros, *et al.* G.E. (2005) Multiple type-B response regulators mediate cytokinin signal transduction in *Arabidopsis*. *The Plant Cell* **17**, 3007–3018.

A.J. Matilla (2000) Ethylene in seed formation and germination. *Seed Science Research* **10**, 111–126.

M.A. Mazzella, M.V. Arana, R.J. Staneloni, *et al.* (2005) Phytochrome control of the *Arabidopsis* transcriptome anticipates seedling exposure to light. *The Plant Cell* **17**, 2507–2516.

D.R. McCarty (1995) Genetic control and integration of maturation and germination pathways in seed development. *Annual Review of Plant Biology* **46**, 71–93.

D.R. McCarty, T. Hattori, C.B. Carson, V. Vasil, M. Lazar and I.K. Vasil (1991) The *viviparous-1* developmental gene of maize encodes a novel transcriptional activator. *Cell* **66**, 895–905.

P. McCourt, S. Lumba, Y. Tsuchiya and S. Gazzarrini (2005) Crosstalk and abscisic acid: the roles of terpenoid hormones in coordinating development. *Physiologia Plantarum* **123**, 147–152.

R.S. McKibbin, M.D. Wilkinson, P.C. Bailey, *et al.* (2002) Transcripts of *Vp-1* homeologues are misspliced in modern wheat and ancestral species. *Proceedings of the National Academy of Sciences of the United States of America* **99**, 10203–10208.

D.W. Meinke (1995) Molecular genetics of plant embryogenesis. *Annual Review of Plant Biology* **46**, 369–394.

D.W. Meinke, L.H. Franzmann, T.C. Nickle and E.C. Yeung (1994) Leafy cotyledon mutants of *Arabidopsis*. *The Plant Cell* **6**, 1049–1064.

A.A. Jacobsen, J.V. Millar, J.J. Ross, *et al.* (2006) Seed dormancy and ABA metabolism in *Arabidopsis* and barley: the role of ABA 8′-hydroxylase. *The Plant Journal* **45**, 942–954.

D.W.S. Mok and M.C. Mok (2001) Cytokinin metabolism and action. *Annual Review of Plant Biology* **52**, 89–118.

M. Monroe-Augustus, B.K. Zolman and B. Bartel (2003) IBR5, a dual-specificity phosphatase-like protein modulating auxin and abscisic acid responsiveness in *Arabidopsis*. *The Plant Cell* **15**, 2979–2991.

A. Nagatani (2004) Light-regulated nuclear localization of phytochromes. *Current Opinion in Plant Biology* **7**, 708–711.

K. Nakaminami, Y. Sawada, M. Suzuki, *et al.* (2003) Deactivation of gibberellin by 2-oxidation during germination of photoblastic lettuce seeds. *Bioscience Biotechnology and Biochemistry* **67**, 1551–1558.

S. Nakamura, T.J. Lynch and R.R. Finkelstein (2001) Physical interactions between ABA response loci of *Arabidopsis*. *The Plant Journal* **26**, 627–635.

E. Nambara, R. Hayama, Y. Tsuchiya, *et al.* (2000) The role of *ABI3* and *FUS3* loci in *Arabidopsis thaliana* on phase transition from late embryo development to germination. *Developmental Biology* **220**, 412–423.

E. Nambara and A. Marion-Poll (2005) Abscisic acid biosynthesis and catabolism. *Annual Review of Plant Biology* **56**, 165–185.

E. Nambara, S. Naito and P. McCourt (1992) A mutant of *Arabidopsis* which is defective in seed development and storage protein accumulation is a new *abi3* allele. *The Plant Journal* **2**, 435–441.

C. Nicolás, G. Nicolás and D. Rodríguez (1998) Transcripts of a gene, encoding a small GTP-binding protein from *Fagus sylvatica*, are induced by ABA and accumulated in the embryonic axis of dormant seeds. *Plant Molecular Biology* **36**, 487–491.

N. Nishimura, N. Kitahata, M. Seki, *et al.* (2005) Analysis of *ABA hypersensitive germination2* revealed the pivotal functions of PARN in stress response in *Arabidopsis*. *The Plant Journal* **44**, 972–984.

J. Normanly (1997) Auxin metabolism. *Physiologia Plantarum* **100**, 431–442.

J. Ogas, J.C. Cheng, Z.R. Sung and C. Somerville (1997) Cellular differentiation regulated by gibberellin in the *Arabidopsis thaliana pickle* mutant. *Science* **277**, 91–94.

J. Ogas, S. Kaufmann, J. Henderson and C. Somerville (1999) *PICKLE* is a CHD3 chromatin-remodeling factor that regulates the transition from embryonic to vegetative development in *Arabidopsis*. *Proceedings of the National Academy of Sciences of the United States of America* **96**, 13839–13844.

M. Ogawa, A. Hanada, Y. Yamauchi, A. Kuwahara, Y. Kamiya and S. Yamaguchi (2003) Gibberellin biosynthesis and response during *Arabidopsis* seed germination. *The Plant Cell* **15**, 1591–1604.

E. Oh, J. Kim, E. Park, J.I. Kim, C. Kang and G. Choi (2004) PIL5, a phytochrome-interacting basic helix-loop-helix protein, is a key negative regulator of seed germination in *Arabidopsis thaliana*. *The Plant Cell* **16**, 3045–3058.

J.J.J. Ooms, K.M. Leon-Kloosterziel, D. Bartels, M. Koornneef and C.M. Karssen (1993) Acquisition of desiccation tolerance and longevity in seeds of *Arabidopsis thaliana* – a comparative study using abscisic acid insensitive *abi3* mutants. *Plant Physiology* **102**, 1185–1191.

F. Ouaked, W. Rozhon, D. Lecourieux and H. Hirt (2003) A MAPK pathway mediates ethylene signaling in plants. *The EMBO Journal* **22**, 1282–1288.

F. Parcy, C. Valon, A. Kohara, S. Miséra and J. Giraudat (1997) The *ABSCISIC ACID-INSENSITIVE3*, *FUSCA3*, and *LEAFY COTYLEDON1* loci act in concert to control multiple aspects of *Arabidopsis* seed development. *The Plant Cell* **9**, 1265–1277.

S. Penfield, E.M. Josse, R. Kannangara, A.D. Gilday, K.J. Halliday and I.A. Graham (2005) Cold and light control seed germination through the bHLH transcription factor SPATULA. *Current Biology* **22**, 1998–2006.

L. Pérez-Flores, F. Carrari, R. Osuna-Fernández, *et al.* (2003) Expression analysis of a GA 20-oxidase in embryos from two sorghum lines with contrasting dormancy: possible participation of this gene in the hormonal control of germination. *Journal of Experimental Botany* **54**, 2071–2079.

J. Phillips, O. Artsaenko, U. Fiedler, *et al.* (1997) Seed-specific immunomodulation of abscisic acid activity induces a developmental switch. *The EMBO Journal* **16**, 4489–4496.

A. Poljakoff-Mayber, I. Popilevski, E. Belausov and Y. Ben-Tal (2002) Involvement of phytohormones in germination of dormant and non-dormant oat (*Avena sativa* L.) seeds. *Plant Growth Regulation* **37**, 7–16.

S. Ramaih, M. Guedira and G.M. Paulsen (2003) Relationship of indoleacetic acid and tryptophan to dormancy and preharvest sprouting of wheat. *Functional Plant Biology* **30**, 939–945.

R.A. Rampey, S. LeClere, M. Kowalczyk, K. Ljung, G. Sandberg and B. Bartel (2004) A family of auxin-conjugate hydrolases that contributes to free indole-3-acetic acid levels during *Arabidopsis* germination. *Plant Physiology* **135**, 978–988.

V. Raz, J.H.W. Bergervoet and M. Koornneef (2001) Sequential steps for developmental arrest in *Arabidopsis* seeds. *Development* **128**, 243–252.

F.A. Razem, A. El-Kereamy, S.R. Abrams and R.D. Hill (2006) The RNA-binding protein FCA is an abscisic acid receptor. *Nature* **439**, 290–294.

F.A. Razem, M. Luo, J.H. Liu, S.R. Abrams and R.D. Hill (2004) Purification and characterization of a barley aleurone abscisic acid-binding protein. *Journal of Biological Chemistry* **279**, 9922–9929.

J.B. Reid, G.M. Symons and J.J. Ross (2004) Regulation of gibberellin and brassinosteroid biosynthesis by genetic, environmental and hormonal factors. In: *Plant Hormones: Biosynthesis, Signal Transduction, Action!* (ed. P.J. Davies), pp. 179–203. Kluwer Academic Publishers, Dordrecht.

C. Ren and A.R. Kermode (1999) Analyses to determine the role of the megagametophyte and other seed tissues in dormancy maintenance of yellow cedar (*Chamaecyparis nootkatensis*) seeds: morphological, cellular and physiological changes following moist chilling and during germination. *Journal of Experimental Botany* **50**, 1403–1419.

S.D. Rider Jr., M.R. Hemm, H.A. Hostetler, H.C. Li, C. Chapple and J. Ogas (2004) Metabolic profiling of the *Arabidopsis pkl* mutant reveals selective derepression of embryonic traits. *Planta* **219**, 489–499.

S.D. Rider, J.T. Henderson, R.E. Jerome, H.J. Edenberg, J. Romero-Severson and J. Ogas (2003) Coordinate repression of regulators of embryonic identity by *PICKLE* during germination in *Arabidopsis*. *The Plant Journal* **35**, 33–43.

M. Riefler, O. Novak, M. Strand and T. Schmulling (2006) *Arabidopsis* cytokinin receptor mutants reveal functions in shoot growth, leaf senescence, seed size, germination, root development, and cytokinin metabolism. *The Plant Cell* **18**, 40–54.

I. Romagosa, D. Prada, M.A. Moralejo, *et al.* (2001) Dormancy, ABA content and sensitivity of a barley mutant to ABA application during seed development and after ripening. *Journal of Experimental Botany* **52**, 1499–1506.

F. Rook, F. Corke, R. Card, G. Munz, C. Smith and M.W. Bevan (2001) Impaired sucrose-induction mutants reveal the modulation of sugar-induced starch biosynthetic gene expression by abscisic acid signalling. *The Plant Journal* **26**, 421–433.

H.S. Saini, E.D. Consolacion, P.K. Bassi and M.S. Spencer (1986) Requirement for ethylene synthesis and action during relief of thermoinhibition of lettuce seed germination by combinations of gibberellic acid, kinetin, and carbon dioxide. *Plant Physiology* **81**, 950–953.

H.S. Saini, D. Evangeline, E.D. Consolacion, K. Pawan, P.K. Bassi and M.S. Spencer (1989) Control processes in the induction and relief of thermoinhibition of lettuce seed germination, actions of phytochrome and endogenous ethylene. *Plant Physiology* **90**, 311–315.

S. Saito, N. Hirai, C. Matsumoto, *et al.* (2004) *Arabidopsis CYP707A*s encode (+)-abscisic acid 8′-hydoxylase, a key enzyme in the oxidative catabolism of abscisic acid. *Plant Physiology* **134**, 1439–1449.

N. Schmitz, S.R. Abrams and A.R. Kermode (2000) Changes in abscisic acid content and embryo sensitivity to (+)-abscisic acid during dormancy termination of yellow-cedar seeds. *Journal of Experimental Botany* **51**, 1159–1162.

N. Schmitz, S.R. Abrams and A.R. Kermode (2002) Changes in ABA turnover and sensitivity that accompany dormancy termination of yellow-cedar (*Chamaecyparis nootkatensis*) seeds. *Journal of Experimental Botany* **53**, 89–101.

J. Schmidt, T. Altmann and G. Adam (1997) Brassinosteroids from seeds of *Arabidopsis thaliana*. *Phytochemistry* **45**, 1325–1327.

N. Schmitz, J.-H. Xia and A.R. Kermode (2001) Dormancy of yellow-cedar seeds is terminated by gibberellic acid in combination with fluridone or with osmotic priming and moist chilling. *Seed Science and Technology* **29**, 331–346.

P. Schopfer and C. Plachy (1985) Control of seed germination by ABA. III. Effect on embryo growth potential (minimum turgor pressure) and growth coefficient (cell wall extensibility) in *Brassica napus* L. *Plant Physiology* **77**, 676–686.

J. Schwachtje and I.T. Baldwin (2004) Smoke exposure alters endogenous gibberellin and abscisic acid pools and gibberellin sensitivity while eliciting germination in the post-fire annual, *Nicotiana attenuata* seed. *Seed Science Research* **14**, 51–60.

S.H. Schwartz, X. Qin and J.A.D. Zeevaart (2003) Elucidation of the indirect pathway of abscisic acid biosynthesis by mutants, genes, and enzymes. *Plant Physiology* **131**, 1591–1601.

A. Schweighofer, H. Hirt and I. Meskiene (2004) Plant PP2C phosphatases: emerging functions in stress signaling. *Trends in Plant Science* **9**, 236–243.

M. Seo, H. Aoki, H. Koiwai, Y. Kamiya, E. Nambara and T. Koshiba (2004) Comparative studies on the *Arabidopsis* aldehyde oxidase (*AAO*) gene family revealed a major role of *AAO3* in ABA biosynthesis in seeds. *Plant and Cell Physiology* **45**, 1694–1703.

M. Seo and T. Koshiba (2002) Complex regulation of ABA biosynthesis in plants. *Trends in Plant Science* **7**, 41–48.

J. Sheen (1998) Mutational analysis of protein phosphatase 2C involved in abscisic acid signal transduction in higher plants. *Proceedings of the National Academy of Sciences of the United States of America* **95**, 975–980.

H. Shen, J. Moon and E. Huq (2005) PIF1 is degraded by light through ubiquitin-mediated proteasome pathway. In: *Abstract, Arabidopsis 2005*, Madison, WI.

T. Shinomura, A. Nagatani, J. Chory and M. Furuya (1994) The induction of seed germination in *Arabidopsis thaliana* is regulated principally by phytochrome B and secondarily by phytochrome A. *Plant Physiology* **104**, 363–371.

T. Shinomura, A. Nagatani, H. Manzawa, M. Kubota, M. Watanabe and M. Furuya (1996) Action spectra for phytochrome A and B-specific photoinduction of seed germination in *Arabidopsis thaliana. Proceedings of the National Academy of Sciences of the United States of America* **93**, 8129–8133.

D.P. Singh, A.M. Jermakow and S.M. Swain (2002) Gibberellins are required for seed development and pollen tube growth in *Arabidopsis. The Plant Cell* **14**, 3133–3147.

G. Siriwitayawan, R.L. Geneve and A.B. Downie (2003) Seed germination of ethylene perception mutants of tomato and *Arabidopsis. Seed Science Research* **13**, 303–314.

J. Smalle, J. Kurepa, P. Yang, *et al.* (2003) The pleiotropic role of the 26S proteasome subunit RPN10 in *Arabidopsis* growth and development supports a substrate-specific function in abscisic acid signaling. *The Plant Cell* **15**, 965–980.

E.M. Söderman, I.M. Brocard, T.J. Lynch and R.R. Finkelstein (2000) Regulation and function of the *Arabidopsis ABA-insensitive4* gene in seed and abscisic acid response signaling networks. *Plant Physiology* **124**, 1752–1765.

W.G. Spollen, M.E. LeNoble, T.D. Samuels, N. Bernstein and R.E. Sharp (2000) Abscisic acid accumulation maintains maize primary root elongation at low water potentials by restricting ethylene production. *Plant Physiology* **122**, 967–976.

C. Steber, S.E. Cooney and P. McCourt (1998) Isolation of the GA-response mutant *sly1* as a suppressor of *ABI1* in *Arabidopsis thaliana. Genetics* **149**, 509–521.

C.M. Steber and P. McCourt (2001) A role for brassinosteroids in germination in *Arabidopsis. Plant Physiology* **125**, 763–769.

S.L. Stone, L.K.M. Kwongm, J. Pelletier, *et al.* (2001) *LEAFY COTYLEDON2* encodes a B3 domain transcription factor that induces embryo development. *Proceedings of the National Academy of Sciences of the United States of America* **98**, 11806–11811.

W. Su and S.H. Howell (1992) A single genetic locus, *ckr1*, defines *Arabidopsis* mutants in which root growth is resistant to low concentrations of cytokinins. *Plant Physiology* **99**, 1569–1574.

J.A. Sullivan and X.W. Deng (2003) From seed to seed: the role of photoreceptors in *Arabidopsis* development. *Developmental Biology* **260**, 289–297.

I.M. Sussex (1975) Growth and metabolism of the embryo and attached seedlings of the viviparous mangrove, *Rhizophora mangle. American Journal of Botany* **62**, 948–953.

M. Suzuki, C.Y. Kao, S. Cocciolone and D.R. McCarty (2001) Maize *VP1* complements *Arabidopsis abi3* and confers a novel ABA/auxin interaction in roots. *The Plant Journal* **28**, 409–418.

M. Suzuki, A.M. Settles, C.-W. Tseung, *et al.* (2006) The maize *viviparous15* locus encodes the molybdopterin synthase small subunit. *The Plant Journal* **45**, 264–274.

U. Sweere, K. Eichenberg, J. Lohrmann, *et al.* (2001) Interaction of the response regulator ARR4 with phytochrome B in modulating red light signaling. *Science* **294**, 1108–1111.

Y. Takeuchi, Y. Omigawa, M. Ogasawara, K. Yoneyama, M. Konnai and A.D. Worsham (1995) Effects of brassinosteroids on conditioning and germination of clover broomrape (*Orobanche minor*) seeds. *Plant Growth Regulation* **16**, 153–160.

B.C. Tan, L.M. Joseph, W.T. Deng, L.J. Liu, Q.B. Li, K. Cline and D.R. McCarty (2003) Molecular characterization of the *Arabidopsis* 9-*cis* epoxycarotenoid dioxygenase gene family. *The Plant Journal* **35**, 44–56.

B.C. Tan, S.H. Schwartz, J.A. Zeevaart and D.R. McCarty (1997) Genetic control of abscisic acid biosynthesis in maize. *Proceedings of the National Academy of Sciences of the United States of America* **94**, 12235–12240.

E. Tillberg (1992) Effect of light on abscisic acid content in photosensitive Scots pine (*Pinus sylvestris* L.) seed. *Plant Growth Regulation* **11**, 147–152.

I. Tiryaki and P.E. Staswick (2002) An *Arabidopsis* mutant defective in jasmonate response is allelic to the auxin-signaling mutant *axr1*. *Plant Physiology* **130**, 887–894.

T. Toyomasu, H. Kawaide, W. Mitsuhashi, Y. Inoue and Y. Kamiya (1998) Phytochrome regulates gibberellin biosynthesis during germination of photoblastic lettuce seeds. *Plant Physiology* **118**, 1517–1523.

T. Toyomasu, H. Tsuji, H. Yamane, *et al.* (1993) Light effects on endogenous levels of gibberellins in photoblastic lettuce seeds. *Journal of Plant Growth Regulation* **12**, 85–90.

H. Ullah, J.G. Chen, S. Wang and A.M. Jones (2002). Role of a heterotrimeric G protein in regulation of *Arabidopsis* seed germination. *Plant Physiology* **129**, 897–907.

T. Urao, B. Yakubov, R. Satoh, *et al.* (1999) A transmembrane hybrid-type histidine kinase in *Arabidopsis* functions as an osmosensor. *The Plant Cell* **11**, 1743–1754.

A.J. van Hengel and K. Roberts (2003) AtAGP30, an arabinogalactan-protein in the cell walls of the primary root, plays a role in root regeneration and seed germination. *The Plant Journal* **36**, 256–270.

B.V. Vardhini and S.S.R. Rao (2001) Effect of brassinosteroids on the seed germination and seedling growth of tomato (*Lycopersicon escultentum* Mill.). *Journal of Plant Biology* **27**, 303–305.

B.V. Vardhini and S.S.R. Rao (2003) Amelioration of osmotic stress by brassinosteroids on seed germination and seedling growth of three varieties of sorghum. *Plant Growth Regulation* **41**, 25–31.

R.D. Vierstra (2003) The ubiquitin/26S proteasome pathway, the complex last chapter in the life of many plant proteins. *Trends in Plant Science* **8**, 135–142.

M. Walker-Simmons (1990) Dormancy in cereals – levels of and response to abscisic acid. In: *Plant Growth Substances* (eds R.P. Pharis and S.B. Rood), pp. 400–406. Springer-Verlag, New York.

M. Wang, S. Heimovaara-Dijkstra and B. Van Duijn (1995) Modulation of germination of embryos isolated from dormant and non-dormant barley grains by manipulation of endogenous abscisic acid. *Planta* **195**, 586–592.

X.Q. Wang, H. Ullah, A.M. Jones and S.M. Assmann (2001). G protein regulation of ion channels and abscisic acid signaling in *Arabidopsis* guard cells. *Science* **292**, 2070–2072.

C.N. White, W.M. Proebsting, P. Hedden and C.J. Riven (2000) Gibberellins and seed development in maize. I. Evidence that gibberellin/abscisic acid balance governs germination versus maturation pathways. *Plant Physiology* **122**, 1081–1088.

C.N. White and C.J. Riven (2000) Gibberellins and seed development in maize. II. Gibberellin synthesis inhibition enhances abscisic acid signaling in cultured embryos. *Plant Physiology* **122**, 1089–1097.

F. Woodger, J.V. Jacobsen and F. Gubler (2004) Gibberellin action in germinating cereal grains. In: *Plant Hormones: Biosynthesis, Signal Transduction, Action!* (ed. P.J. Davies), pp. 221–240. Kluwer Academic Publishers, Dordrecht.

L. Xiong, Z. Gong, C. Rock, *et al.* (2001a) Modulation of abscisic acid signal transduction and biosynthesis by an Sm-like protein in *Arabidopsis*. *Developmental Cell* **1**, 771–781.

L. Xiong, H. Lee, M. Ishitani and J.K. Zhu (2002) Regulation of osmotic stress-responsive gene expression by the *LOS6/ABA1* locus in *Arabidopsis*. *Journal of Biological Chemistry* **277**, 8588–8596.

L. Xiong and J.K. Zhu (2003) Regulation of abscisic acid biosynthesis. *Plant Physiology* **133**, 29–36.

L.M. Xiong, M. Ishitani, H. Lee and J.K. Zhu (2001b) The *Arabidopsis LOS5/ABA3* locus encodes a molybdenum cofactor sulfurase and modulates cold stress- and osmotic stress-responsive gene expression. *The Plant Cell* **13**, 2063–2083.

N. Xu and J.D. Bewley (1991) Sensitivity to abscisic acid and osmoticum changes during embryogenesis of alfalfa (*Medicago sativa*). *Journal of Experimental Botany* **42**, 821–826.

H. Yamada, T. Suzuki, K. Terada, *et al.* (2001) The *Arabidopsis* AHK4 histidine kinase is a cytokinin-binding receptor that transduces cytokinin signals across the membrane. *Plant and Cell Physiology* **42**, 1017–1023.

S. Yamaguchi and Y. Kamiya (2002) Gibberellins and light-stimulated seed germination. *Journal of Plant Growth Regulation* **20**, 369–376.

S. Yamaguchi, Y. Kamiya and T-P. Sun (2001) Distinct cell-specific expression patterns of early and late gibberellin biosynthetic genes during *Arabidopsis* seed germination. *The Plant Journal* **28**, 443–453.

S. Yamaguchi, M.W. Smith, R.G. Brown, Y. Kamiya and T-P. Sun (1998) Phytochrome regulation and differential expression of gibberellin 3β-hydroxylase genes in germinating *Arabidopsis* seeds. *The Plant Cell* **10**, 2115–2126.

T. Yamaguchi, T. Wakizuka, K. Hirai, S. Fujii and A. Fujita (1987) Stimulation of germination in aged rice seeds by pretreatment with brassinolide. *Proceedings of the Plant Growth Regulation Society of America* **14**, 26–27.

Y. Yamauchi, M. Ogawa, A. Kuwahara, A. Hanada, Y. Kamiya and S. Yamaguchi (2004) Activation of gibberellin biosynthesis and response pathways by low temperature during imbibition of *Arabidopsis thaliana* seeds. *The Plant Cell* **16**, 367–378.

T. Yoshioka, T. Endo and S. Satoh (1998) Restoration of seed germination at supraoptimal temperature by fluridone, an inhibitor of abscisic acid biosynthesis. *Plant and Cell Physiology* **39**, 307–312.

T. Yoshida, N. Nishimura, N. Kitahata, *et al.* (2006) *ABA-Hypersensitive Germination3* encodes a protein phosphatase 2C (AtPP2CA) that strongly regulates abscisic acid signaling during germination among *Arabidopsis* protein phosphatase 2Cs. *Plant Physiology* **140**, 115–126.

T.E. Young and D.R. Gallie (2000) Programmed cell death during endosperm development. *Plant Molecular Biology* **44**, 283–301.

T.E. Young, D.R. Gallie and D.A. DeMason (1997) Ethylene mediated programmed cell death during maize endosperm development of *Su* and *sh2* genotypes. *Plant Physiology* **115**, 737–751.

L.I. Zaharia, M.K. Walker-Simmons, C.N. Rodriguez and S.R. Abrams (2005) Chemistry of abscisic acid and abscisic acid catabolites and analogs. *Journal of Plant Growth Regulation* **24**, 274–284.

J.A.D. Zeevaart (1999) Abscisic acid metabolism and its regulation. In: *Biochemistry and Molecular Biology of Plant Hormones* (eds P.J.J. Hooykaas, K.R. Hall and K.R. Libbenga), pp. 189–207. Elsevier Science BV, Amsterdam.

Y. Zeng and A.R. Kermode (2004) A gymnosperm *ABI3* gene functions in a severe abscisic acid-insensitive mutant of *Arabidopsis* (*abi3-6*) to restore the wild-type phenotype and demonstrates a strong synergistic effect with sugar in the inhibition of post-germinative growth. *Plant Molecular Biology* **56**, 731–746.

Y. Zeng, N. Raimondi and A.R. Kermode (2003) Role of an *ABI3* homologue in dormancy maintenance of yellow-cedar seeds and in the activation of storage protein and *Em* gene promoters. *Plant Molecular Biology* **51**, 39–49.

X. Zhang, V. Garreton and N.-H. Chua (2005) The AIP2 E3 ligase acts as a novel negative regulator of ABA signaling by promoting ABI3 degradation. *Genes and Development* **19**, 1532–1543.

9 Regulation of ABA and GA levels during seed development and germination in *Arabidopsis*

Shinjiro Yamaguchi, Yuji Kamiya and Eiji Nambara

9.1 Introduction

Seed germination is regulated by both developmental and environmental factors. Plant hormones are important signaling molecules that transfer the changes of the environment to seeds during germination. Abscisic acid (ABA) and gibberellin (GA) are considered to be major hormones in germinating *Arabidopsis thaliana* seeds. ABA is a sesquiterpene hormone that is necessary during late embryogenesis and inhibits precocious seed germination. ABA also increases the tolerance of plants to stresses such as drought, salinity, and temperature, and induces stomatal closure in leaves (Davies, 2004). Elevated ABA levels are important during mid- and late-embryogenesis to maintain normal seed development. Relatively high levels of ABA in maturing seeds maintain their dormancy and avoid precocious germination. GA is a diterpene hormone that regulates seed development and germination and other aspects of plant growth and development, such as leaf expansion, stem elongation, and flowering. Although some reports suggest that ethylene and brassinosteroids also affect seed germination of *Arabidopsis* (Kucera *et al.*, 2005), we focus here on regulation of ABA and GA levels in seeds. Endogenous levels of active hormones are determined by the relative rates of their formation and conversion into inactive forms. In this chapter, we refer to biosynthesis as production of bioactive forms of a hormone from its precursors, deactivation as conversion of bioactive forms or its precursors to the inactive (or less active) forms, and metabolism as both biosynthesis and deactivation.

Biosynthesis of ABA and GA starts in plastids using a common precursor, geranylgeranyldiphosphate (GGDP), which is synthesized via the 2-*C*-methyl-D-erythritol phosphate (MEP) pathway (Milborrow and Lee, 1998; Hirai *et al.*, 2000; Kasahara *et al.*, 2004). The mevalonate pathway in the cytosol is important for steroid and triterpene biosynthesis, but is not principally involved in the biosynthesis of ABA and GA. Severe alleles of GA-deficient mutants of *Arabidopsis* cannot germinate without exogenous bioactive GAs (Koornneef and van der Veen, 1980). Therefore, GA is an essential hormone leading to germination in *Arabidopsis*. However, some ABA-deficient and -insensitive mutants can germinate in the presence of paclobutrazol (PAC), a GA biosynthesis inhibitor, at the concentration that is sufficient to inhibit germination of wild-type seeds. This suggests that a low level of GA is enough to cause germination when ABA biosynthesis or signaling is impaired.

These observations support the idea that the balance of endogenous ABA and GA levels plays an important role in controlling seed germination.

In this chapter, we first describe the outline of ABA and GA metabolism and the genes involved in the respective pathways. Chemical inhibitors for ABA and GA metabolism enzymes will also be described, because they have contributed significantly to understanding the roles of ABA and GA in seed biology. Finally, we will highlight recent findings on the regulation of ABA and GA levels in developing and germinating seeds. Throughout this chapter, we will focus on *Arabidopsis*, because the regulation of ABA and GA metabolism has been best understood in this model species.

9.2 Biosynthetic and deactivation pathways of ABA and GA

Microarrays have been used to analyze the changes in the transcriptome in *Arabidopsis* seeds (Ogawa *et al.*, 2003; Yamauchi *et al.*, 2004; Nakabayashi *et al.*, 2005). In addition to these published results, a large set of GeneChip data has been made available as public databases such as the AtGenExpress project (http://www.arabidopsis.org/info/expression/ATGenExpress.jsp). We summarized these data to provide a visual overview of the expression profiles of ABA and GA metabolism genes during seed development and germination (Figures 9.1 and 9.2). The identities (AGI [Arabidopsis Genome Initiative] codes) of all ABA and GA metabolism genes are listed in Tables 9.1 and 9.2. Some of these genes and the corresponding loss-of-function mutants are selected to discuss the specific aspects of ABA and GA regulation and to highlight recent findings in the following sections.

9.2.1 ABA biosynthesis

In plants, ABA is synthesized from C_{40} carotenoids via oxidative cleavage (Nambara and Marion-Poll, 2005). Earlier steps in ABA biosynthesis occur in plastids, and the last two steps take place in the cytosol. Zeaxanthin epoxidase (ZEP/AtABA1) is a xanthophyll cycle enzyme that catalyzes two-step epoxidation of zeaxanthin to violaxanthin (Audran *et al.*, 2001; Xiong *et al.*, 2002) (Figure 9.1). Violaxanthin is converted to 9′-*cis*-neoxanthin or 9-*cis*-violaxanthin by unknown mechanisms. 9-*cis*-Epoxycarotenoid dioxygenase (NCED) cleaves the 9-*cis* isomers of neoxanthin and violaxanthin to produce xanthoxin, a C_{15} precursor of ABA (Qin and Zeevaart, 1999; Chernys and Zeevaart, 2000; Iuchi *et al.*, 2001; Schwartz *et al.*, 2003). This reaction is thought to be the most important regulatory step in ABA biosynthesis. In *Arabidopsis*, five NCEDs (AtNCED2, AtNCED3, AtNCED5, AtNCED6, and AtNCED9) are likely to be involved in this reaction (Iuchi *et al.*, 2001; Tan *et al.*, 2003). AtNCEDs have been shown to be chloroplast-targeted *in vitro* (Tan *et al.*, 2003). Xanthoxin is subsequently converted to abscisic aldehyde by short-chain alcohol dehydrogenase/reductase (SDR/AtABA2) and then to abscisic acid by abscisic aldehyde oxidase (AAO3), two cytosolic enzymes (Figure 9.1). Considering the plastid localization of NCEDs, xanthoxin should be transported

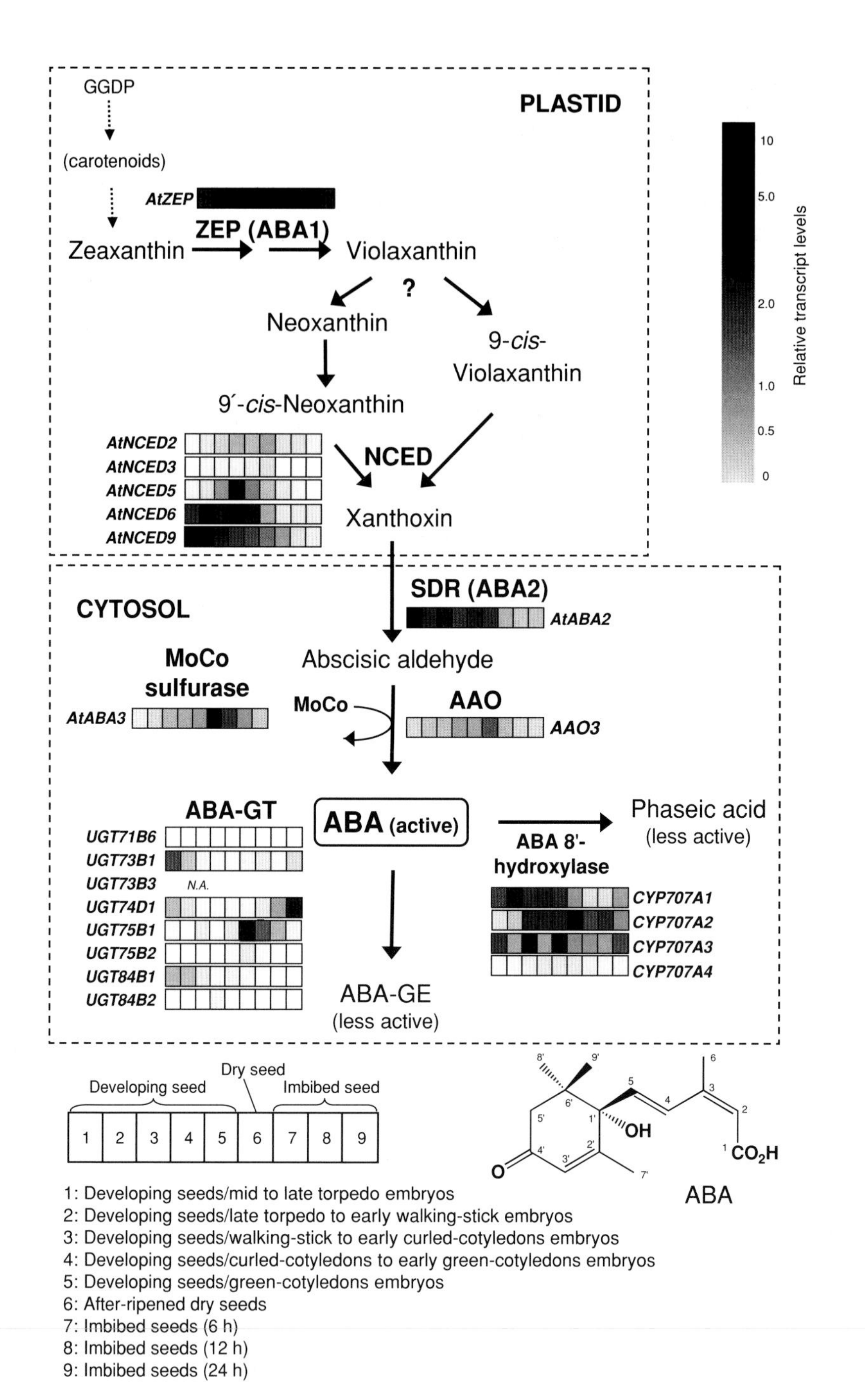

Figure 9.1 The major ABA metabolism pathway in *Arabidopsis* and the expression of associated genes during seed development and germination. ABA precursors, deactivation product and enzymes are shown. Relative transcript levels of the gene(s) responsible for each step were determined by Affymetrix GeneChip microarray and visualized using GeneSpring software (Silicon Genetics, CA, USA). Each bar contains results from nine different seed development/imbibition (*Continued over*)

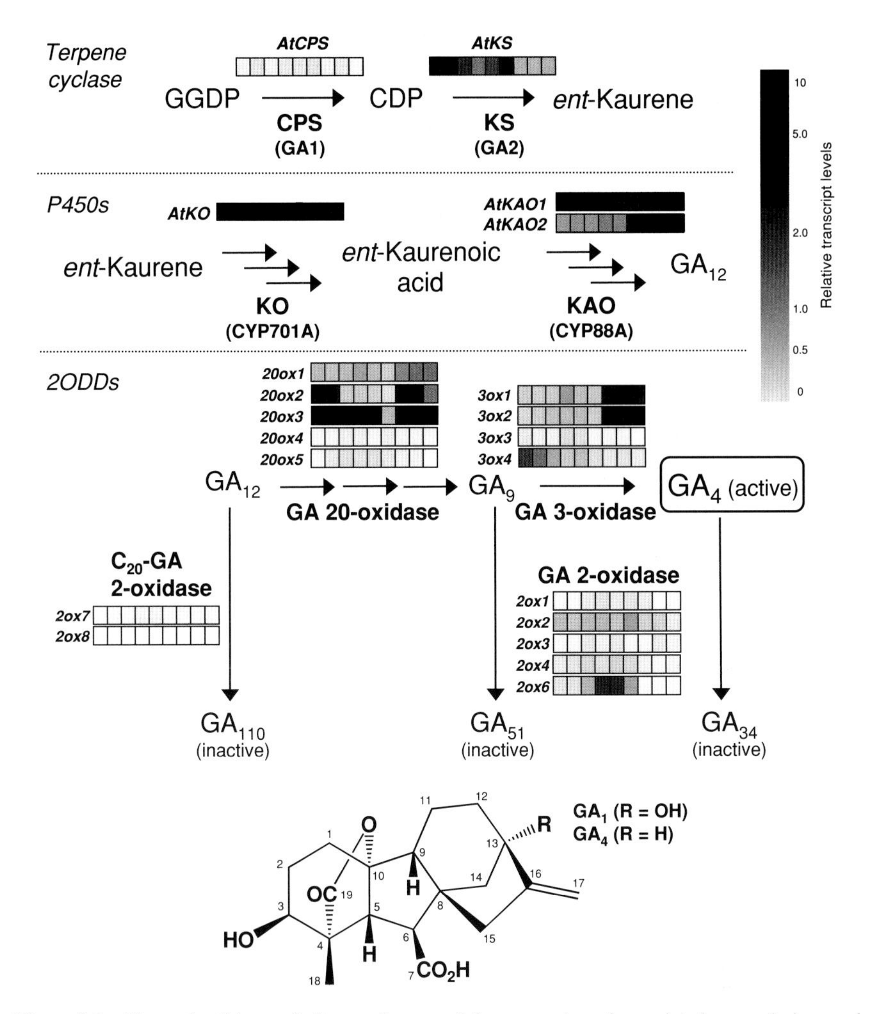

Figure 9.2 The major GA metabolism pathway and the expression of associated genes during seed development and germination. Expression profiles are visualized as detailed in the legend to Figure 9.1. Abbreviations: CPS, *ent*-copalyl diphosphate synthase; CDP, *ent*-copalyl diphosphate; KS, *ent*-kaurene synthase; P450s, cytochrome P450 monooxygenases; KO, *ent*-kaurene oxidase; KAO, *ent*-kaurenoic acid oxidase; 2ODDs, 2-oxoglutarate-dependent dioxygenases. Identities of GA metabolism genes are summarized in Table 9.2.

←

Figure 9.1 (*Continued*) stages as indicated below the pathway. For developing seeds (Stages 1–5), the microarray data were extracted from Expression Atlas of Arabidopsis Development (Schmid *et al.*, 2005), which is available at the AtGeneExpress Web Site (http://arabidopsis.org/servlets/TairObject?type=hyb_descr_collection&id=1006710873). Similar results were obtained among three biological replicates that are available in the database, and one representative set of data is shown. Data for dry (Stage 6) and imbibed seeds (Stages 7–9) were obtained from microarray experiments published by Nakabayashi *et al.* (2005). After-ripened dry seeds were imbibed for 6, 12, or 24 h under continuous white light at 22°C. Similar results were obtained from two biological replicates, and one representative set of data is shown. Identities of ABA metabolism genes are summarized in Table 9.1. Question marks indicate that the corresponding gene has not been identified yet. Abbreviations: GGDP, geranylgeranyl diphosphate; ZEP, zeaxanthin epoxidase; NCED, 9-*cis*-epoxycarotenoid dioxygenase; MoCo, molybdenum cofactor; SDR, short-chain alcohol dehydrogenase/reductase; AAO, *Arabidopsis* aldehyde oxidase; ABA-GT, ABA-glucosyltransferase.

Table 9.1 ABA biosynthesis and deactivation genes in *Arabidopsis*

Enzyme	Gene	AGI code
ZEP	*AtABA1*	At5g67030
NCED	*AtNCED2*	At4g18350
	AtNCED3	At3g14440
	AtNCED5	At1g30100
	AtNCED6	At3g24220
	AtNCED9	At1g78390
Xanthoxin dehydrogenase	*AtABA2*	At1g52340
MoCo sulfurase	*AtABA3*	At1g16540
Abscisic aldehyde oxidase	*AAO3*	At2g27150
ABA 8′-hydroxylase	*CYP707A1*	At4g19230
	CYP707A2	At2g29090
	CYP707A3	At5g45340
	CYP707A4	At3g19270
ABA-GT	*UGT71B6*	At3g21780
	UGT73B1	At4g34138
	UGT73B3	At4g34131
	UGT74D1	At2g31750
	UGT75B1	At1g05560
	UGT75B2	At1g05530
	UGT84B1	At2g23260
	UGT84B2	At2g23250

Table 9.2 Gibberellin biosynthesis and deactivation genes in *Arabidopsis*

Enzyme	Gene	AGI code
CPS	*AtCPS* (*GA1*)	At4g02780
KS	*AtKS* (*GA2*)	At1g79460
KO	*AtKO* (*GA3*)	At5g25900
KAO	*AtKAO1*	At1g05160
	AtKAO2	At2g32440
GA 20-oxidase	*AtGA20ox1* (*GA5*)	At4g25420
	AtGA20ox2	At5g51810
	AtGA20ox3	At5g07200
	AtGA20ox4	At1g60980
	AtGA20ox5	At1g44090
GA 3-oxidase	*AtGA3ox1* (*GA4*)	At1g15550
	AtGA3ox2	At1g80340
	AtGA3ox3	At4g21690
	AtGA3ox4	At1g80330
GA 2-oxidase (Class I and II)[a]	*AtGA2ox1*	At1g78440
	AtGA2ox2	At1g30040
	AtGA2ox3	At2g34555
	AtGA2ox4	At1g47990
	AtGA2ox5	Pseudogene
	AtGA2ox6	At1g02400
GA 2-oxidase (Class III)[b]	*AtGA2ox7*	At1g50960
	AtGA2ox8	At4g21200

[a] GA 2-oxidases of Class I and II use C_{19}-GAs (such as $GA_{1,4,9,20}$) as substrates.
[b] GA 2-oxidases of Class III use C_{20}-GAs (such as $GA_{12,53}$) as substrates.

from plastid to cytosol. To date, the genes encoding these cytosolic enzymes have been identified only from *Arabidopsis*. The *AtABA2* gene encodes a member of the SDR family that catalyzes the conversion of xanthoxin to abscisic aldehyde (Rook *et al.*, 2001; Cheng *et al.*, 2002; González-Guzman *et al.*, 2002). The final step of ABA biosynthesis is oxidation of abscisic aldehyde to ABA, which is catalyzed by aldehyde oxidase (Figure 9.1). Among four *Arabidopsis* aldehyde oxidases (AAO1 to AAO4), AAO3 plays a specific role in ABA biosynthesis *in vivo* (Seo *et al.*, 2000). Aldehyde oxidase is a molybdenum cofactor (MoCo)-requiring enzyme. Therefore, mutations in the genes for MoCo biosynthesis also lead to ABA deficiency. Consistently, mutations in the *AtABA3* gene, which encodes a MoCo sulfurase, show typical ABA-deficient phenotypes, such as reduced seed dormancy and wiltiness (Figure 9.1 and Table 9.1; Bittner *et al.*, 2001; Xiong *et al.*, 2001).

9.2.2 ABA deactivation

ABA is deactivated through several pathways that are triggered by either hydroxylation or conjugation (Nambara and Marion-Poll, 2005). It has been shown that there are three different ABA hydroxylation pathways that oxidize one of the methyl groups (C-7′, C-8′, and C-9′) of the ring structure. The hydroxylation at position C-8′ is thought to be the predominant ABA deactivation pathway in many physiological processes (Cutler and Krochko, 1999; Zeevaart, 1999) (Figure 9.1), and is catalyzed by a cytochrome P450 monooxygenase (P450) (Krochko *et al.*, 1998). The 8′-hydroxy-ABA isomerizes spontaneously to phaseic acid (PA) and is further catabolized to dihydrophaseic acid (DPA) by an unknown soluble reductase (Gillard and Walton, 1976). Recent studies demonstrated that ABA 8′-hydroxylases are encoded by *CYP707A* genes (Kushiro *et al.*, 2004; Saito *et al.*, 2004). It has been shown that 8′-hydroxy-ABA is also oxidized further to 8′-hydroxy-PA or 8′-oxo-PA via several possible intermediates (Zaharia *et al.*, 2004). In addition, ABA is deactivated through conjugation pathways; the most predominant and widespread conjugate is ABA glucose ester (ABA-GE) (Figure 9.1). Conjugated forms of PA and DPA have also been identified. This makes it complicated to evaluate the physiological roles of these pathways based on the levels of the catabolites alone. Loss-of-function mutants of ABA deactivation genes might help elucidate the physiological role of each pathway.

An ABA glucosyltransferase gene (*AOG*) has been identified from adzuki bean (*Vigna angularis*) (Xu *et al.*, 2002). AOG was shown to catalyze the conjugation of UDP-D-glucose to ABA, as well as to 2-*trans*-ABA, (−)-R-ABA, and cinnamic acid, but not to PA. Recently, Lim *et al.* (2005) carried out a comprehensive analysis on 105 *Arabidopsis* glycosyltransferases (GTs) using a heterologous expression system to find those that can attach UDP-D-glucose to ABA (Lim *et al.*, 2005). One GT, UGT71B6, showed conjugation activity specific to natural ABA ((+)-ABA), whereas seven other GTs showed ABA glucosylation activity both with ABA and its analogue (−)-ABA. UGT84B1, which shows the highest activity *in vitro* toward racemic ABA, is known to glucosylate indole acetic acid. Nevertheless, it is still unclear which GTs are involved in ABA deactivation *in vivo* or which GTs catalyze the glycosylation of ABA catabolites, such as hydroxylated ABAs, PA, and DPA.

9.2.3 ABA-deficient mutants and seed germination

Genes involved in ABA biosynthesis and deactivation are summarized in Table 9.1. To date, most ABA biosynthetic genes have been identified in *Arabidopsis* (Figure 9.1). Except for NCED, other known ABA biosynthesis enzymes are encoded by single genes in the *Arabidopsis* genome. Loss of function of such single genes (i.e., *AtABA1/ZEP*, *AtABA2/SDR*, and *AtABA3*) causes severe ABA-deficient phenotypes, such as enhanced germination ability under suboptimal conditions and reduced seed dormancy. Null alleles of these genes are still viable, and so other unknown minor ABA biosynthesis pathways may exist in plants (Barrero *et al.*, 2005). The mild phenotypes of *aao3* null alleles in seeds are attributed to the broad substrate specificity of other AAOs (Seo *et al.*, 2004). In contrast, mutants of *AtNCED* genes display mild and distinct phenotypes, supporting the idea that each member of this gene family regulates ABA biosynthesis in distinct physiological and developmental processes.

During seed development, *AtNCED6* and *AtNCED9* are expressed abundantly in immature seeds (Lefebvre *et al.*, 2006) (Figure 9.1). *AtNCED6* is expressed specifically in immature endosperm, whereas *AtNCED9* is expressed in the embryo and endosperm during seed development. The *nced6* and *nced9* mutants show reduced ABA levels in dry seeds. Germination of these mutants is resistant to PAC. On the other hand, other *AtNCED*s are also expressed weakly or in a particular organ in developing siliques (Tan *et al.*, 2003; Seo *et al.*, 2004). The *nced3* mutant was identified in a screen for enhanced germination on hypertonic media (Ruggiero *et al.*, 2004). This suggests that each AtNCED is involved in the regulation of ABA biosynthesis in seeds under particular conditions. In contrast, ABA deactivation genes *CYP707A1* and *CYP707A2* are abundantly expressed during seed development and after imbibition, respectively (Kushiro *et al.*, 2004). Consistently, the *cyp707a1* and *cyp707a2* mutants exhibited enhanced seed dormancy (Okamoto *et al.*, 2006).

9.2.4 GA biosynthesis

ABA and GAs share GGDP as a common C_{20} biosynthetic precursor (Figures 9.1 and 9.2). Among over one hundred GAs identified from plants (MacMillan, 2002), only a small number of them, such as GA_1 and GA_4, are thought to act as growth hormones (bioactive GAs). Many non-bioactive GAs are precursors for the bioactive forms or deactivated metabolites. The conversions of GGDP into bioactive GAs can be classified into three stages, based on the types of enzymes and their subcellular compartments (Figure 9.2; Hedden *et al.*, 2002; Olszewski *et al.*, 2002).

In the first stage of the pathway, two terpene cyclases, *ent*-copalyl diphosphate synthase (CPS) and *ent*-kaurene synthase (KS), are involved in the conversion of GGDP to *ent*-kaurene, a tetracyclic hydrocarbon (Figure 9.2). These reactions are thought to occur in the plastid, because CPS and KS are localized to this organelle. In the second stage of the pathway, *ent*-kaurene is oxidized to form GA_{12} by two P450s, CYP701A and CYP88A, which are located on the plastid envelope and endoplasmic reticulum, respectively. Each of these P450s catalyzes three consecutive oxidation reactions (Figure 9.2). In the final stages of GA biosynthesis, two classes of 2-oxoglutarate-dependent dioxygenases (2ODDs) are involved in the conversions of GA_{12} to bioactive GA_4. GA 20-oxidase catalyzes multiple oxygenation reactions at C-20 of GA_{12} to produce a C_{19}-GA (due to a loss of C-20 as CO_2), GA_9, which is then converted to bioactive GA_4 by GA 3-oxidase (Figure 9.2). In *Arabidopsis*, GA_4 is thought to be the major bioactive GA, although GA_1 (13-hydroxylated GA_4) is also found at lower levels (Talon *et al.*, 1990; Derkx *et al.*, 1994; Ogawa *et al.*, 2003). When applied exogenously to *Arabidopsis* seeds, GA_4 is approximately 10 times more active than GA_1 (Yang *et al.*, 1995), supporting the idea that GA_4 functions as the main bioactive GA in this species. The gene for GA 13-hydroxylase has not been identified yet in any plant species, and a mutant defective in this enzyme will be necessary to understand the biological meaning of GA 13-hydroxylation.

9.2.5 *GA deactivation*

GA 2-oxidases belong to another group of 2ODDs and are responsible for deactivation of GAs (Figure 9.2). Conventional GA 2-oxidases (recently classified as Class I and II by Lee and Zeevaart, 2005) use bioactive GAs, such as GA_4, GA_1, and their immediate precursors (GA_9 and GA_{20}) as substrates. Recently, a new class of 2ODDs has been shown to function as additional GA 2-oxidases, including AtGA2ox7 and AtGA2ox8 of *Arabidopsis* (Schomburg *et al.*, 2003; Lee and Zeevaart, 2005; Figure 9.2). Unlike the original GA 2-oxidases that utilize C_{19}-GAs as substrates (Thomas *et al.*, 1999; Hedden *et al.*, 2002), the preferred substrates for AtGA2ox7 and AtGA2ox8 are C_{20}-GAs (such as GA_{12} and GA_{53}), earlier intermediates in the pathway (Figure 9.2). These new GA 2-oxidases may play a role in controlling the level of bioactive GAs by reducing the supply of precursor GAs. However, transcript levels of *AtGA2ox7* and *AtGA2ox8* appear consistently low during seed development and germination (Figure 9.2), and the role of these new GA 2-oxidases in seeds is unclear.

GA 2-oxidation has been the only GA deactivation reaction for which the genes have been identified and characterized well so far. Recently, a previously uncharacterized P450, CYP714D1, was shown to be GA 16α,17-epoxidase and act as a new GA-deactivation enzyme in rice internodes (Zhu *et al.*, 2006). This finding indicates the diversity of GA deactivation mechanisms. The role of CYP714 family members (CYP714A1 and CYP714A2) in *Arabidopsis* has yet to be clarified.

9.2.6 *GA-deficient mutants and seed germination*

In *Arabidopsis*, the enzymes involved in the early steps in the GA biosynthetic pathway are encoded by single genes (Table 9.2), and loss-of-function mutants of CPS (*ga1*), KS (*ga2*), or KO (*ga3*) exhibit severe GA-deficient phenotypes, including the inability of seed to germinate in the absence of exogenous GA (Koornneef and van der Veen, 1980). In comparison to early GA biosynthesis enzymes, 2ODDs that catalyze the late steps in the pathway are encoded by small gene families in *Arabidopsis* (Table 9.2). The relative transcript levels of these genes indicate that they are differentially regulated during seed development and germination (Figure 9.2). Among four *AtGA3ox* genes, *AtGA3ox1* and *AtGA3ox2* are expressed at relatively high levels in seeds imbibed in the light. Both *ga3ox1* and *ga3ox2* single mutants germinate normally under continuous white light, but the *ga3ox1 ga3ox2* double mutant fails to germinate (Mitchum *et al.*, 2006). These results indicate that *AtGA3ox1* and *AtGA3ox2* encode major GA 3-oxidases that function redundantly during germination in the light. This is supported by the observation that these two *AtGA3ox* genes show similar cell type-specific expression patterns in seeds imbibed in the light (Yamaguchi *et al.*, 2001). The roles of *AtGA3ox1* and *AtGA3ox2* genes in imbibed seeds will be discussed again below in relation to environmental regulation of seed germination.

9.3 Inhibitors of ABA and GA metabolism: efficacy and side effects of drugs

Chemical inhibitors that block the ABA or GA metabolism pathways have been useful to elucidate the role of these hormones in seeds and to identify important signaling components in hormone-regulated seed germination through genetic screens. We will first describe chemicals that inhibit ABA metabolism enzymes, including those recently designed. Then, we will briefly describe the GA biosynthesis inhibitors that have been commonly used for seed biology research.

Many chemicals are known to block ABA biosynthesis and deactivation in plants. Carotenoid biosynthesis inhibitors are used most frequently in order to reduce endogenous ABA levels (Figure 9.3). In addition, inhibitors of AOs, such as tungstate, have been reported to reduce ABA levels (Figure 9.3; Lee and Milborrow, 1997). On the other hand, some P450 inhibitors are known to increase ABA levels by inhibiting ABA 8′-hydroxylases. Moreover, some ABA analogues have been postulated as specific inhibitors of ABA deactivation enzymes, such as ABA 8′-hydroxylases or ABA-GTs. Cloning of ABA biosynthesis and deactivation enzyme genes allows us to search for better inhibitors or evaluate the modes of action of these chemicals by using heterologous expression systems.

9.3.1 Drugs to reduce endogenous ABA levels

The herbicides fluridone and norflurazon have been widely used to reduce endogenous ABA levels in plants. These chemicals block the activity of phytoene desaturase in carotenoid biosynthesis (Figure 9.3). Because of their high competence of penetration and inhibitory effect on the enzyme activity, these chemicals have been commonly used in seed biology research. Fluridone is known to alleviate primary and secondary dormancy in seeds, such as primary dormancy of tobacco (*Nicotiana plumbaginifolia*) (Grappin *et al.*, 2000), *Arabidopsis* (Ali-Rachedi *et al.*, 2004), and potato (*Solanum tuberosum*) seeds (Alvarado and Bradford, 2005), and

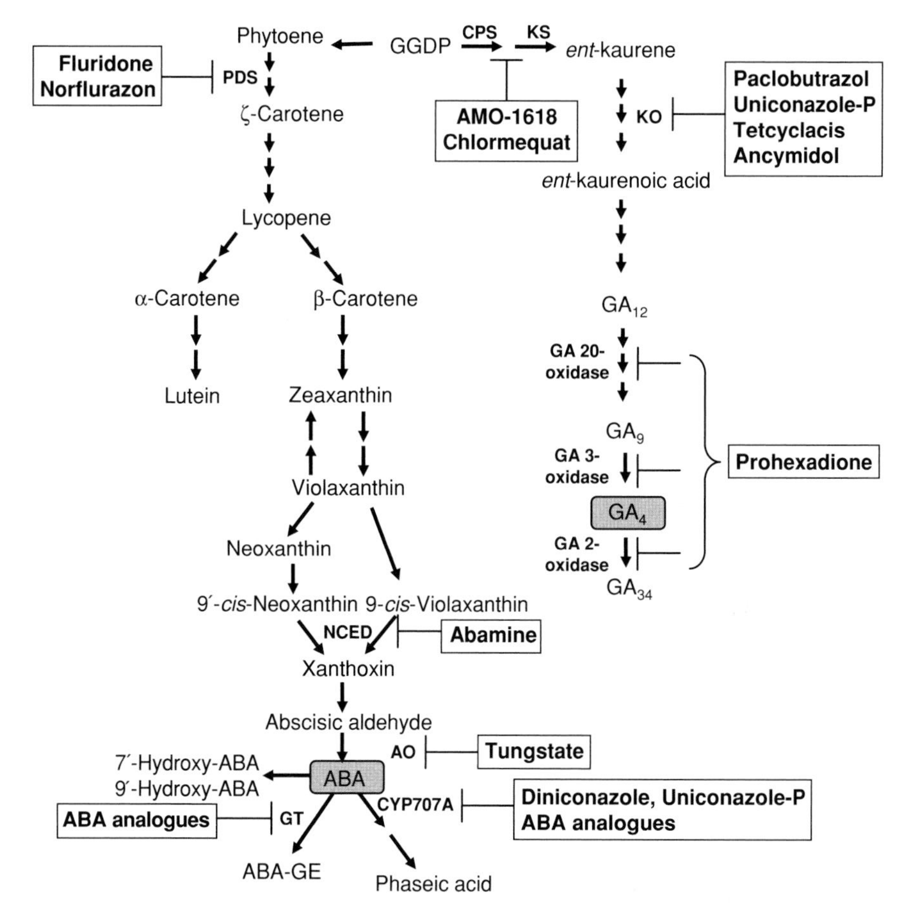

Figure 9.3 Inhibitors and their target steps in ABA and GA metabolic pathways. Target sites of the inhibitors (boxed) used for reducing and increasing ABA and GA levels are shown. Abbreviations: PDS, phytoene desaturase; NCED, 9-*cis*-epoxycarotenoid dioxygenase; AO, aldehyde oxidase; GT, glucosyltransferase; CPS, *ent*-copalyl diphosphate synthase; KS, *ent*-kaurene synthase; KO, *ent*-kaurene oxidase; KAO, *ent*-kaurenoic acid oxidase.

secondary dormancy induced by supraoptimal temperature in lettuce (*Lactuca sativa*) seeds (Yoshioka *et al.*, 1998) and by osmotic stress in oilseeds (*Brassica napus*) (Gulden *et al.*, 2004). Carotenoids are also considered to be precursors of several signal compounds, such as strigolactones (Matusova *et al.*, 2005) and MAX (more axillary)-dependent substance (Sorefan *et al.*, 2003; Booker *et al.*, 2004), an unknown mobile signal produced in roots (Van Norman *et al.*, 2004). Although it remains unknown whether these chemicals are involved in seed physiology, it is recommended to design experiments using several different inhibitors that block different steps in the metabolic pathway.

Fluridone application leads to both reduction in ABA levels and bleaching. Because carotenoids play an essential role in protecting against photooxidative damage, fluridone is lethal to green plants in the light. Recently, an ABA biosynthesis inhibitor that blocks NCED activity was developed (Figure 9.3; Han *et al.*, 2004b). Abamine (ABA biosynthesis inhibitor with an amine moiety) was identified through comprehensive analysis of the derivatives of an uncharacterized ABA biosynthesis inhibitor nordihydroguaiaretic acid, which structurally resembles lignostilbene-α, β-dioxygenase (LSD) inhibitors (Han *et al.*, 2004a,b). LSD catalyzes oxidative cleavage of the central double bond of stilbene in a manner similar to the oxidative cleavage by NCEDs. Abamine-treated spinach (*Spinacia oleracea*), cress (*Lepidium sativum*), and *Arabidopsis* show a reduction in osmotic stress-induced ABA synthesis without bleaching (Han *et al.*, 2004b). This chemical might be a useful tool to investigate ABA physiology, especially in plant species where genetic analysis is difficult.

9.3.2 Drugs to increase endogenous ABA levels

P450 inhibitors have been used as plant growth regulators (Rademacher, 2000). These are largely categorized into two types: ones known as plant growth retardants that inhibit GA biosynthesis and others that act as fungicides by blocking sterol biosynthesis. It is also known that application of some P450 inhibitors leads to enhanced stress tolerance (Fletcher *et al.*, 2000). The inhibitor treatment often causes the alteration of endogenous ABA levels, although it depends on the plant species, organs, and timing of inhibitor applications. The mechanism of ABA accumulation has been extensively examined in the case of tetcyclacis, a norbornanodiazetine-type P450 inhibitor. Radiotracer experiments demonstrated that the conversion of ABA to PA is impaired in tetcyclacis-treated *Xanthium strumarium* leaves (Zeevaart *et al.*, 1990). Likewise, tetcyclacis and a triazole BAS111.W appear to block ABA deactivation in suspension cultures and seedlings of oilseed (*Brassica napus*) (Häuser *et al.*, 1990). Tetcyclacis is shown to inhibit ABA 8′-hydroxylase activity of microsomal fractions from maize (*Zea mays*) suspension cells (Krochko *et al.*, 1998) or from yeast expressing *Arabidopsis* CYP707A1 (Kushiro *et al.*, 2004). Recently, comprehensive analysis was conducted on known P450 inhibitors to investigate their inhibitory effects on ABA deactivation (Kitahata *et al.*, 2005). A fungicide diniconazole was found to inhibit ABA 8′-hydroxylase more effectively than the inhibitors known as GA biosynthesis inhibitors, such as uniconazole, PAC, and tetcyclacis (Figure 9.3). Interestingly, the S-form is more effective as an ABA 8′-hydroxylase inhibitor than the R-form, which is a more potent fungicide. This indicates that the S-form of diniconazole is a useful tool to increase endogenous ABA levels.

Structural analogues of ABA have been synthesized and studied for their structure–activity relationships (Todoroki and Hirai, 2002; Zaharia *et al.*, 2005). These analogues include those exhibiting reduced or increased ABA-like activity and persistent analogues that are resistant to ABA deactivation. Cutler *et al.* (2000) demonstrated that ABA analogues modified at positions C-8′ or C-9′ inhibit ABA 8′-hydroxylase activity in microsomal fractions of maize suspension cells (Cutler *et al.*, 2000). Among them, (+)-8′-acetylene-ABA and (+)-9′-propargyl-ABA were shown to act as suicide substrates and inhibit *Arabidopsis* seed germination more effectively than natural ABA. This suggests that these analogues contain both inhibitory activity toward ABA 8′-hydroxylase and ABA-like biological activity. Ueno *et al.* (2005) conducted a comprehensive analysis of ABA analogues concerning their inhibitory effects on both ABA 8′-hydroxylase and hormone activities (Ueno *et al.*, 2005). Elimination of the 6-methyl and 1′-hydroxy groups had little effect on competitive inhibition of ABA 8′-hydroxylase, but abolished ABA-like biological activity. The authors suggested that it is possible to design specific ABA 8′-hydroxylase inhibitors that do not have ABA-like activity. On the other hand, Priest *et al.* (2005) reported that recognition of ABA by UGT71B6, an *Arabidopsis* GT, does not require the ketone at C-4 and is unaffected by addition of bulky substitution groups at the double bond of the ABA ring (Priest *et al.*, 2005). These designed inhibitors will be powerful tools in the future research on ABA physiology in various plant species.

9.3.3 Drugs to reduce GA levels

As mentioned above, three different classes of enzymes are involved in the conversion of GGDP to bioactive GAs (Figure 9.2). Chemical inhibitors are known for each enzyme class in the GA biosynthesis pathway (Figure 9.3; Rademacher, 2000). The most commonly used GA biosynthesis inhibitors for seed biology are PAC and uniconazole-P, both of which inhibit KO/CYP701A (GA3) and belong to a group of triazole-containing heterocycles. These chemicals also exhibit an inhibitory effect on ABA deactivation enzymes and reduce brassinosteroid levels (Rademacher, 2000). Therefore, it is important to evaluate the specificity as a GA biosynthesis inhibitor by including a control sample to test whether the effect of the chemical is reversed by simultaneous application of exogenous GA. Prohexadione, a structural mimic of 2-oxoglutarate, also inhibits *Arabidopsis* seed germination (Yang *et al.*, 1995). The inhibitory effect of prohexadione on *Arabidopsis* seed germination is efficiently restored by exogenous GA_4, but not by GA_9, consistent with the idea that GA 3-oxidase is a target of this inhibitor (Figure 9.3).

9.3.4 Side effects of drugs

Inhibitors of ABA/GA biosynthesis and deactivation are powerful tools to investigate physiological roles of these hormones in various plant species in which genetic approaches cannot be easily employed. However, we should also consider that many so-called specific inhibitors are not necessarily specific. In addition, the effects and side effects of inhibitors vary when applied by different methods, to different organs or at different concentrations, even in the same plant species. Debeaujon and Koornneef (2000) reported an example of inconsistent results from similar inhibitors. The testa (seed coat) plays an important role in maintaining seed dormancy by protecting embryos both physically and chemically. Thus, *Arabidopsis* mutants having defective testae often show reduced seed dormancy and increased ability to take up chemicals from media (Debeaujon *et al.*, 2000). PAC inhibits germination of all testa mutants more effectively than it inhibits wild-type seeds, but only some of the testa mutants show hypersensitivity to tetcyclacis. This suggests that wild-type seed coats inhibit uptake of such inhibitors. In addition, different effects of PAC and tetcyclacis may be attributed to differences in inhibitory effects on GA biosynthesis or ABA deactivation. Nevertheless, our research is often dependent on such drugs, even though special caution is needed to design such experiments.

9.4 Regulation of ABA and GA levels in *Arabidopsis seeds*

9.4.1 Regulation of ABA and GA levels during seed development

Recent molecular genetic analyses have revealed that the actions of ABA and GAs are modulated directly by developmental regulators that define the 'space' or 'timing' of embryo development. These regulators act as 'balancers' of the metabolism of and sensitivities to these two hormones, which, at least in part, explains the antagonistic actions of ABA and GAs.

9.4.1.1 Roles of ABA and GA during seed development

Both ABA and GA are necessary for embryo growth during seed development. Overexpression of a GA deactivation enzyme gene (pea GA 2-oxidase2) in *Arabidopsis* increases the proportion of seed abortion (Singh *et al.*, 2002). Severe ABA-deficient *aba2* mutants also show an increase in seed abortion compared to the wild type (Cheng *et al.*, 2002). In addition to their roles in embryo growth, ABA and GA antagonistically regulate the degree of seed dormancy during seed development. Precocious germination of *Arabidopsis lec* (*leafy cotyledon*) and *fus* (*fusca*) mutants is suppressed by a GA-deficient mutation, supporting the idea that GA biosynthesis is regulated negatively during seed development to prevent precocious germination (Raz *et al.*, 2001). In contrast, ABA accumulation during seed development is important to induce seed maturation processes and impose seed dormancy. Biphasic ABA accumulation has been observed during seed development (Figure 9.4; Karssen *et al.*, 1983). The first peak of ABA is derived from both zygotic and maternal tissues, while the second peak is derived only from the zygotic tissues. This zygotic-derived ABA is thought to be essential for the induction and maintenance of seed dormancy. In contrast, the maternal-derived ABA is involved in the inhibition of precocious germination and processes of seed maturation (Koornneef *et al.*, 1989; Raz *et al.*, 2001).

9.4.1.2 FUS3, a balancer of ABA and GA levels

The levels of ABA and bioactive GAs are thought to be correlated negatively during seed development (Jacobsen and Chandler, 1987; Batge *et al.*, 1999; White *et al.*, 2000). Recent reports have demonstrated that ABA and GA levels are regulated oppositely by the same transcription factors during *Arabidopsis* seed development. *FUS3* and *LEC2* encode B3 transcription factors (Luerßen *et al.*, 1998; Stone *et al.*, 2001). Loss of function of these genes causes severe defects in seed maturation processes, such as reduced reserve accumulation and desiccation intolerance

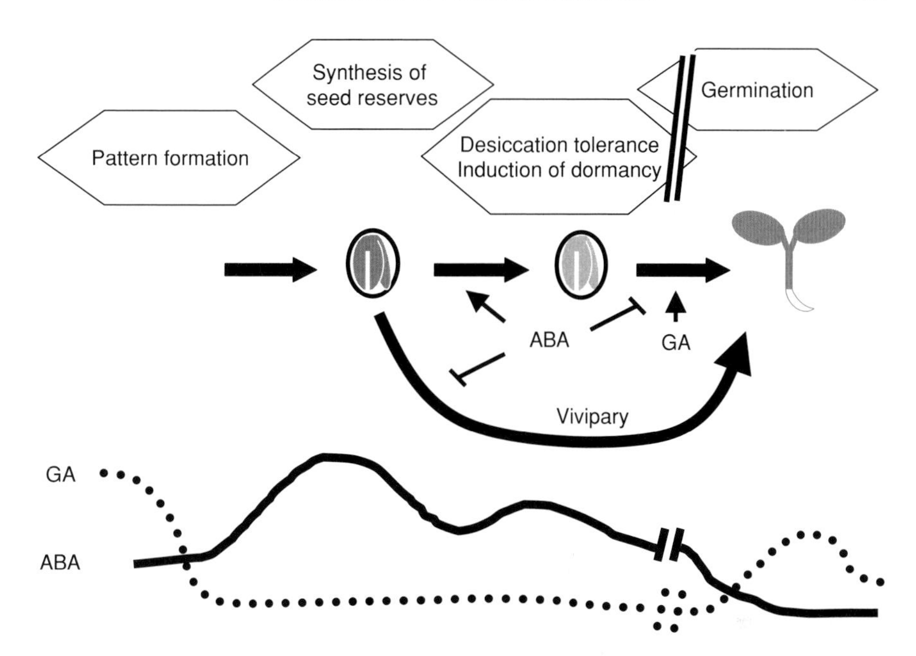

Figure 9.4 Seed developmental phases and hormone levels. Pattern formation occurs during embryogenesis to establish the major tissue types in the seed. Relative hormone levels are indicated at the bottom of the figure. ABA accumulation occurs during mid-maturation stage defined by expression of seed storage protein genes. The late embryo development or post-abscission stage is defined by the expression of *Lea* genes. Acquisition of desiccation tolerance and induction of dormancy occur at this stage, coincident with the second peak of ABA accumulation. ABA deficiency during seed maturation can result in precocious germination or vivipary. Bioactive GA levels during *Arabidopsis* seed development have not been investigated in detail. After seed imbibition, a decrease in the ABA level occurs prior to an increase in the bioactive GA level.

(Meinke *et al.*, 1994). Curaba *et al.* (2004) reported that levels of bioactive GAs increased in immature seeds of the *fus3* and *lec2* mutants (Curaba *et al.*, 2004). The immature embryos of these mutants were found to have higher levels of *AtGA20ox1* and *AtGA3ox2* transcripts than those of wild-type plants. In these mutants, the misexpression of *AtGA3ox2* was observed in the epidermis of embryonic axis and vascular tissues, which coincide with the localization of *FUS3* expression (Tsuchiya *et al.*, 2004). Furthermore, FUS3, but not LEC2, interacts physically with the RY repeats located in the *AtGA3ox2* promoter, presumably to repress its expression (Curaba *et al.*, 2004). Gazzarrini *et al.* (2004) found that the transient induction of *FUS3* resulted in ABA accumulation and repression of the GA biosynthetic genes, *AtGA20ox1* and *AtGA3ox1* (Gazzarrini *et al.*, 2004). These results indicate that FUS3 is a positive regulator of ABA levels and is also a negative regulator of GA levels (Figure 9.5A). Consistent with these roles for FUS3, the *fus3* mutant exhibits delayed ABA accumulation during seed development. Importantly, FUS3 protein levels appear to be regulated by ABA and GA in a positive and negative way, respectively. These regulatory loops support the role of FUS3 as a metabolic switch that

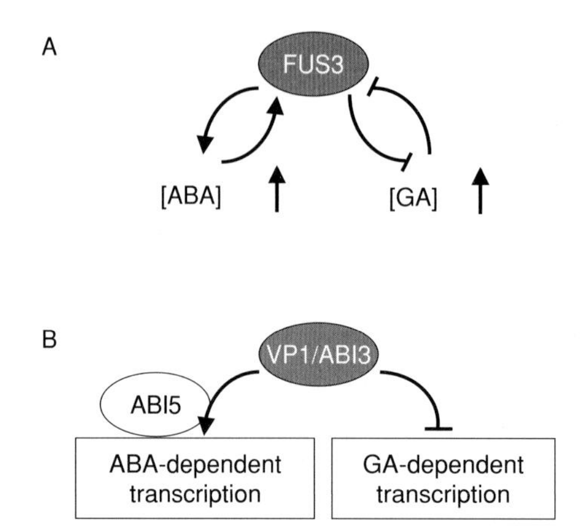

Figure 9.5 The B3 transcription factors regulating ABA and GA metabolism and sensitivity. (A) FUS3 regulates ABA accumulation positively, whereas it regulates GA biosynthesis negatively (Gazzarrini *et al.*, 2004). FUS3 regulation of GA biosynthesis is in part through repression of *AtGA3ox2* gene expression (Curaba *et al.*, 2004). FUS3 protein level is regulated by ABA positively and by GA negatively. Such positive and negative loops might enable FUS3 to function as a metabolic switch regulating ABA and GA levels during seed development. (B) Maize VP1 is proposed to play bifunctional roles in both the enhancement of ABA-dependent gene expression and the repression of GA-dependent gene expression. VP1/ABI3 is thought to be a switch for ABA and GA sensitivities.

regulates ABA and GA levels in an opposing manner. The *fus3* and *lec2* mutants differentially overaccumulate bioactive GAs, GA_1 and GA_4, suggesting that FUS3 and LEC2 regulate GA biosynthesis through distinct mechanisms (Curaba *et al.*, 2004).

In addition to metabolic regulation, sensitivities to ABA and GA are also oppositely regulated by another B3 transcription factor, ABSCISIC ACID-INSENSITIVE3 (ABI3) (Giraudat *et al.*, 1992). The severe *abi3* mutants exhibit defects in both seed reserve accumulation and acquisition of desiccation tolerance (Nambara *et al.*, 1992; Ooms *et al.*, 1993). The maize orthologue VIVIPAROUS1 (VP1) has been shown to be involved in both the activation of ABA-induced transcription and the repression of GA-induced transcription (Hoecker *et al.*, 1995; Suzuki *et al.*, 1997) (Figure 9.5B). ABI3 has also been suggested to play a similar role during seed development (Nambara *et al.*, 2000). Therefore, B3 transcription factors play key roles in balancing both the metabolism of and the sensitivity to ABA and GA during seed development.

9.4.1.3 AGL15, a transcriptional regulator of a GA deactivation gene

AGL15 (AGAMOUS-LIKE15) is a member of the MADS domain family of DNA-binding transcriptional regulators that is expressed at relatively high levels during embryo development (Wang *et al.*, 2002). *AtGA2ox6*, a GA deactivation gene (Figure 9.2), has been identified as a direct target of AGL15 by chromatin immunoprecipitation (Wang *et al.*, 2002). AGL15 bound to the regulatory region of the

AtGA2ox6 gene in a sequence-specific manner *in vitro*, and activated the expression of *AtGA2ox6* in *Arabidopsis* when it was overexpressed. A *ga2ox6* knockdown mutation improved the germination of freshly harvested seeds. In addition, after-ripened *ga2ox6* mutant seeds were more resistant to PAC relative to wild-type seeds (Wang *et al.*, 2004). These results indicate that *AtGA2ox6* plays a role in controlling GA levels during seed germination.

9.4.2 Regulation of ABA metabolism during seed imbibition in Arabidopsis

Reduced seed dormancy of the ABA-deficient mutants might be attributed to ABA deficiency during both seed development and germination. The role of *de novo* ABA biosynthesis after seed imbibition has been investigated using ABA biosynthesis inhibitors. Norflurazon (Figure 9.3) enhances seed germination of *Arabidopsis* wild-type ecotypes, Landsberg *erecta* (L*er*) and Wassilevskija (Ws), which show weak seed dormancy (Debeaujon and Koornneef, 2000). Seed germination of the more dormant Cape Verde Islands (Cvi) ecotype is also induced by fluridone treatment (Ali-Rachedi *et al.*, 2004). During early seed imbibition, a rapid decrease in endogenous ABA levels occurs in both dormant (D) and nondormant (ND) Cvi seeds. D seeds, but not ND seeds, show an increase in ABA levels 3 days after imbibition, which is maintained thereafter. Dormancy of Cvi seeds can be also alleviated by after-ripening, stratification, and nitrate application. In addition, dormancy of Cvi depends on the imbibition temperature. D seeds imbibed at high temperature (20–27°C) maintain dormancy, whereas imbibition of D seeds at lower temperature (such as 13°C) leads to germination and no increase in ABA levels after 3 days. Collectively, these data indicate that maintaining high ABA levels in imbibed seeds might be important for seed dormancy and also suggest that ABA biosynthesis and inactivation are regulated by environmental stimuli after seed imbibition.

ABA deactivation is also involved in regulating ABA levels after seed imbibition. *CYP707A2* mRNA is present in the dry seed, and expression of the gene is induced rapidly after seed imbibition (Kushiro *et al.*, 2004). Consistent with the expression analysis, a *cyp707a2* mutant overaccumulated ABA in dry seeds and high ABA levels were maintained after seed imbibition. Moreover, freshly harvested *cyp707a2* seeds exhibited hyperdormancy. These results indicate that CYP707A2 plays a key role in inactivating ABA during late maturation and germination. On the other hand, transcript levels of other CYP707As remained low in dry and early imbibed seeds. Expression of *CYP707A1* and *CYP707A3* was induced when seeds began to germinate (Kushiro *et al.*, 2004). Consistently, *cyp707a3* mutants did not show distinct phenotypes in seed germination per se, although early seedling growth of these mutants was ABA-hypersensitive (Umezawa *et al.*, 2006).

Ethylene promotes seed germination in many plant species (Kepczynski and Kepczynska, 1997). In agreement, *Arabidopsis* ethylene-insensitive mutants exhibit seed hyperdormancy (Beaudoin *et al.*, 2000; Ghassemian *et al.*, 2000). Simultaneous measurements of hormones and their catabolites indicated that the *etr1-2* (*ethylene receptor1-2*) mutant, a mutant defective in ethylene perception, overaccumulated

ABA in dry seeds (Chiwocha *et al.*, 2005; see Chapter 8). ABA levels of *etr1-2* imbibed seeds decreased immediately after seed imbibition, as occurred in the wild-type seeds, but were then maintained at higher level than in wild type. In contrast, ABA levels in the *etr1-2* imbibed seeds were equivalent to the wild type when seeds were stratified. This stratification-dependent change in ABA levels is consistent with germination phenotypes of the *etr1-2* seed, in which seed hyperdormancy is canceled by the stratification treatment. Moreover, stratified wild-type seeds accumulated a significant amount of ABA-GE, suggesting that the ABA glucosylation pathway is involved in regulating ABA levels during stratification. Recent identification of the putative ABA GT genes might facilitate elucidation of the physiological roles of each ABA inactivation pathway.

9.4.3 Regulation of GA metabolism during seed imbibition in Arabidopsis

GA biosynthesis in imbibed *Arabidopsis* seeds is strictly regulated by environmental conditions. When after-ripened *Arabidopsis* seeds were imbibed under continuous white light at optimal temperature (22°C), the level of bioactive GA_4 increased just prior to radicle emergence (Ogawa *et al.*, 2003). This increase in GA_4 content appears to be achieved through the coordinated regulation of GA biosynthesis and deactivation genes; several GA biosynthesis genes, such as *AtGA20ox3*, *AtGA3ox1*, and *AtGA3ox2*, are upregulated upon imbibition under continuous light, whereas expression of all known *GA 2-oxidase* genes remains at low levels (Figure 9.2; Ogawa *et al.*, 2003). In the following sections, we will briefly describe how GA biosynthesis and deactivation pathways are regulated by light and temperature in imbibed *Arabidopsis* seeds.

9.4.3.1 Regulation of GA biosynthesis by light

Phytochromes play an important role in light-dependent germination of *Arabidopsis* seeds (Shinomura, 1997). Among five phytochromes, PHYB is the major family member that is stored in dry seeds and is responsible for the typical photoreversible response to red (R) and far-red (FR) light shortly after the onset of imbibition (Shinomura *et al.*, 1994, 1996). When imbibed seeds are irradiated with an FR-light pulse to inactivate phytochrome and incubated in the dark, expression of *AtGA3ox1* and *AtGA3ox2* remains at low levels. In such dark-imbibed seeds, transcript levels of *AtGA3ox1* and *AtGA3ox2* are highly elevated by an R pulse, and the effect of R is reversed by subsequent exposure to FR (Yamaguchi *et al.*, 1998). These results indicate that *AtGA3ox1* and *AtGA3ox2* genes are regulated by phytochrome. In the *phyB* mutant, *AtGA3ox2* expression is not increased by R, indicating the primary role of PHYB in mediating the induction of this gene (Yamaguchi *et al.*, 1998). Expression of all *AtGA2ox* genes remained at low levels in imbibed seeds under continuous white light (Figure 9.2; Ogawa *et al.*, 2003). *AtGA2ox2* transcript accumulated at high levels during imbibition in the dark after an FR pulse (Yamauchi *et al.*, 2004). These results suggest that GA deactivation also is modulated by light through regulation of the *AtGA2ox2* gene.

9.4.3.2 Regulation of GA biosynthesis by cold temperature

Exposure of imbibed seeds to cold temperature (cold stratification) often accelerates seed germination (Bewley and Black, 1982). Recently, large-scale expression analysis using GeneChip microarrays revealed that a fraction of GA biosynthesis genes, such as *AtGA3ox1*and *AtGA20ox2*, were upregulated in response to cold treatment in dark-imbibed *Arabidopsis* seeds (Yamauchi *et al.*, 2004). In addition, approximately a quarter of cold-responsive genes were found to be GA-regulated genes, suggesting a role for GA in mediating the cold temperature signal. In agreement with these changes in transcript levels, the amount of GA_4 in dark-imbibed seeds is elevated at 4°C compared to that at 22°C.

Interestingly, *AtGA3ox1* was the only *AtGA3ox* gene induced by cold among the family members (Yamauchi *et al.*, 2004). In the *ga3ox1* mutant, both the increase in the level of GA_4 and cold-stimulated seed germination were not observed, demonstrating the role of this particular family member in mediating the temperature signal. The level of GA_4 at 4°C was significantly higher in comparison with that at 22°C, but the increase was only approximately two-fold (Yamauchi *et al.*, 2004). Nevertheless, the failure of GA-upregulated genes to be induced at 4°C in the *ga3ox1* mutant suggests that the relatively small increase in GA_4 content is necessary to activate the GA response pathway for germination of wild-type seeds. It was previously indicated that cold treatment increases the sensitivity of GA-deficient mutant seeds to exogenously applied GA (Derkx and Karssen, 1993). Therefore, a change in tissue sensitivity to GA might also play a role in the activation of the GA response pathway by the temperature signal.

9.5 Conclusions and perspectives

In this chapter, we have highlighted recent advances in our understanding of ABA and GA metabolism during seed development and germination in *Arabidopsis*. Identification of the majority of ABA and GA metabolism genes in this model species (Hedden *et al.*, 2002; Nambara and Marion-Poll, 2005) has enabled researchers to study how the amounts of these two hormones in seeds are regulated in specific developmental phases and under defined environmental conditions. To fully understand the regulation of ABA and GA levels, it will be necessary to identify additional uncharacterized ABA and GA metabolism enzymes, as these two hormones are likely to be deactivated via multiple pathways. Evidence has now been provided that well-studied developmental regulators, such as FUS3 and LEC2, play direct or indirect roles in modulating hormone levels in developing seeds (Curaba *et al.*, 2004; Gazzarrini *et al.*, 2004). Identification of additional key regulators of ABA and GA levels will increase our knowledge on how cellular concentrations of these antagonistic hormones are balanced during seed development and germination. The application of large-scale expression analysis to seed biology (Ogawa *et al.*, 2003; Nakabayashi *et al.*, 2005) has been a powerful tool to facilitate our understanding of the overall regulation of hormone metabolism pathways. Such genome-wide analyses in combination with molecular genetic studies will allow us to uncover new molecular links among different hormones as well as connections among hormonal and developmental or environmental regulators during seed development and germination.

References

S. Ali-Rachedi, D. Bouinot, M.H. Wagner, *et al.* (2004) Changes in endogenous abscisic acid levels during dormancy release and maintenance of mature seeds: studies with the Cape Verde Islands ecotype, the dormant model of *Arabidopsis thaliana*. *Planta* **219**, 479–488.

V. Alvarado and K.J. Bradford (2005) Hydrothermal time analysis of seed dormancy in true (botanical) potato seeds. *Seed Science Research* **15**, 77–88.

C. Audran, S. Liotenberg, M. Gonneau, *et al.* (2001) Localisation and expression of zeaxanthin epoxidase mRNA in *Arabidopsis* in response to drought stress and during seed development. *Australian Journal of Plant Physiology* **28**, 1161–1173.

J.M. Barrero, P. Piqueras, M.González-Guzman, *et al.* (2005) A mutational analysis of the *ABA1* gene of *Arabidopsis thaliana* highlights the involvement of ABA in vegetative development. *Journal of Experimental Botany* **56**, 2071–2083.

L.S. Batge, J.J. Ross and J.B. Reid, (1999) Abscisic acid levels in seeds of the gibberellin-deficient mutant *lh-2* of pea (*Pisum sativum*). *Physiologia Plantarum* **105**, 485–490.

N. Beaudoin, C. Serizet, F. Gosti and J. Giraudat (2000) Interactions between abscisic acid and ethylene signaling cascades. *The Plant Cell* **12**, 1103–1115.

J.D. Bewley and M. Black (1982) *Physiology and Biochemistry of Seeds: Viability, Dormancy and Environmental Control*, Vol. 2. Springer-Verlag, Berlin.

F. Bittner, M. Oreb and R.R. Mendel (2001) ABA3 is a molybdenum cofactor sulfurase required for activation of aldehyde oxidase and xanthine dehydrogenase in *Arabidopsis thaliana*. *Journal of Biological Chemistry* **276**, 40381–40384.

J. Booker, M. Auldridge, S. Wills, D. McCarty, H. Klee and O. Leyser (2004) MAX3/CCD7 is a carotenoid cleavage dioxygenase required for the synthesis of a novel plant signaling molecule. *Current Biology* **14**, 1232–1238.

W.H. Cheng, A. Endo, L. Zhou, *et al.* (2002) A unique short-chain dehydrogenase/reductase in *Arabidopsis* glucose signaling and abscisic acid biosynthesis and functions. *The Plant Cell* **14**, 2723–2743.

J.T. Chernys and J.A.D. Zeevaart, (2000) Characterization of the 9-*cis*-epoxycarotenoid dioxygenase gene family and regulation of abscisic acid biosynthesis in avocado. *Plant Physiology* **124**, 343–353.

S.D.S. Chiwocha, A.J. Cutler, S.R. Abrams, *et al.* (2005) The *etr1-2* mutation in *Arabidopsis thaliana* affects the abscisic acid, auxin, cytokinin and gibberellin metabolic pathways during maintenance of seed dormancy, moist-chilling and germination. *The Plant Journal* **42**, 35–48.

J. Curaba, T. Moritz, R. Blervaque, *et al.* (2004) *AtGA3ox2*, a key gene responsible for bioactive gibberellin biosynthesis, is regulated during embryogenesis by *LEAFY COTYLEDON2* and *FUSCA3* in *Arabidopsis*. *Plant Physiology* **136**, 3660–3669.

A.J. Cutler and J.E. Krochko (1999) Formation and breakdown of ABA. *Trends in Plant Science* **4**, 472–478.

A.J. Cutler, P.A. Rose, T.M. Squires, *et al.* (2000) Inhibitors of abscisic acid 8′-hydroxylase. *Biochemistry* **39**, 13614–13624.

P.J. Davies (2004) *Plant Hormones: Biosynthesis, Signal Transduction, Action!* Kluwer Academic Publishers, Dordrecht.

I. Debeaujon and M. Koornneef (2000) Gibberellin requirement for *Arabidopsis* seed germination is determined both by testa characteristics and embryonic abscisic acid. *Plant Physiology* **122**, 415–424.

I. Debeaujon, K.M. Léon-Kloosterziel and M. Koornneef (2000) Influence of the testa on seed dormancy, germination, and longevity in *Arabidopsis*. *Plant Physiology* **122**, 403–413.

M.P.M. Derkx and C.M. Karssen (1993) Effects of light and temperature on seed dormancy and gibberellin-stimulated germination in *Arabidopsis thaliana*: studies with gibberellin-deficient and gibberellin-insensitive mutants. *Physiologia Plantarum* **89**, 360–368.

M.P.M. Derkx, E. Vermeer and C.M. Karssen, (1994). Gibberellins in seeds of *Arabidopsis thaliana*: biological activities, identification and effects of light and chilling on endogenous levels. *Plant Growth Regulation* **15**, 223–234.

R.A. Fletcher, A. Gilley, N. Sankhla and T.D. Davis (2000) Triazoles as plant growth regulators and stress protectants. *Horticultural Reviews* **24**, 55–138.

S. Gazzarrini, Y. Tsuchiya, S. Lumba, M. Okamoto and P. McCourt (2004) The transcription factor *FUSCA3* controls developmental timing in *Arabidopsis* through the hormones gibberellin and abscisic acid. *Developmental Cell* **7**, 373–385.

M. Ghassemian, E. Nambara, S. Cutler, H. Kawaide, Y. Kamiya and P. McCourt (2000) Regulation of abscisic acid signaling by the ethylene response pathway in *Arabidopsis*. *The Plant Cell* **12**, 1117–1126.

D.F. Gillard and D.C. Walton (1976) Abscisic acid metabolism by a cell-free preparation from *Echinocystis lobata* liquid endosperm. *Plant Physiology* **58**, 790–795.

J. Giraudat, B.M. Hauge, C. Valon, J. Smalle, F. Parcy and H.M. Goodman (1992) Isolation of the *Arabidopsis ABI3* gene by positional cloning. *The Plant Cell* **4**, 1251–1261.

M. González-Guzman, N. Apostolova, J.M. Bellés, *et al.* (2002) The short-chain alcohol dehydrogenase ABA2 catalyzes the conversion of xanthoxin to abscisic aldehyde. *The Plant Cell* **14**, 1833–1846.

P. Grappin, D. Bouinot, B. Sotta, E. Miginiac and M. Jullien (2000) Control of seed dormancy in *Nicotiana plumbaginifolia*: post-imbibition abscisic acid synthesis imposes dormancy maintenance. *Planta* **210**, 279–285.

R.H. Gulden, S. Chiwocha, S. Abrams, I. McGregor, A. Kermode and S. Shirtliffe (2004) Response to abscisic acid application and hormone profiles in spring *Brassica napus* seed in relation to secondary dormancy. *Canadian Journal of Botany* **82**, 1618–1624.

S.Y. Han, N. Kitahata, T. Saito, *et al.* (2004a) A new lead compound for abscisic acid biosynthesis inhibitors targeting 9-*cis*-epoxycarotenoid dioxygenase. *Bioorganic and Medicinal Chemistry Letters* **14**, 3033–3036.

S.Y. Han, N. Kitahata, K. Sekimata, *et al.* (2004b) A novel inhibitor of 9-*cis*-epoxycarotenoid dioxygenase in abscisic acid biosynthesis in higher plants. *Plant Physiology* **135**, 1574–1582.

P. Hedden, A.L. Phillips, M. Cecilia Rojas, E. Carrera and B. Tudzynski (2002) Gibberellin biosynthesis in plants and fungi: a case of convergent evolution? *Journal of Plant Growth Regulation* **20**, 319–331.

N. Hirai, R. Yoshida, Y. Todoroki and H. Ohigashi (2000) Biosynthesis of abscisic acid by the non-mevalonate pathway in plants, and by the mevalonate pathway in fungi. *Bioscience Biotechnology and Biochemistry* **64**, 1448–1458.

U. Hoecker, I.K. Vasil and D.R. McCarty (1995) Integrated control of seed maturation and germination programs by activator and repressor functions of Viviparous-1 of maize. *Genes and Development* **9**, 2459–2469.

C. Häuser, J. Kwiatkowski, W. Rademacher and K. Grossmann (1990) Regulation of endogenous abscisic acid levels and transpiration in oilseed rape by plant growth retardants. *Journal of Plant Physiology* **137**, 201–207.

S. Iuchi, M. Kobayashi, T. Taji, *et al.* (2001) Regulation of drought tolerance by gene manipulation of 9-*cis*-epoxycarotenoid dioxygenase, a key enzyme in abscisic acid biosynthesis in *Arabidopsis*. *The Plant Journal* **27**, 325–333.

J.V. Jacobsen and P.M. Chandler, (1987) Gibberellin and abscisic acid in germinating cereals. In: *Plant Hormones and Their Role in Plant Growth and Development* (ed. P.J. Davies), pp. 164–193. Martinus Nijhoff, Boston.

C.M. Karssen, D.L.C. Brinkhorst van der Swan, A.E. Breekland and M. Koornneef (1983) Induction of dormancy during seed development by endogenous abscisic acid: studies on abscisic acid deficient genotypes of *Arabidopsis thaliana* (L) Heynh. *Planta* **157** 158–165.

H. Kasahara, K. Takei, N. Ueda, *et al.* (2004) Distinct isoprenoid origins of *cis-* and *trans*-zeatin biosyntheses in *Arabidopsis*. *Journal of Biological Chemistry* **279**, 14049–14054.

J. Kepczynski and E. Kepczynska (1997) Ethylene in seed dormancy and germination. *Physiologia Plantarum* **101**, 720–726.

N. Kitahata, S. Saito and Y. Miyazawa (2005) Chemical regulation of abscisic acid catabolism in plants by cytochrome P450 inhibitors. *Bioscience Biotechnology and Biochemistry* **13**, 4491–4498.

M. Koornneef, C.J. Hanhart, H.W.M. Hilhorst and C.M. Karssen (1989) *In vivo* inhibition of seed development and reserve protein accumulation in recombinants of abscisic acid biosynthesis and responsiveness mutants in *Arabidopsis thaliana*. *Plant Physiology* **90**, 463–469.

M. Koornneef and J.H. van der Veen (1980) Induction and analysis of gibberellin-sensitive mutants in *Arabidopsis thaliana* (L.) Heynh. *Theoretical and Applied Genetics* **58**, 257–263.

J.E. Krochko, G.D. Abrams, M.K. Loewen, S.R. Abrams and A.J. Cutler (1998) (+)-Abscisic acid 8′-hydroxylase is a cytochrome P450 monooxygenase. *Plant Physiology* **118**, 849–860.

B. Kucera, M.A. Cohn and G. Leubner-Metzger (2005) Plant hormone interactions during seed dormancy release and germination. *Seed Science Research* **15**, 281–397.

T. Kushiro, M. Okamoto, K. Nakabayashi, *et al.* (2004) The *Arabidopsis* cytochrome P450 CYP707A encodes ABA 8′-hydroxylases: key enzymes in ABA catabolism. *The EMBO Journal* **23**, 1647–1656.

D.J. Lee and J.A. Zeevaart (2005) Molecular cloning of *GA 2-oxidase3* from spinach and its ectopic expression in *Nicotiana sylvestris*.*Plant Physiology* **138**, 243–254.

H.-S. Lee, and B.V. Milborrow (1997) Endogenous biosynthetic precursors of (+)-abscisic acid. IV. Inhibition by tungstate and its removal by cinchonine shows that xanthoxal is oxidised by a molybdo-aldehyde oxidase. *Australian Journal of Plant Physiology* **24**, 727–732.

V. Lefebvre, H. North and A. Frey, *et al.* (2006) Functional analysis of *Arabidopsis NCED6* and *NCED9* genes indicates that ABA synthesised in the endosperm is involved in the induction of seed dormancy. *The Plant Journal* **45**, 309–319.

E.K. Lim, C.J. Doucet, B. Hou, R.G. Jackson, S.R. Abrams and D.J. Bowles (2005) Resolution of (+)-abscisic acid using an *Arabidopsis* glycosyltransferase. *Tetrahedron: Asymmetry* **16**, 143–147.

K. Lueßben, V. Kirik, P. Herrmann and S. Miséra (1998) *FUSCA3* encodes a protein with a conserved VP1/ABI3-like B3 domain which is of functional importance for the regulation of seed maturation in *Arabidopsis thaliana*. *The Plant Journal* **15**, 755–764.

J. MacMillan (2002) Occurrence of gibberellins in vascular plants, fungi and bacteria. *Journal of Plant Growth Regulation* **20**, 387–442.

R. Matusova, K. Rani, F.W.A. Verstappen, M.C.R. Franssen, M.H. Beale and H.J. Bouwmeester (2005) The strigolactone germination stimulants of the plant-parasitic *Striga* and *Orobanche* spp. are derived from the carotenoid pathway. *Plant Physiology* **139**, 920–934.

D.W. Meinke, L.H. Franzmann, T.C. Nickle and E.C. Yeung (1994) *leafy cotyledon* mutants of *Arabidopsis*. *The Plant Cell* **6**, 1049–1064.

B.V. Milborrow and H.S. Lee (1998) Endogenous biosynthetic precursors of (+)-abscisic acid. VI. Carotenoids and ABA are formed by the 'non-mevalonate' triose–pyruvate pathway in chloroplasts. *Australian Journal of Plant Physiology* **25**, 507–512.

M.G. Mitchum, S. Yamaguchi, A. Hanada, *et al.* (2006) Distinct and overlapping roles of two gibberellin 3-oxidases in *Arabidopsis* development. *The Plant Journal* **45**, 804–818.

K. Nakabayashi, M. Okamoto, T. Koshiba, Y. Kamiya and E. Nambara (2005) Genome-wide profiling of stored mRNA in *Arabidopsis thaliana* seed germination: epigenetic and genetic regulation of transcription in seed. *The Plant Journal* **41**, 697–709.

E. Nambara, R. Hayama, Y. Tsuchiya, *et al.* (2000) The role of *ABI3* and *FUS3* loci in *Arabidopsis thaliana* on phase transition from late embryo development to germination. *Developmental Biology* **220**, 412–423.

E. Nambara and A. Marion-Poll (2005) Abscisic acid biosynthesis and catabolism. *Annual Review of Plant Biology* **56**, 165–185.

E. Nambara, S. Naito and P. McCourt (1992) A mutant of *Arabidopsis* which is defective in seed development and storage protein accumulation is a new *abi3* allele. *The Plant Journal* **2**, 435–441.

M. Ogawa, A. Hanada, Y. Yamauchi, Y. Kuwahara, Y. Kamiya and S. Yamaguchi (2003) Gibberellin biosynthesis and response during *Arabidopsis* seed germination. *The Plant Cell* **15**, 1591–1604.

M. Okamoto, A. Kuwahara, M. Seo *et al.* (2006) CYP707A1 and CYP707A2, which encode ABA 8′-hydroxylases, are indispensable for a proper control of seed dormancy and germination in *Arabidopsis*. *Plant Physiology* **141**, 97–107.

N. Olszewski, T.-P. Sun and F. Gubler (2002) Gibberellin signaling: biosynthesis, catabolism, and response pathways. *The Plant Cell* **14**, S61–S80.

J.J.J. Ooms, K.M. Leonkloosterziel, D. Bartels, M. Koornneef and C.M. Karssen (1993) Acquisition of desiccation tolerance and longevity in seeds of *Arabidopsis thaliana* – a comparative study using abscisic acid-insensitive *abi3* mutants. *Plant Physiology* **102**, 1185–1191.

D.M. Priest, R.G. Jackson, D.A. Ashford, S.R. Abrams and D.J. Bowles (2005) The use of abscisic acid analogues to analyse the substrate selectivity of UGT71B6, a UDP-glycosyltransferase of *Arabidopsis thaliana*. *FEBS Letters* **579**, 4454–4458.

X. Qin and J.A.D. Zeevaart (1999) The 9-*cis*-epoxycarotenoid cleavage reaction is the key regulatory step of abscisic acid biosynthesis in water-stressed bean. *Proceedings of the National Academy of Sciences of the United States of America* **96**, 15354–15361.

W. Rademacher (2000) Growth retardants: effects on gibberellin biosynthesis and other metabolic pathways. *Annual Review of Plant Physiology and Plant Molecular Biology* **51**, 501–531.

V. Raz, J.H.W. Bergervoet and M. Koornneef (2001) Sequential steps for developmental arrest in *Arabidopsis* seeds. *Development* **128**, 243–252.

F. Rook, F. Corke, R. Card, G. Munz, C. Smith and M.W. Bevan (2001) Impaired sucrose-induction mutants reveal the modulation of sugar-induced starch biosynthetic gene expression by abscisic acid signaling. *The Plant Journal* **26**, 421–433.

B. Ruggiero, H. Koiwa, Y. Manabe, *et al.* (2004) Uncoupling the effects of abscisic acid on plant growth and water relations. Analysis of *sto1/nced3*, an abscisic acid-deficient but salt stress-tolerant mutant in *Arabidopsis*. *Plant Physiology* **136**, 3134–3147.

S. Saito, N. Hirai, C. Matsumoto, *et al.* (2004) *Arabidopsis CYP707A*s encode (+)-abscisic acid 8′-hydroxylase, a key enzyme in the oxidative catabolism of abscisic acid. *Plant Physiology* **134**, 1439–1449.

M. Schmid, T.S. Davison, S.R. Henz, *et al.* (2005) A gene expression map of *Arabidopsis thaliana* development. *Nature Genetics* **37**, 501–506.

F.M. Schomburg, C.M. Bizzell, D.J. Lee, J.A.D. Zeevaart and R.M. Amasino (2003) Overexpression of a novel class of gibberellin 2-oxidases decreases gibberellin levels and creates dwarf plants. *The Plant Cell* **15**, 151–163.

S.H. Schwartz, B.C. Tan, D.R. McCarty, W. Welch and J.A.D. Zeevaart (2003) Substrate specificity and kinetics for VP14, a carotenoid cleavage dioxygenase in the ABA biosynthetic pathway. *Biochimica et Biophysica Acta* **1619**, 9–14.

M. Seo, H. Aoki, H. Koiwai, Y. Kamiya, E. Nambara and T. Koshiba (2004) Comparative studies on the *Arabidopsis* aldehyde oxidase (*AAO*) gene family revealed a major role of *AAO3* in ABA biosynthesis in seeds. *Plant and Cell Physiology* **45**, 1694–1703.

M. Seo, A.J.M. Peeters, H. Koiwai, *et al.* (2000) The *Arabidopsis aldehyde oxidase 3* (*AA03*) gene product catalyzes the final step in abscisic acid biosynthesis in leaves. *Proceedings of the National Academy of Sciences of the United States of America* **97**, 12908–12913.

T. Shinomura (1997) Phytochrome regulation of seed germination. *Journal of Plant Research* **110**, 151–161.

T. Shinomura, A. Nagatani, J. Chory and M. Furuya (1994). The induction of seed germination in *Arabidopsis thaliana* is regulated principally by phytochrome B and secondarily by phytochrome A. *Plant Physiology* **104**, 363–371.

T. Shinomura, A. Nagatani, H. Hanzawa, M. Kubota, M. Watanabe and M. Furuya (1996) Action spectra for phytochrome A- and B-specific photoinduction of seed germination in *Arabidopsis thaliana*. *Proceedings of the National Academy of Sciences of the United States of America* **93**, 8129–8133.

D.P. Singh, A.M. Jermakow and S.M. Swain (2002) Gibberellins are required for seed development and pollen tube growth in *Arabidopsis*. *The Plant Cell* **14**, 3133–3147.

K. Sorefan, J. Booker, K. Haurogne, *et al.* (2003) *MAX4* and *RMS1* are orthologous dioxygenase-like genes that regulate shoot branching in *Arabidopsis* and pea. *Genes and Development* **17**, 1469–1474.

S.L. Stone, L.W. Kwong, K.M. Yee, *et al.* (2001) *LEAFY COTYLEDON2* encodes a B3 domain transcription factor that induces embryo development. *Proceedings of the National Academy of Sciences of the United States of America* **98**, 11806–11811.

M. Suzuki, C.Y. Kao and D.R. McCarty (1997) The conserved B3 domain of VIVIPAROUS1 has a cooperative DNA binding activity. *The Plant Cell* **9**, 799–807.

M. Talon, M. Koornneef and J.A.D. Zeevaart (1990) Endogenous gibberellins in *Arabidopsis thaliana* and possible steps blocked in the biosynthetic pathways of the semidwarf *ga4* and *ga5* mutants. *Proceedings of the National Academy of Sciences of the United States of America* **87**, 7983–7987.

B.C. Tan, L.M. Joseph, W.T. Deng, *et al.* (2003) Molecular characterization of the *Arabidopsis* 9-*cis*-epoxycarotenoid dioxygenase gene family. *The Plant Journal* **35**, 44–56.

S.G. Thomas, A.L. Phillips and P. Hedden (1999) Molecular cloning and functional expression of gibberellin 2-oxidases, multifunctional enzymes involved in gibberellin deactivation. *Proceedings of the National Academy of Sciences of the United States of America* **96**, 4698–4703.

Y. Todoroki and N. Hirai (2002) Abscisic acid analogs for probing the mechanism of abscisic acid perception and inactivation. *Studies in Natural Products Chemistry* **27**, 321–360.

Y. Tsuchiya, E. Nambara, S. Naito and P. McCourt (2004) The *FUS3* transcription factor functions through the epidermal regulator *TTG1* during embryogenesis in *Arabidopsis*. *The Plant Journal* **37**, 73–81.

K. Ueno, Y. Araki, N. Hirai, *et al.* (2005) Differences between the structural requirements for ABA 8′-hydroxylase inhibition and for ABA activity. *Bioorganic and Medicinal Chemistry* **13**, 3359–3370.

T. Umezawa, M. Okamoto, T. Kushiro, *et al.* (2006) CYP707A3, a major ABA 8′-hydroxylase involved in dehydration and rehydration response in *Arabidopsis thaliana*. *The Plant Journal* **46**, 171–182.

J.M. Van Norman, R.L. Frederick and L.E. Sieburth (2004) BYPASS1 negatively regulates a root-derived signal that controls plant architecture. *Current Biology* **14**, 1739–1746.

H. Wang, L.V. Caruso, B. Downie and S.E. Perry (2004) The embryo MADS domain protein AGAMOUS-like 15 directly regulates expression of a gene encoding an enzyme involved in gibberellin metabolism. *The Plant Cell* **16**, 1206–1219.

H. Wang, W. Tang, C. Zhu and S.E. Perry (2002) A chromatin immunoprecipitation (ChIP) approach to isolate genes regulated by AGL15, a MADS domain protein that preferentially accumulates in embryos. *The Plant Journal* **32**, 831–843.

C.N. White, W.M. Proebsting, P. Hedden and C.J. Rivin (2000) Gibberellins and seed development in maize. I: Evidence that gibberellin/abscisic acid balance governs germination versus maturation pathways. *Plant Physiology* **122**, 1081–1088.

L.M. Xiong, M. Ishitani, H. Lee and J.K. Zhu (2001) The *Arabidopsis LOS5/ABA3* locus encodes a molybdenum cofactor sulfurase and modulates cold stress- and osmotic stress-responsive gene expression. *The Plant Cell* **13**, 2063–2083.

L.M. Xiong, H.J. Lee, M. Ishitani and J.K. Zhu (2002) Regulation of osmotic stress-responsive gene expression by the *LOS6/ABA1* locus in *Arabidopsis*. *Journal of Biological Chemistry* **277**, 8588–8596.

Z.-J. Xu, M. Nakajima, Y. Suzuki and I. Yamaguchi (2002) Cloning and characterization of abscisic acid-specific glucosyltransferase gene from Adzuki bean seedlings. *Plant Physiology* **129**, 1285–1295.

S. Yamaguchi, Y. Kamiya and T.-P. Sun (2001) Distinct cell-specific expression patterns of early and late gibberellin biosynthetic genes during *Arabidopsis* seed germination. *The Plant Journal* **28**, 443–454.

S. Yamaguchi, M.W. Smith, R.G.S. Brown, Y. Kamiya and T.-P. Sun (1998) Phytochrome regulation and differential expression of gibberellin 3β-hydroxylase genes in germinating *Arabidopsis* seeds. *The Plant Cell* **10**, 2115–2126.

Y. Yamauchi, M. Ogawa, A. Kuwahara, A. Hanada, Y. Kamiya and S. Yamaguchi (2004) Activation of gibberellin biosynthesis and response pathways by low temperature during imbibition of *Arabidopsis* seeds. *The Plant Cell* **16**, 367–378.

Y.Y. Yang, A. Nagatani, Y.J. Zhao, B.J. Kang, R.E. Kentrick and Y. Kamiya (1995) Effects of gibberellins on seed germination of phytochrome-deficient mutants of *Arabidopsis thaliana*. *Plant and Cell Physiology* **36**, 1205–1211.

T. Yoshioka, T. Endo and S. Satoh (1998) Restoration of seed germination at supraoptimal temperatures by fluridone, an inhibitor of abscisic acid biosynthesis. *Plant and Cell Physiology* **39**, 307–312.

L.I. Zaharia, Y. Gai, K.M. Nelson, S.J. Ambrose and S.R. Abrams (2004) Oxidation of 8′-hydroxy abscisic acid in Black Mexican Sweet maize cell suspension cultures. *Phytochemistry* **65**, 3199–3209.

L.I. Zaharia, M.K. Walker-Simmons, C.N. Rodríguez and S.R. Abrams (2005) Chemistry of abscisic acid, abscisic acid catabolites and analogues. *Journal of Plant Growth Regulation* **24**, 274–284.

J.A.D. Zeevaart (1999) Abscisic acid metabolism and its regulation. In: *Biochemistry and Molecular Biology of Plant Hormones* (eds P.J.J. Hooykaas, M.A. Hall and K.R. Libbenga), pp. 189–207. Elsevier, New York.

J.A.D. Zeevaart, D.A. Gage and R.A. Creelman (1990) Recent studies of the metabolism of abscisic acid. In: *Plant Growth Substances 1988* (eds R.P. Pharis and S.B. Rood), pp. 232–240. Springer-Verlag, New York.

Y. Zhu, T. Nomura, Y. Xu, *et al.* (2006) *ELONGATED UPPERMOST INTERNODE* encodes a cytochrome P450 monooxygenase that epoxidizes gibberellins in a novel deactivation reaction in rice. *The Plant Cell* **18**, 442–456.

10 De-repression of seed germination by GA signaling

Camille M. Steber

10.1 Introduction

This chapter explores evidence that the ubiquitin–proteasome pathway plays a role in gibberellin (GA) stimulation of germination via proteolysis of DELLA proteins (named after conserved amino acids). GA stimulates germination, stem elongation, transition to flowering, and fertility. DELLA proteins are repressors of GA responses defined by the presence of conserved DELLA and GRAS (named after GAI [GA-INSENSITIVE], RGA [REPRESSOR OF GA-*INSENSITIVE*], and SCARECROW proteins) domain amino acid sequences. It is clear that the ubiquitin–proteasome pathway relieves DELLA repression of stem elongation in response to GA signaling. The paradigm is that GA stimulates an SCF (named after the Skp1, Cullin, and F-box subunits) E3 ubiquitin ligase complex, which in turn stimulates degradation of the DELLA proteins, negative regulators of GA response, by the proteasome pathway. This induces the downstream events necessary for seed germination, which were under repression in the absence of GA. However, interpretation of the evidence for DELLA regulation of seed germination has been somewhat contentious. Work in *Arabidopsis thaliana* suggests that the DELLA protein RGL2 (RGA-LIKE2) is the main negative regulator of GA response in germination, and that the SCF^{SLY1} E3 ubiquitin ligase complex is required for GA-stimulated disappearance of RGL2.

10.2 Control of germination by GA signaling

The plant hormone GA was first identified in 1926 based on its role in stem elongation (Tamura, 1991; Sun and Gubler, 2004; Thomas *et al.*, 2005). Eiichi Kurosawa identified gibberellin as the agent causing the excessive stem elongation in rice (*Oryza sativa*) seedlings infected with *bakanae* disease. Subsequent work demonstrated that gibberellins are growth regulators in a wide range of plant, algal, and fungal species. GA is not a single plant hormone, but actually a large family of tetracyclic diterpenes. Thus far, 136 naturally occurring GA molecules have been identified in plants and fungi (Thomas *et al.*, 2005). GA acts on many physiological events during plant growth and development (reviewed by Sun and Gubler, 2004; Thomas *et al.*, 2005). GA promotes stem elongation by stimulating both cell elongation and cell division. GA stimulates the transition to flowering in most plant species and appears to be required for normal fertility and embryo development. A wide range of evidence has shown that GA plays a role in seed germination.

By far the strongest evidence for the role of GA in germination comes from studies of mutants defective in GA biosynthesis in dicots. In *Arabidopsis*, mutations in early GA biosynthesis genes result in a requirement for GA application to germinate (Koornneef and van der Veen, 1980). Mutations in the biosynthesis genes *GA1* (copalyl synthase), *GA2* (*ent*-kaurene synthase), and *GA3* (*ent*-kaurene oxidase) result in failure to germinate. Mutations in genes acting late in GA biosynthesis such as *ga4-1* (GA 3-oxidase or GA3ox) and *ga5-1* (GA 20-oxidase or GA20ox) do not require GA application to germinate. The explanation may lie in the fact that both of these genes are part of multigene families in higher plants. These genes show tissue-specific expression. For example, the *Arabidopsis GA3ox2* gene was found to be expressed primarily in germinating seeds and seedlings, while *GA3ox1* was expressed in all growing tissues (Yamaguchi *et al.*, 1998). Future work may use studies of multiple mutants to determine whether these GA biosynthesis genes are also required for germination in *Arabidopsis*. A mutation in *ent*-kaurene synthase called *gib-1* has been identified in tomato (*Lycopersicon esculentum*) and *gib-1* seeds require GA application for germination (Karssen *et al.*, 1989; Benson and Zeevaart, 1990). Thus, the requirement for GA in germination is not unique to *Arabidopsis*.

The plant hormones ABA (abscisic acid) and GA have tightly interwoven roles in the decision to germinate. While GA is often required for germination and for postgerminative reserve mobilization, ABA inhibits germination, promotes nutrient storage, and is required for dormancy and desiccation tolerance during embryo maturation (reviewed by Bentsink and Koornneef, 2002). Seeds are said to be dormant when they are unable to germinate even under favorable conditions. Mutants unable to synthesize ABA have nondormant seeds and a vegetative wilty phenotype. The wilty phenotype results from an inability to close stomates and conserve water in response to drought stress. Thus, in adult tissues ABA is the stress hormone needed for resistance to drought, and is also involved in cold and salt stress. The hormone balance theory postulates that the decision to germinate is a balancing act between ABA and GA signals (Karssen and Lacka, 1986). If ABA biosynthesis or signaling is reduced, a seed is more likely to germinate. If GA biosynthesis or signaling is reduced, a seed is less likely to germinate. A mutation in one pathway can compensate for a mutation in the other. For example, germination of a strong GA biosynthesis mutant can be rescued by mutants causing either reduced ABA biosynthesis or sensitivity (reviewed by Bentsink and Koornneef, 2002). Conversely, the ability of ABA-insensitive mutants to germinate in the presence of high concentrations of ABA is suppressed by reduced GA biosynthesis or signaling (Steber *et al.*, 1998).

By definition, seed germination is complete when the embryo radicle emerges from the seed coat. GA is believed to stimulate germination by inducing hydrolytic enzymes that weaken the seed coat or endosperm cap, by inducing mobilization of seed nutrient storage compounds, and by stimulating expansion of the embryo (Bewley and Black, 1994; see Chapter 11). In tomato, it has been shown that GA stimulates expression of endo-β-mannanase, an enzyme that breaks down cell wall reserves in the endosperm cap (Still and Bradford, 1997; Nonogaki *et al.*, 2000).

During *Arabidopsis* germination, genes expected to be involved in cell wall degradation and induction of cell division are induced by GA (Ogawa *et al.*, 2003). In the cereal grains of rice, barley (*Hordeum vulgare*), and wheat (*Triticum aestivum*), GA induces the expression of genes encoding α-amylase, an enzyme needed for the breakdown of the starchy endosperm (Sun and Gubler, 2004). The mobilization of this nutrient source doubtlessly contributes to embryo growth after germination.

The role of GA biosynthesis and signaling genes in rice germination is currently unclear. An apparent knockout mutation in an *ent*-kaurene synthase gene of rice (*OsKS1*) results in the expected dwarfism and infertility phenotypes, but in contrast to *Arabidopsis* and tomato it does not result in failure to germinate (Sakamoto *et al.*, 2004; Margis-Pinheiro *et al.*, 2005). The fact that *OsKS1* is not expressed in imbibing seeds, however, raises the possibility that one of the five to seven homologues of KS may function in rice seed germination (Margis-Pinheiro *et al.*, 2005). GA-insensitive mutations in the rice F-box gene *GID2* (*GIBBERELLIN-INSENSITIVE DWARF2*) and in the GA receptor *GID1* also result in dwarfism and infertility without compromising germination (Itoh *et al.*, 2003). While the *gid1* mutants fail to induce α-amylase during seedling establishment, the mutants germinate well (Ueguchi-Tanaka *et al.*, 2005; M. Matsuoka, personal communication). There are a number of explanations for these observations. Cereal seeds ('caryopses') including rice have highly specialized seed morphology where the embryo is not entirely enclosed by the endosperm. Tissues covering the embryo may offer relatively little resistance to penetration by the coleorhiza and coleoptile. It is possible that in rice, GA is not required for germination per se but is required only for reserve mobilization. It is also possible that the rice variety in which the *gid1* and *gid2* mutants were identified does not require GA signaling in germination either due to lack of seed dormancy or due to a defect in ABA signaling in that background. The ability of GA mutant rice seeds (caryopses), which have pale testa color to germinate, may be due to reduced dormancy resulting from lack of red pigments in the testa. Red rice has strong seed dormancy compared to pigmentless cultivars (Footitt and Cohn, 1995). Loss of red testa color causes reduced seed dormancy. For example, the reduced dormancy caused by the *Arabidopsis* transparent testa mutant *tt4* suppresses the requirement for GA in germination, allowing the *ga1-1* mutant to germinate (Debeaujon and Koornneef, 2000; Chapter 2). Another possibility is that rice varieties used in these studies have a defect in ABA signaling. For example, it has been shown that all tested wheat cultivars are unable to properly splice the *ZmVP1/AtABI3* (*Zea mays VIVIPAROUS1/ABA-INSENSITIVE3*) homologue of wheat *TaVP1* (McKibbin *et al.*, 2002). Thus, bread wheat has reduced seed dormancy due to reduced ABA signaling. In *Arabidopsis*, mutations causing ABA insensitivity allow both GA biosynthesis and GA-insensitive mutants to germinate. A similar defect in rice ABA signaling might obviate the requirement for GA in germination. The fact that GA biosynthesis is required for germination of the ABA-insensitive *vp5* mutants on the mother plant in maize (*Zea mays*) (White *et al.*, 2000; White and Rivin, 2000) suggests that GA is important for cereal germination during embryo development. Future research will need to address the role of GA in cereal grain germination.

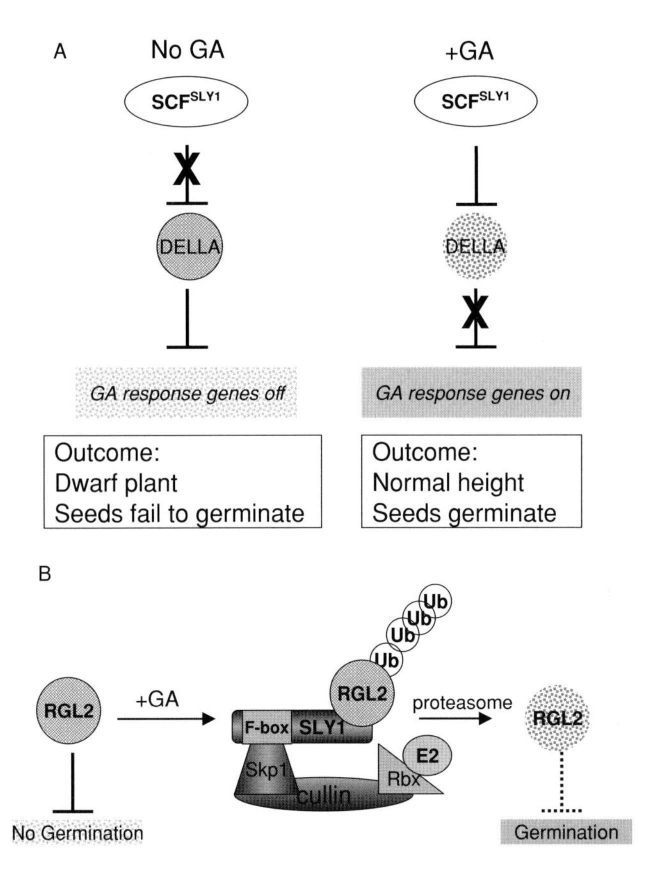

Figure 10.1 Model for control of GA responses by the ubiquitin–proteasome pathway. (A) In the absence of GA, for example in a GA biosynthesis mutant, DELLA proteins repress GA responses such as seed germination and stem elongation. When GA is added, the SCF E3 ubiquitin ligase complex causes disappearance of the DELLA protein, thereby de-repressing GA responses. (B) Destruction of the DELLA protein RGL2 is mediated by SCF^{SLY1} E3 ubiquitin ligase in *Arabidopsis*. In the absence of GA, RGL2 represses seed germination. In the presence of GA, the SCF^{SLY1} complex catalyzes the transfer of ubiquitin from the E2 ubiquitin conjugating enzyme to the target protein RGL2. Addition of at least four ubiquitin moieties allows the target to be recognized and degraded by the 26S proteasome. The configuration of the SCF^{SLY1} complex is based on Zheng *et al.* (2002).

DELLA proteins are negative regulators of GA signaling that form a subfamily of the GRAS family of putative transcription factors (Figure 10.1A). The DELLA proteins are named for a conserved amino acid sequence (D = Asp, E = Glu, L = Leu, L = Leu, A = Ala). The first DELLA genes cloned were *Arabidopsis GAI* and *RGA*. *GAI* was originally identified by Koornneef *et al.* (1985) as a gain-of-function mutation (*gai-1*) causing a GA-insensitive dwarf phenotype similar to the 'green revolution' genes of wheat (Peng *et al.*, 1999a). Loss-of-function mutations in *RGA* were identified based on suppression of the dwarf phenotype of the GA biosynthesis mutant *ga1-3* (Silverstone *et al.*, 1997). Cloning and characterization of *GAI* and *RGA* revealed that they are homologous genes (Peng *et al.*, 1997; Silverstorne *et al.*, 1998). While loss of neither gene alone gives increased stature, the *gai-t6 rga-24* double mutant is taller than wild-type plants, indicating that together they repress stem elongation of *Arabidopsis* (Dill and Sun, 2001; King *et al.*, 2001). This increased height phenotype is reminiscent of the slender phenotype of loss-of-function mutants in DELLA genes of barley (*SLN1*, *SLENDER1*) and rice (*SLR1*, *SLENDER RICE1*) (Ikeda *et al.*, 2001; Chandler *et al.*, 2002). A DELLA protein has been shown to inhibit stem elongation in species as diverse as barley, rice, wheat, maize, grape (*Vitis vinifera*), tobacco (*Nicotiana tabacum*), *Brassica rapa*, and *Arabidopsis* (Peng *et al.*, 1999a; Silverstone *et al.*, 2001; Boss and Thomas, 2002; Gubler *et al.*, 2002; Itoh *et al.*, 2002; Hynes *et al.*, 2003; Dill *et al.*, 2004; Fu *et al.*, 2004; Muangprom *et al.*, 2005).

The current model for GA signaling in stem elongation is widely conserved in the plant kingdom. If the plant never produced GA, as in a GA biosynthesis mutant, repression of stem elongation by DELLA would result in a dwarf phenotype (Figure 10.1A). Addition of GA stimulates cell division and expansion within the meristem by causing disappearance of the DELLA protein. GA seems to cause DELLA protein disappearance by triggering DELLA protein polyubiquitination by an SCF E3 ubiquitin ligase (Itoh *et al.*, 2003). Addition of four or more ubiquitin moieties to a DELLA protein targets it for destruction by the 26S proteasome (Sasaki *et al.*, 2003). The interaction between the DELLA protein and SCF complex may be regulated by DELLA phosphorylation or by interaction between the DELLA protein and the GA receptor GID1 (Fu *et al.*, 2004; Gomi *et al.*, 2004; Ueguchi-Tanaka *et al.*, 2005).

Evidence from work in *Arabidopsis* suggests that GA signaling also stimulates germination via DELLA protein disappearance. There are five DELLA genes in *Arabidopsis*, *RGA*, *GAI*, *RGL1*, *RGL2*, and *RGL3*. *RGA* and *GAI* repress stem elongation, while *RGL1*, *RGA*, and *RGL2* repress the transition to flowering (Itoh *et al.*, 2003). *RGL2* is the main repressor of germination, although *RGL1*, *RGA*, and *GAI* also appear to contribute (Lee *et al.*, 2002; Tyler *et al.*, 2004; Cao *et al.*, 2005). Two questions must be answered in order to understand the role of GA and DELLA proteins in seed germination: (1) Is RGL2 a master negative regulator of seed germination?; and (2) How widely conserved is this mechanism for controlling seed germination? The evidence that DELLA and the SCF E3 ubiquitin ligase complex control germination in *Arabidopsis* will be discussed first, and then evidence in additional plant species will be considered.

10.3 The role of the ubiquitin–proteasome pathway in GA signaling

Several lines of evidence indicate that DELLA proteins are regulated by the ubiquitin–proteasome pathway (Figure 10.1). First, GA treatment causes the disappearance of DELLA proteins in *Arabidopsis*, rice, and barley (Silverstorne *et al.*, 2001; Itoh *et al.*, 2002; Gubler *et al.*, 2002; Dill *et al.*, 2004; Fu *et al.*, 2004). This disappearance requires the F-box subunit of an SCF E3 ubiquitin ligase and the 26S proteasome. The disappearance of DELLA protein in response to GA is blocked by inhibitors of the 26S proteasome (Fu *et al.*, 2002; Hussain *et al.*, 2005) and by mutations in the F-box genes rice *GID2* and *Arabidopsis SLY1* (*SLEEPY1*) (McGinnis *et al.*, 2003; Sasaki *et al.*, 2003). This suggests that GA causes the SCF$^{SLY1/GID2}$ E3 ubiquitin ligase to target the DELLA protein for destruction via the 26S proteasome. Mutations in *GID2* and *SLY1* cause a range of GA-insensitive phenotypes including severe dwarfism, dark green color, and reduced fertility (Itoh *et al.*, 2003). Mutations in *SLY1* also cause increased seed dormancy. SLY1 and GID2 proteins have been shown to interact with DELLA proteins, and ubiquitinated DELLA protein has been detected in rice (Sasaki *et al.*, 2003; Dill *et al.*, 2004; Fu *et al.*, 2004; Gomi *et al.*, 2004).

DELLA proteins are regulated by *Arabidopsis* SCF^{SLY1} and rice SCF^{GID2} E3 ubiquitin ligase complexes (Figure 10.1B; McGinnis *et al.*, 2003; Sasaki *et al.*, 2003). SCF complexes are one of several types of E3 ubiquitin ligases found in plants and in animals (Smalle and Vierstra, 2004). An SCF complex is identified based on the F-box subunit because it is the F-box subunit that binds to a specific target or substrate protein. Based on the crystal structure of mammalian SCF^{Skp2}, the backbone of the SCF complex is cullin (Zheng *et al.*, 2002). Cullin binds a homologue of the ring-finger protein Rbx1 at its C-terminus, and Rbx1 binds the E2 ubiquitin conjugating enzyme. At the N-terminus, cullin binds a homologue of Skp1. Skp1 binds the F-box protein via the conserved F-box domain and tethers the F-box protein to the rest of the complex. The F-box protein is generally composed of an F-box domain at the N-terminus and a protein–protein interaction domain at the C-terminus that binds to a substrate protein such as DELLA. The SCF complex catalyzes transfer of ubiquitin from E2 to the substrate, in this case a DELLA protein.

Ubiquitin is a 76-amino acid protein used in plants and animals as a tag to signal for proteolytic processing, cleavage, or destruction (Smalle and Vierstra, 2004). DELLA proteins appear to be targeted by ubiquitination for destruction by the 26S proteasome. Generally, ubiquitin is first covalently bound to a cysteine of an E1 ubiquitin-activating enzyme via a thioester bond and then transferred to a cysteine of an E2 ubiquitin-conjugating enzyme. Finally, an E3 ubiquitin ligase catalyzes the transfer of ubiquitin from E2 to a lysine residue of the substrate protein. Formation of a chain of four or more ubiquitin moieties on the substrate allows its recognition and proteolysis by the 26S proteasome. The 26S proteasome is a multi-subunit protein complex that can be found both in the nucleus and in the cytosol. It is surmised that DELLA proteins meet their end in the 26S proteasome because the DELLA proteins barley SLN1 and *Arabidopsis* RGL2 were shown to be stabilized by compounds known to inhibit the 26S proteasome (Fu *et al.*, 2002; Hussain *et al.*, 2005). The rice DELLA SLR1 was shown to be subjected to ubiquitination by western analysis using an antibody to ubiquitin (Sasaki *et al.*, 2003). Finally, DELLA proteins in a wide range of plant species have been shown to be regulated by a highly conserved SCF E3 ubiquitin ligase (Itoh *et al.*, 2003).

Protein destruction is a recurrent theme in plant signal transduction (Smalle and Vierstra, 2004). There are almost 700 F-box proteins in the *Arabidopsis* genome. It is already clear that F-box proteins and the ubiquitin–proteasome pathway play a crucial role in plant hormone signaling. The ubiquitin–proteasome pathway was implicated in ABA and cytokinin signaling when mutations in subunits of the 26S proteasome were shown to result in changes in hormone sensitivity (Smalle *et al.*, 2002, 2003). Recently, the ABA signaling protein ABI3 was found to be regulated by the AIP2 (AB13-INTERACTING PROTEIN2) E3 ubiquitin ligase (Zhang *et al.*, 2005). In ethylene signaling, the transcription factors EIN3 (ETHYLENE INSENSITIVE3) and EIL1 (EIN3-LIKE) are regulated by the homologous F-box proteins EBF1 (EIN3-BINDING F-BOX) and EBF2 (Guo and Ecker, 2003; Potuschak *et al.*, 2003; Gagne *et al.*, 2004). EBF1 and EBF2 F-box proteins contain the leucine-rich repeat (LRR) type of protein–protein interaction domain at the C-terminus. The SLY1 F-box protein contains no consensus protein–protein interaction domain at the C-terminus. However, based on mutation studies, the C-terminus is required both for function and for interaction with DELLA proteins (McGinnis *et al.*, 2003; Dill *et al.*, 2004; Fu *et al.*, 2004). The COI1 (CORONATINE INSENSITIVE1) F-box protein required for jasmonic acid signaling contains an LRR protein–protein interaction domain at the C-terminus (Xu *et al.*, 2002). This is similar to the LRRs found in the C-terminus of the TIR1 (TRANSPORT INHIBITOR RESPONSE1) F-box protein (Gray *et al.*, 1999). TIR1 was recently shown to be an auxin receptor (Dharmasiri *et al.*, 2005; Kepinski and Leyser, 2005).

Auxin signaling has a number of similarities to GA signaling. Auxin triggers auxin responses by targeting the IAA/AUX family of transcriptional repressors for destruction via the 26S proteasome (Gray *et al.*, 2001). The IAA/AUX proteins repress auxin responses by binding to the ARF (AUXIN RESPONSE FACTOR) transcription factors. DELLA proteins repress GA responses by an unknown mechanism. Auxin binds the F-box protein TIR1, thus enabling TIR1 to bind to the IAA/AUX proteins and polyubiquitinate them, thereby targeting them for destruction by the 26S proteasome (Dharmasiri *et al.*, 2005, Kepinski and Leyser, 2005). The DELLA proteins have been shown to be regulated by the F-box proteins *Arabidopsis* SLY1 or rice GID2. However, it is still unclear precisely how GA controls the interaction between the F-box protein and the DELLA protein. While interactions have been detected between F-box proteins SLY1/GID2 and DELLA proteins, no interaction between GA and SLY1/GID2 has yet been reported. However, it has been reported that GA binds to the rice GID1 protein, a GA receptor (Ueguchi-Tanaka *et al.*, 2005). GID1 shows GA-dependent interaction with the rice DELLA protein SLR1. Mutations in *GID1* lead to overaccumulation of DELLA protein, suggesting that *GID1* is required for destruction of the DELLA protein by SCF^{GID2}. Whether GID1 controls the proteolysis of DELLA proteins via direct protein–protein interaction between GID1 and GID2 F-box protein needs to be investigated.

While DELLA proteins are known to be negative regulators of GA responses, their precise function remains unknown. They are considered likely transcription factors based on amino acid homology and on yeast two-hybrid data showing that they can activate transcription (Peng *et al.*, 1999a; Itoh *et al.*, 2002). The DELLA protein family is defined by conserved amino acid sequences (Sun and Gubler, 2004). The C-terminus of DELLA proteins contains homology to the GRAS family of proteins, including a nuclear localization sequence, the VHIID domain, leucine heptad repeats (LHR), and SH2 domains. The SH2 domain of metazoan STAT (signal transducers and activators of transcription) transcription factors is involved in phosphotyrosine signaling (Peng *et al.*, 1999a). The DELLA subfamily of the GRAS family contains conserved domains at the N-terminus including DELLA and VHYNP. Based on mutation studies, it appears that the C-terminus is the functional domain while the N-terminus is a regulatory domain needed for response to GA (Sun and Gubler, 2004). However, this work is based on stem elongation as a GA response. Future work will need to define the protein domains needed for negative regulation of seed germination by DELLA proteins.

10.4 Is RGL2 a 'master regulator' of seed germination?

The evidence that DELLA proteins are negative regulators of germination came from studies of RGL2 in *Arabidopsis* (summarized in Figure 10.2; Lee *et al.*, 2002; Tyler *et al.*, 2004; Cao *et al.*, 2005). The first study by Lee *et al.* (2002) showed that loss of *RGL2* function restores germination of the strong GA biosynthesis mutant *ga1-3* and restores germination in the presence of GA biosynthesis inhibitor paclobutrazol. This result was confirmed by Tyler *et al.* (2004). These genetic studies are the most compelling evidence that the DELLA protein RGL2 is normally needed to repress seed germination and that GA acts by alleviating RGL2 repression of seed germination.

The initial study by Lee *et al.* (2002) suggested that RGL2 activity is controlled at the level of transcript accumulation. Northern blot analysis showed that *RGL2* transcript is induced within the first 24 h of seed imbibition, the process during which dry seeds take up water. *RGL2* mRNA levels remained high as long as the seeds were incubated at 4°C, a temperature at which seeds did not germinate. *RGL2* mRNA levels greatly decreased and germination took place within 48 h of transfer to 23°C. This correlative evidence suggested that germination is associated with a decrease in *RGL2* mRNA. This correlation between *RGL2* mRNA accumulation and failure

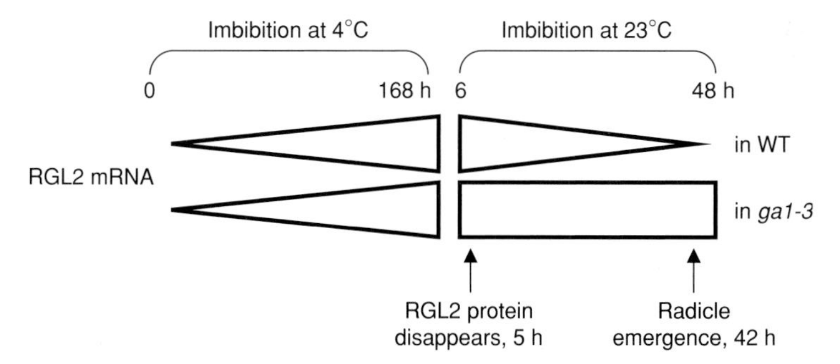

Figure 10.2 Schematic summary of the pattern of *RGL2* expression during imbibition and germination in wild type (WT) and *ga1-3* mutant seeds. Diagram shows increasing *RGL2* mRNA accumulation during imbibition in the cold (4°C). *RGL2* mRNA levels decrease over 48 h of transfer to 23°C in wild-type, but remain high in the GA biosynthesis mutant *ga1-3* (Lee *et al.*, 2002). However, some *RGL2* mRNA remains at the time of radicle emergence in wild-type seeds (Bassel *et al.*, 2004). Radicle emergence begins within 30 h of transfer to higher temperature and is complete within 42 h (Bassel *et al.*, 2004). In the *ga1-3* mutant, RGL2 protein disappears within 5 h of adding GA to imbibing seeds (Tyler *et al.*, 2004).

to germinate was further supported by data showing that the strong GA biosynthesis mutant *ga1-3* accumulates high levels of *RGL2* mRNA during imbibition both at 4°C and at 23°C. *RGL2* mRNA abundance decreases within 48 h of addition of GA to imbibing *ga1-3* seeds. This suggests that one way that GA stimulates germination of *ga1-3* seeds is by decreasing the level of *RGL2* transcript. It is important to note, however, that the *RGL2* transcript did not completely disappear within 48 h of adding GA. Thus Lee *et al.* (2002) concluded that downregulation of *RGL2* mRNA level is consistent with the notion that it is a negative regulator of seed germination, but could not conclude that complete absence of *RGL2* mRNA is required for germination to occur.

A subsequent study by Bassel *et al.* (2004) sought to determine whether downregulation of DELLA transcript is required for seed germination in *Arabidopsis*. This study very carefully examined the level of *RGL2* mRNA relative to percent germination. They found that even after *Arabidopsis* seed germination is complete, *RGL2* transcript can still be detected (Bassel *et al.*, 2004). This study clearly indicates that disappearance of *RGL2* mRNA cannot be the sole mechanism for stimulating germination. However, it does leave open the possibility that *RGL2* is a negative regulator of germination subject to posttranscriptional regulation, such as translation, nuclear localization, protein stability, or posttranslational modification of RGL2 protein.

A study by Tyler *et al.* (2004) suggests that RGL2 function is regulated by proteolysis and that GA-dependent disappearance of RGL2 protein is regulated by the SLY1 F-box protein. Western analysis of RGL2 protein accumulation in the *ga1-3* mutant background showed that RGL2 protein disappears within 5 h after GA application to the *ga1-3* seeds, but does not disappear in the absence of GA. This correlation between GA-triggered disappearance of RGL2 protein and GA stimulation of germination suggests that disappearance of RGL2 protein is required for germination in *Arabidopsis*. The rapid disappearance of RGL2 protein would be consistent with a role during germination proper (prior to radicle emergence), rather than a role in postgermination events such as seedling establishment. Taken together with the fact that loss of *RGL2* function rescues *ga1-3* seed germination, this study strongly suggests that RGL2 is a negative regulator of germination in *Arabidopsis*. Proof that RGL2 is a negative regulator of seed germination requires demonstration that germination fails to occur if RGL2 protein persists during germination.

RGL2 protein does persist after GA treatment of *sly1* mutant seeds (Tyler *et al.*, 2004). This suggests that the SCF^{SLY1} E3 ubiquitin ligase normally targets RGL2 for destruction in response to GA. Mutations in the *SLY1* gene result in an increase in seed dormancy and in an increase in sensitivity to ABA in germination (Steber *et al.*, 1998; Steber and McCourt, 2001; Strader *et al.*, 2004). However, *sly1* mutants do eventually after-ripen, and some seed lots do germinate (C.M. Steber, unpublished results). This suggests that the disappearance of RGL2 protein is important for germination, but is not the sole mechanism regulating germination in *Arabidopsis*. It is possible that there are additional regulators of seed germination in *Arabidopsis*, or that there are additional mechanisms for posttranslational regulation of RGL2.

Recent work by Hussain *et al.* (2005) in the tobacco BY2 cell line indicates that RGL2 protein is phosphorylated *in planta* and suggests that RGL2 protein is stabilized by phosphorylation at multiple sites. Future work will need to investigate whether posttranslational modification alters RGL2 activity in seed germination.

10.5 *Sleepy1* is a positive regulator of seed germination in *Arabidopsis*

Mutations in the *SLY1* gene were recovered in two screens that looked for reduced ability to germinate (Figure 10.3). The first was a screen for suppressors of the ability of the ABA-insensitive mutant *ABI1-1* to germinate in the presence of 3 μM ABA (Steber *et al.*, 1998). *ABI1-1* is a semi-dominant mutation that permits seeds to germinate at up to 100 μM ABA, while germination of wild-type *Arabidopsis* seeds is inhibited by 1.2 μM ABA. Suppressor mutations were isolated in *SLY1* and in the GA biosynthesis gene *GA1*. All mutant isolates of *SLY1* from this screen were a single frameshift mutation causing loss of the last 40 amino acids, and are now referred to as *sly1-2* for historical reasons (McGinnis *et al.*, 2003). An additional allele, *sly1-10*, was recovered in a screen for brassinosteroid-dependent germination (Figure 10.3) and is a complex rearrangement resulting in loss of the last 8 amino acids (Steber and McCourt, 2001). The *sly1-2* mutation appears to result in stronger phenotypes than does the *sly1-10* mutation.

The *SLY1* gene is defined as a positive regulator of germination, because loss of function alleles lead to reduced germination and gain of function leads to increased ability to germinate. Loss of *SLY* function leads to GA-insensitive phenotypes including reduced ability to germinate, as well as dwarfism and reduced fertility. The F2 seeds resulting from crosses of *sly1-2* to wild-type plants segregated for failure

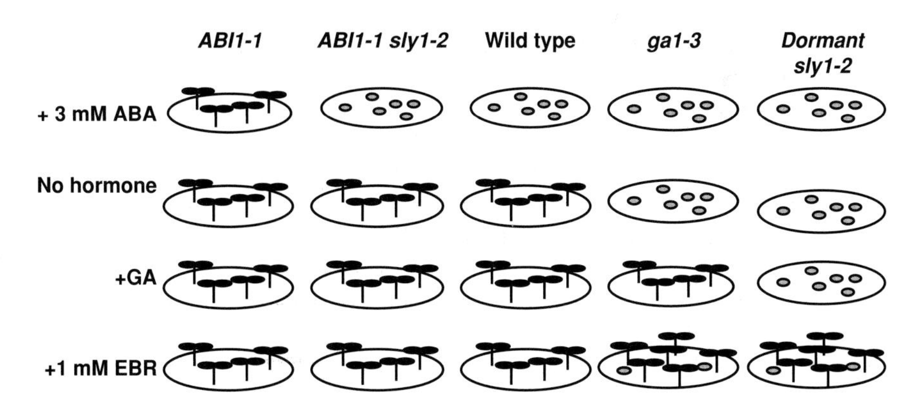

Figure 10.3 Schematic of rationale for screens that isolated mutations in the *SLY1* gene. The *sly1-2* allele was isolated as a suppressor of *ABI1-1*. *ABI1-1* is able to germinate in the presence of 3 μM ABA (Steber *et al.*, 1998). Either *sly1* or *ga1* mutations in the *ABI1-1* background cause failure to germinate on 3 μM ABA. The *sly1-10* allele was isolated based on brassinosteroid-dependent germination. Both *sly1* and *ga1* mutants are unable to germinate in the absence of hormone. Addition of GA fully rescues *ga1-3* germination, but not *sly1*. Epibrassinolide (EBR) is able to partly rescue both *ga1* and *sly1* germination.

to germinate. This suggested that even in the absence of the *ABI1-1* mutation, the recessive *sly1-2* mutation has a strong germination phenotype (Steber *et al.*, 1998). Most fresh (unaged) *sly1-2* seed lots germinate between 0 and 20% (Steber and McCourt, 2001; C.M. Steber unpublished results). Such seed stocks can require up to 3 years to after-ripen. Some rare seed lots germinate 80–100% soon after harvest. However, such seed lots show increased sensitivity to ABA in germination, suggesting that they also have a reduced capacity to germinate (Strader *et al.*, 2004). Further work is needed to determine how environmental conditions during seed development and during seed storage influence the degree of seed dormancy in *sly1* mutants. Germination of *sly1* mutants can be rescued by cutting the seed coat, suggesting that the germination phenotype is the result of seed coat-imposed dormancy (McGinnis *et al.*, 2003; see Chapters 2 and 11). The fact that *sly1-10* mutants also show reduced capacity to germinate indicates that multiple alleles of *sly1* decrease the germination efficiency of *Arabidopsis* seeds. The notion that *SLY1* is a positive regulator of seed germination is further supported by the fact that *SLY1* overexpression and a gain-of-function mutation in *SLY1*, *sly1-gar2* (*gai revertant2*), both cause increased resistance to the GA biosynthesis inhibitor paclobutrazol in germination (Peng *et al.*, 1999b; Fu *et al.*, 2004).

Evidence that the reduced germination in the *sly1* mutant is caused by failure of RGL2 protein to disappear in response to GA is strictly correlative at this point. It is not yet known whether a mutation in *RGL2* can suppress the germination phenotype of the *sly1* mutant in the same way that it suppresses the *ga1-3* germination phenotype. As described above, it is known that whereas RGL2 protein disappears within 5 h of adding GA to *ga1-3* mutant seeds, it does not disappear within 5 h of adding GA to imbibing seeds of the GA-insensitive mutant *sly1-10* (Tyler *et al.*, 2004). Thus *SLY1* is required for the GA-induced disappearance of RGL2 protein in seed germination. Since the *sly1-10* mutant has reduced ability to germinate, it is possible that this germination phenotype is due to over-accumulation of RGL2 protein. This suggests that the SCF^{SLY1} complex is responsible for relieving RGL2 repression of germination, just as it relieves DELLA repression of stem elongation (Figure 10.1). It is not yet known, however, whether *sly1* mutant seed lots that germinate well accumulate less RGL2 protein than *sly1* seed lots that do not germinate well. If RGL2 protein levels correlate with the severity of the *sly1* germination phenotype, it is possible that other ubiquitin ligases can target RGL2 protein for destruction in the absence of *SLY1*. Candidates for ubiquitin ligases that might have overlapping function with SLY1 protein in germination include the U-box protein PHOR1 (PHOTOPERIOD-RESPONSIVE1) that acts in GA signaling in potato and the F-box protein SNEEZY (SNE), which is a homologue of SLY1 in *Arabidopsis* (Amador *et al.*, 2001; Fu *et al.*, 2004; Strader *et al.*, 2004).

10.6 Do DELLA proteins have a conserved role in seed germination?

Both the *SLY1* gene and DELLA gene family are highly conserved among plant species. Homologues of *SLY1* are known in soybean (*Glycine max*), *Medicago truncatula*, sunflower (*Helianthus annuus*), cotton (*Gossypium arboretum*), tomato, rice, barley, and wheat (Itoh *et al.*, 2003, McGinnis *et al.*, 2003). The *SLY1* homologue of rice, *GID2*, clearly functions in GA signaling in stem elongation, flowering, and fertility. However, *GID2* is not required for rice seed germination. This suggests that GA may not play an essential role in the seed germination of all plant species. Members of the DELLA gene family have also been identified in numerous plant species. A single DELLA has been identified in barley (*SLN1*; Gubler *et al.*, 2002), maize (*d8* or *Dwarf8*; Peng *et al.*, 1999a), wine grape (Boss and Thomas, 2002), and tomato (Bassel *et al.*, 2004). Two DELLA genes have been identified in Hawaiian silversword (*Dubautia arborea*) (Remington and Purugganan, 2002), soybean (Bassel *et al.*, 2004), and hexaploid bread wheat (Peng *et al.*, 1999a).

While it is known that DELLA proteins exist in other plant species, it is not yet known whether they function similarly to *Arabidopsis* RGL2 in seed germination. Bassel *et al.* (2004) took the first steps toward exploring the role of DELLA homologues in seed germination of other plant species by identifying and characterizing the mRNA expression of DELLA homologues in tomato and in soybean. Soybean was chosen because, unlike tomato and *Arabidopsis*, it does not have seed coat-imposed dormancy. The two soybean DELLA homologues, *GmGAI1* and *GmGAI2*, identified by RT-PCR were not induced until after germination was complete. Thus, it is possible that GA and DELLA proteins are involved only in seedling establishment rather than in seed germination in this case (Bassel *et al.*, 2004). However, future work will have to determine whether these are the only DELLA or GRAS family genes expressed during soybean seed germination.

Like *Arabidopsis*, tomato seeds have seed coat-imposed dormancy and require GA to germinate. A single DELLA gene, *LeGAI*, was recovered from tomato (Bassel *et al.*, 2004). Similarly to *AtRGL2*, *LeGAI* transcript accumulated at very low levels in dry seeds and was strongly induced within 24 h of imbibition. If *LeGAI* behaved like *RGL2*, *LeGAI* mRNA should be present at high levels during imbibition of the strong GA biosynthesis mutant *gib-1*, but disappear upon imbibition in the presence of GA. In contrast, *LeGAI* mRNA failed to disappear upon imbibition in GA. This suggests either that *LeGAI* activity is not regulated at the level of transcript accumulation, or that *LeGAI* is not a negative regulator of seed germination in tomato. The main limitation in this study is that they were unable to examine LeGAI protein levels. Moreover, *LeGAI* was identified as a potential regulator of seed germination based mainly on sequence homology to *RGL2* and on transcriptional induction during imbibition. It is possible that another GRAS protein plays the role of *AtRGL2* in tomato germination. What is needed is evidence that LeGAI is functionally equivalent to *Arabidopsis RGL2*. For example, does loss of *LeGAI* function rescue *gib-1* mutant seed germination?

The example of *Arabidopsis* emphasizes the importance of genetics in determining whether a DELLA gene functions in seed germination. *RGL2* transcript is not the only DELLA transcript that accumulates during seed imbibition. *RGL3*, *RGA*, and *GAI* transcripts also accumulate to high levels during the first 24 h of imbibition (Tyler *et al.*, 2004). Loss of RGL3 does not enhance the ability of *rgl2* mutants to rescue *ga1-3* germination. However, mutations in *RGL1*, *RGA*, and *GAI* appear to enhance *rgl2* mutant suppression of the *ga1-3* germination phenotype (Cao *et al.*, 2005). Thus, induction during imbibition alone is not a sufficient criterion for determining whether homologues of *RGL2* share *RGL2* function. Future research on DELLA or GRAS gene function will need to determine whether homologues are expressed during seed imbibition and whether loss of function causes reduced requirement for GA in germination.

10.7 Future directions

It is clear that GA stimulated de-repression of germination via proteolysis of the DELLA protein RGL2 is an important component of the decision to germinate in *Arabidopsis* seeds. Further work is needed to determine whether RGL2 is a 'master regulator' of seed germination in *Arabidopsis*. Future work will also need to clarify whether GA is required for germination in other plant species, and if so, whether DELLA repression of germination is conserved within the plant kingdom. Some plant species may use other plant hormones known to stimulate germination, such as ethylene or brassinosteroids, to control seed germination (Karssen *et al.*, 1989; Leubner-Metzger, 2001; Steber and McCourt, 2001). It will be important to determine whether plant hormones that stimulate germination destabilize DELLA proteins, while hormones that inhibit germination stabilize DELLA proteins. It is known that ethylene, a hormone that stimulates germination, can delay GA-induced disappearance of DELLA proteins in seedlings (Achard *et al.*, 2003). While this is the reverse of what would be expected, the experiment needs to be performed with germinating seeds. It is possible that the role of DELLA proteins in seed germination is not as highly conserved as their role in stem elongation. It is also possible that species with seed coat-imposed dormancy like tomato and *Arabidopsis* use DELLA proteins, while species that have different mechanisms for seed dormancy such as embryo dormancy use a completely different mechanism for controlling seed germination (Bassel *et al.*, 2004).

References

P. Achard, W.H. Vriezen, D. Van Der Straeten and N.P. Harberd (2003) Ethylene regulates *Arabidopsis* development via the modulation of DELLA protein growth repressor function. *The Plant Cell* **15**, 2816–2825.

V. Amador, E. Monte, J.L. Garcia-Martinez and S. Prat (2001) Gibberellins signal nuclear import of PHOR1, a photoperiod-responsive protein with homology to *Drosophila* armadillo. *Cell* **106**, 343–354.

G.W. Bassel, E. Zielinska, R.T. Mullen and J.D. Bewley (2004) Down-regulation of *DELLA* genes is not essential for germination in tomato, soybean, and *Arabidopsis* seeds. *Plant Physiology* **136**, 2782–2789.

J. Benson and J.A. Zeevaart (1990) Comparison of *ent*-kaurene synthase A and B activities in cell-free extracts from young tomato fruits of wild-type and *gib-1*, *gib-2*, and *gib-3* tomato plants. *Journal of Plant Growth Regulation* **9**, 237–242.

L. Bentsink and M. Koornneef (2002) Seed dormancy and germination. In: *The Arabidopsis Book* (eds C.R. Somerville and E.M. Meyerowitz). American Society of Plant Biologists, Rockville, MD. http://www.aspb.org/publications/arabidopsis/, pp. 1–18.

J.D. Bewley and M. Black (1994) *Seeds: Physiology of Development and Germination*. Plenum Press, New York.

P.K. Boss and M.R. Thomas (2002) Association of dwarfism and floral induction with a grape 'green revolution' mutation. *Nature* **416**, 847–850.

D. Cao, A. Hussain, H. Cheng and J. Peng (2005) Loss of function of four DELLA genes leads to light- and gibberellin-independent seed germination in *Arabidopsis*. *Planta*, **223**, 105–113.

P.M. Chandler, A. Marion-Poll, M. Ellis and F. Gubler (2002) Mutants at the *Slender1* locus of barley cv Himalaya. Molecular and physiological characterization. *Plant Physiology* **129**, 181–190.

I. Debeaujon and M. Koornneef (2000) Gibberellin requirement for *Arabidopsis* seed germination is determined both by testa characteristics and embryonic abscisic acid. *Plant Physiology* **122**, 415–424.

N. Dharmasiri, S. Dharmasiri and M. Estelle (2005) The F-box protein TIR1 is an auxin receptor. *Nature* **435**, 441–445.

A. Dill and T. Sun (2001) Synergistic derepression of gibberellin signaling by removing *RGA* and *GAI* function in *Arabidopsis thaliana*. *Genetics* **159**, 777–785.

A. Dill, S.G. Thomas, J. Hu, C.M. Steber and T.P. Sun (2004) The *Arabidopsis* F-box protein SLEEPY1 targets gibberellin signaling repressors for gibberellin-induced degradation. *The Plant Cell* **16**, 1392–1405.

S. Footitt and M.A. Cohn (1995) Seed dormancy in red rice (*Oryza sativa*). IX: Embryo fructose-2,6-bisphosphate during dormancy breaking and subsequent germination. *Plant Physiology* **107**, 1365–1370.

X. Fu, D.E. Richards, T. Ait-Ali, *et al.* (2002) Gibberellin-mediated proteasome-dependent degradation of the barley DELLA protein SLN1 repressor. *The Plant Cell* **14**, 3191–3200.

X. Fu, D.E. Richards, B. Fleck, D. Xie, N. Burton and N.P. Harberd (2004) The *Arabidopsis* mutant sleepy1^{gar2-1} protein promotes plant growth by increasing the affinity of the SCF^{SLY1} E3 ubiquitin ligase for DELLA protein substrates. *The Plant Cell* **16**, 1406–1418.

J.M. Gagne, J. Smalle, D.J. Gingerich, *et al.* (2004) *Arabidopsis* EIN3-binding F-box 1 and 2 form ubiquitin-protein ligases that repress ethylene action and promote growth by directing EIN3 degradation. *Proceedings of the National Academy of Sciences of the United States of America* **101**, 6803–6808.

K. Gomi, A. Sasaki, H. Itoh, *et al.* (2004) GID2, an F-box subunit of the SCF E3 complex, specifically interacts with phosphorylated SLR1 protein and regulates the gibberellin-dependent degradation of SLR1 in rice. *The Plant Journal* **37**, 626–634.

W.M. Gray, J.C. del Pozo, L. Walker, *et al.* (1999) Identification of an SCF ubiquitin-ligase complex required for auxin response in *Arabidopsis thaliana*. *Genes and Development* **13**, 1678–1691.

W.M. Gray, S. Kepinski, D. Rouse, O. Leyser and M. Estelle (2001) Auxin regulates SCF (TIR1)-dependent degradation of AUX/IAA proteins. *Nature* **414**, 271–276.

F. Gubler, P.M. Chandler, R.G. White, D.J. Llewellyn and J.V. Jacobsen (2002) Gibberellin signaling in barley aleurone cells. Control of SLN1 and GAMYB expression. *Plant Physiology* **129**, 191–200.

H. Guo and J.R. Ecker (2003) Plant responses to ethylene gas are mediated by SCF (EBF1/EBF2)-dependent proteolysis of EIN3 transcription factor. *Cell* **115**, 667–677.

A. Hussain, D. Cao, H. Cheng, Z. Wen and J. Peng (2005) Identification of the conserved serine/threonine residues important for gibberellin-sensitivity of *Arabidopsis* RGL2 protein. *The Plant Journal* **44**, 88–99.

L.W. Hynes, J. Peng, D.E. Richards and N. Harberd (2003) Transgenic expression of the *Arabidopsis* DELLA proteins GAI and gai confers altered gibberellin response in tobacco. *Transgenic Research* **12**, 707–714.

A. Ikeda, M. Ueguchi-Tanaka, Y. Sonoda, *et al.* (2001) *slender rice*, a constitutive gibberellin response mutant, is caused by a null mutation of the *SLR1* gene, an ortholog of the height-regulating gene *GAI/RGA/RHT/D8*. *The Plant Cell* **13**, 999–1010.

H. Itoh, M. Matsuoka and C.M. Steber (2003) A role for the ubiquitin–26S-proteasome pathway in gibberellin signaling. *Trends in Plant Science* **8**, 492–497.

H. Itoh, M. Ueguchi-Tanaka, Y. Sato, M. Ashikari and M. Matsuoka (2002) The gibberellin signaling pathway is regulated by the appearance and disappearance of SLENDER RICE1 in nuclei. *The Plant Cell* **14**, 57–70.

C.M. Karssen and E. Lacka (1986) A revision of the hormone balance theory of seed dormancy: studies on gibberellin- and/or abscisic acid-deficient mutants of *Arabidopsis thaliana*. In: *Plant Growth Substances 1985* (ed. M. Bopp), pp. 315–323. Springer-Verlag, Heidelberg.

C.M. Karssen, S. Zagorski, J. Kepczynski and S.P.C. Groot (1989) Key role for endogenouse gibberellins in the control of seed germination. *Annals of Botany* **63**, 71–80.

S. Kepinski and O. Leyser (2005) The *Arabidopsis* F-box protein TIR1 is an auxin receptor. *Nature* **435**, 446–451.

K.E. King, T. Moritz and N.P. Harberd (2001) Gibberellins are not required for normal stem growth in *Arabidopsis thaliana* in the absence of *GAI* and *RGA*. *Genetics* **159**, 767–776.

M. Koornneef, A. Elgersma, C.J. Hanhart, E.P. van Loenen-Martinet, L. van Rijn and J.A.D. Zeevaart (1985) A gibberellin insensitive mutant of *Arabidopsis thaliana*. *Plant Physiology* **65**, 33–39.

M. Koornneef and J.H. van der Veen (1980) Induction and analysis of gibberellin sensitive mutants in *Arabidopsis thaliana* (L.) Heynh. *Theoretical and Applied Genetics* **58**, 257–263.

S. Lee, H. Cheng, K.E. King, *et al.* (2002) Gibberellin regulates *Arabidopsis* seed germination via *RGL2*, a *GAI/RGA*-like gene whose expression is up-regulated following imbibition. *Genes and Development* **16**, 646–658.

G. Leubner-Metzger (2001) Brassinosteroids and gibberellins promote tobacco seed germination by distinct pathways. *Planta* **213**, 758–763.

M. Margis-Pinheiro, X.R. Zhou, Q.H. Zhu, E.S. Dennis and N.M. Upadhyaya (2005) Isolation and characterization of a Ds-tagged rice (*Oryza sativa* L.) GA-responsive dwarf mutant defective in an early step of the gibberellin biosynthesis pathway. *Plant Cell Reports* **23**, 819–833.

K.M. McGinnis, S.G. Thomas, J.D. Soule, *et al.* (2003) The *Arabidopsis SLEEPY1* gene encodes a putative F-box subunit of an SCF E3 ubiquitin ligase. *The Plant Cell* **15**, 1120–1130.

R.S. McKibbin, M.D. Wilkinson, P.C. Bailey, *et al.* (2002) Transcripts of *Vp-1* homeologues are misspliced in modern wheat and ancestral species. *Proceedings of the National Academy of Sciences of the United States of America* **99**, 10203–10208.

A. Muangprom, S.G. Thomas, T. Sun and T.C. Osborn (2005) A novel dwarfing mutation in a green revolution gene from *Brassica rapa*. *Plant Physiology* **123**, 1235–1246.

H. Nonogaki, O.H. Gee and K.J. Bradford (2000) A germination-specific endo-β-mannanase gene is expressed in the micropylar endosperm cap of tomato seeds. *Plant Physiology* **123**, 1235–1246.

M. Ogawa, A. Hanada, Y. Yamauchi, A. Kuwahara, Y. Kamiya and S. Yamaguchi (2003) Gibberellin biosynthesis and response during *Arabidopsis* seed germination. *The Plant Cell* **15**, 1591–1604.

J. Peng, P. Carol, D.E. Richards, *et al.* (1997) The *Arabidopsis GAI* gene defines a signaling pathway that negatively regulates gibberellin responses. *Genes and Development* **11**, 3194–3205.

J. Peng, D.E. Richards, N.M. Hartley, *et al.* (1999a) 'Green revolution' genes encode mutant gibberellin response modulators. *Nature* **400**, 256–261.

J. Peng, D.E. Richards, T. Moritz, A. Cano-Delgado and N.P. Harberd (1999b) Extragenic suppressors of the *Arabidopsis gai* mutation alter the dose–response relationship of diverse gibberellin responses. *Plant Physiology* **119**, 1199–1208.

T. Potuschak, E. Lechner, Y. Paramentier, *et al.* (2003) EIN3-dependent regulation of plant ethylene hormone signaling by two *Arabidopsis* F-box proteins: EBF1 and EBF2. *Cell* **115**, 679–689.

D.L. Remington and M.D. Purugganan (2002) *GAI* homologues in the Hawaiian silversword alliance (Asteraceae-Madiinae): molecular evolution of growth regulators in a rapidly diversifying plant lineage. *Molecular Biology and Evolution* **19**, 1563–1574.

T. Sakamoto, K. Miura, H. Itoh, *et al.* (2004) An overview of gibberellin metabolism enzyme genes and their related mutants in rice. *Plant Physiology* **134**, 1642–1653.

A. Sasaki, H. Itoh, K. Gomi, *et al.* (2003) Accumulation of phosphorylated repressor for gibberellin signaling in an F-box mutant. *Science* **299**, 1896–1898.

A.L. Silverstone, C.N. Ciampaglio and T.P. Sun (1998) The *Arabidopsis* RGA gene encodes a transcriptional regulator repressing the gibberellin signal transduction pathway. *The Plant Cell* **10**, 155–169.

A.L. Silverstone, H.S. Jung, A. Dill, H. Kawaide, Y. Kamiya and T.P. Sun (2001) Repressing a repressor: gibberellin-induced rapid reduction of the RGA protein in *Arabidopsis*. *The Plant Cell* **13**, 1555–1566.

A.L. Silverstone, P.Y.A. Mak, E.C. Martinez and T.-P. Sun (1997) The new RGA locus encodes a negative regulator of gibberellin response in *Arabidopsis thaliana*. *Genetics* **146**, 1087–1099.

J. Smalle, J. Kurepa, P. Yang, E. Babiychuk, S. Kushnir and R.D. Vierstra (2002) Cytokinin growth responses in *Arabidopsis* involve the 26S proteasome subunit RPN12. *The Plant Cell* **14**, 17–32.

J. Smalle, J. Kurepa, P. Yang *et al.* (2003) The pleiotropic role of the 26S proteasome subunit RPN10 in *Arabidopsis* growth and development supports a substrate-specific function in abscisic acid signaling. *The Plant Cell* **15**, 965–980.

J. Smalle and R.D. Vierstra (2004) The ubiquitin 26S proteasome pathway. *Annual Review of Plant Biology* **55**, 555–590.

C.M. Steber, S.E. Cooney and P. McCourt (1998) Isolation of the GA-response mutant *sly1* as a suppressor of *ABI1-1* in *Arabidopsis thaliana*. *Genetics* **149**, 509–521.

C.M. Steber and P. McCourt (2001) A role for brassinosteroids in germination in *Arabidopsis*. *Plant Physiology* **125**, 763–769.

D.W. Still and K.J. Bradford (1997) Endo-β-mannanase activity from individual tomato endosperm caps and radicle tips in relation to germination rates. *Plant Physiology* **113**, 21–29.

L.C. Strader, S. Ritchie, J.D. Soule, K.M. McGinnis and C.M. Steber (2004) Recessive-interfering mutations in the gibberellin signaling gene *SLEEPY1* are rescued by overexpression of its homologue, *SNEEZY*. *Proceedings of the National Academy of Sciences of the United States of America* **101**, 12771–12776.

T.P. Sun and F. Gubler (2004) Molecular mechanism of gibberellin signaling in plants. *Annual Review of Plant Biology* **55**, 197–223.

S. Tamura (1991) Historical aspects of gibberellins. In: *Gibberellins* (eds N. Takahashi, B.O. Phinney and J. MacMillan), pp. 1–8. Springer-Verlag, New York.

S.G. Thomas, I. Rieu and C.M. Steber (2005) Gibberellin metabolism and signaling. In: *Vitamins and Hormones*, Vol. 72 (ed. G. Litwack), pp. 289–337. Elsevier, London.

L. Tyler, S.G. Thomas, J. Hu, *et al.* (2004) DELLA Proteins and gibberellin-regulated seed germination and floral development in *Arabidopsis*. *Plant Physiol*ogy **135**, 1008–1019.

M. Ueguchi-Tanaka, M. Ashikari, M. Nakajima, *et al.* (2005) *GIBBERELLIN INSENSITIVE DWARF1* encodes a soluble receptor for gibberellin. *Nature* **437**, 693–698.

C.N. White, W.M. Proebsting, P. Hedden and C.J. Rivin (2000) Gibberellins and seed development in maize. I. Evidence that gibberellin/abscisic acid balance governs germination versus maturation pathways. *Plant Physiology* **122**, 1081–1088.

C.N. White and C.J. Rivin (2000) Gibberellins and seed development in maize. II. Gibberellin synthesis inhibition enhances abscisic acid signaling in cultured embryos. *Plant Physiology* **122**, 1089–1097.

L. Xu, F. Liu, E. Lechner, *et al.* (2002) The SCF (COI1) ubiquitin-ligase complexes are required for jasmonate response in *Arabidopsis*. *The Plant Cell* **14**, 1919–1935.

S. Yamaguchi, M.W. Smith, R.G. Brown, Y. Kamiya and T. Sun (1998) Phytochrome regulation and differential expression of gibberellin 3β-hydroxylase genes in germinating *Arabidopsis* seeds. *The Plant Cell* **10**, 2115–2126.

X. Zhang, V. Garreton and N.H. Chua (2005) The AIP2 E3 ligase acts as a novel negative regulator of ABA signaling by promoting ABI3 degradation. *Genes and Development* **19**, 1532–1543.

N. Zheng, B.A. Schulman, L. Song, *et al.* (2002) Structure of the Cul1-Rbx1-Skp1-F boxSkp2 SCF ubiquitin ligase complex. *Nature* **416**, 703–709.

11 Mechanisms and genes involved in germination *sensu stricto*

Hiroyuki Nonogaki, Feng Chen and Kent J. Bradford

11.1 Introduction

The term 'germination' has a surprisingly large number of meanings. In general, the term is applied to seeds, spores, and pollen to indicate when these quiescent structures reinitiate growth. Physiologists often consider seed germination to be the period from imbibition up to the point that embryo growth is initiated and the embryo protrudes through any covering tissues. Seed quality analysts, on the other hand, do not consider germination to be complete until the seedling has grown sufficiently to be able to observe and evaluate the root, hypocotyl, and cotyledons (ISTA, 1999). Growers often consider germination to be completed only when the seedling emerges through the soil. For this chapter, we will consider germination in the strict sense (*sensu stricto*; Perino and Côme, 1991) to be the period from the start of imbibition of a dry seed until the embryo (usually the radicle) first emerges from any tissues enclosing it. Processes that are associated uniquely with germination, such as storage reserve mobilization and de-etiolation, still occur after this point, but most physiological processes in seedlings are similar to those occurring during plant growth. Gene expression patterns during germination, however, are quite distinct from those in other stages of the plant life cycle (Czechowski *et al.*, 2005; Ma *et al.*, 2005).

In addition to restricting discussion to germination *sensu stricto*, we will also focus primarily on processes associated mechanistically with the initiation and completion of embryo emergence through any covering tissues. Other chapters in this book look in depth at the regulatory and signaling pathways involved in germination and dormancy, so we will not repeat that material except to discuss the regulation of the genes and processes directly involved in embryo emergence. We also will not address reserve mobilization in any depth, as the majority of that process occurs after germination *sensu stricto*. Instead, our focus will be on the mechanisms and genes that are involved in the progress of a seed from imbibition through radicle emergence.

11.2 Imbibition and water relations of seed germination

Beginning in most cases from an air-dry state, seeds must first imbibe water in order to activate metabolic processes. Seeds having impermeable seed coats that prevent water uptake (termed 'hard seeds') remain quiescent until weathering or scarification of the seed coat allows water to penetrate. The initial uptake of water is a physical process driven by the matric potential of the seed constituents and occurs in both living and dead seeds. In most cases, viable seeds show a three-phase pattern of water uptake, where the initial rapid uptake is followed by a plateau phase of variable duration that ends with the resumption of water uptake associated with embryo growth (Figure 11.1). As seed water content increases during Phase I of imbibition (Figure 11.1), initially strong binding sites, then weak binding sites, and finally multimolecular binding sites for water become saturated (Vertucci and Farrant, 1995; Walters *et al.*, 2002). Swelling of the tissues, particularly of the

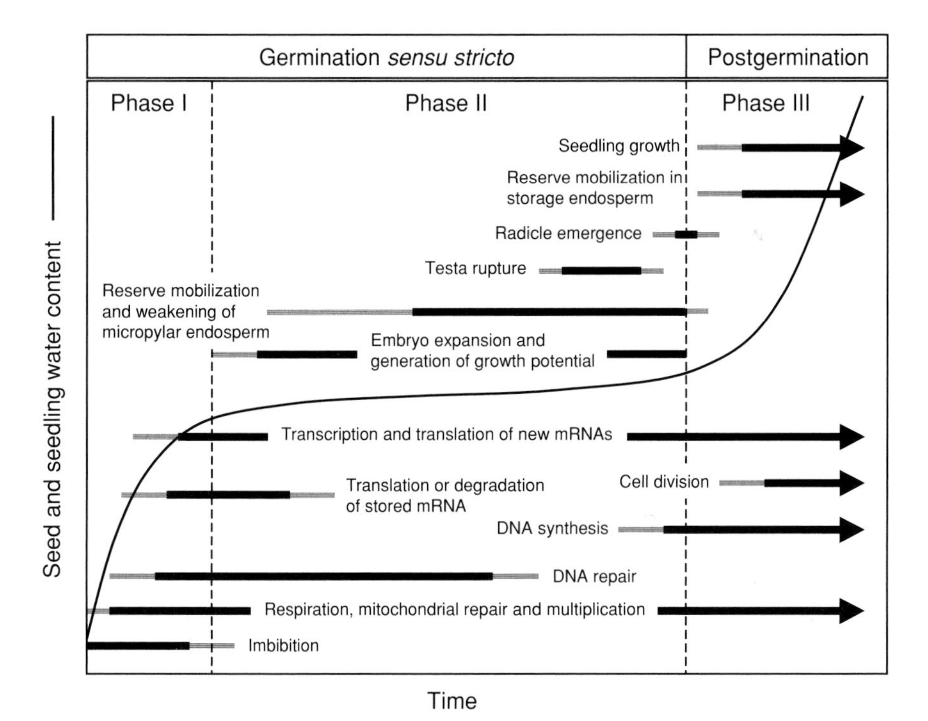

Figure 11.1 Time courses of imbibition and of some important events associated with germination and seedling growth. Initial absorption of water during Phase I is primarily a physical process, but physiological activities such as respiration, protein synthesis, and DNA repair are activated well before all tissues are completely hydrated. During Phase II of imbibition seed water content is relatively constant or only slowly increasing and metabolic activities transition from those characteristic of a developing seed to those associated with germination. Some stored mRNAs are translated, but most are degraded as new genes are transcribed. Storage reserves in the micropylar endosperm are mobilized and enzymes associated with endosperm weakening and embryo growth are synthesized. The embryo develops the turgor required to penetrate tissues enclosing it. In some seeds, testa rupture precedes endosperm rupture. Radicle emergence through the endosperm and testa marks the end of germination *sensu stricto* and the beginning of seedling growth. Although DNA synthesis may begin in the meristem prior to radicle emergence, in most cases cell division does not occur until after radicle emergence. (Modified from Bewley *et al.*, 2000.)

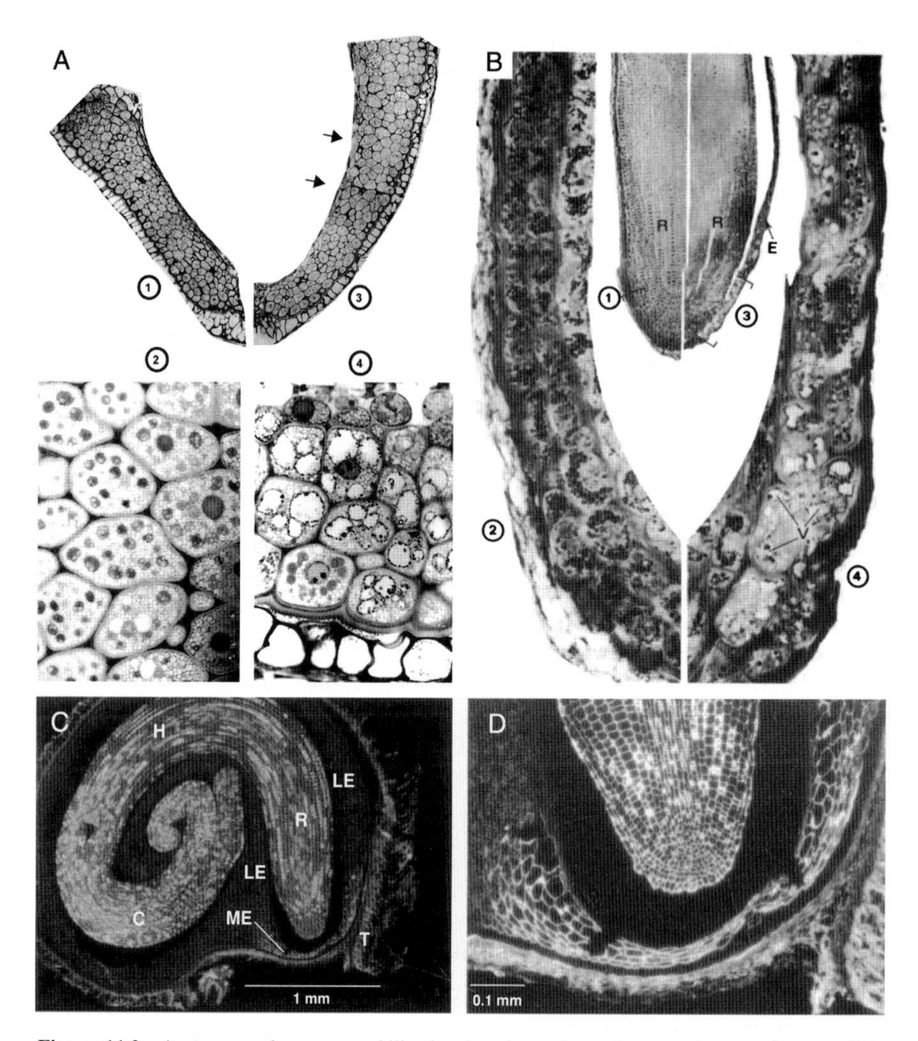

Figure 11.2 Anatomy and reserve mobilization in micropylar endosperm tissues prior to radicle emergence. (A) Endosperm tissues from *Datura ferox* seeds that have been imbibed for 48 h and treated with far-red (sections 1 and 2) or red (sections 3 and 4) light are shown in median longitudinal sections. The arrows shown in section 3 indicate a zone with extensive degradation of protein bodies and vacuolization, which can be seen in the enlarged view in 4. In contrast, a similar section of tissue from 1 (far-red treated) does not show protein body degradation or vacuolization (2). (Reprinted from Mella *et al.*, 1995, with permission from NRC Research Press.) (B) Median longitudinal sections of a lettuce embryo and endosperm imbibed in water (1) or GA (~1.5 μM) (3) for 12 h. Panels 2 and 4 show enlarged views of the two-cell-layered endosperm from the regions bracketed in subpanels 1 and 3. Note the vacuolization and loss of protein bodies from the endosperm in response to GA in subpanel 4 relative to subpanel 2. (E, endosperm; R, radicle; V, vacuoles) (Reprinted from Psaras and Georghiou, 1983, with permission from Elsevier.) (C) Median longitudinal section of a tomato seed stained with calcofluor white, which stains cellulose. Note that the embryo and testa cells are brightly stained relative to the lateral endosperm cells, indicating differences in their cell wall structure or composition. (C, cotyledons; H, hypocotyl; LE, lateral endosperm; ME, micropylar endosperm; R, radicle; T, testa) (Previously unpublished photo by B. Downie.)

protein components, accompanies this hydration (Leopold, 1983). Additional water is absorbed by the cells osmotically, due to the solutes present within the cytoplasm contained by the plasmamembrane and cell walls.

Rates of water uptake into dry seeds are controlled initially by the permeability of the testa, which may contain lignified cells and a waxy covering limiting water penetration. Water enters most rapidly through the most permeable regions of the testa, which in cereals and many other seeds is in the micropylar region (McDonald *et al.*, 1988, 1994), and then spreads throughout the seed tissues. Proton nuclear magnetic resonance (NMR) microimaging of intact seeds has revealed further details of the time course of water uptake and distribution during seed imbibition and germination (Hou *et al.*, 1997; Spoelstra, 2002; Manz *et al.*, 2005; Terskikh *et al.*, 2005). These studies confirm earlier dissection studies indicating that water distribution is not uniform among seed tissues. In particular, in tomato (*Lycopersicon esculentum*), tobacco (*Nicotiana tabacum*), and pine (*Pinus sylvestris*) seeds, water density (i.e., water per unit tissue volume) increased in the micropylar region and in the radicle tip prior to and at the time of radicle emergence. It has been suggested that these areas could represent regions of greater water availability to support metabolic processes or embryo expansion (Spoelstra, 2002; Manz *et al.*, 2005). However, the relative water densities as revealed by NMR are also dependent upon the types and amounts of cellular contents. Cells filled with lipid bodies, for example, will have lower water densities than will vacuolated cells with few lipid bodies, even if both cells are at the same total water potential (ψ). Storage body breakdown and vacuolization occur first in the same tissues where NMR reveals higher water densities (Figure 11.2), perhaps accounting for the differences observed among tissues. It remains to be tested whether this water is more available than water in other seed tissues to support metabolism or expansion. As discussed below, multiple genes are induced exclusively in the micropylar region of seeds. However, even at the later stages of imbibition when seed water content has reached a plateau, the rest of the endosperm (termed 'lateral endosperm'), which has also fully hydrated by this time, does not express these genes. Therefore, greater water availability alone cannot explain the differential metabolic activation of various seed tissues. Nonetheless, the nondestructive visualization of water (and lipid) distribution inside intact seeds by proton NMR microimaging is a highly revealing and valuable approach to study imbibition and germination.

Figure 11.2 (*Continued*) (D) Enlarged view of the micropylar region of the tomato seed in panel C. The seed had been imbibed until just prior to radicle emergence before being dehydrated and sectioned, causing the radicle tip to pull away from the micropylar endosperm, against which it would have been appressed in the hydrated seed. Note that the micropylar endosperm cells are essentially empty of storage materials compared to the adjacent lateral endosperm cells (left side of panel) and cracks are appearing. They also have thinner walls than lateral endosperm cells, even prior to imbibition (Haigh, 1988). The difference in structure between the lateral and micropylar endosperm cell walls is evident in the differential staining by calcofluor white, which could be due to degradation of the galactomannan storage polysaccharides only in the micropylar region.

Water uptake may also be influenced by the presence of water channel proteins (aquaporins) in either the seed coat or the endosperm/embryo tissues (Maurel *et al.*, 1997). Certain types of aquaporins, or α-Tonoplast Intrinsic Proteins (α-TIPs), are expressed exclusively in seeds and may be involved in the regulation of water uptake as internal membranes and vacuoles re-form during imbibition (Hofte *et al.*, 1992; Gao *et al.*, 1999). Two *TIP*-related genes are expressed during germination of rice (*Oryza sativa*) seeds, although primarily during seedling growth in association with reserve mobilization (Takahashi *et al.*, 2004). *Plasmamembrane Intrinsic Protein* (*PIP*) gene family members are also expressed in seeds, including one expressed in both the seed coat and the cotyledons of pea whose mRNA is present throughout seed development and remains abundant in dry seeds, suggesting a possible role in seed hydration (Schuurmans *et al.*, 2003). Whether membrane intrinsic proteins or aquaporins have a specific role in seed imbibition or are primarily associated with inter- and intracellular water transport during seed development and germination remains an intriguing but largely unexplored question. Genetic manipulation of aquaporin expression combined with NMR imaging of water uptake and distribution could be a powerful approach to test the role of aquaporins in seed hydration.

Following the initial rapid water uptake during Phase I, water uptake rate slows and seed water content either is constant or slowly increases during Phase II of imbibition (Figure 11.1). There are two ways that further water uptake and swelling are prevented during Phase II. The embryo cells can develop turgor, an internal hydraulic pressure that offsets the osmotic gradient for water uptake. In this case, the hydraulic pressure would be contained by the cell walls of the living tissues of the seed, preventing further expansion until the walls relax. Alternatively, the tissues external to the embryo can restrict the expansion of the embryo. That is, as the embryo expands due to water uptake, it presses against the more rigid enclosing tissues (either endosperm or testa), which then limit further expansion. The embryo still develops hydraulic pressure due to water uptake, but this pressure would be contained by the enclosing tissues, rather than by the embryo cell walls. If the embryo cell walls are sufficiently relaxed initially, this pressure against the enclosing tissues would develop during Phase II. If the embryonic cell walls are not sufficiently relaxed, the turgor pressure would be contained within the embryo and little pressure would be exerted on the external tissues. Thus, both embryo cell wall extensibility and the resistance of enclosing tissues to expansion can be involved in determining the water content achieved during Phase II of imbibition. Since dead embryos have lost membrane integrity and cannot develop turgor, they often absorb more water than do viable seeds (Hill and Taylor, 1989; Terskikh *et al.*, 2005), but do not enter Phase III. Imbibed dormant seeds also remain in Phase II until dormancy is released, as they do not progress into growth.

Seed water content can also be controlled by the ψ of the water supply available to the seed. Reducing the ψ of the imbibition solution will also reduce the internal hydrostatic pressure and seed water content, and, if the ψ is sufficiently low, prevent radicle emergence (Bradford, 1995; Toorop *et al.*, 1998). This is at least partly a physical phenomenon related to the requirement for the embryo to exert sufficient pressure to penetrate the enclosing tissues (Figure 11.3). However, reduced ψ also

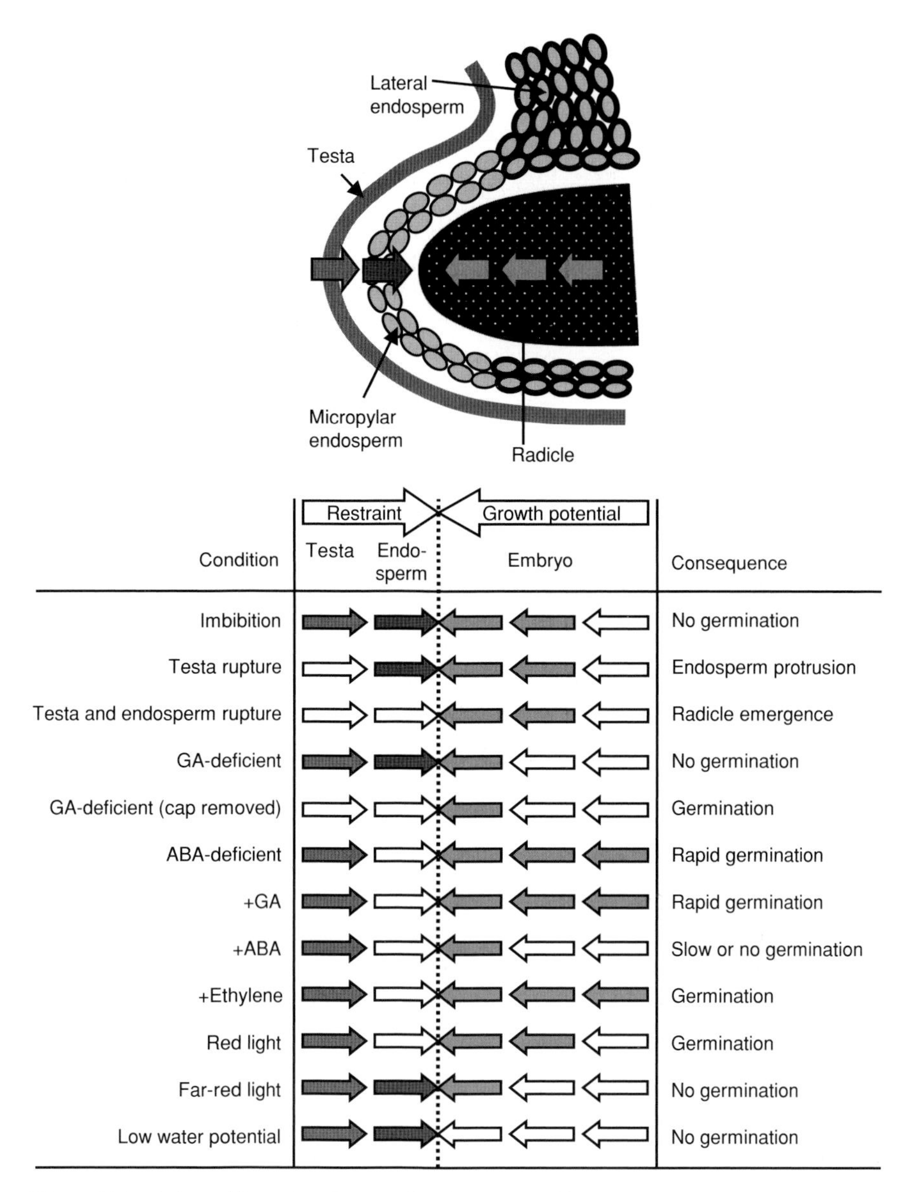

Figure 11.3 Scheme illustrating the opposing forces controlling germination of many seeds. The embryo is always enclosed within a testa (and in some cases the pericarp), and many seeds also contain endosperm tissue surrounding the radicle tip. To complete germination, the radicle must penetrate through these tissues. The table illustrates how various hormonal and environmental factors influence either the strength of these enclosing tissues or the growth potential of the embryo to exert pressure on them. Filled arrows indicate greater restraint or growth potential, while open arrows indicate weakening of restraint or reduction in embryo growth potential.

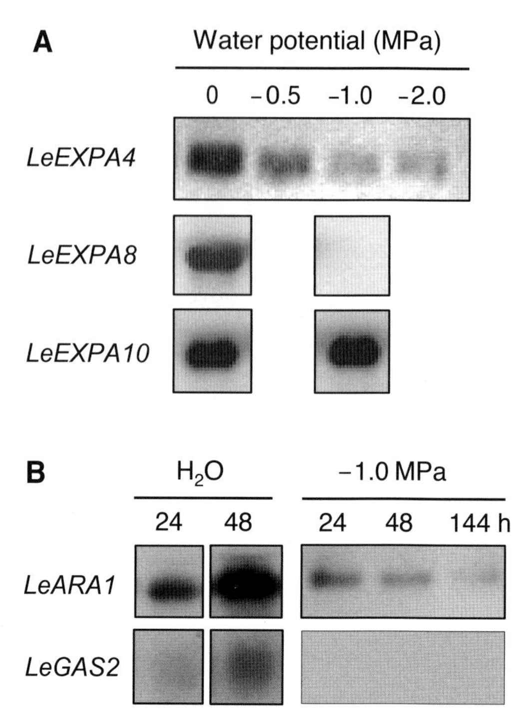

Figure 11.4 Effects of reduced water potential on gene expression during germination *sensu stricto*. (A) Three expansin genes show differential expression responses to reduced ψ. *LeEXPA4* and *LeEXPA8*, both of which are expressed during germination, are inhibited by reduced ψ during imbibition (using solutions of polyethylene glycol 8000 (PEG 8000)), while *LeEXPA10*, which is expressed in both seed development and germination, is unaffected. (Data of RNA gel blot hybridization from Chen and Bradford, 2000, and Chen *et al.*, 2001; copyright American Society of Plant Biologists.) (B) Expression of two tomato genes exhibiting high homology to arabinofuranosidases (*LeARF1*) and to a GA-stimulated transcript (*LeGAS2*) of unknown function is sensitive to reduced ψ (imbibition in –1.0 MPa PEG solution). (P. Dahal and K.J. Bradford, previously unpublished results.)

extends the duration of Phase II of imbibition even at ψ at which radicle emergence will eventually be completed. Analysis of seed germination using the hydrotime concept suggests that the rate of metabolic progress toward radicle emergence may also be sensitive to ψ (Bradford and Trewavas, 1994; Bradford, 1995; see Chapter 4). This is supported by the effects of reduced ψ on gene expression during germination *sensu stricto*. Reduced ψ inhibited the expression of several (but not all) genes whose mRNAs normally increase in tomato seeds in association with germination (Figure 11.4). For the majority of these genes, expression was not inhibited by abscisic acid (ABA), indicating that reduced ψ can act on gene expression independently of ABA (Table 11.1). Reduced ψ can also alter the germination response of *Datura ferox* seeds to light, apparently by preventing weakening of the endosperm (de Miguel and Sánchez, 1992; Sánchez *et al.*, 2002).

Table 11.1 Some genes associated with germination mechanisms that are expressed in seeds during germination *sensu stricto*

Enzyme or protein	Gene	Tissue localization	GA	ABA	Low ψ	FR	Species	References
α–L-Arabinofuranosidase	*LeARA1*	LAT	+	O	–	–	Tomato	P. Dahal, C. Itai, J.D. Bewley, K.J. Bradford, unpublished
Chitinase	*LeChi9*	CAP, RT	+	O			Tomato	Wu *et al.*, 2001; Wu and Bradford, 2003
Endo-β-mannanase	*LeMAN1*	LAT	+	O	O		Tomato	Bewley *et al.*, 1997
	LeMAN2	CAP	+	O			Tomato	Nonogaki *et al.*, 2000
	CamanA		–				Coffee	Marraccini *et al.*, 2001
	CamanB		–				Coffee	Marraccini *et al.*, 2001
	LsMan1	CAP, LAT					Lettuce	Wang *et al.*, 2004
	DfMan	EMB, CAP	+			–	Datura	Arana *et al.*, 2006
Expansin	*LeEXPA4*	CAP	+	O	–	–	Tomato	Chen and Bradford, 2000
	LeEXPA8	EMB	+	O	–		Tomato	Chen *et al.*, 2001
	LeEXPA10	EMB	+	O	O		Tomato	Chen *et al.*, 2001
	DfExpa1	EMB, CAP	+	O		–	Datura	Mella *et al.*, 2004; Arana *et al.*, 2006
α–Galactosidase	*LeaGal1*	CAP, LAT, EMB					Tomato	Feurtado *et al.*, 2001
β-1,3-Glucanase	*NtGluB*	CAP	+	–			Tobacco	Leubner-Metzger *et al.*, 1995
	LeGluB	CAP	+	–			Tomato	Wu *et al.*, 2001
β-1,4-Glucanase (cellulase)	*LeCel55*	CAP, RT, ROS	+	O	–	–	Tomato	B. Downie, unpublished
β-Mannosidase	*LeMside1*	CAP, LAT, EMB					Tomato	Mo and Bewley, 2002
Polygalacturonase	*LeXPG1*	CAP, RT		O		–	Tomato	Sitrit *et al.*, 1999
Xyloglucan endotransglycosylase/hydrolase	*LeXET4*	CAP	+	O			Tomato	Chen *et al.*, 2002
β-D-Xylosidase	*LeXYL2*	EMB	+				Tomato	C. Itai, J.D. Bewley, unpublished

Enzymes or proteins thought to have a mechanistic role in the initiation of germination that have been identified as being expressed in seeds prior to radicle emergence are listed on the left with the corresponding gene. If known, the tissue localization of expression is indicated (CAP, endosperm cap; EMB, embryo; LAT, lateral endosperm; RT, radicle tip; ROS, rest of seed except micropylar tip). The qualitative effects of GA, ABA, low water potential (Low ψ), and far-red light (FR), if known, are also indicated (+, promotes expression; O, no effect on expression; –, inhibits expression; blank, information not available). See text for details. For related genes expressed in *Arabidopsis* seeds, see Figure 11.7.

Imbibing seeds at reduced ψ is the basis of seed priming, a process utilized in the seed industry to enhance germination speed and break thermo- and photodormancy (Halmer, 2004). Microarray and proteomic studies in *Arabidopsis* and *Brassica oleracea* have found that the majority of genes whose expression is altered (either increased or decreased) during germination *sensu stricto* show similar changes during priming at reduced ψ (Gallardo *et al.*, 2001; Soeda *et al.*, 2005). However, specific subsets of genes whose expression increases during germination increased less or not at all during priming, consistent with the results discussed above. A number of genes associated with stress tolerance or pathogen defense were among this group, suggesting that these genes may be expressed only when radicle emergence is imminent (Soeda *et al.*, 2005). Thus, reduced ψ has specific effects on gene expression and protein synthesis in addition to lowering seed water content.

11.3 Testa/endosperm restraint and embryo growth potential

Testa mutants exhibiting reduced or enhanced dormancy characteristics demonstrate that the presence and/or properties of tissues external to the embryo can affect the timing and completion of germination (Debeaujon *et al.*, 2000; Downie *et al.*, 2004; see Chapter 2). In some seeds, such as those of the the Brassicaceae, the testa and endosperm tissues enclosing the embryo are thin and relatively weak, and may split during imbibition, therefore presenting little resistance to penetration by the embryo (Figure 11.3; Schopfer *et al.*, 1979). In other cases, such as tobacco, the testa may rupture during imbibition, but germination may still be controlled by an endosperm layer enclosing the embryo (Manz *et al.*, 2005). In tomato, pepper (*Capsicum annuum*), *Datura*, or lettuce (*Lactuca sativa*), the primary regulator of germination appears to be the endosperm tissues enclosing the embryo (Watkins and Cantliffe, 1983b; Groot and Karssen, 1987; Sánchez *et al.*, 1990; Sung *et al.*, 1998), although the testa may still contribute to a lesser extent (Hilhorst and Downie, 1996). Even in the absence of an external restraint, the embryo tissues must expand in order to protrude, so processes related to cell growth (embryo growth potential) can also regulate germination (de Miguel and Sánchez, 1992; Toorop *et al.*, 1998). It is therefore the balance between the restraint exerted by enclosing tissues and the force or growth potential that can be generated by the embryo that in the end determines whether radicle emergence will occur (Figure 11.3). Here we will review the roles of each of these components and possible mechanisms underlying their involvement in germination.

11.3.1 Testa and pericarp

Seeds are enclosed in a testa that develops from the integuments of the ovule. The testa is a multifunctional organ. During seed development, the testa plays an important role in embryo nutrition, as the phloem ends in the testa, and all sugars, amino acids, and mineral ions transported to the developing seed are unloaded in the testa and move apoplastically to the embryo and endosperm (Patrick and Offler,

2001). In mature seeds, the testa protects the seed against detrimental agents from the environment. As a result of this protective function, the testa may impose dormancy on the seeds due to its impermeability to water and/or oxygen, its mechanical strength that resists radicle protrusion, or both. The function of the testa in seed dormancy has been extensively studied using mutants in *Arabidopsis* and tomato (Hilhorst and Downie, 1996; Debeaujon *et al.*, 2000; Downie *et al.*, 2003, 2004) and transgenic modification in tobacco (Leubner-Metzger, 2002, 2005; see Chapter 2). Some seeds (i.e., dispersal units) also contain pericarp tissues, which develop from the ovary wall. Such seeds are actually dry fruits in which the pericarp and testa are fused, as in grass ('caryopses'), Compositae ('achenes'), and Umbelliferae ('schizocarps') species. As an embryo-covering tissue, the pericarp may also regulate seed germination by restricting radicle penetration (Ogawa and Iwabuchi, 2001).

Because of the integumental origin of the testa, its properties are mainly controlled by the genetics of the maternal plant. However, communication between the maternal testa and the embryo and/or endosperm can also affect testa attributes in tomato (Downie *et al.*, 2003). After seed maturity and desiccation, the testa is generally dead and considered to be physiologically inactive, although in some seeds, such as tobacco, a living cell layer that is interposed between the dead outer testa and the endosperm has been identified (Matzke *et al.*, 1993). Testa rupture is associated, at least partly, with an increase of water uptake during imbibition that leads to embryo swelling and initial growth. In some seeds, splitting or cracking of the testa can occur prior to, and distinct from, radicle emergence through endosperm envelopes inside the testa (see below). Interestingly, such testa cracking generally does not occur when dormant seeds are imbibed (Leubner-Metzger, 2002; Penfield *et al.*, 2004; Liu *et al.*, 2005a), indicating that release of dormancy is associated with either weakening of the testa or relaxation of the constraints on expansion of the embryo. Such after-ripening-mediated testa rupture can be replaced by overexpression of a class I β-1,3-glucanase gene in the seed-covering layers (Leubner-Metzger, 2002). Thus, the extent of embryo swelling and/or testa strength during Phase II of imbibition seems to be under physiological control.

11.3.2 Endosperm

Endosperm is formed from fusion of two haploid maternal nuclei and one haploid paternal nucleus during double fertilization. Endosperm cells are rich in food reserves such as carbohydrates (polysaccharides), oils, and proteins. In some seeds, absorption of the food reserves from the endosperm by the embryo is completed early in seed development. In these seeds, such as bean or pea, the endosperm has disappeared or is reduced to a few cell layers at seed maturity. In contrast, some seeds, such as cereals, contain a large dead endosperm and a few living cell layers of endosperm (called the aleurone layer) at maturity. In these seeds, the embryo is not covered by the endosperm tissue. The dead endosperm in the latter types of seeds serves solely as a food reserve for early seedling growth. The degradation of the cereal endosperm has been studied intensively, especially the synthesis and secretion of amylases and proteases from the aleurone layer and regulation of their synthesis by gibberellins (GA) and abcisic acid (ABA) (Jacobsen *et al.*, 2002; Gubler *et al.*, 2005).

In contrast, the seeds of many plant species at maturity contain living endosperm tissues that completely enclose the embryo. Examples of such seeds include tomato, tobacco, pepper, and *Datura* (all of which are in the Solanaceae family); lettuce (Compositae); muskmelon (*Cucumis melo*, Cucurbitaceae); and coffee (*Coffea arabica*, Rubiaceae). In these seeds, the endosperm possesses two biological functions: first, as in other types of seeds, to provide reserves for seed germination and seedling growth; and second, to control the timing of germination. Because of its complete enclosure of the embryo, the endosperm serves as a physical constraint that has to be overcome in order for the radicle tip to emerge. The endosperm in such seeds functions as a major determinant of 'coat-imposed' dormancy.

The regulatory role of the endosperm in germination is realized mainly through the micropylar endosperm. In tomato and other endospermic seeds, the micropylar endosperm forms a cone that encloses the radicle tip. It has therefore been called the 'endosperm cap'. The micropylar endosperm can be anatomically distinguished from the remainder of the endosperm, or the lateral endosperm. In tomato seeds, for example, the micropylar endosperm is composed of smaller cells with thinner cell walls compared to the lateral endosperm (Figure 11.2; Haigh, 1988).

The micropylar endosperm undergoes dramatic changes during germination *sensu stricto*. The inner surface of the endosperm cell walls is initially smooth and becomes increasingly degraded following imbibition (Sánchez *et al.*, 1990; Nonogaki *et al.*, 1992; Welbaum *et al.*, 1995). Similar changes occur in the micropylar region of the megagametophyte that encloses the embryo of spruce (*Picea glauca*) seeds (Downie *et al.*, 1997a). The storage reserves in the micropylar cells are degraded and the cells become highly vacuolated prior to radicle emergence, while these changes do not occur in adjacent cells of the lateral endosperm until after radicle emergence (Figure 11.2). The division between the two types of endosperm is sharp, with adjacent cells exhibiting either a micropylar or a lateral endosperm type of response. These cellular changes are associated with physical weakening of the endosperm cap tissue. Using radicle-sized probes and force analyzers, several groups have measured the physical force that is needed to break through the isolated micropylar endosperms (or megagametophyte/nucellar tissue) of tomato, muskmelon, pepper, coffee, and spruce seeds (Watkins and Cantliffe, 1983b; Groot and Karssen, 1987; Welbaum *et al.*, 1995; Downie *et al.*, 1997b; Chen and Bradford, 2000; Wu *et al.*, 2001; da Silva *et al.*, 2004). In all cases, the micropylar endosperm physically weakened following imbibition and prior to radicle emergence (Figure 11.5).

Endosperm weakening is a prerequisite for radicle protrusion in such seeds. In tomato, this conclusion was drawn based on a study with GA-deficient mutant (*gib-1*) seeds that fail to complete germination without exogenous GA. When the micropylar endosperm of *gib-1* seeds is removed, however, embryo expansion can occur without GA (Figure 11.4) (Groot and Karssen, 1987). Application of GA results in weakening of the endosperm cap and allows radicle protrusion (Figure 11.5). In some seeds, the endosperm cap region stretches and protrudes through the testa, forming a protuberance as the embryo expands until the cap is penetrated by the radicle tip (Watkins and Cantliffe, 1983a; de Miguel and Sánchez, 1992; da Silva

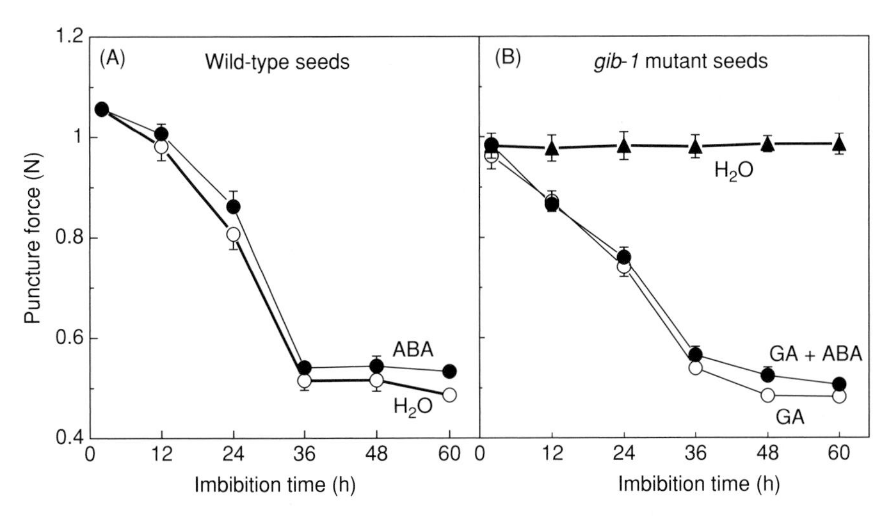

Figure 11.5 Weakening of the micropylar endosperm cap tissue of tomato seeds. The force required to penetrate the excised endosperm caps of (A) wild-type cv. Moneymaker or (B) gibberellin-deficient *gib-1* mutant tomato seeds was measured using a texture analyzer essentially as described by Groot and Karssen (1987). Endosperm weakening was not prevented by ABA (100 μM) in wild-type seeds (A); nor did ABA counteract the promotive action of GA_{4+7} (100 μM) on weakening in *gib-1* seeds (B). (Data from Chen and Bradford, 2000.)

et al., 2004; Manz *et al.*, 2005). The effect of ABA, a seed germination inhibitor, on endosperm weakening is less clear. In imbibed tomato seeds, ABA did not prevent the initial weakening of the endosperm and did not counteract the stimulatory effect of GA on weakening (Figure 11.5) (Chen and Bradford, 2000; Toorop *et al.*, 2000; Wu *et al.*, 2001). However, it does inhibit the completion of germination, and it has been proposed that there is a second phase of endosperm weakening that is sensitive to inhibition by ABA (Toorop *et al.*, 2000). While not all data from tomato show a second phase of weakening (Wu *et al.*, 2001), an ABA-sensitive second phase of weakening is clearly evident in coffee (da Silva *et al.*, 2004). It is also likely that ABA reduces the growth potential of the embryo, so that it is unable to penetrate even the weakened endosperm; i.e., the second phase of weakening actually may be due to pressure from the expanding embryo (da Silva *et al.*, 2004). ABA has been reported to reduce embryo growth potential (Schopfer and Plachy, 1984, 1985; Ni and Bradford, 1992; da Silva *et al.*, 2004). Seeds of a tomato mutant deficient in ABA synthesis (sit^w) are able to germinate at lower ψ (Groot and Karssen, 1992; Ni and Bradford, 1993; Hilhorst and Downie, 1996), even though endosperm plus testa strength is not initially any less in the majority of mutant seeds compared to wild type (Downie *et al.*, 1999). Differential effects of GA and ABA on the endosperm and embryo are also evident in their regulation of gene expression (see below).

Weakening of the micropylar endosperm can be influenced by environmental factors as well. For example, weakening of the tomato micropylar endosperm is inhibited by low ψ and far-red light (Chen and Bradford, 2000). The *Datura ferox* system has been particularly valuable in identifying the important role of phytochrome in regulating endosperm weakening in that species (de Miguel and Sánchez, 1992; de Miguel *et al.*, 2000; Sánchez *et al.*, 2002; Mella *et al.*, 2004; Arana *et al.*, 2006).

The effects of the various factors discussed above on the restraint offered by the testa and endosperm or on the growth potential of the embryo are summarized in Figure 11.3. Not all of these situations are applicable to all species, but the overall scheme illustrates how different factors can affect either the envelope restraint or the ability of the embryo to exert the pressure required for emergence, resulting in rapid, normal, slow, or no germination.

11.3.3 Cell wall proteins and hydrolases involved in weakening of covering tissues

The mechanisms leading to endosperm weakening have been of interest to seed biologists since at least the early 1960s, when Ikuma and Thimann (1963) suggested, based on their studies of lettuce seed germination, that the endosperm is weakened through enzyme-mediated degradation. Because cell walls determine the shape and strength of cells, cell wall hydrolases have been the focus of many subsequent studies. In the past two decades, the biochemical and molecular mechanisms responsible for endosperm cell wall modification have been extensively studied. A number of cell wall proteins and hydrolases and the genes encoding them have been identified that are likely to have a mechanistic role in endosperm weakening (summarized in Table 11.1 and Figure 11.6).

11.3.3.1 Expansins

Expansins are novel cell wall proteins that have the ability to cause extension of isolated cell walls. Expansin genes are highly conserved in higher plants and are often present as gene families with multiple members (Cosgrove *et al.*, 2002). *In silico* analyses of the *Arabidopsis* and rice genomes revealed the existence of 26 and 33 expansin genes (α-type) in the two genomes, respectively, and 14 expansin genes have been reported in tomato (Cosgrove, 2005). Multiple expansin genes often are expressed in association with developmental events such as root hair initiation or fruit growth (Brummell *et al.*, 1999; Harrison *et al.*, 2001; Cho and Cosgrove, 2002). They are also involved in processes such as fruit ripening and abscission where cell wall modification occurs without expansion (Rose *et al.*, 1997; Cho and Cosgrove, 2000). Thus, it is likely that expansins may be involved in both embryo growth and endosperm weakening during germination.

At least three expansin genes are expressed in tomato seeds during germination. One of them, *LeEXPA4* (previously termed *LeEXP4*; Kende *et al.*, 2004), was specifically expressed in the micropylar endosperm cap region within 12 h of imbibition (Figure 11.6), a time when endosperm weakening had just begun (Figure 11.5; Chen and Bradford, 2000). GA induced the expression of *LeEXPA4* in GA-deficient *gib-1* mutant seeds, and in wild-type seeds, *LeEXPA4* mRNA accumulation was blocked by far-red light and decreased by low water potential. Although ABA prevented radicle emergence, it did not inhibit *LeEXPA4* expression in wild-type seeds or in *gib-1* seeds in the presence of GA, consistent with the inability of ABA to block

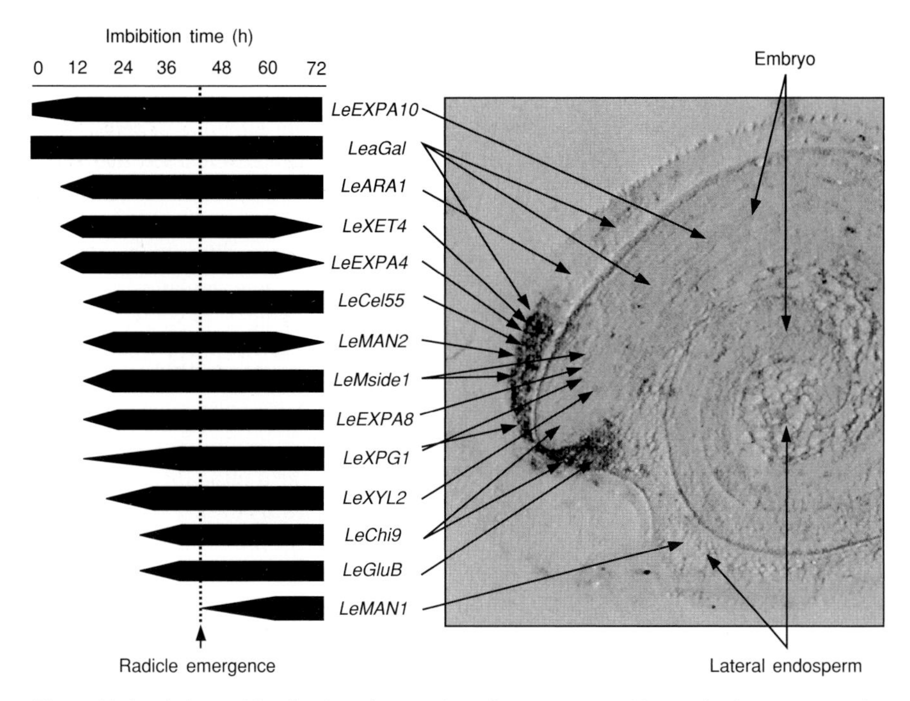

Figure 11.6 Timing and localization of expression of genes expressed in germinating tomato seeds. The chart at left indicates the times at which mRNAs of the indicated genes become detectable or disappear. The vertical dotted line indicates the time when radicle emergence occurs (~42 h). The image at the right shows a tissue print of a tomato seed expressing a gene specifically in the micropylar endosperm cap (stained region). The arrows from the genes to the image indicate the tissues in which expression is detected (the endosperm cap, the lateral endosperm, or the embryo). For gene identifications, see Table 11.1.

endosperm weakening (Figure 11.5). An expansin transcript was also detected in the micropylar endosperms of germinating seeds of *Datura* prior to radicle protrusion (Mella *et al.*, 2004). Expression of this gene, which has 90% nucleotide identity with *LeEXPA4*, was under the control of the low fluence (red/far-red reversible) response of phytochrome and was stimulated by GA but not repressed by ABA, as in tomato. The close parallels between specific expansin expression and endosperm weakening suggest that *LeEXPA4* in tomato or its homologues in other species are involved in this process during germination.

11.3.3.2 Xyloglucan endotransglycosylase/hydrolases

Xyloglucan endotransglycosylase/hydrolases (XTHs) modify xyloglucans, which are major components of primary cell walls in dicots (Carpita and Gibeaut, 1993). XTHs play divergent physiological roles as indicated by their involvement in many developmental processes, including cell growth, fruit ripening, and reserve mobilization following germination in xyloglucan-storing seeds (Tine *et al.*, 2000; Fry, 2004). An *XTH* gene, *LeXET4* (XTHs were previously termed xyloglucan endotransglycosylases, or XETs), was isolated from germinating tomato seeds using a homology-based polymerase chain reaction (PCR) approach combined with cDNA library screening (Chen *et al.*, 2002). *LeXET4* was not seed specific, as its expression could be detected also in stems, although at a lower level. During seed germination, *LeXET4* mRNA was detectable 12 h after imbibition, with maximal expression at 24 h. Tissue prints showed that *LeXET4* mRNA was localized exclusively to the endosperm cap region (Figure 11.6). Like *LeEXPA4*, expression of *LeXET4* was GA dependent but was not inhibited by ABA. *LeXET4* mRNA disappeared after radicle emergence, even though degradation of the lateral endosperm cell walls continued. This suggests that the function of *LeXET4* is specific to the micropylar endosperm. It is possible that other *XTH* genes remain to be discovered that are expressed in the lateral endosperm and/or embryo, analogous to the situation for endo-β-mannanase (Nonogaki *et al.*, 2000). In *Arabidopsis* seeds, three *XTH* genes (*AtXTH5*, *AtXTH9*, and *AtXTH31*) were upregulated within 6 h and two more (*AtXTH3* and *AtXTH33*) after 12 h of imbibition (Figure 11.7) (Ogawa *et al.*, 2003).

Xyloglucan, the substrate of XTH, does not seem to be a major hemicellulosic polysaccharide in endosperm cell walls of tomato seeds based on the low amounts of xylose detected (Dahal *et al.*, 1997). This raises a question about the significance of *LeXET4* in endosperm cap weakening. The cellulose/xyloglucan network in dicots is the major 'loading–bearing' association that determines the mechanical properties of primary cell walls (Scheible and Pauly, 2004). Weakening of this structure is

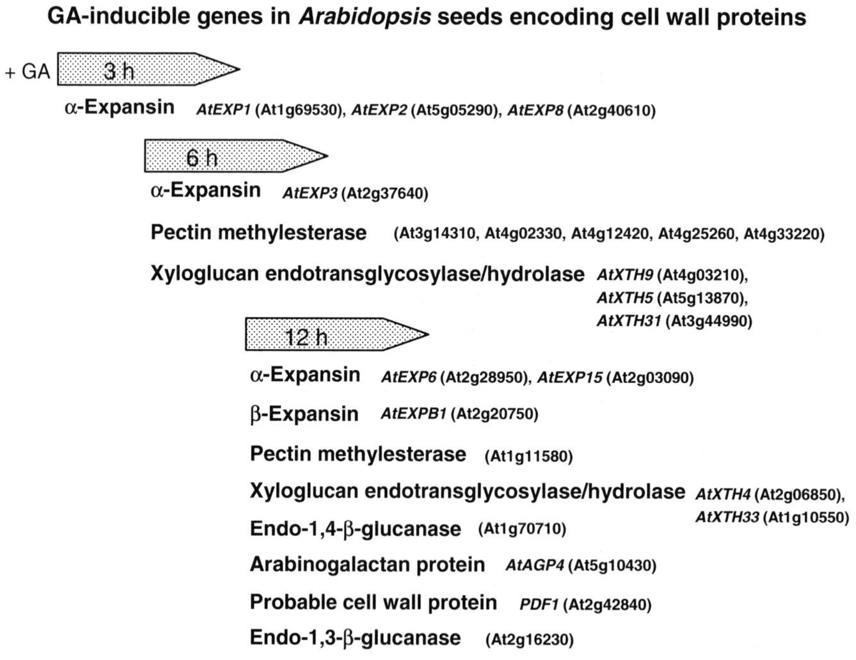

GA-inducible genes in *Arabidopsis* seeds encoding cell wall proteins

+ GA 3 h

α-Expansin *AtEXP1* (At1g69530), *AtEXP2* (At5g05290), *AtEXP8* (At2g40610)

6 h

α-Expansin *AtEXP3* (At2g37640)

Pectin methylesterase (At3g14310, At4g02330, At4g12420, At4g25260, At4g33220)

Xyloglucan endotransglycosylase/hydrolase *AtXTH9* (At4g03210), *AtXTH5* (At5g13870), *AtXTH31* (At3g44990)

12 h

α-Expansin *AtEXP6* (At2g28950), *AtEXP15* (At2g03090)

β-Expansin *AtEXPB1* (At2g20750)

Pectin methylesterase (At1g11580)

Xyloglucan endotransglycosylase/hydrolase *AtXTH4* (At2g06850), *AtXTH33* (At1g10550)

Endo-1,4-β-glucanase (At1g70710)

Arabinogalactan protein *AtAGP4* (At5g10430)

Probable cell wall protein *PDF1* (At2g42840)

Endo-1,3-β-glucanase (At2g16230)

Figure 11.7 Timing of expression of cell-wall-related enzymes and proteins during early imbibition of *Arabidopsis* seeds. Based on microarray analyses of mRNA abundance by Ogawa *et al.* (2003).

believed to be essential for loosening primary cell walls during growth. Although present in low amounts, if xyloglucan is integral to the structure of endosperm cell walls, its degradation could still be important for the disassembly of the overall structure. Further studies of XTH enzyme activity, localization, and substrates are needed to determine the mechanistic role of XTH *in vivo* during germination (Fry, 2004).

11.3.3.3 Endo-β-mannanase, α-galactosidase, and β-mannosidase

Mannans are hemicellulosic polysaccharides present in the endosperm of seeds of a large number of plant species. Some seeds such as carob (*Ceratonia siliqua*), guar (*Cyamopsis tetragonolobus*), and fenugreek (*Trigonella foenum graecum*) have high percentages of galactose (20–45%) attached to the mannan chain, resulting in their gummy or mucilaginous properties (Reid and Edwards, 1995). The endosperm cell walls of tomato, tobacco, or *Datura ferox* contain ~60% mannose and ~10% galactose as galactomannans (Groot *et al.*, 1988; Sánchez *et al.*, 1990; Dahal *et al.*, 1997; Reid *et al.*, 2003). The lower percentage of galactose in the galactomannans of these seeds results in a rigid structure with considerable mechanical strength. The even lower percentage of galactose in coffee mannan (~2% or less) results in the hard and brittle properties of its endosperm (Reid and Edwards, 1995). Degradation of mannan polymers involves endo-β-mannanase, galactosidase, and mannosidase, all of which have been identified in germinating seeds.

Endo-β-mannanases randomly hydrolyze the internal β-1,4-D-mannopyranosyl linkage in the backbone of mannan polymers. In the seeds of a number of plant species, endo-β-mannanase activity shows a dramatic increase during germination. Such activity is often contributed by different electrophoretic isoforms of endo-β-mannanase that are produced in a temporally- and spatially-specific manner (Dirk *et al.*, 1995). In germinating and germinated tomato seeds, for example, endo-β-mannanases are expressed sequentially in different parts of the tomato endosperm, initially in the micropylar endosperm (germinative expression) and subsequently in the remaining lateral endosperm (postgerminative expression) surrounding the rest of the embryo (Nonogaki *et al.*, 1998, 2000). These different isoforms of endo-β-mannanase are produced by the expression of different genes, as revealed by the isolation of multiple endo-β-mannanse genes that show distinct temporal- and spatial specificities in the germinating seeds (Figure 11.6; Nonogaki *et al.*, 2000).

Two endo-β-mannanase genes, *LeMAN1* and *LeMAN2*, have been isolated from germinating tomato seeds. *LeMAN1* was isolated based on the partial amino acid sequences of a purified endo-β-mannanase (Bewley *et al.*, 1997) and *LeMAN2* was isolated by screening a cDNA library of germinating tomato seeds, using *LeMAN1* as a probe (Nonogaki *et al.*, 2000). *LeMAN1* and *LeMAN2* displayed different timing of expression during tomato seed germination. While *LeMAN2* mRNAs were detectable by 12–18 h after imbibition, *LeMAN1* was not expressed until germination was completed at about 48 h. The two genes also showed distinct tissue specificity: expression of *LeMAN2* was restricted to the endosperm cap, while *LeMAN1* transcripts were localized to the lateral endosperm (Figure 11.6). LeMAN1 is likely responsible for reserve mobilization in the lateral endosperm that provides resources for early seedling growth following germination *sensu stricto*. When expressed in *Escherichia coli*, LeMAN2 protein exhibited endo-β-mannanase activity *in vitro* and caused the release of mannose when incubated with isolated endosperm cap cell walls, implying that LeMAN2 is involved in degradation of galactomannans in the endosperm cap. *LeMAN2* expression was induced by GA in GA-deficient *gib-1* mutant seeds, but was not inhibited by ABA in wild-type seeds (Nonogaki *et al.*, 2000), consistent with endosperm cap weakening under both conditions (Figure 11.5). While correlative evidence is strong that LeMAN2 is involved in the endosperm-cap-weakening process, specific suppression of its expression during germination is needed to definitively confirm that it is required for endosperm weakening.

In *Datura ferox* seeds, red light promotes expression of an endo-β-mannanase gene (*DfMan*) in the micropylar endosperm and an increase in enzyme activity, and also increases the response of the gene to GA (Arana *et al.*, 2006). The presence of the embryo enhances *DfMan* expression in the endosperm, an effect that can be replaced by a cytokinin (zeatin) (Arana *et al.*, 2006). Expression of homologues (*DfMYB* and *DfPHOR*) of two genes involved in GA signal transduction in barley (*GAMYB*) and potato (*Solanum tuberosum*) (*PHOR1*), respectively (Gubler *et al.*, 1995; Amador *et al.*, 2001), is also modulated by phytochrome in the micropylar endosperm of *Datura ferox*. Expression of a third germination-related gene (*DfSPY*, a homologue of barley *SPY*; Jacobsen *et al.*, 1996), was not affected by light, the embryo, or zeatin (Arana *et al.*, 2006). Thus, some components of the GA-signaling pathway identified in other plant tissues appear to be involved in the response of the micropylar endosperm to the hormone and in the modulation of that response by the embryo.

Lettuce seeds are another important model for studying the function of endo-β-mannanses in germination. Several isoforms of endo-β-mannanase, including three major ones, were detected in germinating lettuce seeds, using isoelectric focusing (Nonogaki and Morohashi, 1999; Wang *et al.*, 2004). Endo-β-mannanase activity during lettuce seed germination is promoted by GA but suppressed by ABA (Halmer *et al.*, 1976; Dulson *et al.*, 1988). High temperature (>25°C) during imbibition can delay or completely inhibit lettuce seed germination, which can be mitigated by ethylene treatment. Endo-β-mannanase activity was significantly increased in germinating lettuce seeds by ethylene treatment (Nascimento *et al.*, 2000), leading to the hypothesis that the promotion of seed germination after ethylene treatment is due to the endosperm weakening resulting from elevated endo-β-mannanase production (Nascimento *et al.*, 2004). An endo-β-mannanase gene (*LsMan1*) was isolated from lettuce seeds (Wang *et al.*, 2004). This gene was endosperm specific and its expression was detected only after lettuce seeds had germinated, consistent with enzyme assays and activity tissue prints that also did not show endo-β-mannanase activity prior to radicle emergence (Nonogaki and Morohashi, 1999). On the other hand, lettuce endosperm cell walls were capable of autohydrolysis when isolated prior to radicle emergence, possibly due to an endo-β-mannanase activity tightly bound to the walls (Dutta *et al.*, 1997). As multiple endo-β-mannanase genes were suggested by Southern blots of lettuce DNA (Wang *et al.*, 2004), it is possible that some genes in this family may be ethylene responsive or expressed at low levels prior to radicle emergence. The preponderance of data, however, suggests that endo-β-mannanase in lettuce endosperm is associated with reserve mobilization closely following radicle emergence rather than with endosperm weakening prior to radicle emergence. This is consistent with the inhibition of *LsMan1* expression by ABA (Wang *et al.*, 2004), as expression of *LeMAN1* in the lateral endosperm of tomato is inhibited by ABA (H. Nonogaki and Y. Morohashi, unpublished results), while the endosperm-cap-specific expression of *LeMAN2* is not (Nonogaki *et al.*, 2000).

Endo-β-mannanase genes have also been cloned from coffee seeds. As in tomato, two distinct endo-β-mannanase genes, *manA* and *manB*, were isolated from germinating coffee grains (Marraccini *et al.*, 2001). Both genes appeared to be expressed only in germinated seeds, as no expression was detected in other tissues, including maturing seeds. Expression of the two genes was first detected coincident with the initiation of radicle emergence, then peaked and declined after radicle emergence. Surprisingly, the expression of *manA* and *manB* as well as endo-β-mannanase activity was reported to be inhibited by exogenous GA_3 (Marraccini *et al.*, 2001). On the other hand, exogenous GA_{4+7} also inhibits germination of coffee seeds and causes embryo death, while endogenous GAs nonetheless seem to be required for germination (da Silva *et al.*, 2005). The latter study also found a marked stimulation of endo-β-mannanase activity by GA_{4+7}, while GA-synthesis inhibitors (paclobutrazol and tetcyclasis) inhibited the appearance of activity. Endo-β-mannanase activity was present in the coffee endosperm cap, but unlike in tomato, the increase in its activity was strongly suppressed by ABA (da Silva *et al.*, 2004). Apparently, while some endogenous GA is required for endosperm weakening and embryo growth, excess GA may 'desynchronize' these processes and in some way damage the embryos of coffee (da Silva *et al.*, 2005).

α-Galactosidases remove the galactose side-chains attached to the mannan backbone of galactomannans. α-Galactosidase activity has been detected in the micropylar and lateral endosperm and in the embryo of germinating tomato seeds, and a gene encoding an α-galactosidase, *LeaGal*, was cloned by screening a germinating seed cDNA library using the coffee bean galactosidase cDNA as a probe (Feurtado *et al.*, 2001). *LeaGal* transcripts were present in both the micropylar and the lateral endosperm, and also in the embryo to a lesser extent (Figure 11.6), suggesting that *LeaGal* is responsible for the α-galactosidase activities identified in various tissues of germinating tomato seeds (Feurtado *et al.*, 2001). Activity was essentially constitutive, being present throughout imbibition and after radicle emergence (Figure 11.6). This, and its localization to multiple tissues, suggests that the LeaGal1 α-galactosidase plays a wider role in the plant in addition to its involvement in micropylar endosperm weakening.

β-Mannosidases hydrolyze the terminal, nonreducing β-D-mannose residues in the oligo-mannans released by the action of endo-β-mannanases and α-galactosidases. β-Mannosidase activity has been detected in the micropylar endosperm of a number of seeds, including tomato (Mo and Bewley, 2002), *Datura* (de Miguel *et al.*, 2000), and coffee (da Silva *et al.*, 2005). In both tomato and coffee seeds, the activity of β-mannosidase increased in a similar temporal pattern as that of endo-β-mannanase (Mo and Bewley, 2002; da Silva *et al.*, 2005). The β-Mannosidase in tomato seed was purified to homogeneity and partially sequenced, allowing the gene encoding it (*LeMside1*) to be cloned (Mo and Bewley, 2002). *LeMside1* transcripts were detected in the micropylar endosperm 24 h after imbibition, increased after 36 h, and remained high until 48 h. *LeMside1* expression was also detected in the lateral endosperm, where the expression remained high even 72 h after imbibition (Figure 11.6). After germination, *LeMside1* mRNA was also present in the embryo (Mo and Bewley, 2002).

As endo-β-mannanase, α-galactosidase, and β-mannosidase are all required for breakdown of galactomannans, their coordinate expression during germination might be expected. However, α-galactosidase is largely constitutive, and differences in endo-β-mannanase and β-mannosidase expression have also been reported. In *Datura ferox* seeds in which micropylar endosperm weakening and germination in response to red light were reduced by low ψ, β-mannosidase activity was inhibited by low ψ while endo-β-mannanase activity was not (Sánchez *et al.*, 2002). Thus, while endo-β-mannanase and β-mannosidase expression and activity are often coordinated during germination, they are independently regulated. This conclusion was supported by an analysis of β-mannosidase and endo-β-mannanase activities in individual tomato seeds. Surprisingly, when assayed on an individual seed basis, no close correlation was found between the activities of these two enzymes (Mo and Bewley, 2003). Thus, the details of cooperation among mannan-degrading enzymes during germination require further investigation. Genetic evidence, using either mutants or transgenics in which the expression of individual genes is altered, will help clarify whether and under what conditions these mannan-degrading enzymes are involved in endosperm weakening that controls seed germination.

11.3.3.4 Cellulase, arabinosidase, xylosidase

The key enzyme for degrading cellulose is endo-β-1,4-glucanase, or cellulase. Cellulase activity has been detected in germinating seeds of a number of plant species. In tomato, cellulase activity was detected in germinating seeds but was not closely related to germination timing (Leviatov *et al.*, 1995). In contrast, cellulase activity in *Datura* seeds was closely linked with phytohormone-dependent germination (Sánchez *et al.*, 1986). An increase in cellulase activity also occurred during the initial phase of micropylar endosperm weakening in germinating coffee seeds, while an elevated level of endo-β-mannanase occurred during the second phase of weakening (da Silva *et al.*, 2004). Puncture force analysis showed that ABA had no effect on the first phase of weakening or on cellulase activity, but did inhibit both the activity of endo-β-mannanse and the second phase of weakening, implicating the involvement of cellulase and endo-β-mannanse in the first and second phases of micropylar endosperm weakening, respectively. A cDNA that shared high homology to known cellulase genes (*LeCel55*) was isolated from germinating tomato seeds (Bradford *et al.*, 2000). Its mRNA abundance increased in tomato radicles and micropylar endosperm prior to the completion of germination (Figure 11.6), and expression was reduced by low ψ, far-red light, or dormancy, consistent with the effects of these factors on germination. However, whether the cellulase encoded by *LeCel55* plays a mechanistic role in germination has not been confirmed.

Tomato endosperm cell walls contain up to 10% arabinose but little xylose, while the reverse is true for embryo cell walls (Groot *et al.*, 1988; Dahal *et al.*, 1997). A cDNA encoding a putative α-L-arabinofuranosidase (*LeARA1*) was isolated by differential display screening from germinating tomato seeds (Bradford *et al.*, 2000), and the same gene was also isolated from developing and ripening tomato fruit (termed *LeARF1*) (Itai *et al.*, 2003). Although its mRNA appeared within 12 h after imbibition, expression of *LeARA1* was restricted to the lateral endosperm of tomato seeds (Figure 11.6; P. Dahal and K.J. Bradford, unpublished results). A β-D-xylosidase gene (*LeXYL2*) was also cloned from tomato fruits (Itai *et al.*, 2003), and it was expressed almost exclusively in the embryos of imbibed tomato seeds (Figure11.6; A. Itai and J.D. Bewley, personal communication, 12 May 2004). Although both genes are expressed well prior to radicle emergence, their localization suggests that they may primarily be involved in postgermination events.

11.3.3.5 Polygalacturonase and pectin methylesterase

According to a widely accepted model (Carpita and Gibeaut, 1993), pectin functions as a matrix in which cellulose microfibrils and hemicellulosic polymers are embedded in plant primary cell walls. Pectin is especially abundant in the middle lamella and is important for maintaining cell-to-cell cohesion. Disassembly of pectin (polygalacturonic acid) by polygalacturonases (PGs) is associated with many developmental processes and has been particularly well studied in fruit ripening (Hadfield and Bennett, 1998). In germinating tomato seeds, a calcium-dependent exo-PG activity was detected and a gene (*LeXPG1*) was cloned from germinating seeds using degenerate primers corresponding to two conserved regions of known *PG* genes (Sitrit *et al.*, 1999). *LeXPG1* mRNA abundance was low during seed development, increased during imbibition, and was high in seeds that completed germination. Expression of *LeXPG1* during germination was detected in both the micropylar endosperm and the radicle tip (Figure 11.6), suggesting a role for *LeXPG1* in both tissues.

The degree of esterification of pectin, regulated by pectin methylesterase (PME), plays an important role in its rigidity. De-esterification of pectin catalyzed by PME has been suggested to be involved in cell wall weakening of the megagametophyte of yellow cedar (*Chamaecyparis nootkatensis*) seeds during dormancy breakage and germination (Ren and Kermode, 2000). In germinating tomato seeds, PME activity was detected in the micropylar endosperm, although the majority of activity was in the embryo (Downie *et al.*, 1998). A number of *PME* genes were upregulated by GA within 6–12 h of imbibition in *Arabidopsis* seeds as well (Figure 11.7; Ogawa *et al.*, 2003).

11.3.3.6 β-1,3-Glucanase and chitinase

Class I β-1,3-glucanases (βGLU) are well-known pathogenesis-related (PR) proteins involved in plant defense against fungal infection and other physiological processes (Leubner-Metzger, 2003). The involvement of βGLU in seed germination was first revealed in transgenic tobacco seeds containing a chimeric GUS (glucuronidase) reporter gene fused with the promoter of a tobacco βGLU (*GluB*) (Vogeli-Lange *et al.*, 1994). During germination, GUS activity was highly induced and specifically localized to the micropylar endosperm tissue prior to radicle emergence (Leubner-Metzger *et al.*, 1995). Expression analysis showed that *Gluβ* mRNA accumulation and enzyme activity increased just prior to endosperm rupture. Chitinases are another group of PR proteins whose expression often accompanies the expression of βGLU during plant defense responses (van Loon, 1997). In germinating tobacco seeds, chitinase activity is very low (Leubner-Metzger *et al.*, 1995), but in tomato seeds, genes encoding both βGLU and chitinase are expressed in the micropylar endosperm just prior to radicle emergence (Figure 11.6; Wu *et al.*, 2001).

The role of βGLU in germinating tobacco seeds has been studied using a transgenic approach. Overexpression of βGLU in tobacco seeds was achieved by transforming a *βGLU* gene under the control of an ABA-inducible *Cat1* promoter from castor bean (*Ricinus communis*) into tobacco (Leubner-Metzger and Meins, 2000). Thus, in addition to expression during seed development and higher initial levels of βGLU, seeds containing *Cat1::βGLU* also increase their expression of *βGLU* in response to ABA, in contrast to the inhibition (delay in expression) of the endogenous *βGLU* gene by ABA (Leubner-Metzger and Meins, 2000). Immunolocalization analysis showed that βGLU in transgenic tobacco seeds is mainly localized in the micropylar endosperm and in an inner layer of living cells in the testa (Leubner-Metzger, 2002). Overexpression of *βGLU* had no effect on the rate of testa rupture in either the presence or absence of ABA or on the time to endosperm rupture in the absence of ABA, but reduced the time to endosperm rupture by about 10–20% in the presence of ABA (Leubner-Metzger and Meins, 2000). Expression of antisense constructs of *βGLU*, on the other hand, had a delaying effect on the time to testa rupture, even though there was no effect on the expression of βGLU associated with endosperm rupture (Leubner-Metzger and Meins, 2001). Freshly harvested seeds are normally dormant when imbibed in the dark and lose this dormancy during dry storage (after-ripening). Sense expression of *βGLU* was able to partially replace this requirement for after-ripening (Leubner-Metzger and Meins, 2000), while antisense expression extended the dormancy period (Leubner-Metzger and Meins, 2001). Reciprocal crosses confirmed the maternal inheritance of the effect of sense-*βGLU* expression on germination speed, indicating that it was due to expression in the testa (Leubner-Metzger, 2002). These studies led to the conclusion that in tobacco, *βGLU* expression promotes both testa and endosperm rupture, with only the latter being sensitive to ABA (Leubner-Metzger, 2002).

Somewhat contrasting results were obtained for βGLU and chitinase expression in germinating tomato seeds. As in tobacco, one gene encoding a basic (type I) form of βGLU was expressed specifically in the micropylar endosperm in tomato just prior to radicle emergence (Figure 11.6; Wu *et al.*, 2001), and its expression is GA-dependent but suppressed by ABA. In contrast to tobacco, a chitinase gene was expressed coordinately with *βGLU*, and was also localized to the tomato endosperm cap (with lower expression in the radicle tip) but was not inhibited by ABA (Wu *et al.*, 2001). Interestingly, chitinase (but not *βGLU*) expression in the endosperm cap was also regulated by wounding and jasmonic acid (Wu and Bradford, 2003). Thus, while chitinase and *βGLU* genes are normally expressed simultaneously and are co-localized in the endosperm cap tissue of tomato seeds (Figure 11.6), they are differentially regulated by ABA, jasmonate and wounding. In addition, a different βGLU enzyme was reported to increase in tomato seeds following germination (Morohashi and Matsushima, 2000). βGLU hydrolyzes β-1,3-glucan polymers, or callose, which is present in the covering tissues of some seeds, such as muskmelon (Yim and Bradford, 1998). However, callose was not detected by staining in tomato seeds (Beresniewicz *et al.*, 1995), and when βGLU expressed in *E. coli* was incubated with isolated tomato endosperm cell walls, no hydrolytic activity was detected, suggesting that substrates for βGLU may not be present in the endosperm cell walls (Wu *et al.*, 2001). Similarly, chitins are not known to be present in the endosperm, although other substrates for chitinase, such as glucosamine-containing arabinogalactan proteins, could be present (Gomez *et al.*, 2002). Thus, it remains an open question whether βGLU and chitinase are involved in the mechanism of germination *per se* in tomato or whether they are components of a defense response to delay or prevent entry of fungi into the lateral endosperm following radicle emergence (Wu and Bradford, 2003). βGLU could be playing a role other than in endosperm weakening, as synthesis and degradation of callose is associated with closing and opening of plasmodesmata that are correlated with imposition and loss of dormancy and in intercellular signaling in shoot apical meristems (Rinne and van der Schoot, 2003).

11.3.3.7 Concerted action of cell wall hydrolases and expansins

The presence of multiple cell wall hydrolases and expansins in the endosperm during seed germination suggests that these enzymes and proteins work together to cause endosperm cell wall disassembly. Potential cooperation between expansins and XTHs has been studied in several biological processes. Whitney *et al.* (2000) found that expansin protein was active on polymer composites containing xyloglucan but not on those made with mannan polymers. From this and related studies they concluded that the cellulose/xyloglucan matrix is the site of expansin action (Whitney *et al.*, 1999, 2000). The concerted expression of an expansin gene (*LeEXP2*) and an *XTH* gene (*LeEXT1*) in early tomato fruit development and in auxin-treated tomato hypocotyl elongation supports this proposal (Catala *et al.*, 1997, 2000). Similarly, *LeEXP4* and *LeXET4* showed concerted expression in the endosperm cap during tomato seed germination and displayed similar responses to environmental and hormonal factors (Figure 11.6; Table 11.1), suggesting that the enzymes encoded by *LeEXPA4* and *LeXET4* may work cooperatively towards endosperm weakening.

Such a cooperative action may be extended to other cell wall hydrolases during different stages of cell wall disassembly. During fruit ripening, two stages of wall structure alteration are apparent (Rose *et al.*, 1998; Bennett, 2002). The first stage is associated with softening, in which expansins and XTHs are likely essential. The second stage is associated with tissue disruption and decay, for which other cell wall hydrolases such as PGs are key players. A two-stage process may also be the case for micropylar endosperm weakening. *LeEXPA4*, *LeXET4* and *LeMAN2* are all expressed specifically in the micropylar endosperm, but *LeMAN2* mRNA appears somewhat later than that of the other two (Figure 11.6). One hypothesis is that the action of *LeEXP4* and *LeXET4*, expressed at an early stage of imbibition, causes loosening of the cellulose-xyloglucan network, which in turn makes the wall matrix accessible to other cell wall hydrolases, such as endo-β-mannanase. However, it should be noted that this two-stage cell wall modification does not explain the secondary weakening hypothesis, since all above mentioned expansin and hemicellulase genes are associated with the drastic reduction in the mechanical resistance of the endosperm cap during the first (ABA-insensitive) stage (Figure 11.5).

Hypotheses concerning the functions of the individual cell wall hydrolases in endosperm weakening are mainly based on the association between enzyme activity/gene expression and endosperm weakening/seed germination. While the correlative and circumstantial evidence is quite strong, only for βGLU is there genetic evidence supporting a specific biological function for cell wall enzymes in germination. Similar evidence is desired for all other cell wall hydrolases and expansins. With antisense or RNAi strategies, the expression of individual genes can be selectively eliminated in the seed to test whether they play essential roles, and if so, what those roles are. GA appears to be a key regulator of most of these genes (Table 11.1). Interestingly, many of these genes expressed during germination *sensu stricto* are insensitive to exogenous ABA, while other typical GA-inducible genes in postgermination, such as α-amylase or *LeMAN1*, are repressed by ABA. This suggests that regulatory mechanisms differ between germinative and postgerminative gene expression patterns. The identification of the regulatory genes and transcription factors that connect hormonal signals to their targeted genes will help understand the concerted action of cell wall hydrolases.

11.3.4 Embryo growth potential

While weakening of enclosing tissues is a prerequisite for germination in many seeds, embryo growth is essential for germination of all seeds. Seeds display a wide range of embryo morphologies at seed maturity, from rudimentary to fully developed. In some seeds, such as celery (*Apium graveolens*), considerable embryo growth and development must occur within the seed prior to completion of germination (van der Toorn and Karssen, 1992). In others, only expansion of the existing embryonic structures is required. In the latter cases, DNA replication and cell cycle activity may commence prior to radicle emergence, but cell division in the meristematic tissues is generally delayed until at or after radicle emergence (de Castro *et al.*, 2000; Barrôco *et al.*, 2005). Cell cycle initiation may play an additive role in *Arabidopsis* seed germination, but its effect was relatively small and not essential, as germination was delayed but not prevented by cell cycle inhibitors or mutations in some cyclin D genes (Masubelele *et al.*, 2005). Here, we will address only the expansion of the embryo that occurs during Phase II and transition to Phase III of imbibition (Figure 11.1) and results in completion of germination *sensu stricto*.

11.3.4.1 Generation of embryo growth potential

Plant growth is dependent upon cell expansion due to water uptake. The driving force for water uptake is the osmotic or solute potential (ψ_s) of the cytoplasm and vacuole, reduced by the hydrostatic pressure or turgor (ψ_p) generated by resistance to expansion by the cell walls. Thus, at ψ equilibrium for a non-expanding cell in water the values of ψ_s and ψ_p are equal and opposite in value and net water flux into or out of the cells would be nil. For a seed imbibed in water, Phase II represents a period of near equilibrium with the external water supply, as net water uptake is either zero or only slowly increasing (Figure 11.1). Thus, either the cell walls of the embryo are not expansive and contain the pressure generated, or they are relaxed and the hydraulic pressure is exerted on the external enclosing tissues (endosperm and testa). The former is the case with true embryo dormancy, where removal of the embryo from any enclosing tissues does not result in embryo growth. As mentioned previously, even the extent of embryo swelling during Phases I and II of imbibition seems to be under physiological control in these cases, as dormant seeds often do not split their enclosing envelopes while non-dormant seeds do, even though embryo growth *per se* will not commence in the latter until considerably later. In cases where the endosperm protrudes prior to breaking under pressure from expansion of the embryo or due to further weakening, it is evident that the embryo walls are relaxed and the internal pressure generated by water uptake into the embryo is being transferred to the external envelopes.

This situation complicates the measurement of ψ and its components ψ_s and ψ_p in embryos during germination *sensu stricto*, and therefore estimation of what the embryo growth potential is. Measurement of ψ requires excision of the tissue and incubation in a psychrometer for a period of time for equilibration (30 min to an hour). In general, and particularly for seeds in which embryo expansion is limited by external tissues, such measurements have reported quite low values for embryo ψ and ψ_s (–0.6 to –6 MPa) and ψ_p values often near 0 MPa (Haigh and Barlow, 1987; Welbaum *et al.*, 1998; da Silva *et al.*, 2004, 2005). These low values are likely due to the loss of external restraint and therefore release of the pressure component of ψ in the embryo, for if the embryo was actually pushing against its enclosing tissues, its own cell walls were relaxed. Upon excision, embryo ψ would fall to near ψ_s and turgor would be low unless additional water was provided, for which there would now be a large gradient for uptake. This was observed for tomato embryos, for example, which very rapidly increased in water content once the radicle penetrated the endosperm cap or when the embryos were excised and placed in water (Haigh and Barlow, 1987; Dahal and Bradford, 1990). No such increase occurred in lettuce embryos, suggesting that the endosperm exerts less physical constraint on germination in this species (Bradford, 1990). Care should be exercised in interpreting direct ψ measurements of excised embryos, particularly from seeds having strong tissues surrounding the embryo. In cases where an external restraining tissue is present, measurement of low ψ, ψ_s and ψ_p values of embryos excised from seeds in Phase II of imbibition in water ($\psi \sim 0$ MPa) may actually indicate that the embryo was exerting high pressure against the external restraint prior to excision.

In cases where the embryo cell walls are relaxed, the determinant of embryo growth potential is ψ_s, which creates the driving force for water uptake. Results are mixed with respect to changes in ψ_s in embryos during germination *sensu stricto*, with some reports finding solute accumulation prior to radicle emergence and others finding little change in ψ_s (see Welbaum *et al.*, 1998 for details). Reserve mobilization in the endosperm may contribute to embryo growth potential. Even though the

Arabidopsis endosperm is only a single cell layer, removing it or blocking reserve mobilization in it reduced subsequent hypocotyl growth (Penfield *et al.*, 2004). Thus, the early mobilization of reserves, as is evident in micropylar endosperms (Figure 11.2) may provide solutes to the embryo to increase its growth potential.

Due to the difficulties in measuring the actual pressure that the embryo can exert to penetrate any covering tissues, embryo growth potential is often measured as the ability of the embryo to either complete germination or elongate at a range of external water potentials when intact or with the endosperm excised. One approach to this is to determine the base or minimum water potential (ψ_b) that allows germination of a given fraction of the seed population (Dahal and Bradford, 1990; Bradford and Somasco, 1994). This approach is described in detail in Chapter 4, and integrates both embryo growth potential and the restraint of external tissues. Alternatively, the ability of excised embryos to grow in water or at a range of ψ after excision can give an indication of the embryo growth potential. This approach has shown, for example, that red light and GA increase the growth potential of the embryo of *Datura ferox* while also stimulating endosperm weakening (de Miguel and Sánchez, 1992; Arana *et al.*, 2006). The *Datura* embryo also produces a signal (possibly a cytokinin) in response to red light that enhances the sensitivity of the micropylar endosperm to GA (Arana *et al.*, 2006).

11.3.4.2 Gene expression associated with embryo growth

In comparison with the number of genes known to be expressed in endosperm tissues during germination *sensu stricto*, relatively few genes have been shown to be specifically expressed in the embryo in association with germination *sensu stricto*. Two expansin genes (*LeEXPA8* and *LeEXPA10*) were expressed in tomato embryos prior to radicle emergence (Figure 11.6; Chen *et al.*, 2001). *LeEXPA10* was expressed throughout the embryo during both embryogenesis and germination, while *LeEXPA8* was specific to germination and was expressed only in the elongation zone of the radicle. Lack of expression in GA-deficient *gib-1* mutant seeds unless supplied with GA indicates that expression of both genes is dependent on the hormone. ABA did not prevent the initial elevation of transcript abundance of both genes in wild-type seeds, although with extended incubation in ABA without germination, transcript abundance declined to low levels by 96 h of imbibition. Low ψ had differential effects on expression: expression of *LeEXPA8* was prevented, while that of *LeEXPA10* was unaffected initially, then declined with extended incubation. This pattern, along with the tissue localization, suggests that *LeEXPA8* may be specifically associated with expansion of the elongation zone of the radicle during germination, while *LeEXPA10* may be involved more generally in growth throughout embryogenesis and during seedling growth. Expansin transcripts were also expressed in the embryo during germination of *Datura ferox* seeds (Mella *et al.*, 2004). In addition to similar regulation by GA and ABA as described above, expression was also reduced by far-red light, which is known to lower embryo growth potential in this species (de Miguel and Sánchez, 1992).

In *Arabidopsis*, expansins were among the earliest genes up-regulated by GA in GA-deficient (*ga1-3*) seeds following imbibition (Figure 11.7; Ogawa *et al.*, 2003). Transcripts for three α-expansins (*AtEXP1*, *AtEXP2* and *AtEXP8*) were more abundant within 3 h of imbibition, *AtEXP3* appeared by 6 h, and *AtEXP6* and *AtEXP15* after 12 h. As noted previously, cell wall hydrolases were also up-regulated by GA, including pectin methylesterases and xyloglucan endotransglycosylases by 6 h and endo-1,4-glucanases (cellulases) and endo-1,3-β-glucanases by 12 h of imbibition (Figure 11.7; Ogawa *et al.*, 2003). *In situ* analysis showed that one of the *XTH* genes (*AtXTH5*) is expressed exclusively in the embryonic axis, suggesting that this enzyme is potentially involved in cell wall loosening associated with embryo growth (Ogawa *et al.*, 2003). Pectin methylesterase activity was also expressed in the embryo during tomato seed germination, primarily in the radicle tip and in a discrete band just distal to the elongation zone after germination (Downie *et al.*, 1998). Thus, as for the endosperm cap, co-expression of expansins and cell wall hydrolases in the embryo accompanies germination and comparable expression patterns are detected across different species.

11.4 Approaches to identify additional genes involved in germination

Much of the research on the mechanisms and biochemistry of germination has focused on a few model species such as tomato, tobacco, *Datura* and lettuce where the relationships between the embryo and the endosperm in controlling germination have been investigated. While having the advantage of being large enough for dissection of seed parts and application of single-seed analyses (Still and Bradford, 1997; Nascimento *et al.*, 2001; Mo and Bewley, 2003), identification and analysis of genes and proteins involved in germination has proceeded primarily on a one-by-one candidate basis in these species. The availability of GA- and ABA-deficient mutants in tomato has greatly facilitated this work (Koornneef *et al.*, 1990; Hilhorst and Karssen, 1992), but useful germination mutants are limited in number. In contrast, a large number of mutations affecting seed development, dormancy and germination have been identified in *Arabidopsis* that have advanced our knowledge of the genetic basis of seed biology (Bentsink and Koornneef, 2002; Finkelstein *et al.*, 2002; Chapters 2, 5, 8). In addition to seed germination mutants, the extensive mutation libraries, cDNA microarrays, proteomics and bioinformatics resources make *Arabidopsis* the system of choice for fundamental genetic seed research. Methods such as activation tagging and enhancer trapping are also being applied to identify additional candidates involved in germination. As these tools and genomic resources become more available in other species, information developed in *Arabidopsis* can be translated back to agricultural species. Here, we describe how these integrated approaches are being applied to understand the mechanisms and genes involved in seed germination.

11.4.1 Transcriptome and proteome analyses

Application of high-throughput approaches such as transcriptome and proteome analyses to seed research is still limited, although there are good examples of seed development research using these techniques (Girke *et al.*, 2000; Ruuska *et al.*, 2002; Hajduch *et al.*, 2005). Ogawa *et al.* (2003) provided the first comprehensive gene expression profiles during *Arabidopsis* seed germination using oligonucleotide-based DNA microarrays. In this study, *ga1-3* mutant seeds that do not complete germination without exogenous GA (Koornneef and van der Veen, 1980) were used with or without GA application to identify GA up- and down-regulated genes. The authors pre-stratified *ga1-3* seeds at 4°C in the dark for 48 h and then incubated for 24 h under continuous white light before GA application, because the stratification treatment was known to increase tissue sensitivity to exogenous GA (Derkx and Karssen, 1993). They detected 230 and 127 GA up- and down-regulated genes, respectively, which exhibited more than four-fold changes (Ogawa *et al.*, 2003). The complete data are available from the journal website (http://www.plantcell.org/cgi/content/full/15/7/1591/DC1), providing valuable information for seed germination researchers.

The gene expression profiles reported by Ogawa *et al.* (2003) not only confirmed the involvement of cell wall proteins in early seed germination (described previously) but also identified the expression of genes encoding regulatory proteins such as MYB transcription factors, homeobox-leucine zipper proteins, basic helix-loop-helix (bHLH) transcription factors, DOF and GATA zinc finger proteins. These proteins could be upstream regulators of the cell wall proteins that potentially determine their tissue and stage specificity during seed germination. Elucidation of the biological functions of these potential upstream regulators will provide detailed information on the mechanisms of GA regulation of gene expression in germinating seeds.

Another important implication from the microarray data is that GA may be associated with ethylene and auxin biosynthesis during germination. GA up-regulates the gene encoding ACO (1-aminocyclopropane-1-carboxylic acid oxidase) that catalyzes ethylene formation, and ethylene-inducible genes such as *HOOKLESS* (*HLS1*) (Lehman *et al.*, 1996) and an ethylene receptor gene *ERS1* (*ETHYLENE RESPONSE SENSOR 1*) (Hua *et al.*, 1995, 1998), suggesting that GA activates ethylene biosynthesis and/or response. *CYP79B2* and *CYP79B3*, which potentially enhance auxin biosynthesis, were also up-regulated by GA (Ogawa *et al.*, 2003). Auxin is not typically considered to be a germination-associated hormone; however, recent studies have indicated the possibility of cross-talk between auxin and GA (Ogawa *et al.*, 2003) or ethylene (Chiwocha *et al.*, 2005) in seed germination (see Chapter 8).

Nakabayashi *et al.* (2005) applied microarrays to compare gene expression in dry and germinating *Arabidopsis* seeds. This analysis indicated that stored mRNAs from more than 12 000 different genes are present in dry seeds. It is known that the maternal mRNA stored in the unfertilized egg of sea urchin supports most of the protein synthesis in the early embryo (Davidson *et al.*, 1982). Although the biological function of stored mRNA in seeds is still largely unknown (see below), the extensive transcriptome of dry seeds implies potential involvement of stored mRNA in seed germination.

Recently, microarrays were used to characterize changes in the transcriptome of seeds of the dormant Cape Verde Islands (Cvi) accession of *Arabidopsis* in different states of dormancy (Cadman *et al.*, 2006). These studies indicated that transcriptional activity differs among seeds in primary or secondary dormancy or non-dormant states and tended to support the concept that changes in the balance of ABA and

GA synthesis/catabolism and action is associated with dormancy transitions (see Chapter 3).

Microarrays have also been used to characterize gene expression in seeds of *Brassica oleracea* (cabbage, broccoli, kale and other related vegetables). Soeda *et al.* (2005) prepared cDNA microarrays using the libraries of *B. napus* developing, germinating and germinated seeds and used them for hybridization with *B. oleracea* mRNAs from seeds at various stages of maturation, germination or priming. Since there is high sequence homology between *Brassica* and *Arabidopsis* (average sequence identity in the coding region between *Brassica* and *Arabidopsis* species is 87%; http://ukcrop.net/brassica.html), they were able to classify the unique genes from the libraries. Using this approach, the authors found that a similar set of genes were activated during seed priming and germination. This supports the hypothesis that germination-associated events are initiated during priming, some of which are completed during treatment resulting in fast and uniform germination of primed seeds upon rehydration. Although the gene expression profiles of *B. oleracea* are not as comprehensive as *Arabidopsis* due to the limited number of genes spotted on the arrays, this type of approach applied in additional species will be very informative regarding the universal molecular mechanisms underlying seed germination.

While the transcriptome provides useful information on transcription and accumulation of mRNA in seeds, some important processes for induction of germination are likely to be regulated at the protein level. Modification of existing proteins including phosphorylation and glycosylation could also play critical roles in seed germination. The importance of regulation at the protein level is also well exemplified by the degradation of DELLA proteins by the ubiquitin-26S proteasome pathway in seed dormancy breakage (see Chapter 10). Therefore, proteome analysis is another important approach for seed germination research.

About 1,300 total seed proteins extracted from germinating *Arabidopsis* seeds were resolved in two-dimensional gels and polypeptides specifically changing in abundance during germination *sensu stricto* have been identified (Gallardo *et al.*, 2001). For example, Actin 7, a fundamental component of the cytoskeleton that participates in a number of cellular processes, such as cytoplasmic streaming, cell division and elongation, and tip growth (Kost *et al.*, 1999), was expressed during germination. The amount of a WD-40 repeat protein showing high identity to the receptor of activated C kinase encoded by alfalfa *Msgb1* (*Medicago sativa Gβ-like1*) that functions in signal transduction and hormone-controlled cell division (McKhann *et al.*, 1997) also increased during germination. The WD-40 protein is possibly involved in the resumption of cell cycle activity which occurs during germination (Gallardo *et al.*, 2001). More comprehensive proteome analyses were conducted using *Arabidopsis ga1* mutant seeds and also wild-type seeds treated with GA biosynthesis inhibitor paclobutrazol, both of which were tested with or without exogenous GA. In this study also, proteins similar to the ones previously found were induced during germination *sensu stricto* (Gallardo *et al.*, 2002).

As indicated by the presence of stored mRNA in dry seeds (Nakabayashi *et al.*, 2005), pre-existing transcripts might contribute to *de novo* protein synthesis during germination. Rajjou *et al.* (2004) advanced this concept further by conducting seed germination and proteome analyses using α-amanitin, a transcriptional inhibitor of RNA polymerase II. The authors showed that radicle protrusion was not inhibited by α-amanitin, while it did block subsequent seedling growth. This was not due to impermeability of the testa, since *transparent testa 2-1* seeds that have a highly permeable testa were used for this experiment. In contrast, the translational inhibitor cycloheximide completely blocked radicle protrusion, indicating that de novo protein synthesis is indispensable for germination. Although there is general agreement about the importance of protein synthesis for seed germination, it seems still controversial whether germination can be completed without transcription of any genes. Certainly, many genes are normally expressed early in seed germination (Ogawa *et al.*, 2003). As Rajjou *et al.* (2004) proposed, microarray analyses of seed germination in the presence or absence of α-amanitin would be very informative and could lead to integration of the information from both transcriptome and proteome analyses.

11.4.2 Activation tagging and enhancer trapping

Activation tagging is an efficient approach in functional genomics (Walden *et al.*, 1994). An activation tagging vector uses four copies of an enhancer element from the constitutively active cauliflower mosaic virus 35S (CaMV 35S) promoter. When the T-DNA is inserted into the plant genome, the multiple copies of enhancers activate the transcription of nearby genes, which results in a gain-of-function mutation. This is a powerful approach, especially for large gene families with redundancy in their function. A large population of *Arabidopsis* activation tagging lines has been created. In this population, the majority of over-expressed genes were found to be immediately adjacent to the enhancers. The CaMV 35S enhancers primarily enhance an endogenous expression pattern and do not induce constitutive ectopic expression (Weigel *et al.*, 2000). Activation tagging has been used successfully to identify genes involved in flowering (Kardailsky *et al.*, 1999; Lee *et al.*, 2000), phenylpropanoid biosynthesis (Borevitz *et al.*, 2000), disease resistance (Xia *et al.*, 2004), and hormone biosynthesis (Zhao *et al.*, 2001), catabolism (Busov *et al.*, 2003) and signaling (Neff *et al.*, 1999). Interesting mutants that exhibited seedless fruit development were also isolated using this approach (Ito and Meyerowitz, 2000). Recent successful outcomes from activation tagging are the discovery of mutants over-expressing microRNA genes (Aukerman and Sakai, 2003; Palatnik *et al.*, 2003).

Salaita *et al.* (2005) successfully applied the activation tagging approach for seed germination research. The authors screened 52 000 independent activation tagging lines for seeds that germinated faster than wildtype at 10°C and recovered the three dominant mutants termed *ctg* (*cold temperature germination*) *10*, *ctg41*, and *ctg144*. The recessive mutants *ctg156* and *ctg225* were also isolated from this screen, which probably result simply from the disruption of genes responsible for germination by the activation tagging T-DNA (Salaita *et al.*, 2005). This powerful approach, as shown by research on other aspects of plant development, has not been extensively used for seed germination research and could be applied to screen for genes involved in other traits such as high temperature germination or secondary dormancy.

Another efficient tool in *Arabidopsis* functional genomics is enhancer trapping. The enhancer-trap vector contains a minimal promoter and a reporter gene (Campisi *et al.*, 1999). The minimal promoter alone drives very low or no expression of the reporter gene. When the T-DNA containing the minimal promoter::reporter construct is inserted in the vicinity of the promoter of a gene in the plant genome, the reporter gene is expressed, being driven by the native plant promoter and reflecting the tissue- and stage-specific expression of the gene. Since this approach starts with identification of the reporter gene expression in the tissue of interest at the chosen timing, it is ideal for research into germination *sensu stricto*.

Liu *et al.* (2005a) screened an *Arabidopsis* enhancer-trap population for GUS reporter gene expression in germinating seeds. A technical problem for GUS activity staining in germinating *Arabidopsis* seeds is the impermeability of testa to the GUS substrate before radicle emergence. For the analysis of genes associated with seed germination *sensu stricto*, GUS staining needs to be detected in seeds before radicle emergence. However, the substrate will not penetrate into intact *Arabidopsis* seeds when the testa covers the endosperm and the embryo. The authors focused on the period between testa and endosperm rupture. At this time, the testa has split, allowing entrance of GUS substrate into the endosperm and the embryo, but the radicle tip is still enclosed by the endosperm. This approach allowed GUS detection in *Arabidopsis* seeds before the completion of germination and over 120 enhancer-trap lines were isolated having GUS expression in seeds. These lines, termed '*Seed-GUS-Expression*' library, are available from the *Arabidopsis* Biological Resource Center for the seed research community.

More than half of these *Arabidopsis* enhancer-trap lines exhibited micropylar-specific GUS expression in germinating seeds (Liu *et al.*, 2005a). As described previously, multiple cell wall proteins are expressed exclusively in the micropylar region of the endosperm (Figure 11.5), and degradation of protein and lipid bodies which results in disappearance of cell inclusions also occurs exclusively in this region (Figure 11.2; H. Nonogaki, unpublished results), indicating that a specific activation of the micropylar tissue occurs during seed germination. The results from the enhancer-trap lines support the hypothesis that specific gene activation in the micropylar region is a widespread phenomenon in seeds of diverse species.

The T-DNA insertion sites in some of these enhancer-trap lines have been identified using genome-walking PCR. A line carrying a T-DNA insertion in the 5′ upstream region of a GATA-type zinc finger gene has been characterized (Liu *et al.*, 2005b). This gene is expressed in *Arabidopsis* seeds during cold stratification. Although the gene is not completely cold-dependent, as it can be induced at 22°C in partially dormant seeds, expression of the gene is always initiated before radicle emergence. Freshly harvested seeds of plants in which this gene has been knocked out exhibited deeper dormancy than wild-type seeds, which was rescued by exogenous GA. This is interesting because exogenous GA is usually less effective than pre-chilling treatment in terms of breaking dormancy of many flower and vegetable seeds. This indicates that the GATA zinc finger protein could be associated with GA biosynthesis. In the knockout seeds, expression of *GA20ox3* and *GA3ox1*, which are known to be induced in *Arabidopsis* seeds by cold stratification (Yamauchi *et al.*,

2004), was down-regulated, supporting the hypothesis that the GATA zinc finger protein might be an upstream regulator of GA biosynthesis. Characterization of the promoter region of the GATA zinc finger protein revealed that this gene is specifically activated in the embryonic axis of germinating seeds that are still enclosed by the endosperm. Thus, the zinc finger protein could be involved in generation of embryo growth potential to overcome mechanical resistance of the endosperm and complete radicle emergence (Liu *et al.*, 2005b). Another enhancer-trap line has identified expression of a PWWP (proline-tryptophan-tryptophan-proline) protein whose function is largely unknown in plants (T.M. Homrichhausen, P.-P. Liu and H. Nonogaki, unpublished results). Further characterization of the '*Seed-GUS-Expression*' lines will discover more genes playing critical roles in seed germination. Non-destructive, green fluorescent protein-based enhancer-trap lines have also been created for *Arabidopsis* (Engineer *et al.*, 2005) and rice (Johnson *et al.*, 2005) that can be used for seed research.

11.4.3 Potential involvement of microRNAs in seed germination

MicroRNAs (miRNAs) are naturally occurring small (21∼24 nucleotides) single-stranded RNAs that transcriptionally and post-transcriptionally regulate target genes sharing their sequences (Llave *et al.*, 2002; Bao *et al.*, 2004). In plants, the miRNA targets are computationally predicted based on the high sequence complementarity between miRNAs and their targets, most of which are genes encoding transcription factors or other regulatory proteins (Reinhart *et al.*, 2002; Rhoades *et al.*, 2002). miRNAs are involved in many aspects of plant growth and development including root formation, leaf morphogenesis, flower induction, determination of tissue identity in flower, anther development and embryogenesis (Chen, 2005).

miRNA microarrays are not commercially available for plants yet, although there are successful applications of miRNA microarrays to plant research (Axtell and Bartel, 2005). Martin *et al.* (2005) have developed techniques to extract and purify small RNAs from seeds and to screen miRNAs expressed in seeds. The screening method, though not high-throughput like microarrays, is rapid and efficient and allows simultaneous detection with multiple (∼30) miRNA probes at one time. Since only about 50 families of miRNAs are known for plants, several hybridizations can examine expression of all known miRNAs with replication. Using this method, Martin *et al.* (2005) detected major miRNAs expressed in *Arabidopsis* and tomato seeds. Interestingly, germinating seeds of these two species showed similar miRNA expression profiles, suggesting that a specific set of miRNAs plays roles in seed germination. In contrast, the authors found a shift in the expression profiles between germinative and postgerminative stages of tomato seeds.

Seed-expressed miRNAs are being identified and, as mentioned above, their target genes are computationally predicted. The next phase of research is the characterization of the target genes expressed during seed germination. miRNA-resistant target genes can be created by mutagenizing the miRNA complementary sequences in the target genes and transforming them into *Arabidopsis* (Mallory *et al.*, 2005;

Millar and Gubler, 2005). This will cause over-accumulation of the target genes that are usually repressed by miRNAs. Using this 'de-repression' approach, the biological functions of miRNA and their target genes in seed germination will be elucidated.

References

V. Amador, E. Monte, J.-L. Garcia-Martinez and S. Prat (2001) Gibberellins signal nuclear import of PHOR1, a photoperiod-responsive protein with homology to *Drosophila* armadillo. *Cell* **106**, 343–354.

M.V. Arana, L.C. de Miguel and R.A. Sánchez (2006) A phytochrome-dependent embryonic factor modulates gibberellin responses in the embryo and micropylar endosperm of *Datura ferox* seeds. *Planta* **223**, 847–857.

M.J. Aukerman and H. Sakai (2003) Regulation of flowering time and floral organ identity by a microRNA and its APETALA2-like target genes. *The Plant Cell* **15**, 2730–2741.

M.J. Axtell and D.P. Bartel (2005) Antiquity of microRNAs and their targets in land plants. *The Plant Cell* **17**, 1658–1673.

N. Bao, K.W. Lye and M.K. Barton (2004) MicroRNA binding sites in *Arabidopsis* class III HD-ZIP mRNAs are required for methylation of the template chromosome. *Developmental Cell* **7**, 653–662.

R.M. Barrôco, K. Van Pounke, J.H.W. Bergervoet, *et al.* (2005) The role of the cell cycle machinery in resumption of postembryonic development. *Plant Physiology* **137**, 127–140.

A.B. Bennett (2002) Biochemical and genetic determinants of cell wall disassembly in ripening fruit: a general model. *Hortscience* **37**, 447–450.

L. Bentsink and M. Koornneef (2002) Seed dormancy and germination. In: *The Arabidopsis Book* (eds C.R. Somerville and E.M., Meyerowitz). American Society of Plant Biologists, Rockville, MD. (http://www.aspb.org/publications/arabidopsis/), pp. 1–18.

M.M. Beresniewicz, A.G. Taylor, M.C. Goffinet and W.D. Koeller (1995) Chemical nature of a semipermeable layer in seed coats of leek, onion (Liliaceae), tomato and pepper (Solanaceae). *Seed Science and Technology* **23**, 135–145.

J.D. Bewley, R.A. Burton, Y. Morohashi and G.B. Fincher (1997) Molecular cloning of a cDNA encoding a (1→4)-β-mannan endohydrolase from the seeds of germinated tomato (*Lycopersicon esculentum*). *Planta* **203**, 454–459.

J.D. Bewley, F.D. Hempel, S. McCormick and P. Zambryski (2000) Reproductive development. In: *Biochemistry & Molecular Biology of Plants* (eds B. Buchanan, W. Gruissem and R. Jones), pp. 988–1043. American Society of Plant Biologists, Rockville, MD.

J.O. Borevitz, Y. Xia, J. Blount, R.A. Dixon and C. Lamb (2000) Activation tagging identifies a conserved MYB regulator of phenylpropanoid biosynthesis. *The Plant Cell* **12**, 2383–2394.

K.J. Bradford (1990) A water relations analysis of seed germination rates. *Plant Physiology* **94**, 840–849.

K.J. Bradford (1995) Water relations in seed germination. In: *Seed Development and Germination* (eds J. Kigel and G. Galili), pp. 351–396. Marcel Dekker, Inc., New York.

K.J. Bradford, F. Chen, M.B. Cooley, *et al.* (2000) Gene expression prior to radicle emergence in imbibed tomato seeds. In: *Seed Biology: Advances and Applications* (eds M. Black, K.J. Bradford and J. Vazquez-Ramos), pp. 231–251. CABI, Wallingford, U.K.

K.J. Bradford and O.A. Somasco (1994) Water relations of lettuce seed thermoinhibition. I. Priming and endosperm effects on base water potential. *Seed Science Research* **4**, 1–10.

K.J. Bradford and A.J. Trewavas (1994) Sensitivity thresholds and variable time scales in plant hormone action. *Plant Physiology* **105**, 1029–1036.

D.A. Brummell, M.H. Harpster and P. Dunsmuir (1999) Differential expression of expansin gene family members during growth and ripening of tomato fruit. *Plant Molecular Biology* **39**, 161–169.

V.B. Busov, R. Meilan, D.W. Pearce, C. Ma, S.B. Rood and S.H. Strauss (2003) Activation tagging of a dominant gibberellin catabolism gene (GA 2-oxidase) from poplar that regulates tree stature. *Plant Physiology* **132**, 1283–1291.

C.S.C. Cadman, P.E. Toorop, H.W.M. Hilhorst and W.E. Finch-Savage (2006) Gene expression profiles of *Arabidopsis* Cvi seeds during cycling through dormant and non-dormant states indicate a common underlying dormancy control mechanism. *The Plant Journal* **46**, 805–822.

L. Campisi, Y. Yang, Y. Yi, *et al.* (1999) Generation of enhancer trap lines in *Arabidopsis* and characterization of expression patterns in the inflorescence. *The Plant Journal* **17**, 699–707.

N.C. Carpita and D.M. Gibeaut (1993) Structural models of primary cell walls in flowering plants - consistency of molecular structure with the physical properties of the walls during growth. *The Plant Journal* **3**, 1–30.

C. Catala, J.K.C. Rose and A.B. Bennett (1997) Auxin regulation and spatial localization of an endo-1,4-β-D-glucanase and a xyloglucan endotransglycosylase in expanding tomato hypocotyls. *The Plant Journal* **12**, 417–426.

C. Catala, J.K.C. Rose and A.B. Bennett (2000) Auxin-regulated genes encoding cell wall-modifying proteins are expressed during early tomato fruit growth. *Plant Physiology* **122**, 527–534.

F. Chen and K.J. Bradford (2000) Expression of an expansin is associated with endosperm weakening during tomato seed germination. *Plant Physiology* **124**, 1265–1274.

F. Chen, P. Dahal and K.J. Bradford (2001) Two tomato expansin genes show divergent expression and localization in embryos during seed development and germination. *Plant Physiology* **127**, 928–936.

F. Chen, H. Nonogaki and K.J. Bradford (2002) A gibberellin-regulated xyloglucan endotransglycosylase gene is expressed in the endosperm cap during tomato seed germination. *Journal of Experimental Botany* **53**, 215–223.

X. Chen (2005) microRNA biogenesis and function in plants. *FEBS Letters* **579**, 5923–5931.

S.D. Chiwocha, A.J. Cutler, S.R. Abrams, *et al.* (2005) The *etr1-2* mutation in *Arabidopsis thaliana* affects the abscisic acid, auxin, cytokinin and gibberellin metabolic pathways during maintenance of seed dormancy, moist-chilling and germination. *The Plant Journal* **42**, 35–48.

H.T. Cho and D.J. Cosgrove (2000) Altered expression of expansin modulates leaf growth and pedicel abscission in *Arabidopsis thaliana*. *Proceedings of the National Academy of Sciences United States of America* **97**, 9783–9788.

H.T. Cho and D.J. Cosgrove (2002) Regulation of root hair initiation and expansin gene expression in *Arabidopsis*. *The Plant Cell* **14**, 3237–3253.

D.J. Cosgrove (2005) Expansin web page. (http://www.bio.psu.edu/expansins/), date of access 1 October 2005.

D.J. Cosgrove, L.C. Li, H.T. Cho, S. Hoffmann-Benning, R.C. Moore and D. Blecker (2002) The growing world of expansins. *Plant and Cell Physiology* **43**, 1436–1444.

T. Czechowski, M. Stitt, T. Altmann, M.K. Udvardi and W.-R. Scheible (2005) Genome-wide identification and testing of superior reference genes for transcript normalization in *Arabidopsis*. *Plant Physiology* **139**, 5–17.

E.A.A. da Silva, P.E. Toorop, J. Nijsse, J.D. Bewley and H.W.M. Hilhorst (2005) Exogenous gibberellins inhibit coffee (*Coffea arabica* cv. Rubi) seed germination and cause cell death in the embryo. *Journal of Experimental Botany* **56**, 1029–1038.

E.A.A. da Silva, P.E. Toorop, A.C. van Aelst and H.W.M. Hilhorst (2004) Abscisic acid controls embryo growth potential and endosperm cap weakening during coffee (*Coffea arabica* cv. Rubi) seed germination. *Planta* **220**, 251–261.

P. Dahal and K.J. Bradford (1990) Effects of priming and endosperm integrity on seed germination rates of tomato genotypes. II. Germination at reduced water potential. *Journal of Experimental Botany* **41**, 1441–1453.

P. Dahal, D.J. Nevins and K.J. Bradford (1997) Relationship of endo-β-D-mannanase activity and cell wall hydrolysis in tomato endosperm to germination rates. *Plant Physiology* **113**, 1243–1252.

E.H. Davidson, B.R. Hough-Evans and R.J. Britten (1982) Molecular biology of the sea urchin embryo. *Science* **217**, 17–26.

I. Debeaujon, K.M. Leon-Kloosterziel and M. Koornneef (2000) Influence of the testa on seed dormancy, germination, and longevity in *Arabidopsis*. *Plant Physiology* **122**, 403–413.
R.D. de Castro, A.A.M. van Lammeren, S.P.C. Groot, R.J. Bino and H.W.M. Hilhorst (2000) Cell division and subsequent radicle protrusion in tomato seeds are inhibited by osmotic stress but DNA synthesis and formation of microtubular cytoskeleton are not. *Plant Physiology* **122**, 327–335.
L. de Miguel, M.J. Burgin, J.J. Casal and R.A. Sánchez (2000) Antagonistic action of low-fluence and high-irradiance modes of response of phytochrome on germination and β-mannanase activity in *Datura ferox* seeds. *Journal of Experimental Botany* **51**, 1127–1133.
L. de Miguel and R.A. Sánchez (1992) Phytochrome-induced germination, endosperm softening and embryo growth-potential in *Datura ferox* seeds. Sensitivity to low water potential and time to escape to Fr reversal. *Journal of Experimental Botany* **43**, 969–974.
M.P.M. Derkx and C.M. Karssen (1993) Effects of light and temperature on seed dormancy and gibberellin stimulated germination in *Arabidopsis thaliana*: studies with gibberellin-deficient and -insensitive mutants. *Physiologia Plantarum* **89**, 360–368.
L.M.A. Dirk, A.M. Griffen, B. Downie and J.D. Bewley (1995) Multiple isozymes of endo-β-D-mannanase in dry and imbibed seeds. *Phytochemistry* **40**, 1045–1056.
B. Downie, L.M.A. Dirk, K.A. Hadfield, T.A. Wilkins, A.B. Bennett and K.J. Bradford (1998) A gel diffusion assay for quantification of pectin methylesterase activity. *Analytical Biochemistry* **264**, 149–157.
A.B. Downie, L.M.A. Dirk, Q.L. Xu, *et al.* (2004) A physical, enzymatic, and genetic characterization of perturbations in the seeds of the brownseed tomato mutants. *Journal of Experimental Botany* **55**, 961–973.
B. Downie, S. Gurusinghe and K.J. Bradford (1999) Internal anatomy of individual tomato seeds: relationship to abscisic acid and germination physiology. *Seed Science Research* **9**, 117–128.
B. Downie, S. Gurusinghe, K.J. Bradford, C.G. Plopper, J.S. Greenwood and J.D. Bewley (1997a) Embryo root cap cells adhere to the megagametophyte and sheathe the radicle of white spruce (*Picea glauca* [Moench.] Voss.) seeds following germination. *International Journal of Plant Sciences* **158**, 738–746.
B. Downie, H.W.M. Hilhorst and J.D. Bewley (1997b) Endo-β-mannanase activity during dormancy alleviation and germination of white spruce (*Picea glauca*) seeds. *Physiologia Plantarum* **101**, 405–415.
A.B. Downie, D.Q. Zhang, L.M.A. Dirk, *et al.* (2003) Communication between the maternal testa and the embryo and/or endosperm affect testa attributes in tomato. *Plant Physiology* **133**, 145–160.
J. Dulson, J.D. Bewley and R.N. Johnston (1988) Abscisic acid is an endogenous inhibitor in the regulation of mannanase production by isolated lettuce (*Lactuca sativa* cv Grand Rapids) endosperms. *Plant Physiology* **87**, 660–665.
S. Dutta, K.J. Bradford and D.J. Nevins (1997) Endo-β-mannanase activity present in cell wall extracts of lettuce endosperm prior to radicle emergence. *Plant Physiology* **113**,155–161.
C.B. Engineer, K.C. Fitzsimmons, J.J. Schmuke, S.B. Dotson and R.G. Kranz (2005) Development and evaluation of a Gal4-mediated LUC/GFP/GUS enhancer trap system in *Arabidopsis*. *BMC Plant Biology* 2005, **5**:9 doi:10.1186/1471-2229-5-9.
J.A. Feurtado, M. Banik and J.D. Bewley (2001) The cloning and characterization of α-galactosidase present during and following germination of tomato (*Lycopersicon esculentum* Mill.) seed. *Journal of Experimental Botany* **52**, 1239–1249.
R.R. Finkelstein, S.S.L. Gampala and C.D. Rock (2002) Abscisic acid signaling in seeds and seedlings. *The Plant Cell* **14**, S15–S45.
S.C. Fry (2004) Primary cell wall metabolism: tracking the careers of wall polymers in living plant cells. *New Phytologist* **161**, 641–675.
K. Gallardo, C. Job, S.P. Groot, *et al.* (2002) Proteomics of *Arabidopsis* seed germination. A comparative study of wild-type and gibberellin-deficient seeds. *Plant Physiology* **129**, 823–837.
K. Gallardo, C. Job, S.P.C. Groot, *et al.* (2001) Proteomic analysis of *Arabidopsis* seed germination and priming. *Plant Physiology* **126**, 835–848.

Y.P. Gao, L. Young, P. Bonham-Smith and L.V. Gusta (1999) Characterization and expression of plasma and tonoplast membrane aquaporins in primed seed of *Brassica napus* during germination under stress conditions. *Plant Molecular Biology* **40**, 635–644.

T. Girke, J. Todd, S. Ruuska, J. White, C. Benning and J. Ohlrogge (2000) Microarray analysis of developing *Arabidopsis* seeds. *Plant Physiology* **124**, 1570–1581.

L. Gomez, I. Allona, R. Casado and C. Aragoncillo (2002) Seed chitinases. *Seed Science Research* **12**, 217–230.

S.P.C. Groot and C.M. Karssen (1987) Gibberellins regulate seed germination in tomato by endosperm weakening–a study with gibberellin-deficient mutants. *Planta* **171**, 525–531.

S.P.C. Groot and C.M. Karssen (1992) Dormancy and germination of abscisic acid-deficient tomato seeds. Studies with the *sitiens* mutant. *Plant Physiology* **99**, 952–958.

S.P.C. Groot, B. Kieliszewskarokicka, E. Vermeer and C.M. Karssen (1988) Gibberellin-induced hydrolysis of endosperm cell walls in gibberellin-deficient tomato seeds prior to radicle protrusion. *Planta* **174**, 500–504.

F. Gubler, R. Kalla, J.K. Roberts and J.V. Jacobsen (1995) Gibberellin-regulated expression of a *myb* gene in barley aleurone cells: evidence for *myb* transactivation of a high-pI α-amylase gene promoter. *The Plant Cell* **7**, 1879–1891.

F. Gubler, A.A. Millar and J.V. Jacobsen (2005) Dormancy release, ABA and pre-harvest sprouting. *Current Opinion in Plant Biology* **8**, 183–187.

K.A. Hadfield and A.B. Bennett (1998) Polygalacturonases: many genes in search of a function. *Plant Physiology* **117**, 337–343.

A.M. Haigh (1988) *Why do tomato seeds prime? Physiological Investigations into the Control of Tomato Seed Germination and Priming*. Ph.D. Dissertation, School of Biological Sciences, Macquarie University, North Ryde, NSW, Australia.

A.M. Haigh and E.W.R. Barlow (1987) Water relations of tomato seed germination. *Australian Journal of Plant Physiology* **14**, 485–492.

M. Hajduch, A. Ganapathy, J.W. Stein and J.J. Thelen (2005) A systematic proteomic study of seed filling in soybean. Establishment of high-resolution two-dimensional reference maps, expression profiles, and an interactive proteome database. *Plant Physiology* **137**, 1397–1419.

P. Halmer (2004) Methods to improve seed performance in the field. In: *Handbook of Seed Physiology. Applications to Agriculture* (eds R.L. Bench-Arnold and R.A. Sánchez), pp. 125–166. Food Products Press, New York.

P. Halmer, J.D. Bewley and T.A. Thorpe (1976) Enzyme to degrade lettuce endosperm cell walls – appearance of a mannanase following phytochrome-induced and gibberellin-induced germination. *Planta* **130**, 189–196.

E.P. Harrison, S.J. McQueen-Mason and K. Manning (2001) Expression of six expansin genes in relation to extension activity in developing strawberry fruit. *Journal of Experimental Botany* **52**, 1437–1446.

H.W.M. Hilhorst and B. Downie (1996) Primary dormancy in tomato (*Lycopersicon esculentum* cv Moneymaker): studies with the *sitiens* mutant. *Journal of Experimental Botany* **47**, 89–97.

H.W.M. Hilhorst and C.M. Karssen (1992) Seed dormancy and germination – the role of abscisic acid and gibberellins and the importance of hormone mutants. *Plant Growth Regulation* **11**, 225–238.

H.J. Hill and A.G. Taylor (1989) Relationship between viability, endosperm integrity, and imbibed lettuce seed density and leakage. *Hortscience* **24**, 814–816.

H. Hofte, L. Hubbard, J. Reizer, D. Ludevid, E.M. Herman and M.J. Chrispeels (1992) Vegetative and seed-specific forms of tonoplast intrinsic protein in the vacuolar membrane of *Arabidopsis thaliana*. *Plant Physiology* **99**, 561–570.

J. Hou, E. Kendall and G. Simpson (1997) Water uptake and distribution in non-dormant and dormant wild oat (*Avena fatua* L.) caryopses. *Journal of Experimental Botany* **48**, 683–692.

J. Hua, C. Chang, Q. Sun and E.M. Meyerowitz (1995) Ethylene insensitivity conferred by *Arabidopsis ERS* gene. *Science* **269**, 1712–1714.

J. Hua, H. Sakai, S. Nourizadeh, *et al.* (1998) *EIN4* and *ERS2* are members of the putative ethylene receptor gene family in *Arabidopsis*. *The Plant Cell* **10**, 1321–1332.

H. Ikuma and K.V. Thimann (1963) The role of the seed-coats in germination of photosensitive lettuce seeds. *Plant and Cell Physiology* **4**, 169–185.

ISTA (1999) International rules for seed testing. *Seed Science and Technology* **27**, supplement.

A. Itai, K. Ishihara and J.D. Bewley (2003) Characterization of expression, and cloning, of β-D-xylosidase and α-L-arabinofuranosidase in developing and ripening tomato (*Lycopersicon esculentum* Mill.) fruit. *Journal of Experimental Botany* **54**, 2615–2622.

T. Ito and E.M. Meyerowitz (2000) Overexpression of a gene encoding a cytochrome P450, *CYP78A9*, induces large and seedless fruit in *Arabidopsis*. *The Plant Cell* **12**, 1541–1550.

S.E. Jacobsen, K.A. Binkowski and N.E. Olszewski (1996) SPINDLY, a tetratricopeptide repeat protein involved in gibberellin signal transduction in *Arabidopsis*. *Proceedings of the National Academy of Sciences United States of America* **93**, 9292–9296.

J.V. Jacobsen, D.W. Pearce, A.T. Poole, R.P. Pharis and L.N. Mander (2002) Abscisic acid, phaseic acid and gibberellin contents associated with dormancy and germination in barley. *Physiologia Plantarum* **115**, 428–441.

A.A. Johnson, J.M. Hibberd, C. Gay, *et al.* (2005) Spatial control of transgene expression in rice (*Oryza sativa* L.) using the GAL4 enhancer trapping system. *The Plant Journal* **41**, 779–789.

I. Kardailsky, V.K. Shukla, J.H. Ahn, *et al.* (1999) Activation tagging of the floral inducer FT. *Science* **286**, 1962–1965.

H. Kende, K.J. Bradford, D.A. Brummell, *et al.* (2004) Nomenclature for members of the expansin superfamily of genes and proteins. *Plant Molecular Biology* **55**, 311–314.

M. Koornneef, T.D.G. Bosma, C.J. Hanhart, J.H. Vanderveen and J.A.D. Zeevaart (1990) The isolation and characterization of gibberellin-deficient mutants in tomato. *Theoretical and Applied Genetics* **80**, 852–857.

M. Koornneef and J.H. van der Veen (1980) Induction and analysis of gibberellin sensitive mutants in *Arabidopsis thaliana* (L.) Heynh. *Theoretical Applied Genetics* **58**, 257–263.

B. Kost, J. Mathur and N.H. Chua (1999) Cytoskeleton in plant development. *Current Opinion in Plant Biology* **2**, 462–470.

H. Lee, S.S. Suh, E. Park, *et al.* (2000) The AGAMOUS-LIKE 20 MADS domain protein integrates floral inductive pathways in *Arabidopsis*. *Genes and Development* **14**, 2366–2376.

A. Lehman, R. Black and J.R. Ecker (1996) HOOKLESS1, an ethylene response gene, is required for differential cell elongation in the *Arabidopsis* hypocotyl. *Cell* **85**, 183–194.

A.C. Leopold (1983) Volumetric components of seed imbibition. *Plant Physiology* **73**, 677–680.

G. Leubner-Metzger (2002) Seed after-ripening and over-expression of class I β-1,3-glucanase confer maternal effects on tobacco testa rupture and dormancy release. *Planta* **215**, 959–968.

G. Leubner-Metzger (2003) Functions and regulation of β-1,3-glucanases during seed germination, dormancy release and after-ripening. *Seed Science Research* **13**, 17–34.

G. Leubner-Metzger (2005) β-1,3-glucanase gene expression in low-hydrated seeds as a mechanism for dormancy release during tobacco after-ripening. *Plant Journal* **41**, 133–145.

G. Leubner-Metzger, C. Frundt, R. Vogeli-Lange and F. Meins (1995) Class-I β-1,3-glucanases in the endosperm of tobacco during germination. *Plant Physiology* **109**, 751–759.

G. Leubner-Metzger and F. Meins (2000) Sense transformation reveals a novel role for class I β-1,3-glucanase in tobacco seed germination. *Plant Journal* **23**, 215–221.

G. Leubner-Metzger and F. Meins (2001) Antisense-transformation reveals novel roles for class I β-1,3-glucanase in tobacco seed after-ripening and photodormancy. *Journal of Experimental Botany* **52**, 1753–1759.

S. Leviatov, O. Shoseyov and S. Wolf (1995) Involvement of endomannanase in the control of tomato seed germination under low temperature conditions. *Annals of Botany* **76**, 1–6.

P.-P. Liu, N. Koizuka, T.M. Homrichhausen, J.R. Hewitt, R.C. Martin and H. Nonogaki (2005a) Large-scale screening of *Arabidopsis* enhancer-trap lines for seed germination-associated genes. *Plant Journal* **41**, 936–944.

P.-P. Liu, N. Koizuka, R.C. Martin and H. Nonogaki (2005b) The *BME3* (*Blue Micropylar End 3*) GATA zinc finger transcription factor is a positive regulator of *Arabidopsis* seed germination. *Plant Journal* **44**, 960–971.

C. Llave, Z. Xie, K.D. Kasschau and J.C. Carrington (2002) Cleavage of *Scarecrow-like* mRNA targets directed by a class of *Arabidopsis* miRNA. *Science* **297**, 2053–2056.

L. Ma, N. Sun, X. Liu, Y. Jiao, H. Zhao and X.W. Deng (2005) Organ-specific expression of *Arabidopsis* genome during development. *Plant Physiology* **138**, 80–91.

A.C. Mallory, D.P. Bartel and B. Bartel (2005) MicroRNA-directed regulation of *Arabidopsis AUXIN RESPONSE FACTOR17* is essential for proper development and modulates expression of early auxin response genes. *The Plant Cell* **17**, 1360–1375.

B. Manz, K. Muller, B. Kucera, F. Volke and G. Leubner-Metzger (2005) Water uptake and distribution in germinating tobacco seeds investigated *in vivo* by nuclear magnetic resonance imaging. *Plant Physiology* **138**, 1538–1551.

P. Marraccini, W.J. Rogers, C. Allard, *et al.* (2001) Molecular and biochemical characterization of endo-β-mannanases from germinating coffee (*Coffea arabica*) grains. *Planta* **213**, 296–308.

R.C. Martin, P.-P. Liu and H. Nonogaki (2005) Simple purification of small RNAs from seeds and efficient detection of multiple microRNAs expressed in *Arabidopsis thaliana* and tomato (*Lycopersicon esculentum*) seeds. *Seed Science Research* **15**, 319–328.

R.C. Martin, P.-P. Liu and H. Nonogaki (2006) MicroRNAs in seeds – modified detection techniques and potential applications. *Canadian Journal of Botany* **84**, 189–198.

N.H. Masubelele, W. Dewitte, M. Menges, *et al.* (2005) D-type cyclins activate division in the root apex to promote seed germination in *Arabidopsis*. *Proceedings of the National Academy of Sciences United States of America* **102**, 15694–15699.

A.J.M. Matzke, E.M. Stoger and M.A. Matzke (1993) A zein gene promoter fragment drives Gus expression in a cell layer that is interposed between the endosperm and the seed coat. *Plant Molecular Biology* **22**, 553–554.

C. Maurel, M. Chrispeels, C. Lurin, *et al.* (1997) Function and regulation of seed aquaporins. *Journal of Experimental Botany* **48**, 421–430.

M.B. McDonald, J. Sullivan and M.J. Lauer (1994) The pathway of water-uptake in maize seeds. *Seed Science and Technology* **22**, 79–90.

M.B. McDonald, C.W. Vertucci and E.E. Roos (1988) Soybean seed imbibition – water-absorption by seed parts. *Crop Science* **28**, 993–997.

H.I. McKhann, F. Frugier, G. Petrovics, *et al.* (1997) Cloning of a WD-repeat-containing gene from alfalfa (*Medicago sativa*): a role in hormone-mediated cell division? *Plant Molecular Biology* **34**, 771–780.

R.A. Mella, M.J. Burgin and R.A. Sánchez (2004) Expansin gene expression in *Datura ferox* L. seeds is regulated by the low-fluence response, but not by the high-irradiance response, of phytochromes. *Seed Science Research* **14**, 61–71.

R.A. Mella, S. Maldonado and R.A. Sánchez (1995) Phytochrome-induced structural changes and protein degradation prior to radicle protrusion in *Datura ferox* seeds. *Canadian Journal of Botany* **73**, 1371–1378.

A.A. Millar and F. Gubler (2005) The *Arabidopsis GAMYB-like* genes, *MYB33* and *MYB65*, are microRNA-regulated genes that redundantly facilitate anther development. *The Plant Cell* **17**, 705–721.

B.X. Mo and J.D. Bewley (2002) β-Mannosidase (EC 3.2.1.25) activity during and following germination of tomato (*Lycopersicon esculentum* Mill.) seeds. Purification, cloning and characterization. *Planta* **215**, 141–152.

B. Mo and J.D. Bewley (2003) The relationship between β-mannosidase and endo-β-mannanase activities in tomato seeds during and following germination: a comparison of seed populations and individual seeds. *Journal of Experimental Botany* **54**, 2503–2510.

Y. Morohashi and H. Matsushima (2000) Development of β-1,3-glucanase activity in germinated tomato seeds. *Journal of Experimental Botany* **51**, 1381–1387.

K. Nakabayashi, M. Okamoto, T. Koshiba, Y. Kamiya and E. Nambara (2005) Genome-wide profiling of stored mRNA in *Arabidopsis thaliana* seed germination: epigenetic and genetic regulation of transcription in seed. *The Plant Journal* **41**, 697–709.

W.M. Nascimento, D.J. Cantliffe and D.J. Huber (2000) Thermotolerance in lettuce seeds: association with ethylene and endo-β-mannanase. *Journal of the American Society for Horticultural Science* **125**, 518–524.

W.M. Nascimento, D.J. Cantliffe and D.J. Huber (2001) Endo-β-mannanase activity and seed germination of thermosensitive and thermotolerant lettuce genotypes in response to seed priming. *Seed Science Research* **11**, 255–264.

W.M. Nascimento, D.J. Cantliffe and D.J. Huber (2004) Ethylene evolution and endo-β-mannanase activity during lettuce seed germination at high temperature. *Scientia Agricola* **61**, 156–163.

M.M. Neff, S.M. Nguyen, E.J. Malancharuvil, *et al.* (1999) *BAS1*: A gene regulating brassinosteroid levels and light responsiveness in *Arabidopsis*. *Proceedings of the National Academy of Sciences United States of America* **96**, 15316–15323.

B.R. Ni and K.J. Bradford (1992) Quantitative models characterizing seed germination responses to abscisic acid and osmoticum. *Plant Physiology* **98**, 1057–1068.

B.R. Ni and K.J. Bradford (1993) Germination and dormancy of abscisic acid-deficient and gibberellin-deficient mutant tomato seeds. Sensitivity of germination to abscisic acid, gibberellin, and water potential. *Plant Physiology* **101**, 607–617.

H. Nonogaki, O.H. Gee and K.J. Bradford (2000) A germination-specific endo-β-mannanase gene is expressed in the micropylar endosperm cap of tomato seeds. *Plant Physiology* **123**, 1235–1245.

H. Nonogaki, H. Matsushima and Y. Morohashi (1992) Galactomannan hydrolyzing activity develops during priming in the micropylar endosperm tip of tomato seeds. *Physiologia Plantarum* **85**, 167–172.

H. Nonogaki and Y. Morohashi (1999) Temporal and spatial pattern of the development of endo-β-mannanase activity in germinating and germinated lettuce seeds. *Journal of Experimental Botany* **50**, 1307–1313.

H. Nonogaki, M. Nomaguchi, N. Okumoto, Y. Kaneko, H. Matsushima and Y. Morohashi (1998) Temporal and spatial pattern of the biochemical activation of the endosperm during and following imbibition of tomato seeds. *Physiologia Plantarum* **102**, 236–242.

M. Ogawa, A. Hanada, Y. Yamauchi, A. Kuwalhara, Y. Kamiya and S. Yamaguchi (2003) Gibberellin biosynthesis and response during *Arabidopsis* seed germination. *The Plant Cell* **15**, 1591–1604.

K. Ogawa and M. Iwabuchi (2001) A mechanism for promoting the germination of *Zinnia elegans* seeds by hydrogen peroxide. *Plant and Cell Physiology* **42**, 286–291.

J.F. Palatnik, E. Allen, X. Wu, *et al.* (2003) Control of leaf morphogenesis by microRNAs. *Nature* **425**, 257–263.

J.W. Patrick and C.E. Offler (2001) Compartmentation of transport and transfer events in developing seeds. *Journal of Experimental Botany* **52**, 551–564.

S. Penfield, E.L. Rylott, A.D. Gilday, S. Graham, T.R. Larson and I.A. Graham (2004) Reserve mobilization in the *Arabidopsis* endosperm fuels hypocotyl elongation in the dark, is independent of abscisic acid, and requires *PHOSPHOENOLPYRUVATE CARBOXYKINASE1*. *The Plant Cell* **16**, 2705–2718.

C. Perino and D. Côme (1991) Physiological and metabolic study of the germination phases in apple embryo. *Seed Science and Technology* **19**, 1–14.

G. Psaras and K. Georghiou (1983) Gibberellic acid-induced structural alterations in the endosperm of germinating *Lactuca sativa* L. achenes. *Zeitschrift Fur Pflanzenphysiologie* **112**, 14–19.

L. Rajjou, K. Gallardo, I. Debeaujon, J. Vandekerckhove, C. Job and D. Job (2004) The effect of α-amanitin on the *Arabidopsis* seed proteome highlights the distinct roles of stored and neosynthesized mRNAs during germination. *Plant Physiology* **134**, 1598–1613.

J.S.G. Reid and M. Edwards (1995) Galactomannans and other cell wall storage polysaccharides in seeds. In: *Food Polysaccharides and Their Applications* (ed. A.M. Stephen), pp. 155–186. Marcel Dekker, New York.

J.S.G. Reid, M.E. Edwards, C.A. Dickson, C. Scott and M.J. Gidley (2003) Tobacco transgenic lines that express fenugreek galactomannan galactosyltransferase constitutively have structurally altered galactomannans in their seed endosperm cell walls. *Plant Physiology* **131**, 1487–1495.

B.J. Reinhart, E.G. Weinstein, M.W. Rhoades, B. Bartel and D.P. Bartel (2002) MicroRNAs in plants. *Genes and Development* **16**, 1616–1626.

C.W. Ren and A.R. Kermode (2000) An increase in pectin methyl esterase activity accompanies dormancy breakage and germination of yellow cedar seeds. *Plant Physiology* **124**, 231–242.

M.W. Rhoades, B.J. Reinhart, L.P. Lim, C.B. Burge, B. Bartel and D.P. Bartel (2002) Prediction of plant microRNA targets. *Cell* **110**, 513–520.

P.L.H. Rinne and C. van der Schoot (2003) Plasmodesmata at the crossroads between development, dormancy, and defense. *Canadian Journal of Botany* **81**, 1182–1197.

J.K.C. Rose, K.A. Hadfield, J.M. Labavitch and A.B. Bennett (1998) Temporal sequence of cell wall disassembly in rapidly ripening melon fruit. *Plant Physiology* **117**, 345–361.

J.K.C. Rose, H.H. Lee and A.B. Bennett (1997) Expression of a divergent expansin gene is fruit-specific and ripening-regulated. *Proceedings of the National Academy of Sciences United States of America* **94**, 5955–5960.

S.A. Ruuska, T. Girke, C. Benning and J.B. Ohlrogge (2002) Contrapuntal networks of gene expression during *Arabidopsis* seed filling. *The Plant Cell* **14**, 1191–206.

L. Salaita, R.K. Kar, M. Majee and A.B. Downie (2005) Identification and characterization of mutants capable of rapid seed germination at 10°C from activation-tagged lines of *Arabidopsis thaliana*. *Journal of Experimental Botany* **56**, 2059–2069.

R.A. Sánchez, L. de Miguel, C. Lima and R.M. de Lederkremer (2002) Effect of low water potential on phytochrome-induced germination, endosperm softening and cell-wall mannan degradation in *Datura ferox* seeds. *Seed Science Research* **12**, 155–163.

R.A. Sánchez, L. de Miguel and O. Mercuri (1986) Phytochrome control of cellulase activity in *Datura ferox* L. seeds and its relationship with germination. *Journal of Experimental Botany* **37**, 1574–1580.

R.A. Sánchez, L. Sunell, J.M. Labavitch and B.A. Bonner (1990) Changes in the endosperm cell walls of two *Datura* species before radicle protrusion. *Plant Physiology* **93**, 89–97.

W.R. Scheible and M. Pauly (2004) Glycosyltransferases and cell wall biosynthesis: novel players and insights. *Current Opinion in Plant Biology* **7**, 285–295.

P. Schopfer, D. Bajracharya and C. Plachy (1979) Control of seed germination by abscisic acid. I. Time course of action in *Sinapis alba* L. *Plant Physiology* **64**, 822–827.

P. Schopfer and C. Plachy (1984) Control of seed germination by abscisic acid. II. Effect on embryo water uptake in *Brassica napus* L. *Plant Physiology* **76**, 155–160.

P. Schopfer and C. Plachy (1985) Control of seed germination by abscisic acid. III. Effect on embryo growth potential (minimum turgor pressure) and growth coefficient (cell wall extensibility) in *Brassica napus* L. *Plant Physiology* **77**, 676–686.

J. Schuurmans, J.T. van Dongen, B.P.W. Rutjens, A. Boonman, C.M.J. Pieterse and A.C. Borstlap (2003) Members of the aquaporin family in the developing pea seed coat include representatives of the PIP, TIP, and NIP subfamilies. *Plant Molecular Biology* **53**, 655–667.

Y. Sitrit, K.A. Hadfield, A.B. Bennett, K.J. Bradford and A.B. Downie (1999) Expression of a polygalacturonase associated with tomato seed germination. *Plant Physiology* **121**, 419–428.

Y. Soeda, M. Konings, O. Vorst, *et al.* (2005) Gene expression programs during *Brassica oleracea* seed maturation, osmopriming, and germination are indicators of progression of the germination process and the stress tolerance level. *Plant Physiology* **137**, 354–368.

P. Spoelstra (2002) *Germination and dormancy of single tomato seeds. A Study Using Non-Invasive Molecular and Biophysical Techniques*. Ph.D. Disseration, Wageningen University, The Netherlands.

D.W. Still and K.J. Bradford (1997) Endo-β-mannanase activity from individual tomato endosperm caps and radicle tips in relation to germination rates. *Plant Physiology* **113**, 21–29.

Y. Sung, D.J. Cantliffe and R. Nagata (1998) Using a puncture test to identify the role of seed coverings on thermotolerant lettuce seed germination. *Journal of the American Society for Horticultural Science* **123**, 1102–1106.

H. Takahashi, M. Rai, T. Kitagawa, S. Morita, T. Masumura and K. Tanaka (2004) Differential localization of tonoplast intrinsic proteins on the membrane of protein body type II and aleurone grain in rice seeds. *Bioscience Biotechnology and Biochemistry* **68**, 1728–1736.

V.V. Terskikh, J.A. Feurtado, C. Ren, S.R. Abrams and A.R. Kermode (2005) Water uptake and oil distribution during imbibition of seeds of western white pine (*Pinus monticola* Dougl. ex D. Don) monitored *in vivo* using magnetic resonance imaging. *Planta* **221**, 17–27.

M.A.S. Tine, A.L. Cortelazzo and M.S. Buckeridge (2000) Xyloglucan mobilization in cotyledons of developing plantlets of *Hymenaea courbaril* L. (Leguminosae-Cesalpinoideae). *Plant Science* **154**, 117–126.

P.E. Toorop, A.C. van Aelst and H.W.M. Hilhorst (1998) Endosperm cap weakening and endo-β-mannanase activity during priming of tomato (*Lycopersicon esculentum* cv. Moneymaker) seeds are initiated upon crossing a threshold water potential. *Seed Science Research* **8**, 483–491.

P.E. Toorop, A.C. van Aelst and H.W.M. Hilhorst (2000) The second step of the biphasic endosperm cap weakening that mediates tomato (*Lycopersicon esculentum*) seed germination is under control of ABA. *Journal of Experimental Botany* **51**, 1371–1379.

P. van der Toorn and C.M. Karssen (1992) Analysis of embryo growth in mature fruits of celery (*Apium graveolens*). *Physiologia Plantarum* **84**, 593–599.

L.C. van Loon (1997) Induced resistance in plants and the role of pathogenesis-related proteins. *European Journal of Plant Pathology* **103**, 753–765.

C.W. Vertucci and J.M. Farrant (1995) Acquisition and loss of desiccation tolerance. In: *Seed Development and Germination* (eds J. Kigel and G. Galili), pp. 237–271. Marcel Dekker, Inc., New York.

R. Vogeli-Lange, C. Frundt, C.M. Hart, R. Beffa, F. Nagy and F. Meins (1994) Evidence for a role of β-1,3-glucanase in dicot seed germination. *The Plant Journal* **5**, 273–278.

R. Walden, K. Fritze, H. Hayashi, E. Miklashevichs, H. Harling and J. Schell (1994) Activation tagging: a means of isolating genes implicated as playing a role in plant growth and development. *Plant Molecular Biology* **26**, 1521–1528.

C. Walters, J.M. Farrant, N.W. Pammenter and P. Berjak (2002) Desiccation stress and damage. In: *Desiccation and Survival in Plants. Drying without Dying* (eds M. Black and H.W. Pritchard), pp. 263–291. CABI Publishing, Wallingford, Oxon, UK.

A.X. Wang, J.R. Li and J.D. Bewley (2004) Molecular cloning and characterization of an endo-β-mannanase gene expressed in the lettuce endosperm following radicle emergence. *Seed Science Research* **14**, 267–276.

J.T. Watkins and D.J. Cantliffe (1983a) Hormonal control of pepper seed germination. *Hortscience* **18**, 342–343.

J.T. Watkins and D.J. Cantliffe (1983b) Mechanical resistance of the seed coat and endosperm during germination of *Capsicum annuum* at low temperature. *Plant Physiology* **72**, 146–150.

D. Weigel, J.H. Ahn, M.A. Blazquez, *et al.* (2000) Activation tagging in *Arabidopsis*. *Plant Physiology* **122**, 1003–1013.

G.E. Welbaum, K.J. Bradford, K.O. Yim, D.T. Booth and M.O. Oluoch (1998) Biophysical, physiological and biochemical processes regulating seed germination. *Seed Science Research* **8**, 161–172.

G.E. Welbaum, W.J. Muthui, J.H. Wilson, R.L. Grayson and R.D. Fell (1995) Weakening of muskmelon perisperm envelope tissue during germination. *Journal of Experimental Botany* **46**, 391–400.

S.E.C. Whitney, M.J. Gidley and S.J. McQueen-Mason (2000) Probing expansin action using cellulose/hemicellulose composites. *The Plant Journal* **22**, 327–334.

S.E.C. Whitney, M.G.E. Gothard, J.T. Mitchell and M.J. Gidley (1999) Roles of cellulose and xyloglucan in determining the mechanical properties of primary plant cell walls. *Plant Physiology* **121**, 657–663.

C.T. Wu and K.J. Bradford (2003) Class I chitinase and β-1,3-glucanase are differentially regulated by wounding, methyl jasmonate, ethylene, and gibberellin in tomato seeds and leaves. *Plant Physiology* **133**, 263–273.

C.T. Wu, G. Leubner-Metzger, F. Meins and K.J. Bradford (2001) Class I β-1,3-glucanase and chitinase are expressed in the micropylar endosperm of tomato seeds prior to radicle emergence. *Plant Physiology* **126**, 1299–1313.

Y. Xia, H. Suzuki, J. Borevitz, *et al.* (2004) An extracellular aspartic protease functions in *Arabidopsis* disease resistance signaling. *EMBO Journal* **23**, 980–988.

Y. Yamauchi, M. Ogawa, A. Kuwahara, A. Hanada, Y. Kamiya and S. Yamaguchi (2004) Activation of gibberellin biosynthesis and response pathways by low temperature during imbibition of *Arabidopsis thaliana* seeds. *The Plant Cell* **16**, 367–378.

K.O. Yim and K.J. Bradford (1998) Callose deposition is responsible for apoplastic semipermeability of the endosperm envelope of muskmelon seeds. *Plant Physiology* **118**, 83–90.

Y. Zhao, S.K. Christensen, C. Fankhauser, *et al.* (2001) A role for flavin monooxygenase-like enzymes in auxin biosynthesis. *Science* **291**, 306–309.

12 Sugar and abscisic acid regulation of germination and transition to seedling growth

Bas J.W. Dekkers and Sjef C.M. Smeekens

12.1 Introduction

Carbohydrates serve diverse functions in plants ranging from energy sources, storage molecules, and structural components to intermediates for the synthesis of other organic molecules. Sugars also act as signaling molecules in both prokaryotes and eukaryotes including plants. Plants are autotrophic organisms and it is important to adjust sugar consumption to sugar production and to properly allocate these resources between source and sink tissues. Sugars are potent regulators of these processes and sugar-induced signal transduction has been shown to control gene expression in plants via diverse mechanisms that include transcription, translation, and modification of mRNA and protein stability. Consequently, many physiological and developmental processes in plants are affected by sugars. Sugar signaling is intimately related with plant hormone signaling, in particular abscisic acid (ABA) signaling.

Here we review advances mainly in *Arabidopsis* research concerning sugar and ABA signaling and their interactions with particular focus on two early developmental stages, namely seed germination and seedling establishment. Sugar and ABA signaling affect both germination (i.e., radicle emergence) and the switch from embryonic to vegetative growth (i.e., seedling development).

12.2 ABA signaling during germination and early seedling growth

12.2.1 ABA response mutants isolated in germination-based screens

ABA is involved in several developmental processes including seed development and phase transitions from embryonic to germinative growth. Furthermore, ABA plays an integral part in responses to biotic and abiotic stresses, regulation of stomatal aperture, and nitrate and sugar signaling (Leung and Giraudat, 1998; Finkelstein *et al.*, 2002). In *Arabidopsis* alone, about 50 genes are known to be involved in ABA responses (Finkelstein *et al.*, 2002). Different genetic approaches were used to identify many mutants with aberrant responses to ABA. The 'classic' screen for mutants in ABA signaling is a germination-based assay initially performed by Koornneef *et al.* (1984). The ABA-insensitive (*abi*) mutants are able to germinate in the presence of otherwise inhibitory concentrations of ABA. Such germination-based screens proved to be remarkably successful and have identified many important mutants. These include mutants insensitive or hypersensitive to ABA and that are affected in diverse ABA-regulated processes such as seed dormancy and stomatal closure. The genes affected in *abi* mutants were identified and found to encode protein phosphatases (*ABI1* and *ABI2*) and a B3 class transcription factor (*ABI3*) (Giraudat *et al.*, 1992; Leung *et al.*, 1994; Meyer *et al.*, 1994; Leung *et al.*, 1997). These three mutants lack normal seed dormancy, and *abi1* and *abi2* are defective in regulation of stomatal closure as well (Koornneef *et al.*, 1984). AtP2C-HA/HAB1 phosphatase 2C, which is homologous to ABI1 and ABI2, was also shown to be involved in ABA signaling (Leonhardt *et al.*, 2004; Saez *et al.*, 2004). Three additional *abi* mutants, *abi4*, *abi5*, and *abi8*, were isolated. *ABI4* and *ABI5* encode AP2 (APETALA2) and bZIP family transcription factors, respectively, whereas *ABI8* encodes a protein of unknown function that is also known as KOBITO1 or ELONGATION DEFECTIVE1 (Finkelstein, 1994; Finkelstein *et al.*, 1998; Finkelstein and Lynch, 2000a; Soderman *et al.*, 2000; Brocard-Gifford *et al.*, 2004). Using a nonnatural (−)-(*R*)-ABA analogue instead of the natural (+)-(*S*)-ABA, Nambara *et al.* (2002) isolated many new alleles of known *abi* mutants but also two new loci named *CHOTTO1* and *CHOTTO2*. Genetic analysis suggests that CHOTTO1 may act downstream of ABI4 in parallel with a pathway containing ABI3 and ABI5 (E. Nambara, personal communication).

In addition to screens for ABA-insensitive mutants, germination assays have been used to search also for ABA-hypersensitive mutants, which resulted in isolation of the *enhanced response to ABA* (*era*) and *ABA-hypersensitive* (*abh*) mutants. Analysis of these mutants revealed the involvement of a subunit of a farnesyl transferase (ERA1), an ethylene signal transduction protein ETHYLENE INSENSITIVE2 (EIN2/ERA3), and an mRNA cap binding protein (ABH1) in ABA signaling (Cutler *et al.*, 1996; Ghassemian *et al.*, 2000; Hugouvieux *et al.*, 2001). Recently, four novel *ABA-hypersensitive germination* (*ahg*) mutants have been characterized (Nishimura *et al.*, 2004). These mutants also show enhanced sensitivity to salt and osmotic stress during germination. However, the responsiveness to sucrose and glucose during germination is different in each *ahg* mutant (Nishimura *et al.*, 2004). In addition, screens for suppressors and enhancers of *abi1-1* revealed the importance of gibberellin (GA) and ethylene in the regulation of ABA responses (Steber *et al.*, 1998; Beaudoin *et al.*, 2000).

12.2.2 ABA inhibition of seed germination is suppressed by sugars

Exogenous ABA inhibits *Arabidopsis* seed germination, but this inhibition can be relieved by the addition of metabolizable sugars or peptone (Garciarrubio *et al.*, 1997; Finkelstein and Lynch, 2000b). These studies showed that glucose or peptone do not affect ABA uptake or repress ABA signaling in general since (i) mobilization of storage proteins in seeds, which is blocked by ABA treatment, still does not occur in the presence of both ABA and glucose; and (ii) ABA-induced gene expression is not affected by glucose treatment. Thus, glucose (or peptone) rescues radicle emergence of ABA-treated seeds but not seedling growth, indicating that this antagonistic effect of sugar on ABA signaling is specific to the process of radicle emergence only.

Since ABA treatment inhibits the breakdown of storage proteins, Garciarrubio *et al.* (1997) hypothesized that ABA inhibits germination by restricting the availability of nutrients and metabolites. This metabolic restriction could be relieved by sugar treatment. Finkelstein and Lynch (2000b) hypothesized that the glucose rescue of ABA inhibition is not solely nutritional but may involve sugar signaling events. This is based on several observations that (i) sugar application suppresses only ABA inhibition of radicle emergence and not of seedling growth; (ii) the sugar response was enhanced by light; and (iii) a relatively low concentration (35 mM) of glucose is the optimum to suppress the ABA inhibition of germination while higher concentrations of glucose are less effective. However, glucose also delays germination at low concentrations (5–55 mM) (Price *et al.*, 2003; Dekkers *et al.*, 2004). This glucose-induced delay is dose dependent and is enhanced by ABA. Therefore, a separate process may be involved in the lower effectiveness of higher glucose concentrations in the recovery of ABA-inhibited germination observed by Finkelstein and Lynch (2000b). Using glucose analogues did not provide clear evidence for glucose signaling in this response (Finkelstein and Gibson, 2001), possibly because such analogues inhibit seed germination as well (Pego *et al.*, 1999; Dekkers *et al.*, 2004).

Interestingly, ABA-treated *Arabidopsis* seeds are still able to mobilize their lipid storage reserves, albeit at lower rates (Pritchard *et al.*, 2002; see Chapter 6). This mobilization of lipid storage reserves in ABA-treated seeds results in increased levels of sucrose and glucose. Puzzlingly, this internal increase in sugar concentration by itself does not provoke the ABA-treated seeds to germinate, in contrast to exogenously supplied metabolizable sugars. However, lipid storage in *Arabidopsis* seeds is not restricted to the embryo. The endosperm tissue contains about 10% of the total seed storage lipids (Penfield *et al.*, 2004). These lipid reserves in the endosperm are important to fuel hypocotyl elongation in the dark. GA is required for storage lipid mobilization in both embryo and endosperm. Normally, endosperm-derived carbon is transported to the embryo to fuel growth. However, ABA blocks storage lipid breakdown in the embryo but not in the endosperm tissues (Penfield *et al.*, 2004). Possibly, this differential regulation of storage lipid breakdown in the endosperm and embryo by ABA results in carbon limitation in the embryo only, which can somehow be overcome by exogenous sugar treatment (Penfield *et al.*, 2004). ABA might affect carbon allocation between these two seed tissues as well. Thus far, however, it is not yet established whether this differential storage lipid breakdown indeed results in different sugar levels between endosperm and embryo.

12.2.3 ABA blocks the transition from embryonic to vegetative growth

Lopez-Molina and Chua (2000) used an adapted version of the classic germination screen in their search for ABA-insensitive mutants. Seeds were plated on ABA media containing 1% sucrose. Such treatment allows seeds to germinate but seedling growth (i.e., root growth, greening of the cotyledons, and leaf formation) is arrested immediately after germination. Mutants whose seedling growth was not blocked by ABA were selected. One such *growth-insensitive to ABA* mutant (*gia1*) was isolated and was shown to be allelic to *abi5* (Lopez-Molina and Chua, 2000). ABI5 mRNA and protein accumulate in seeds in response to ABA and also in response to NaCl and drought stress (Lopez-Molina *et al.*, 2001). Interestingly, seedling development can be blocked by exogenous ABA only in a restricted time frame after germination, and only during this ABA-sensitive phase does ABI5 protein accumulate. Such a postgerminative arrested state can be maintained for at least a month and protects the seedling against the effects of drought stress (Lopez-Molina *et al.*, 2001). This arrested seedling state is relieved as soon as growth conditions become favorable. It has been hypothesized by Lopez-Molina *et al.* (2001) that seedlings may use this postgermination time frame to monitor the osmotic environment before the shift to vegetative growth. Collectively, different mechanisms controlling germination and seedling establishment by developmental as well as environmental cues optimize seedling survival by allowing seed germination and seedling establishment only in favorable conditions.

Interestingly, ABI5 protein level is enhanced in the *rpn10* (*regulatory particle non-ATPase subunit* [of 265 proteasome]) mutant, which is ABA-, sucrose-, and NaCl-hypersensitive, among other phenotypes (Smalle *et al.*, 2003). *rpn10* is mutated in a locus encoding a 26S subunit of the proteasome, consistent with the finding that an ABI5-interacting protein, AFP, acts as a negative regulator of ABA signaling by stimulating ABI5 protein degradation (Lopez-Molina *et al.*, 2003). ABI5 appears to be an important regulator of early seedling growth arrest due to ABA, with protein turnover of ABI5 via the 26S proteasome pathway being involved in regulating ABI5 protein levels.

Another ABA-insensitive mutant, *abi3-1*, was also shown to have a *gia* phenotype (Lopez-Molina *et al.*, 2002). Interestingly, ABI3 protein levels are also controlled by the ubiquitin–26S proteasome pathway via binding to an ABI3-INTERACTING PROTEIN, AIP2 (Zhang *et al.*, 2005). Genetic analysis suggests that ABI5 acts downstream of ABI3 to arrest early seedling development in response to ABA.

In addition, HYPONASTIC LEAVES1 (HYL1) is a dsRNA-binding protein involved in microRNA-mediated gene regulation that affects ABA signaling (Lu and Fedoroff, 2000; Han *et al.*, 2004; Vazquez *et al.*, 2004). The *hyl1* mutant is hypersensitive to ABA, glucose, NaCl, and osmotic stress (Lu and Fedoroff, 2000; Han *et al.*, 2004). Using the *hyl1* mutant, Lu *et al.* (2002) showed that the ABA-induced seedling arrest is mediated by MAPK (mitogen-activated protein kinase) signaling. In this mutant, MAPK activity and ABI5 protein accumulation are enhanced in response to ABA.

Both radicle emergence and seedling development are inhibited by ABA (Koornneef *et al.*, 1984). Perhaps it is not surprising that common mutants are identified in the *abi* and *gia* screens because the original *abi* mutants were screened for plantlets developing on ABA-containing media. Thus *abi1-1*, *abi2-1*, *abi3-1*, *abi4*, and *abi5* are insensitive to ABA during radicle emergence as well as early seedling development (Koornneef *et al.*, 1984; Finkelstein, 1994; Lopez-Molina *et al.*, 2002).

12.3 Sugar signaling represses germination and the transition to vegetative growth

12.3.1 Plant sugar signaling and the identification of sugar-response mutants

Carbohydrates have many important functions in plants. It is not surprising that the addition of sugar to plant growth media affects many developmental processes and gene expression (Gibson, 2005). Microarray experiments show that glucose is a potent regulator of transcription and that hundreds of genes are responsive to glucose treatment (Price *et al.*, 2004; Villadsen and Smith, 2004). Responsive genes include those involved in carbohydrate and nitrogen metabolism, signal transduction, and stress responses. Starvation conditions (famine response) upregulate genes associated with photosynthesis and resource mobilization while genes related to storage and utilization functions are repressed. Sugar excess (feast response) induces the opposite effects on gene expression (Koch, 1996). Such sugar-regulated gene expression adjusts cellular activity to nutrient availability. In this way sugars control plant resource allocation and adaptive developmental changes resulting in enhanced survival and/or competitiveness (Koch, 1996). Potentially all sugars or intermediary metabolites may serve as signaling molecules, but so far there is evidence for the existence of separate glucose, sucrose, and trehalose signaling pathways (Smeekens, 2000).

Factors involved in sugar signaling in plants have been identified using genetic approaches in *Arabidopsis*. A large number of mutants in different sugar-regulated processes were isolated and studied. Several review papers summarize and discuss the different screens performed and the mutants and genes identified (Smeekens, 2000; Gazzarrini and McCourt, 2001; Finkelstein and Gibson, 2001; Rolland *et al.*, 2002; Leon and Sheen, 2003; Rook and Bevan, 2003; Gibson, 2005). Four screens are of particular interest for this review since these were performed at germination and early seedling growth and have resulted in a significant number of identified mutant genes (Table 12.1). These mutant screens can be divided into two classes that are discussed in more detail below.

The first class of mutant screens used a sugar-regulated promoter fused to a reporter gene. Mutants were identified that showed an aberrant regulation of the reporter gene when challenged with sugar. One such example is a screen for *sucrose uncoupled* (*sun*) mutants. In this screen, the promoter of the plastocyanin (*PC*) gene was used. When seeds are germinated in the dark, the seedlings show transiently increased expression of several photosynthesis genes including *PC* (Brusslan and Tobin, 1992; Dijkwel *et al.*, 1996). This transient increase in expression of photosynthesis genes is developmentally orchestrated and is independent of light. Such increase in expression might be important for seedlings to be prepared for the utilization of light for photosynthesis once it becomes available. Dijkwel *et al.* (1996) showed that this transient activation of *PC* could be repressed by addition of sucrose to the germination medium. Mutagenized transgenic plants containing a *PC* promoter::luciferase (*LUC*) fusion were screened for mutants that lack the

Table 12.1 Overview of *sun*, *isi*, *gin*, and *sis* screens

Screen	Screening method	Developmental stage	Mutants identified	References
Sucrose uncoupled (*sun*)	Expression of PC promoter fused to LUC reporter gene in dark-grown seedlings in response to 88 mM sucrose	Dark grown, 2–3-day-old seedlings	*sun1* = unknown *sun2* = unknown *sun3* = unknown *sun4* = unknown *sun6* = *abi4*, AP2 transcription factor *sun7–sun16* = unknown	Dijkwel *et al.* (1997); Huijser *et al.* (2000)
Impaired sucrose induction (*isi*)	Expression of Apl3 promoter fused to a negative selection marker in response to 100 mM sucrose	Light grown, 1-week-old seedlings	*isi1* = novel gene, vascular expressed related to sink-to-source transition *isi2* = unknown *isi3* = *abi4*, AP2 transcription factor *isi4* = *aba2*, short chain dehydrogenase/reductase	Rook *et al.* (2001) Rook and Bevan (2003)
Glucose insensitive (*gin*)	Developmental arrest on 330 mM glucose	Light grown, germinated seedlings	*gin1* = *aba2*, short chain dehydrogenase/reductase *gin2* = *hxk1*, hexokinase/ sugar sensor *gin3* = unknown *gin4* = *ctr1*, RAF-like protein kinase *gin5* = *aba3*, molybdenum cofactor sulfurase *gin6* = *abi4*, AP2 transcription factor	Zhou *et al.* (1998); Arenas-Huertero *et al.* (2000); Cheng *et al.* (2002); Moore *et al.* (2003)
Sugar insensitive (*sis*)	Developmental arrest on 300 mM sucrose	Light grown, germinated seedlings	*sis1* = *ctr1*, RAF-like protein kinase *sis4* = *aba2*, short chain dehydrogenase/reductase *sis5* = *abi4*, AP2 transcription factor	Laby *et al.* (2000); Gibson *et al.* (2001)

transcriptional downregulation of the *PC* promoter. Thus far, 16 *sun* mutants have been isolated, one of which (*sun6*) was shown to be allelic to *abi4* (Dijkwel *et al.*, 1997; Huijser *et al.*, 2000).

Another screen of this class is the *impaired sucrose induction* (*isi*) screen, which is based on the observation that sucrose induces the *ApL3* gene encoding a large subunit of the ADP-glucose pyrophosphorylase (AGPase) complex involved in starch biosynthesis. A bacterial cytochrome P450 was used as a reporter gene in this system. This enzyme metabolizes the nontoxic proherbicide R7402 compound to a highly phytotoxic form. Expression of the P450 gene in the presence of R7402 results in lethality, serving as a negative selection marker (O'Keefe *et al.*, 1994). Transgenic seeds containing the *ApL3* promoter fused to the P450 marker gene were grown on media containing sucrose and R7402. Since the *ApL3* promoter is activated by sucrose and causes lethality, only mutants that lacked the response to sucrose survived this treatment. Four *isi* mutants were identified, two of which had lesions in the *ABA2* and *ABI4* genes (Rook *et al.*, 2001).

The second class of mutant screens is based on the inhibitory effect of high sugar concentrations on early seedling development. On control media, seedlings develop expanded, green cotyledons and produce leaves. In contrast, when seedlings are germinated on media containing elevated sugar concentrations, seedling development is arrested quickly after germination, i.e., no greening, cotyledon expansion, and leaf formation are observed. This developmental block is not evoked by osmotic control treatments with sorbitol or mannitol. Both high glucose (>330 mM) as well as high sucrose (>300 mM) concentrations have been used to screen for sugar-insensitive mutants. *Glucose-insensitive* (*gin*) and *sugar-insensitive* (*sis*) mutants are resistant to high glucose and high sucrose concentrations, respectively (Zhou *et al.*, 1998; Laby *et al.*, 2000). These mutants develop green cotyledons and leaves when grown under high sugar conditions. Interestingly, Gibson *et al.* (2001) showed that the sucrose-induced developmental arrest could be triggered only in a limited time window after sowing.

Remarkably, all four different screening conditions used (i.e., *gin*, *sis*, *isi*, *sun*) yielded *abi4* alleles. In addition, *aba2* mutants were identified in *gin*, *sis*, and *isi* screens and the ethylene signaling mutant *ctr1* (*constitutive triple response1*) was isolated in both *gin* and *sis* screens. On the other hand, there are differences among the mutants identified in these screens. Two *isi* mutants (*isi1* and *isi2*) do not show a *gin* phenotype (Rook *et al.*, 2001). Furthermore, whereas ABA and elevated glucose and sucrose concentrations induce a similar early seedling developmental arrest, sucrose but not ABA is able to induce the *ApL3* gene. Probably, ABA modulates the sensitivity of *ApL3* gene regulation to sucrose (Rook *et al.*, 2001).

More recent studies have isolated novel mutants that are sugar hypersensitive as well (Leon and Sheen, 2003; Baier *et al.*, 2004). Identification and characterization of such mutants and the affected genes will advance our understanding of sugar signaling.

12.3.2 The glucose-insensitive response pathway

Seeds plated on media containing elevated glucose concentrations (>6%) are developmentally arrested immediately following germination. This early seedling arrest is characterized by the absence of cotyledon expansion, greening, and leaf formation. Seedling growth of *gin* mutants is not blocked by such glucose concentrations but show greening of the cotyledons and leaf formation. Mutants exhibiting the glucose-insensitive response can be subdivided into three groups namely, *hexokinase1* (*gin2/hxk1*), ABA-related, and ethylene-related mutants. An overview of the glucose-insensitive response and factors involved in it is presented in Figure 12.1.

Glucose analogues such as 2-deoxyglucose, which can be phosphorylated by hexokinase (HXK) but not used in general metabolism, were shown to induce several sugar responses. Such studies suggested that HXK substrate phosphorylation suffices to induce sugar responses (Jang and Sheen, 1994). Evidence for the

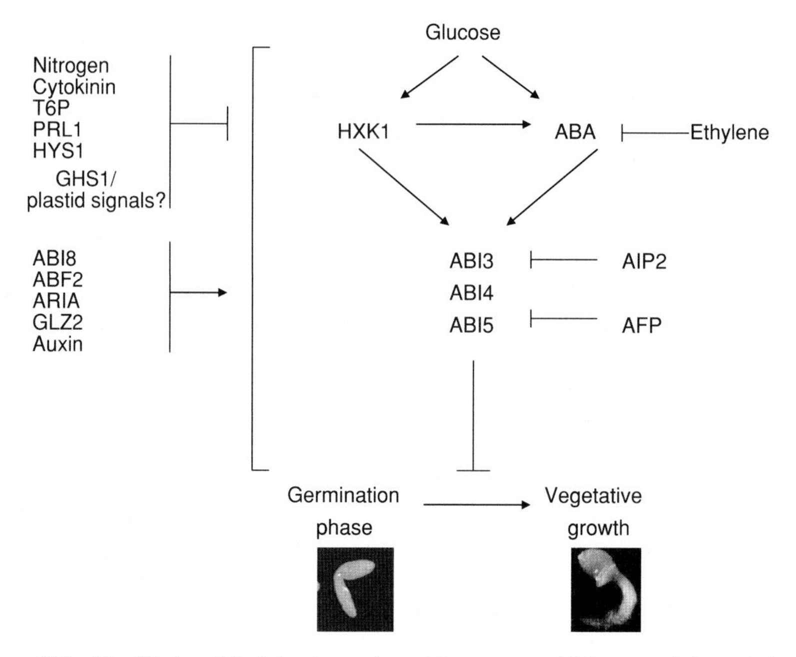

Figure 12.1 Simplified model of the glucose-insensitive response. ABA accumulation and the expression of downstream effectors (ABI3, ABI4, and ABI5) are enhanced in glucose-treated seedlings (Arenas-Huertero *et al.*, 2000; Cheng *et al.*, 2002). AIP2 and AFP induce ABI3 and ABI5 degradation (Lopez-Molina *et al.*, 2003; Zhang *et al.*, 2005) and are putative negative regulators of glucose signalling as well. On the left negative regulators (indicated with a bar) and positive regulators (indicated with an arrow) of the glucose-insensitive response are indicated. See text for details. It is largely unknown how these factors impinge on the glucose response pathway. Abbreviations: ABI, ABA INSENSITIVE; ABF2, ABRE-BINDING FACTOR2; AFP, ABI5-INTERACTING PROTEIN; AIP2, ABI3-INTERACTING PROTEIN; ARIA, ARM REPEAT PROTEIN INTERACTING WITH ABF2; GHS1, GLUCOSE HYPERSENSITIVE1; GLZ2, GAOLAOZHUANGREN2; HXK1, HEXOKINASE1; HYS1, HYPERSENESCENCE1; PRL1, PLEIOTROPIC RESPONSE LOCUS1; T6P, TREHALOSE-6-PHOSPHATE.

involvement of HXK in sugar responses was shown by work of Xiao *et al.* (2000) using antisense and overexpressing *HXK1* lines. Antisense *HXK1* lines showed reduced sugar responsiveness whereas *HXK1* overexpressing lines were glucose hypersensitive. These studies suggested a role for HXK1 in sugar responses, but it remained unclear whether HXK catalytic activity is required for its signaling function. A *gin2/hxk1* mutant was isolated in the glucose-insensitive screen. This mutant is glucose insensitive and shows reduced growth under high light conditions, reduced seed set, hypersensitivity to cytokinin, and a reduced sensitivity to auxin (Moore *et al.*, 2003). Remarkably, introduction of catalytically inactive versions of HXK1 into the *gin2/hxk1* mutant suppresses several of its phenotypes, such as the growth defects under high light, reduced sugar sensitivity, and altered hormone sensitivity (Moore *et al.*, 2003). This indicates that the catalytically inactive HXK1 protein supports several important signaling functions.

The second group of *gin* mutants (*gin1/aba2*, *gin5/aba3*, and *gin6/abi4*) has defects in either ABA biosynthesis or ABA response (Arenas-Huertero *et al.*, 2000; Cheng *et al.*, 2002). *ABI4* gene expression is enhanced in sugar-treated, developmentally arrested seedlings compared to untreated, nonarrested seedlings (Arenas-Huertero *et al.*, 2000; Cheng *et al.*, 2002). However, the response to sugar is dependent on the developmental stage; i.e., older seedlings lose their sugar sensitivity and do not show *ABI4* gene induction (Gibson *et al.*, 2001; Arroyo *et al.*, 2003). The *abi4* mutant was originally identified in a germination-based screen for ABA-insensitive mutants (Finkelstein, 1994). It encodes an AP2 domain transcription factor. In total, 145 ERF (ETHYLENE RESPONSE FACTOR)/AP2 domain proteins have been reported that are subdivided into five subfamilies (Sakuma *et al.*, 2002). ABI4 is a member of the DREB (Dehydration-Responsive Element Binding Protein) subfamily, which is further divided into six classes. ABI4 forms one such DREB class by itself and no close homologues are present in the *Arabidopsis* genome (Finkelstein *et al.*, 1998; Soderman *et al.*, 2000; Sakuma *et al.*, 2002). Moreover, the AP2 domain of AtABI4 is more similar to the AP2 domains of its maize and rice orthologues, ZmABI4 and OsABI4 respectively, than to its closest homologues in the *Arabidopsis* genome (Niu *et al.*, 2002). The maize *ZmABI4* gene is able to complement the *Arabidopsis abi4-1* mutant and binds the CE1 (Coupling Element1)-like sequences in ABA- and sugar-regulated promoters (Niu *et al.*, 2002). Recently, Acevedo-Hernandez *et al.* (2005) showed that the S-box (containing a CE1-like element) in the *RBCS* (ribulose-1,5-bisphosphate carboxylase small subunit) promoter and ABI4 binding to this motif are essential for sugar and ABA repression of *RBCS* transcription. Experiments using an S-box-less *RBCS* promoter::reporter construct indicated that this promoter is no longer subject to repression by sugar and ABA, suggesting the importance of this element in the downregulation of *RBCS* by sugar and ABA. Also, reporter analysis using the authentic S-box-containing promoter in an *abi4-1* background indicated that the S-box does not function in the absence of ABI4.

In addition to its role in ABA and sugar regulation, ABI4 is also involved in osmotic and salt responses during germination, lateral root outgrowth in response to nitrate, and primed callose deposition in β-amino butyric acid-induced resistance against necrotrophic pathogens (Quesada *et al.*, 2000; Signora *et al.*, 2001; Carles *et al.*, 2002; Ton and Mauch-Mani, 2004).

ABA and the ABI4 transcription factor are essential for several sugar responses, raising the question of whether other ABA response loci are important in sugar signaling as well. Several papers reported the early seedling developmental phenotypes of *abi1*, *abi2*, *abi3*, *abi5*, and *abi8* mutants grown in the presence of glucose and sucrose. In general, *abi1*, *abi2*, and *abi3* seeds showed wild-type sensitivity to high sugar concentrations, while *abi5* and *abi8* seeds were insensitive (Leon and Sheen, 2003; Brocard-Gifford *et al.*, 2004). In contrast, overexpression studies with *ABI3* and *ABI5* suggest that both genes affect sugar responses (Brocard *et al.*, 2002; Finkelstein *et al.*, 2002; Zeng and Kermode, 2004). Moreover, *abi3* mutants were insensitive to glucose in combination with ABA (Nambara *et al.*, 2002). Clearly, this issue needs further work.

The third group of *gin* mutants exhibits altered ethylene biosynthesis or responses. Zhou *et al.* (1998) found that the glucose insensitivity of the *gin1* mutant could be phenocopied by treatment of wild-type seeds with the ethylene precursor 1-aminocyclopropane carboxylic acid (ACC). Moreover, mutants that overproduce ethylene or have a constitutive ethylene response show a *gin* phenotype. In two screens (*gin* and *sis*) alleles of constitutive ethylene response (*ctr*) mutants were identified as sugar insensitive (Gibson *et al.*, 2001; Cheng *et al.*, 2002; Leon and Sheen, 2003). Vice versa, *ethylene-insensitive* (*ein*) mutants were found to be hypersensitive to glucose-induced seedling arrest (Zhou *et al.*, 1998; Leon and Sheen, 2003). Interestingly, Yanagisawa *et al.* (2003) showed that glucose promotes EIN3 turnover via a HXK1-dependent pathway. EIN3 regulation is an important step in ethylene signaling. EIN3 is degraded via the ubiquitin–proteasome pathway which involves the F-box proteins EBF1 (EIN3-BINDING F-BOX) and EBF2. Upon ethylene treatment, EIN3 protein accumulates, dependent on several genes of the ethylene response pathway including *ETR1* (ETHYLENE RECEPTOR1), *CTR1*, *EIN2*, and *EIN6* (Guo and Ecker, 2003; Potuschak *et al.*, 2003).

Both ABA and ethylene are regulators of sugar responses and also interact in their signaling pathways. Constitutive ethylene signaling decreases ABA sensitivity and ethylene-insensitive mutants are ABA hypersensitive (Beaudoin *et al.*, 2000; Ghassemian *et al.*, 2000). Genetic analysis of glucose-induced seedling arrest using ethylene signaling *etr1* and *ein2* mutants and *gin1/aba2* demonstrated that *gin1/aba2* acts downstream of ethylene signaling (Zhou *et al.*, 1998; Cheng *et al.*, 2002). Possibly, ethylene affects sugar responses by modifying ABA levels or ABA signaling, as was shown for germination (Beaudoin *et al.*, 2000; Ghassemian *et al.*, 2000) and for stomatal closure (Tanaka *et al.*, 2005).

12.3.3 Other factors affecting the glucose response during early seedling development

The analysis of *gin* mutants provides an important basis for our knowledge of glucose signaling during early seedling development. However, other nutrients, metabolites, hormones, and mutations are known to affect the sugar response during early seedling development. For example, the *gin2/hxk1* mutant is hypersensitive to cytokinin and has a reduced sensitivity to auxin (Moore *et al.*, 2003). Interestingly, addition of cytokinin to developing seedlings results in a reduced sugar sensitivity as does the addition of ACC (Zhou *et al.*, 1998; Moore *et al.*, 2003). The effect of cytokinin is independent from ethylene signaling. In addition, auxin signaling mutants, such as *tir1* (*transport inhibitor response1*), show a *gin* phenotype (Moore *et al.*, 2003).

Both carbon (C) and nitrogen (N) are important nutrients for plant growth and development. C and N signaling intersect and several responses are more responsive to C/N ratio rather than to C concentration alone (Martin *et al.*, 2002; for reviews see Coruzzi and Bush, 2001; Coruzzi and Zhou, 2001). Microarray analyses of C, N, or C + N responsive genes supported the hypothesis that plants possess such C/N ratio-responsive mechanisms (Palenchar *et al.*, 2004). Furthermore, this study showed that C/N ratio controls transcription via multiple mechanisms. Increasing N concentration in the growth medium decreases responsiveness of seedlings to sugar, while lowering nitrogen level increases the sensitivity of seedlings to sugar (Martin *et al.*, 2002; Moore *et al.*, 2003). Photosynthetic gene expression and lipid mobilization in response to sugar are affected by C/N ratio as well (Martin *et al.*, 2002).

Trehalose metabolism also affects sugar utilization during early seedling growth. Genes for trehalose metabolism were found in many plants, but with the exception of some resurrection plants, trehalose levels are very low to undetectable (Eastmond and Graham, 2003). Also in *Arabidopsis* the amount of trehalose is very low but, nevertheless, trehalose metabolism was found to be essential for *Arabidopsis* development and carbon metabolism (Eastmond *et al.*, 2002; Schluepmann *et al.*, 2003). Transgenic plants with enhanced levels of trehalose-6-phosphate (T6P) show improved growth on media supplemented with sugars, while transgenic lines with decreased levels of T6P develop slower and exhibit reduced cotyledon expansion and greening (Schluepmann *et al.*, 2003; Avonce *et al.*, 2004). Thus, T6P levels determine the responses of seedlings to exogenously supplied sugar.

Recently several mutants that have an altered *gin* response were described. Mutants in the ABRE-binding factor ABF2 and its interacting partner ARIA (ARM REPEAT PROTEIN INTERACTING WITH ABF2) result in a *gin* phenotype, and the opposite result was obtained when both genes were overexpressed (Kim *et al.*, 2004a,b). Also a genetic lesion in *GAOLAOZHUANGREN2* (*GLZ2*) has been identified as a *gin* mutant (Chen *et al.*, 2004). On the other hand, mutations in genes like *PLEIOTROPIC RESPONSE LOCUS1* (*PRL1*), *HYPERSENESCENCE1* (*HYS1*)/*CONSTITUTIVE EXPRESSOR OF PATHOGENESIS-RELATED GENES* (*CPR5*), and *GLUCOSE HYPERSENSITIVE1* (*GHS1*) result in an enhanced response to sugar during early seedling growth (Nemeth *et al.*, 1998; Yoshida *et al.*, 2002; Morito-Yamamuro *et al.*, 2004). Interestingly, *ghs1* has a lesion in plastid 30S ribosomal protein S21. This mutant has reduced levels of RBCL (RBC large subunit) and impaired chloroplast development (Morita-Yamamuro *et al.*, 2004). Possibly, a translation-dependent plastidic signal is necessary for proper sugar signaling, although alternative chloroplast defects cannot be ruled out (Morita-Yamamuro *et al.*, 2001). Interestingly, treatment of dark-grown seedlings with lincomycin, an inhibitor of plastid translation, inhibits the developmentally regulated PC expression and mimics sugar treatment of dark-grown seedlings (Dijkwel, 1997).

12.3.4 Sugar delays seed germination in Arabidopsis

The storage reserves in seeds are consumed during germination and seedling growth to sustain cellular processes and growth. It is, therefore, remarkable that the addition of metabolizable sugars such as glucose delays seed germination in *Arabidopsis* (To *et al.*, 2002; Ullah *et al.*, 2002; Price *et al.*, 2003; Dekkers *et al.*, 2004). An inhibition of germination by sugars is also reported for tomato (*Lycopersicon esculentum*), squash (*Cucurbita pepo*), and watermelon (*Citrullus lanatus*) seeds (Martin and Blackburn, 2003). In contrast, addition of metabolizable sugar relieves the inhibition of *Arabidopsis* seed germination by mannose (see below) and ABA (Garciarrubio *et al.*, 1997; Pego *et al.*, 1999; Finkelstein and Lynch, 2000a). In *Arabidopsis* the delay of germination is provoked by low concentrations of glucose. This glucose-induced germination delay is dose dependent but independent of the osmotic stress, because equal concentrations of sorbitol or mannitol do not cause a similar delay in germination (To *et al.*, 2002; Price *et al.*, 2003; Dekkers *et al.*, 2004).

The glucose analog mannose is a potent inhibitor of *Arabidopsis* seed germination and is effective at low concentrations (5–10 mM) (Pego *et al.*, 1999). The mannose-induced inhibition of germination depends on a HXK-mediated step, and *abi4* mutants have a reduced sensitivity for such treatment. Surprisingly, *aba* mutants show normal sensitivity to mannose during germination (Pego *et al.*, 1999; Huijser *et al.*, 2000; Laby *et al.*, 2000). Glucose and mannose both inhibit seed germination but this occurs via separate pathways, as the effect of mannose can be counteracted by the addition of 60 mM glucose (Pego *et al.*, 1999; Dekkers *et al.*, 2004). Originally, it was hypothesized that mannose signaling restricts carbon mobilization necessary for germination. Analysis of a mutant in *AtSTP1* (*Arabidopsis thaliana SUGAR TRANSPORT PROTEIN1*), a monosaccharide/proton symporter in *Arabidopsis*, showed that this is a major sugar transporter in seeds and seedlings (Sherson *et al.*, 2000). Glucose and mannose uptake is reduced up to 60% in the *stp1* mutant seedlings. The *stp1* seeds exhibit *mannose-insensitive germination* (*mig*; Pego *et al.*, 1999), indicating that mannose transport is important for the germination inhibition. Sherson *et al.* (2000) therefore suggested that the rescue of mannose-inhibited germination by glucose is possibly caused by reduced mannose uptake due to competition with excess glucose.

As discussed earlier, the ability of elevated sugar concentrations to inhibit early seedling growth was used to screen for sugar-insensitive mutants. The *hxk1*, *aba2*, *abi4*, and *ctr1* mutants lacked this postgerminative developmental arrest when challenged with elevated sugar concentrations (Leon and Sheen, 2003). Such sugar-insensitive mutants were predicted to be insensitive also to the sugar-induced germination delay. Strikingly, *hxk1*, *abi4*, and *ctr1* mutants turned out to be sensitive to the sugar-induced germination delay. Only ABA-deficient mutants were insensitive to sugar to some extent. These results show that the sugar-induced delay of germination and sugar-induced postgerminative developmental arrest are two distinct responses (Price *et al.*, 2003; Dekkers *et al.*, 2004).

Other plant hormones were investigated for their roles in the sugar-induced germination delay. Sugar treatment delayed ABA degradation (Price *et al.*, 2003); however, it is unclear whether this is the *cause* of the delay of germination or is a *consequence* of the glucose-triggered delay of germination. Mutants with a reduced ABA sensitivity such as *abi1*, *abi2*, *abi4*, *abi5*, and *ctr1* exhibit normal sensitivity to glucose during germination (Price *et al.*, 2003; Dekkers *et al.*, 2004). These results support the hypothesis that the delayed ABA degradation is a *consequence* rather than the *cause* of the germination delay. It is unclear why ABA-deficient mutants show a reduced sensitivity to glucose. Possibly, ABA levels are important for regulating the glucose sensitivity in seeds. ABA concentrations that are not inhibitory to germination synergistically enhance the repression of germination by glucose (Dekkers *et al.*, 2004). This enhancement of glucose repression of germination by ABA is not due to osmotic signaling and is independent of *ABI4* (Figure 12.2). Thus, ABI4 is involved neither in the glucose response during germination nor in

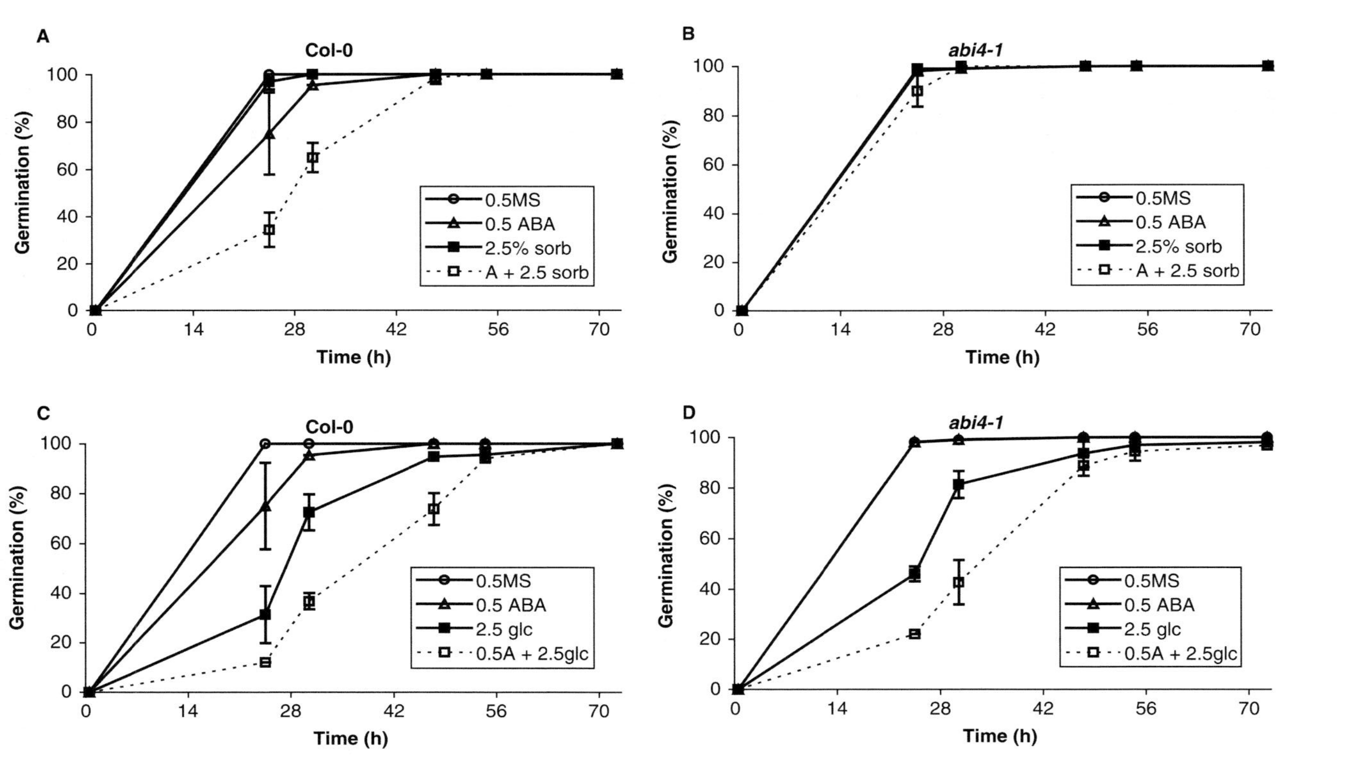

Figure 12.2 ABI4 is required neither for the glucose response during germination nor its modulation by ABA. (A, B) Col-0 (wild-type) and *abi4-1* seeds were germinated on 0.5 MS media, 0.5 MS containing 0.5 μM ABA, 2.5% sorbitol or a combination of 0.5 μM ABA + 2.5% sorbitol (A + 2.5sorb). The osmotic response is synergistically enhanced by low concentrations of ABA in Col-0, but *abi4-1* is insensitive to this combination. (C, D) Col-0 and *abi4-1* seeds are germinated on 0.5 MS, or 0.5 MS containing 0.5 μM ABA, 2.5% glucose or a combination of 0.5 μM ABA + 2.5% glucose (0.5A + 2.5glc). Both Col-0 and *abi4-1* have enhanced glucose sensitivity in combination with ABA compared to glucose treatment alone. Apparently, based on the differential response of *abi4-1* to ABA + sorbitol compared to ABA + glucose, the enhanced response to the combination of ABA and glucose is not due to osmotic signaling. Seeds were stratified for 3 days at 4°C before transfer to the growth chamber (16 h/8 h light/dark, 22°C). The results are the mean of single experiment performed *in duplo*. Error bars indicate SE (SE smaller than 2 is not indicated). The experiment was repeated twice with similar results.

the modulation of this response by ABA. Furthermore, the addition of hormones such as GA, brassinosteroids, and ethylene (via ACC) that are known to stimulate germination did not suppress the glucose-induced germination delay (Dekkers *et al.*, 2004).

Another mutant with an enhanced sensitivity to sugar during germination is *gpa1* (Ullah *et al.*, 2002). *GPA1* (*G PROTEIN α-SUBUNIT*) is a gene encoding the alpha subunit of the G protein in *Arabidopsis*. Inhibition of carotenoid synthesis by fluridone (which causes an inhibition of ABA synthesis) suppresses the hypersensitivity of *gpa1* seeds to glucose (Ullah *et al.*, 2002). The *gpa1* mutants are also less sensitive to GA and lack the response to brassinosteroids.

An impaired germination phenotype is also observed in mutants that have defects in fatty acid metabolism, such as *ketoacyl-CoA thiolase2* (*kat2*), *comatose* (*cts*), and *acyl-CoA oxidase1 acyl-CoA oxidase2* (*acx1 acx2*) double mutant. A recent study of Pinfield-Wells *et al.* (2005) showed that seed germination of fatty acid breakdown mutants cannot be rescued by sucrose addition (see Chapter 6). This suggests that peroxisomal β-oxidation activity has another essential function in germination in addition to the provision of carbon to support growth. Possibly peroxisomal β-oxidation activity is necessary for the generation of metabolites or signaling molecules that induce germination or, alternatively, is involved in the degradation of a germination inhibitor (Pinfield-Wells *et al.*, 2005). Studies on the fatty acid transport mutant *cts* show that its 'forever' dormant phenotype is suppressed in certain genetic backgrounds. Combining *cts* with mutations such as *aba1*, *abi3*, *lec1* (*leafy cotyledon1*), and *fus3* (*fusca3*) rescues the germination phenotype (Russell *et al.*, 2000). However, *cts* germination phenotype is not rescued by *abi1-1* or *reduced dormancy2* (*rdo2*), both of which show a reduced dormancy phenotype (Footitt *et al.*, 2002). Notably, *aba*, *abi3-5*, *lec1*, and *fus3* have reduced glucose sensitivity during germination but *abi1-1* and *rdo2* are sensitive to sugar during germination (Brocard-Gifford *et al.*, 2003; Price *et al.*, 2003; Dekkers *et al.*, 2004; our unpublished results). Moreover, germination of the fatty acid breakdown mutants *acx1, acx2* and *cts* is not significantly improved by exogenous GA (Russell *et al.*, 2000; Pinfield-Wells *et al.*, 2005). Similarly, the glucose-induced delay of germination is not rescued by GA (Dekkers *et al.*, 2004). A correlation might exist between different genotypes that specify glucose sensitivity and their sensitivity to a dysfunctional peroxisomal β-oxidation pathway during germination. Moreover, sugar addition delayed storage lipid breakdown in *Arabidopsis* seeds (Martin *et al.*, 2002; To *et al.*, 2002). One could suggest that glucose addition to seeds delays germination by transiently repressing peroxisomal β-oxidation activity, which is essential for proper germination. However, this hypothesis seems unlikely, since the sugar-insensitive mutant *abi4* is insensitive to glucose repression of storage lipid mobilization but is sensitive to glucose-induced germination delay (To *et al.*, 2002). Moreover, nitrogen concentrations do not affect the glucose-induced delay of seed germination but C/N ratio does influence storage lipid breakdown (Martin *et al.*, 2002). Finally, the glucose analogue 3-O-methylglucose delays germination but does not affect storage lipid breakdown (To *et al.*, 2002). These observations uncouple the inhibitory effects of glucose on storage lipid mobilization and on germination.

12.3.5 *Imbibed seeds rapidly lose sensitivity for the glucose-induced germination delay*

The temporal window of seed glucose sensitivity was investigated by transferring non-cold-stratified seeds from sugar-free media to glucose-containing media at different times after imbibition. The sensitivity to glucose rapidly declined during imbibition prior to radicle emergence, in contrast to the sensitivity to ABA, which remained essentially constant during this time period (Figures 12.3A and 12.3B). Since cold stratification (imbibed chilling at 4°C) promotes loss of dormancy in *Arabidopsis* seeds, it was of interest to test whether the sensitivity to glucose would also be lost during cold stratification. When *Arabidopsis* seeds were treated with 2.5% glucose during cold stratification and subsequent incubation at 22°C, germination was significantly delayed (Dekkers *et al.*, 2004). In contrast, when seeds were exposed to 2.5% glucose only immediately after cold stratification, they had lost their sensitivity to glucose and exhibited no delay in germination (B.J.W. Dekkers and J.C.M. Smeekens, unpublished results). Moreover, seeds were resistant to even higher glucose concentrations (6%) after cold stratification, while the response to 6% sorbitol was not affected by cold stratification (Figures 12.3C and 12.3D). Germination of seeds that were treated with glucose after cold stratification mimics the germination curves of sorbitol-treated seeds (Figures 12.3C and 12.3D). On the other hand, continuous glucose treatment during cold stratification and germination resulted in a clear germination delay (Figures 12.3C and 12.3D).

As discussed previously, metabolizable sugars antagonize ABA signaling during germination and promote germination of ABA-treated seeds (Garciarrubio *et al.*, 1997; Finkelstein and Lynch, 2000b). This rescue of germination of ABA-treated seeds by glucose is more pronounced when glucose is added to the medium after cold stratification (Figures 12.3E and 12.3F), which is consistent with the observation that glucose is inhibitory when present during cold stratification. This finding shows that the stimulatory effect of glucose on germination is exerted after cold stratification. It appears that the inhibitory and stimulatory (in the case of ABA treatment) actions of glucose during germination are separable in time. Mutants with disrupted seed development such as *lec1*, *fus3*, and *abi3-5* are insensitive to sugar during germination as well (Brocard-Gifford *et al.*, 2003; B.J.W. Dekkers and J.C.M. Smeekens, unpublished results). The temporal separation of glucose sensitivity of seeds between cold stratification and germination periods might explain the insensitivity of these mutants; i.e., the developmental modes of *lec1*, *fus3* and *abi3-5* may be already shifted to a sugar-insensitive phase due to their heterochronic nature.

Thus far, little is known about the physiological or molecular basis of the glucose response during seed germination. In *Arabidopsis* many genes involved in dormancy and germination have been described using mutant or natural variation analyses (Bentsink and Koornneef, 2002; see Chapter 5). Genes such as *IMBIBITION-INDUCIBLE1* (*IMB1*), *PIF3-LIKE5* (*PIL5*), *RGA-LIKE2* (*RGL2*), *DELAY OF GERMINATION* (*DOG*), and *REDUCED DORMANCY* (*RDO1/4*) are important regulators of germination (Leon-Kloosterziel *et al.*, 1996; Bentsink and Koornneef, 2002; Lee *et al.*, 2002; Alonso-Blanco *et al.*, 2003, Duque and Chua, 2003; Oh *et al.*,

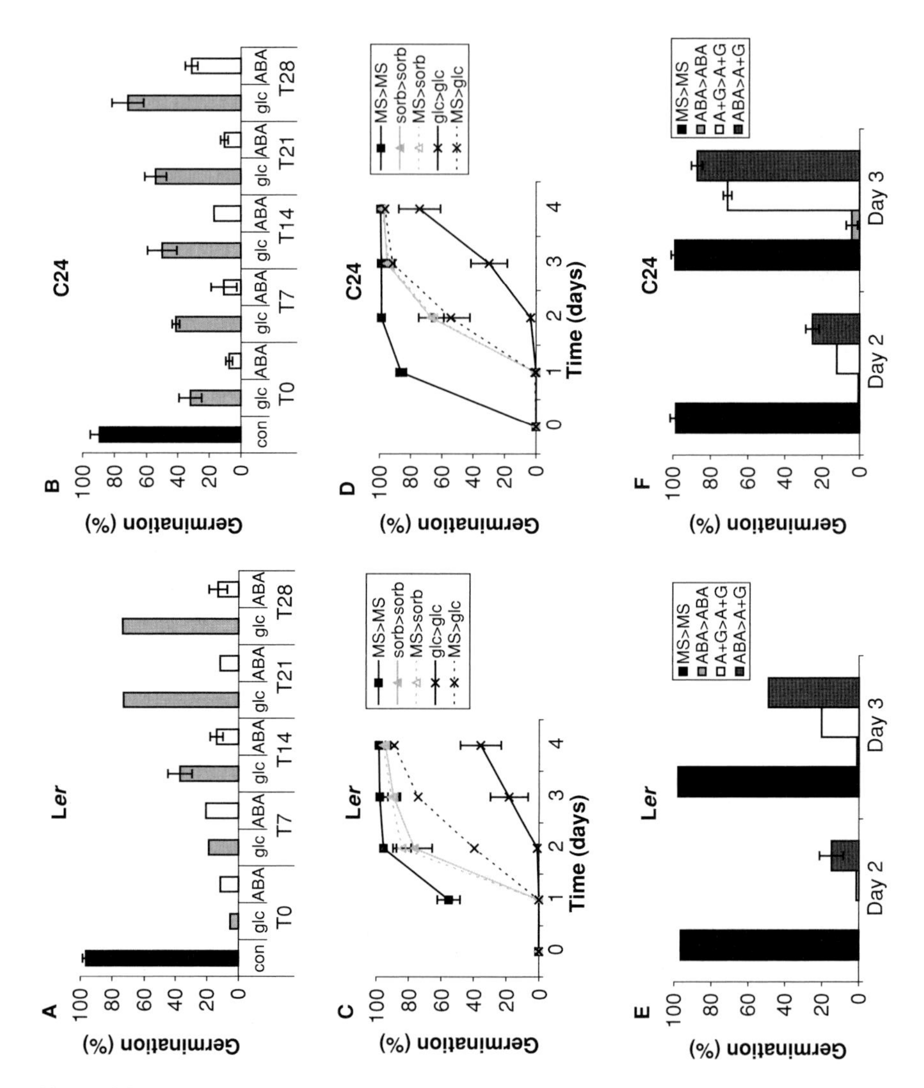

Figure 12.3 Imbibed seeds rapidly lose sensitivity to glucose during germination. (A, B) Non-stratified seeds were germinated at 22°C on 0.5 MS (con, control), on 0.5 MS containing 2.5% glucose (glc), on 0.5 MS + 3 μM ABA (ABA) at $t = 0$ or on 2.5% glucose or 3 μM ABA after incubation for 7, 14, 21, or 28 h on 0.5 MS (T0 to T28). At 28 h the seeds were not germinated but germination started within 7 h thereafter. Germination was scored on days 2 or 3 for 0.5 MS and glucose treatments and on days 5–7 for ABA treatments depending on the ecotype and experiment. In both L*er* and C24 (wild-type) ecotypes, glucose sensitivity was rapidly lost during imbibition while the sensitivity for ABA was unaffected. (C, D) Seeds were stratified for three days at 4°C on 0.5 MS and transferred after stratification to 0.5 MS (MS > MS), 6% sorbitol (MS > sorb) or 6% glucose (MS > glc), or stratified on 6% sorbitol and transferred after stratification to 6% sorbitol (sorb > sorb), or stratified on 6% glucose and transferred to 6% glucose (glc > glc). In both L*er* and C24 ecotypes the sensitivity to glucose during germination is suppressed by stratification in the absence of glucose. (E, F) Seeds were stratified on 0.5 MS and transferred after stratification to 0.5 MS (MS > MS), stratified on 0.5 MS + 10 μM ABA and transferred

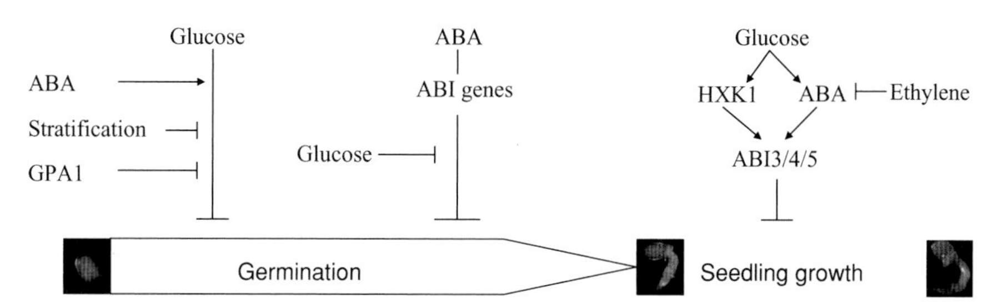

Figure 12.4 Overview of three different responses with alternating synergistic and antagonistic ABA and glucose interactions during *Arabidopsis* seed germination and early seedling growth.

2004). Possibly, sugars delay germination by affecting expression of such genes or activities of their products. Interestingly, Bradford *et al.* (2003) showed that the tomato *SUCROSE NON-FERMENTING4* (*LeSNF4*) gene is differently regulated by ABA and GA in germinating tomato seeds. SNF4 is part of an SnRK1 (SNF1-related protein kinase) complex involved in the regulation of sugar metabolism. Bradford *et al.* (2003) suggest that regulation of *LeSNF4* may act as a potential link between hormonal and sugar signals during germination.

12.4 Conclusions

Sugar and ABA signaling are intimately related in their effects on seed germination and the switch from germinative to seedling growth stages. Thus far, three different phases encompassing germination and early seedling growth with alternating synergistic or antagonistic sugar and ABA interactions can be distinguished (Figure 12.4).

First, the sugar-induced delay of germination is an early response. During imbibition the sensitivity to glucose rapidly decreases while the sensitivity to ABA is unaffected. This glucose response during germination is largely repressed when glucose is added after the stratification period. Small amounts of ABA, which do not affect germination, enhance the sugar-induced delay of germination. In this case, ABA and sugar act synergistically to repress germination. This synergistic response is also observed in *abi4*. Thus, ABI4 has no function in either the glucose-induced delay of germination or the synergistic action between sugar and ABA.

Figure 12.3 (*Continued*) to 0.5 MS + 10 μM ABA (ABA > ABA) or 0.5 MS + 10 μM ABA + 2.5% glucose (ABA > A+G), or stratified on 0.5 MS + 10 μM ABA + 2.5% glucose and transferred to 0.5 MS + 10 μM ABA + 2.5% glucose (A+G > A+G). The addition of glucose rescues germination of ABA-treated seeds of both L*er* and C24, but is more pronounced when glucose is added after stratification. Data points are the means of a single experiment performed *in duplo*. Error bars indicate SE (SE smaller than 2 is not indicated). The experiment was repeated one to three times with similar results. Control experiments showed that seed transfer did not affect their germination behavior (data not shown). Similar results were found as well using the Col-0 ecotype.

Second, ABA inhibits seed germination and the addition of sugar rescues the germination of ABA-treated seeds. This antagonistic interaction of metabolizable sugar with ABA is observed only with respect to radicle emergence. Early seedling growth is still inhibited by ABA in the presence of sugar. The rescue of germination by sugar is improved when sugar is applied after stratification.

Third, both ABA and elevated sugar concentrations block early seedling development. The response to sugar depends on the presence of ABA, as shown by the fact that several ABA biosynthesis mutants are glucose insensitive. Also the ABA response loci *ABI4* and *ABI5* are important for a proper glucose response. Thus, sugar and ABA signaling cooperate to induce early seedling developmental arrest.

Screens for sugar-insensitive mutant phenotypes based on inhibition of seedling development identified many important mutants in sugar signaling. Escape from early seedling developmental arrest induced by elevated sugar concentrations is a rapid assay for selecting sugar-sensitivity mutants. Detailed analysis of these sugar-response mutants is important for our understanding of sugar signaling. Seed germination is affected by multiple factors including growth conditions during seed development, seed age, and storage conditions. Such factors are likely to affect sugar sensitivity during germination and early seedling development as well. We have observed that aged seeds are often less responsive to sugar. Therefore, in sugar response assays it is necessary to use wild-type and mutant seed batches derived from plants grown under similar conditions and that had been stored and aged identically.

Sugar responses during germination and during the transition to seedling growth can be distinguished (Price *et al.*, 2003; Dekkers *et al.*, 2004). Thus, seedling developmental arrest is the result of two distinct sugar responses, one affecting germination and the other affecting seedling development. Mutants or transgenic lines that are insensitive to sugar during germination possibly grow faster than wild type during seedling development when challenged with sugar. Thus, potentially such mutants might appear as *gin* or *sis* mutants, especially when screening conditions are less stringent and seedling development is delayed instead of blocked. The glucose-induced germination delay is largely suppressed when sugars are added after stratification. Therefore, adding sugars after stratification might constitute a better way of analyzing sugar sensitivity of seedling growth, as the interference of sugars during germination is separated from sugar-induced developmental arrest of seedlings.

Acknowledgments

We thank Jolanda Schuurmans, Leonie Bentsink and Shanna Bastiaan-Net for critically reading the manuscript and Eiji Nambara for sharing unpublished results.

References

G.J. Acevedo-Hernandez, P. Leon and L.R. Herrera-Estrella (2005) Sugar and ABA responsiveness of a minimal RBCS light-responsive unit is mediated by direct binding of ABI4. *The Plant Journal* **43**, 506–519.

C. Alonso-Blanco, L. Bentsink, C.J. Hanhart, H. Blankestijn-de Vries and M. Koornneef (2003) Analysis of natural allelic variation at seed dormancy loci of *Arabidopsis thaliana. Genetics* **164**, 711–729.

F. Arenas-Huertero, A. Arroyo, L. Zhou, J. Sheen and P. Leon (2000) Analysis of *Arabidopsis* glucose insensitive mutants, *gin5* and *gin6*, reveals a central role of the plant hormone ABA in the regulation of plant vegetative development by sugar. *Genes and Development* **14**, 2085–2096.

A. Arroyo, F. Bossi, R.R. Finkelstein and P. Leon (2003) Three genes that affect sugar sensing (*abscisic acid insensitive 4*, *abscisic acid insensitive 5*, and *constitutive triple response 1*) are differentially regulated by glucose in *Arabidopsis. Plant Physiology* **133**, 231–242.

N. Avonce, B. Leyman, J.O. Mascorro-Gallardo, P. Van Dijck, J.M. Thevelein and G. Iturriaga (2004) The *Arabidopsis* trehalose-6-P synthase *AtTPS1* gene is a regulator of glucose, abscisic acid, and stress signaling. *Plant Physiology* **136**, 3649–3659.

M. Baier, G. Hemmann, R. Holman, *et al.* (2004) Characterization of mutants in *Arabidopsis* showing increased sugar-specific gene expression, growth, and developmental responses. *Plant Physiology* **134**, 81–91.

N. Beaudoin, C. Serizet, F. Gosti and J. Giraudat (2000) Interactions between abscisic acid and ethylene signaling cascades. *The Plant Cell* **12**, 1103–1116.

L. Bentsink and M. Koornneef (2002) Seed dormancy and germination. In: *The Arabidopsis Book* (eds C.R. Somerville and E.M. Meyerowitz). American Society of Plant Biologists, Rockville, MD. http//:www.aspb.org/publications/arabidopsis/, pp. 1–18.

K.J. Bradford, A.B. Downie, O.H. Gee, V. Alvarado, H. Yang and P. Dahal (2003) Abscisic acid and gibberellin differentially regulate expression of genes of the SNF1-related kinase complex in tomato seeds. *Plant Physiology* **132**, 1560–1576.

I.M. Brocard, T.J. Lynch and R.R. Finkelstein (2002) Regulation and role of the *Arabidopsis ABSCISIC ACID-INSENSITIVE 5* gene in abscisic acid, sugar, and stress response. *Plant Physiology* **129**, 1533–1543.

I.M. Brocard-Gifford, T.J. Lynch and R.R. Finkelstein (2003) Regulatory networks in seeds integrating developmental, abscisic acid, sugar, and light signalling. *Plant Physiology* **131**, 78–92.

I. Brocard-Gifford, T.J. Lynch, M.E. Garcia, B. Malhotra and R.R. Finkelstein (2004) The *Arabidopsis thaliana ABSCISIC ACID-INSENSITIVE 8* locus encodes a novel protein mediating abscisic acid and sugar responses essential for growth. *The Plant Cell* **16**, 406–421.

J.A. Brusslan and E.M. Tobin (1992) Light-independent developmental regulation of *cab* gene expression in *Arabidopsis thaliana. Proceedings of the National Academy of Sciences of the United States of America* **89**, 7791–7795.

C. Carles, N. Bies-Etheve, L. Aspart, *et al.* (2002) Regulation of *Arabidopsis thaliana Em* genes: role of ABI5. *The Plant Journal* **30**, 373–383.

M. Chen, X. Xia, H. Zheng, Z. Yuan and H. Huang (2004) The *GAOLAOZHUANGREN2* gene is required for normal glucose response and development of *Arabidopsis. Journal of Plant Research* **117**, 473–476.

W.H. Cheng, A. Endo, L. Zhou, *et al.* (2002) A unique short-chain dehydrogenase/reductase in *Arabidopsis* glucose signaling and abscisic acid biosynthesis and functions. *The Plant Cell* **14**, 2723–2743.

G. Coruzzi and D.R. Bush (2001) Nitrogen and carbon nutrient and metabolite signaling in plants. *Plant Physiology* **125**, 61–64.

G.M. Coruzzi and L. Zhou (2001) Carbon and nitrogen sensing and signaling in plants: emerging matrix effects. *Current Opinion in Plant Biology* **4**, 247–253.

S. Cutler, M. Ghassemian, D. Bonetta, S. Cooney and P. McCourt (1996) A protein farnesyl transferase involved in abscisic acid signal transduction in *Arabidopsis. Science* **273**, 1239–1241.

B.J.W. Dekkers, J.A.M.J. Schuurmans and S.C. Smeekens (2004) Glucose delays seed germination in *Arabidopsis thaliana. Planta* **218**, 579–588.

P.P. Dijkwel, P.A.M. Kock, R. Bezemer, P.J. Weisbeek and S.C.M. Smeekens (1996) Sucrose represses the developmentally controlled transient activation of the plastocyanin gene in *Arabidopsis thaliana* seedlings. *Plant Physiology* **110**, 455–463.

P.P. Dijkwel (1997) *Metabolic Control of Light Signalling in* Arabidopsis thaliana. PhD Thesis, University of Utrecht.

P.P. Dijkwel, C. Huijser, P.J. Weisbeek, N.H. Chua and S.C.M. Smeekens (1997) Sucrose control of phytochrome A signaling in *Arabidopsis*. *The Plant Cell* **9**, 583–595.

P. Duque and N.H. Chua (2003) IMB1, a bromodomain protein induced during seed imbibition, regulates ABA- and phyA-mediated responses of germination in *Arabidopsis*. *The Plant Journal* **35**, 787–799.

P.J. Eastmond, A.J.H. van Dijken, M. Spielman, *et al.* (2002) Trehalose-6-phosphate synthase 1, which catalyses the first step in trehalose synthesis, is essential for *Arabidopsis* embryo maturation. *The Plant Journal* **29**, 225–235.

P.J. Eastmond and I.A. Graham (2003) Trehalose metabolism: a regulatory role for trehalose-6-phosphate? *Current Opinion in Plant Biology* **6**, 231–235.

R.R. Finkelstein (1994) Mutations at two new *Arabidopsis* ABA response loci are similar to the *abi3* mutations. *The Plant Journal* **5**, 765–771.

R.R. Finkelstein, S.S.L. Gampala and C.D. Rock (2002) Abscisic acid signalling in seeds and seedlings. *The Plant Cell* **14**, S15–S45.

R.R. Finkelstein and S.I. Gibson (2001) ABA and sugar interactions regulating development: cross-talk or voices in a crowd? *Current Opinion in Plant Biology* **5**, 26–32.

R.R. Finkelstein and T.J. Lynch (2000a) The *Arabidopsis* abscisic acid response gene *ABI5* encodes a basic leucine zipper transcription factor. *The Plant Cell* **12**, 599–610.

R.R. Finkelstein and T.J. Lynch (2000b) Abscisic acid inhibition of radicle emergence but not seedling growth is suppressed by sugars. *Plant Physiology* **122**, 1179–1186.

R.R. Finkelstein, M.L. Wang, T.J. Lynch, S. Rao and H.M. Goodman (1998) The *Arabidopsis* abscisic acid response locus *ABI4* encodes an APETALA 2 domain protein. *The Plant Cell* **10**, 1043–1054.

S. Footitt, S.P. Slocombe, V. Larner, *et al.* (2002) Control of germination and lipid mobilization by *COMATOSE*, the *Arabidopsis* homologue of human ALDP. *The EMBO Journal* **21**, 2912–2922.

A. Garciarrubio, J.P. Legaria and A.A. Covarrubias (1997) Abscisic acid inhibits germination of mature *Arabidopsis* seeds by limiting the availability of energy and nutrients. *Planta* **203**, 182–187.

S. Gazzarrini and P. McCourt (2001) Genetic interactions between ABA, ethylene and sugar signalling pathways. *Current Opinion in Plant Biology* **4**, 387–391.

M. Ghassemian, E. Nambara, S. Cutler, H. Kawaide, Y. Kamiya and P. McCourt (2000) Regulation of abscisic acid signaling by the ethylene response pathway in *Arabidopsis*. *The Plant Cell* **12**, 1117–1126.

S.I. Gibson (2005) Control of plant development and gene expression by sugar signaling. *Current Opinion in Plant Biology* **8**, 93–102.

S.I. Gibson, R.J. Laby and D. Kim (2001) The *sugar-insensitive1* (*sis1*) mutant of *Arabidopsis* is allelic to *ctr1*. *Biochemical and Biophysical Research Communications* **280**, 196–203.

J. Giraudat, B.M. Hauge, C. Valon, J. Smalle, F. Parcy and H.M. Goodman (1992) Isolation of the *Arabidopsis ABI3* gene by positional cloning. *The Plant Cell* **4**, 1251–1261.

H. Guo and J.R. Ecker (2003) Plant responses to ethylene gas are mediated by SCFEBF1/EBF2-dependent proteolysis of EIN3 transcription factor. *Cell* **115**, 667–677.

M.H. Han, S. Goud, L. Song and N. Fedoroff (2004) The *Arabidopsis* double-stranded RNA-binding protein HYL1 plays a role in microRNA-mediated gene regulation. *Proceedings of the National Academy of Sciences of theUnited States of America* **101**, 1093–1098.

V. Hugouvieux, J.M. Kwak and J.I. Schroeder (2001) A mRNA cap binding protein, ABH1, modulates early abscisic acid signal transduction in *Arabidopsis*. *Cell* **106**, 477–487.

C. Huijser, A. Kortstee, J. Pego, P. Weisbeek, E. Wisman and S. Smeekens (2000) The *Arabidopsis SUCROSE UNCOUPLED-6* gene is identical to *ABSCISIC ACID INSENSITIVE-4*: involvement of abscisic acid in sugar responses. *The Plant Journal* **23**, 577–585.

J.C. Jang and J. Sheen (1994) Sugar sensing in higher plants. *The Plant Cell* **6**, 1665–1679.

S. Kim, H. Choi, H.J. Ryu, J.H. Park, M.D. Kim and S.Y. Kim (2004a) ARIA, an *Arabidopsis* arm repeat protein interacting with a transcriptional regulator of abscisic acid-responsive gene expression, is a novel abscisic acid signaling component. *Plant Physiology* **136**, 3639–3648.

S. Kim, J. Kang, D.I. Cho, J.H. Park and S.Y. Kim (2004b) ABF2, an ABRE-binding bZIP factor, is an essential component of glucose signaling and its overexpression affects multiple stress tolerance. *The Plant Journal* **40**, 75–87.

K.E. Koch (1996) Carbohydrate-modulated gene expression in plants. *Annual Review of Plant Physiology and Plant Molecular Biology* **47**, 509–540.

M. Koornneef, G. Reuling and C.M. Karssen (1984) The isolation and characterization of abscisic acid-insensitive mutants of *Arabidopsis thaliana*. *Physiologia Plantarum* **61**, 377–383.

R.J. Laby, M.S. Kincaid, D. Kim and S.I. Gibson (2000) The *Arabidopsis* sugar-insensitive mutants *sis4* and *sis5* are defective in abscisic acid synthesis and response. *The Plant Journal* **23**, 587–596.

S. Lee, H. Cheng, K.E. King, *et al.* (2002) Gibberellin regulates *Arabidopsis* seed germination via *RGL2*, a *GAI/RGA-like* gene whose expression is up-regulated following imbibition. *Genes and Development* **16**, 646–658.

P. Leon and J. Sheen (2003) Sugar and hormone connections. *Trends in Plant Science* **8**, 110–116.

K.M. Leon-Kloosterziel, G.A. van de Bunt, J.A.D. Zeevaart and M. Koornneef (1996) *Arabidopsis* mutants with a reduced seed dormancy. *Plant Physiology* **110**, 233–240.

N. Leonhardt, J.M. Kwak, N. Robert, D. Waner, G. Leonhardt and J. Schroeder (2004) Microarray expression analysis of *Arabidopsis* guard cells and isolation of a recessive abscisic acid hypersensitive protein phosphatase 2C mutant. *The Plant Cell* **16**, 596–615.

J. Leung, M. Bouvier-Durand, P.-C. Morris, D. Guerrier, F. Chefdor and J. Giraudat (1994) *Arabidopsis* ABA response gene *ABI1*: features of a calcium-modulated protein phosphatase. *Science* **264**, 1448–1452.

J. Leung and J. Giraudat (1998) Abscisic acid signal transduction. *Annual Review of Plant Physiology and Plant Molecular Biology* **49**, 199–222.

J. Leung, S. Merlot and J. Giraudat (1997) The *Arabidopsis ABSCISIC ACID-INSENSITIVE2* (*ABI2*) and *ABI1* genes encode homologous protein phosphatases 2C involved in abscisic acid signal transduction. *The Plant Cell* **9**, 759–771.

L. Lopez-Molina and N.H. Chua (2000) A null mutation in a bZIP factor confers ABA-insensitivity in *Arabidopsis thaliana*. *Plant and Cell Physiology* **41**, 541–547.

L. Lopez-Molina, S. Mongrand and N.H. Chua (2001) A postgermination developmental arrest checkpoint is mediated by abscisic acid and requires the ABI5 transcription factor in *Arabidopsis*. *Proceedings of the National Academy of Sciences of the United States of America* **98**, 4782–4787.

L. Lopez-Molina, S. Mongrand, N. Kinoshita and N.H. Chua (2003) AFP is a novel negative regulator of ABA signaling that promotes ABI5 protein degradation, *Genes and Development* **17**, 410–418.

L. Lopez-Molina, S. Mongrand, D.T. McLachlin, B.T. Chait and N.H. Chua (2002) ABI5 acts downstream of ABI3 to execute an ABA-dependent growth arrest during germination. *The Plant Journal* **32**, 317–328.

C. Lu and N. Fedoroff (2000) A mutation in the *Arabidopsis HYL1* gene encoding a dsRNA binding protein affects responses to abscisic acid, auxin, and cytokinin. *The Plant Cell* **12**, 2351–2366.

C. Lu, M.H. Han, A. Guevara-Garcia and N.V. Fedoroff (2002) Mitogen-activated protein kinase signaling in postgermination arrest of development by abscisic acid. *Proceedings of the National Academy of Sciences of the United States of America* **99**, 15812–15817.

P.A.W. Martin and M. Blackburn (2003) Inhibition of seed germination by extracts of bitter Hawkesbury watermelon containing cucurbitacin, a feeding stimulant for corn rootworm (Coleoptera: Chrysomelidae). *Journal of Economic Entomology* **96**, 441–445.

T. Martin, O. Oswald and I.A. Graham (2002) *Arabidopsis* seedling growth, storage lipid mobilization, and photosynthetic gene expression are regulated by carbon:nitrogen availability. *Plant Physiology* **128**, 472–481.

K. Meyer, M. Leube and E. Grill (1994) A protein phosphatase 2C involved in ABA signal transduction in *Arabidopsis thaliana*. *Science* **264**, 1452–1455.

B. Moore, L. Zhou, F. Rolland, *et al.* (2003) Role of the *Arabidopsis* glucose sensor HXK1 in nutrient, light, and hormonal signaling. *Science* **300**, 332–336.

C. Morito-Yamamuro, T. Tsutsui, A. Tanaka and J. Yamaguchi (2004) Knock-out of the plastid ribosomal protein S21 causes impaired photosynthesis and sugar response during germination and seedling development in *Arabidopsis thaliana*. *Plant and Cell Physiology* **45**, 781–788.

E. Nambara, M. Suzuki, S. Abrams, D.R. McCarty, Y. Kamiya and P. McCourt (2002) A screen for genes that function in abscisic acid signaling in *Arabidopsis thaliana*. *Genetics* **161**, 1247–1255.

K. Nemeth, K. Salchert, P. Putnoky, *et al.* (1998) Pleiotropic control of glucose and hormone responses by PRL1, a nuclear WD protein, in *Arabidopsis. Genes and Development* **12**, 3059–3073.

N. Nishimura, T. Yoshida, M. Murayama, T. Asami, K. Shinozaki and T. Hirayama (2004) Isolation and characterization of novel mutants affecting the abscisic acid sensitivity of *Arabidopsis* germination and seedling growth. *Plant and Cell Physiology* **45**, 1485–1499.

X. Niu, T. Helentjaris and N.J. Bate (2002) Maize ABI4 binds Coupling Element1 in abscisic acid and sugar response genes. *The Plant Cell* **14**, 2565–2575.

E. Oh, J. Kim, E. Park, J.I. Kim, C. Kang and G. Choi (2004) PIL5, a phytochrome-interacting basic helix-loop-helix protein, is a key negative regulator of seed germination in *Arabidopsis thaliana. The Plant Cell* **16**, 3045–3058.

D.P. O'Keefe, J.M. Tepperman, C. Dean, K.J. Leto, D.L. Erbes and J.T. Odell (1994) Plant expression of a bacterial cytochrome P450 that catalyzes activation of a sulfonylurea pro-herbicide. *Plant Physiology* **105**, 473–482.

P. Palenchar, A. Kouranov, L. Lejay and G. Coruzzi (2004) Genome-wide patterns of carbon and nitrogen regulation of gene expression validate the combined carbon and nitrogen (CN)-signaling hypothesis in plants. *Genome Biology* **5**, R91.

J.V. Pego, P.J. Weisbeek and S.C.M. Smeekens (1999) Mannose inhibits *Arabidopsis* germination via a hexokinase-mediated step. *Plant Physiology* **119**, 1017–1024.

S. Penfield, E.L. Rylott, A.D. Gilday, S. Graham, T.R. Larson and I.A. Graham (2004) Reserve mobilization in the *Arabidopsis* endosperm fuels hypocotyl elongation in the dark, is independent of abscisic acid, and requires *PHOSPHOENOLPYRUVATE CARBOXYKINASE1. The Plant Cell* **16**, 2705–2718.

H. Pinfield-Wells, E.L. Rylott, A.D. Gilday, *et al.* (2005) Sucrose rescues seedling establishment but not germination of *Arabidopsis* mutants disrupted in peroxisomal fatty acid catabolism. *The Plant Journal* **43**, 861–872.

T. Potuschak, E. Lechner, Y. Parmentier, *et al.* (2003) EIN3-dependent regulation of plant ethylene hormone signaling by two *Arabidopsis* F box proteins: EBF1 and EBF2. *Cell* **115**, 679–689.

J. Price, A. Laxmi, S.K. St. Martin and J.C. Jang (2004) Global transcription profiling reveals multiple sugar signal transduction mechanisms in *Arabidopsis. The Plant Cell* **16**, 2128–2150.

J. Price, T.C. Li, S.G. Kang, J.K. Na and J.C. Jang (2003) Mechanisms of glucose signaling during germination of *Arabidopsis. Plant Physiology* **132**, 1424–1438.

S.L. Pritchard, W.L. Charlton, A. Baker and I.A. Graham (2002) Germination and storage reserve mobilization are regulated independently in *Arabidopsis. The Plant Journal* **31**, 639–647.

V. Quesada, M.R. Ponce and J.L. Micol (2000) Genetic analysis of salt-tolerant mutants in *Arabidopsis thaliana. Genetics* **154**, 421–436.

F. Rolland, B. Moore and J. Sheen (2002) Sugar sensing and signalling in plants. *The Plant Cell* **14**, S185–S205.

F. Rook and M.W. Bevan (2003) Genetic approaches to understanding sugar-response pathways. *Journal of Experimental Botany* **54**, 495–501.

F. Rook, F. Corke, R. Card, G. Munz, C. Smith and M.W. Bevan (2001) Impaired sucrose-induction mutants reveal the modulation of sugar-induced starch biosynthetic gene expression by abscisic acid signalling. *The Plant Journal* **26**, 421–433.

L. Russell, V. Larner, S. Kurup, S. Bougourd and M. Holdsworth (2000) The *Arabidopsis COMATOSE* locus regulates germination potential. *Development* **127**, 3759–3767.

A. Saez, N. Apostolova, M. Gonzalez-Guzman, *et al.* (2004) Gain-of-function and loss-of-function phenotypes of the protein phophatase 2C *HAB1* reveal its role as a negative regulator of abscisic acid signalling. *The Plant Journal* **37**, 354–369.

Y. Sakuma, Q. Liu, J.G. Dubouzet, H. Abe, K. Shinozaki and K. Yamaguchi-Shinozaki (2002) DNA-binding specificity of the ERF/AP2 domain of *Arabidopsis* DREBs, transcription factors involved in dehydration- and cold-inducible gene expression. *Biochemical and Biophysical Research Communications* **290**, 998–1009.

H. Schluepmann, T. Pellny, A. van Dijken, S. Smeekens and M. Paul (2003) Trehalose 6-phosphate is indispensable for carbohydrate utilization and growth in *Arabidopsis thaliana. Proceedings of the National Academy of Sciences of the United States of America* **100**, 6849–6854.

S.M. Sherson, G. Hemmann, G. Wallace, *et al.* (2000) Monosaccharide/proton symporter AtSTP1 plays a major role in uptake and response of *Arabidopsis* seeds and seedlings to sugars. *The Plant Journal* **24**, 849–857.

L. Signora, I. De Smet, C.H. Foyer and H. Zhang (2001) ABA plays a central role in mediating the regulatory effects of nitrate on root branching in *Arabidopsis. The Plant Journal* **28**, 655–662.

J. Smalle, J. Kurepa, P. Yang, *et al.* (2003) The pleiotropic role of the 26S proteasome subunit RPN10 in *Arabidopsis* growth and development supports a substrate-specific function in abscisic acid signaling. *The Plant Cell* **15**, 965–980.

S. Smeekens (2000) Sugar-induced signal transduction in plants. *Annual Review of Plant Physiology and Plant Molecular Biology* **51**, 49–81.

E.M. Soderman, I.M. Brocard, T.J. Lynch and R.R. Finkelstein (2000) Regulation and function of the *Arabidopsis ABA-INSENSITIVE 4* gene in seed and abscisic acid response signaling networks. *Plant Physiology* **124**, 1752–1765.

C.M. Steber, S.E. Cooney and P. McCourt (1998) Isolation of the GA-response mutant *sly1* as a suppressor of *abi1-1* in *Arabidopsis thaliana. Genetics* **149**, 509–521.

Y. Tanaka, T. Sano, M. Tamaoki, N. Nakajima, N. Kondo and S. Hasezawa (2005) Ethylene inhibits abscisic acid-induced stomatal closure in *Arabidopsis. Plant Physiology* **138**, 2337–2343.

J. To, W.D. Reiter and S. Gibson (2002) Mobilization of seed storage lipid by *Arabidopsis* seedlings is retarded in the presence of exogenous sugars. *BMC Plant Biology* **2**, 4.

J. Ton and B. Mauch-Mani (2004) β-Amino-butyric acid-induced resistance against necrotrophic pathogens is based on ABA-dependent priming for callose. *The Plant Journal* **38**, 119–130.

H. Ullah, J.G. Chen, S. Wang and A.M. Jones (2002) Role of a heterotrimeric G protein in regulation of *Arabidopsis* seed germination. *Plant Physiology* **129**, 897–907.

F. Vazquez, V. Gasciolli, P. Crete and H. Vaucheret (2004) The nuclear dsRNA binding protein HYL1 is required for microRNA accumulation and plant development, but not posttranscriptional transgene silencing. *Current Biology* **14**, 346–351.

D. Villadsen and S. Smith (2004) Identification of more than 200 glucose-responsive *Arabidopsis* genes none of which responds to 3-O-methylglucose or 6-deoxyglucose. *Plant Molecular Biology* **55**, 467–477.

W. Xiao, J. Sheen and J.C. Jang (2000) The role of hexokinase in plant sugar signal transduction and growth and development. *Plant Molecular Biology* **44**, 451–461.

S. Yanagisawa, S.D. Yoo and J. Sheen (2003) Differential regulation of EIN3 stability by glucose and ethylene signalling in plants. *Nature* **425**, 521–525.

S. Yoshida, M. Ito, I. Nishida and A. Watanabe (2002) Identification of a novel gene *HYS1/CPR5* that has a repressive role in the induction of leaf senescence and pathogen-defence responses in *Arabidopsis thaliana. The Plant Journal* **29**, 427–437.

Zeng and A.R. Kermode (2004) A gymnosperm *ABI3* gene functions in a severe abscisic acid-insensitive mutant of *Arabidopsis* (*abi3-6*) to restore the wild-type phenotype and demonstrates a strong synergistic effect with sugar in the inhibition of post-germinative growth. *Plant Molecular Biology* **56**, 731–746.

X. Zhang, V. Garreton and N.H. Chua (2005) The AIP2 E3 ligase acts as a novel negative regulator of ABA signaling by promoting ABI3 degradation. *Genes and Development* **19**, 1532–1543.

L. Zhou, J.C. Jang, T.L. Jones and J. Sheen (1998) Glucose and ethylene signal transduction crosstalk revealed by an *Arabidopsis* glucose-insensitive mutant. *Proceedings of the National Academy of Sciences of the United States of America* **95**, 10294–10299.

Index